AN INTRODUCTION TO

Mathematical Statistics and Its Applications

AN INTRODUCTION TO
Mathematical Statistics and Its Applications

Second Edition

RICHARD J. LARSEN
Vanderbilt University

MORRIS L. MARX
University of Mississippi

Prentice-Hall, Englewood Cliffs, New Jersey 07632

Library of Congress Cataloging-in-Publication Data

Larsen, Richard J.
   An introduction to mathematical statistics and its
applications

   Bibliography: p.
   Includes index.
   1. Mathematical statistics.   I. Marx, Morris L.
II. Title.
QA276.L314   1986   519.5   85–19330
ISBN 0–13–487174–X

Editorial/production supervision and
   interior design: Maria McColligan
Cover design: Karen A. Stephens
Manufacturing buyer: John B. Hall

Printed in the United States of America

10   9   8   7   6   5   4   3

ISBN 0-13-487174-X   01

Prentice-Hall International (UK) Limited, *London*
Prentice-Hall of Australia Pty. Limited, *Sydney*
Prentice-Hall Canada Inc., *Toronto*
Prentice-Hall Hispanoamericana, S.A., *Mexico*
Prentice-Hall of India Private Limited, *New Delhi*
Prentice-Hall of Japan, Inc., *Tokyo*
Prentice-Hall of Southeast Asia Pte. Ltd., *Singapore*
Editora Prentice-Hall do Brasil, Ltda., *Rio de Janeiro*
Whitehall Books Limited, *Wellington, New Zealand*

Contents

Contents

Preface

Changes in the second edition have been made with the intention of improving the readability of the text without compromising its mathematical integrity or application orientation. We have streamlined our treatment of random variables, introduced a more convenient numbering system for theorems and examples, simplified some of the notation, and moved review exercises from the end of the chapters to the sections to which they apply. Answers to all odd-numbered exercises are given at the end of the book; in many cases, hints and intermediate steps are also provided.

Our coverage of combinatorics and classical probability has been expanded and new material has been added to the regression chapter. Expected values and higher moments have been dealt with at greater length. Whenever possible, examples have been updated; others have been deleted and new ones added.

The intended audience for this text remains the same—sophomore, junior, and senior math majors as well as upper level or graduate students in related disciplines. The recommended prerequisite is three semesters of calculus, although a strong two-semester course would suffice for much of what is covered. The book can be adapted to courses lasting one semester, two quarters, or two semesters.

Our overriding motivation in deciding which topics to present—and in what order—derived from our objective of writing a book that emphasizes the interrelationship between probability theory, mathematical statistics, and data analysis. We believe that for all three to be presented more or less simultaneously is vitally important, particularly so for those students who take only one statistics course during their college careers. Our experience in the classroom has certainly strengthened our faith in this approach: Each of the three comes more alive and generates greater enthusiasm when viewed in the context of the other two.

The second edition, like its predecessor, contains a fair amount of historical information. Apart from their inherent interest, comments about the intellectual heritage of modern probability and statistics are effective in unifying the two subjects. The merits of this tack are particularly evident in Chapter 4, which profiles the chronological and mathematical evolution of several important probability distributions, and in Chapter 10, which examines some of the original problems that eventually led to the development of regression analysis.

We have tried hard to make this book as "teachable" as possible. Motivation, in particular, has been given high priority. Chapter introductions describe at some length the nature of the material to come, how it relates to what has already been covered, and to what use it will ultimately be put. An extensive set of exercises is provided, ranging in difficulty from routine numerical applications to theoretical results not taken up in the text. The index is very detailed and extensively cross-referenced. The examples, though, will be the most help to the teacher in presenting this material. The bibliography lists almost two hundred references, the majority of which are the original sources for the many case studies and examples used throughout the book. With these as a backdrop, it is easy for the student to see the relevance and applicability of statistics.

We would like to acknowledge the following sources of the chapter opening photographs and epigraphs: Chapter 1 from *Natural Inheritance* by Francis Galton (1908); Chapter 2 courtesy of the History of Science Collections, University of Oklahoma; Chapters 3, 4, and 6 from *Studies in the History of Statistical Method* by Helen M. Walker (Williams & Wilkins, Baltimore, 1929); Chapters 5, 11, and 12 from *R. A. Fisher: The Life of a Scientist* by Joan Fisher Box (John Wiley & Sons, New York, 1978); Chapter 8 from the *Annals of Eugenics*, vol. 9, 1939; Chapter 9 from *Karl Pearson: An Appreciation of Some Aspects of His Life and Work* by Egon Sharpe Pearson (Cambridge University Press, Cambridge, 1938); Chapter 10 from *The Life, Letters and Labours of Francis Galton*, vol. IIIA (Cambridge University Press, Cambridge, 1930); and Chapter 13 from "Effect of Non-normality on the Power Function of the Sign Test" by Jean D. Gibbons (*Journal of the American Statistical Association*, vol. 59, 1964). We especially wish to thank Marcia Goodman of the History of Science Collections, University of Oklahoma Libraries for her assistance in obtaining the photographs.

We would like to express our indebtedness to the editors at J. R. Geigy, Biometrika, and McGraw-Hill for letting us use the tables that appear in the Appendix and to all the many researchers whose data we have used for examples. We also wish to thank the numerous individuals and organizations who gave us the approval to reprint tables and figures. Acknowledgments for these permissions are incorporated in the bibliography. For their detailed comments, criticisms, and suggestions we thank the reviewers: Professor Donald Myers, Dept. of Mathematics, University of Arizona, Tucson, AZ; Professor Galen Shorack, Dept. of Statistics, University of Washington, Seattle, WA; Professor Franklin Sheehan, Dept. of Mathematics, San Francisco State University, San Francisco, CA.

Finally, we would like to express our thanks for the support and encouragement given by Bob Sickles, who guided this endeavor from its inception, and to Maria McColligan, who was the production editor for this edition. Their counsel on matters of form and content was sincerely appreciated.

RJL
Nashville, Tennessee

MLM
Oxford, Mississippi

1

Introduction

Francis Galton

Some people hate the very name of statistics, but I find them full of beauty and interest. Whenever they are not brutalized, but delicately handled by the higher methods, and are warily interpreted, their power of dealing with complicated phenomena is extraordinary. They are the only tools by which an opening can be cut through the formidable thicket of difficulties that bars the path of those who pursue the Science of man.

1.1 A BRIEF HISTORY

The astragalus, a cube-shaped bone in the heel of dogs, sheep, and horses, has been found far more often than other types of bones in a number of archeological excavations. Hypotheses of ancient beasts with many legs, or of wild pagan foot feasts, come quickly to mind, but the correct, albeit less dramatic, explanation can be read off the Egyptian tomb inscriptions: the bones were used as primitive dice—for gambling! While this implies that some knowledge of probability, however rudimentary, was every bit as old as human socialization itself, it is nevertheless true that the formal mathematization of chance events—what this book is all about—is a much more recent development. We will begin our story in seventeenth-century France.

There are many early writings on the art of gambling, some of them even mathematical, but the date when the so-called calculus of probability was first applied to games of chance is 1654. In that year began a correspondence between the renowned mathematicians Blaise Pascal and Pierre de Fermat. The French nobleman, Antoine Gombaud, Chevalier de Méré, brought to Pascal's attention two old problems, one concerning odds in dice games and the other relating to the equitable division of stakes in case a game is interrupted before completion and one competitor holds an advantage. Pascal was intrigued by the problems and communicated his solutions to Fermat, thus prompting an exchange of letters that some consider to be the genesis of mathematical probability.

A year later a young Dutchman, Christiaan Huygens (best known for his work in optics), visited Paris and was told of the work of Pascal and Fermat. Not able to meet with Pascal, who had withdrawn to religious and philosophical contemplation, Huygens elected to pursue the subject on his own. Two years later, in 1657, his efforts produced the first published treatise on probability, *De Ratiociniis in Aleae Ludo* (*Calculations in Games of Chance*).

The newly conceived science began to receive serious consideration in the years after 1657. The birth process can be thought of as completed with the posthumous publication of Jacob Bernoulli's *Ars Conjectandi* (*The Art of Conjecture*) in 1713. Among the contents of *Ars Conjectandi* was the first published theorem on the limiting behavior of probability in repetitions of a simple chance experiment. Variations on this idea are still topics of research today.

Let us leave probability for a moment and turn to the other subject of our inquiry, *statistics*, a word denoting collections of data as well as the methodology for drawing inferences from data. To be sure, lists of data are as old as the art of writing, but the first comprehensive attack on the numerical interpretation of biological and social phenomena is due to John Graunt. In 1662 Graunt published *Natural and Political Observations Made upon the Bills of Mortality*, a book based on birth and death records gathered in plague-stricken England. Shortly thereafter, Sir William Petty began developing what became known as Political Arithmetic, "the art of reasoning by figures upon things relating to government." Finally, by 1700, the work of Edmund Halley (of Halley's Comet fame) on mortality tables and life expectancy had established statistical inference as a science both viable and eminently worthwhile.

By the time of the publication of *Ars Conjectandi*, the disciplines of probability and statistics were both well recognized but for the most part separate. Over the next century and a half, though, they were amalgamated into a single theory of mathematical statistics. A number of the important contributions to that union are discussed in later chapters, so we will limit ourselves here to just a few brief comments.

The scholarship of the eighteenth century in statistics and probability culminated in the work of Pierre-Simon Laplace. In 1812 his influential book, *Theorie Analytique des Probabilities*, presented all his previous contributions, unified the basic research in the field, and discussed a number of applications of probability, particularly to the theory of observational errors. By the end of the first quarter of the nineteenth century it was well established that the theory of probability could be profitably applied to empirical data in the physical sciences.

About this same time, Adolphe Quetelet, a most remarkable Belgian academician, began to demonstrate the use of probability models in describing social and biological phenomena. Quetelet had studied briefly with Laplace, and the latter's influence on the Belgian was unmistakable. Quetelet traveled widely throughout Europe in the ensuing years, spreading with fervor the statistical " gospel."

By the 1870s "modern" mathematical statistics was poised and ready for its debut. Among its leaders at this point was Sir Francis Galton, the great British scientist. Galton's story and the rapid evolution of the subject near the turn of the century will be told later. We look now at four examples that raise some of the sorts of questions we will eventually be considering.

1.2 SOME EXAMPLES

Do stock and commodity markets rise and fall randomly? Is there a common element in the aesthetic standards of the ancient Greeks and the Shoshoni Indians? Can external forces, such as phases of the moon, affect admissions to mental hospitals? What kind of relationship exists between exposure to radiation and cancer mortality?

Contentwise, these questions are quite diverse, but they have some important traits in common. They are all, for example, difficult or impossible to study in a laboratory, and none of them admits any self-evident axioms from which one may reason deductively. Indeed, such questions are usually attacked by collecting data, making guesses about the processes generating those data, and then testing those guesses. As it turns out, this approach leads to still another similarity—the element of chance or uncertainty that affects each data point recorded.

The goal of this text is to develop the mathematical tools and concepts necessary for incorporating the chance element into the methodology for describing or making predictions about real-world phenomena. The examples that follow are cases in point.

Each evening the radio and TV reporters offer an often bewildering array of averages, indices, and the like that presumably indicate the state of the stock market. But do they? Are these numbers conveying any really useful information? Some financial analysts would say "No," arguing that speculative markets tend to rise and fall randomly, much as though some hidden roulette wheel were spinning out the figures. In this example we attempt to examine quantitatively the reasonableness of this "random-movement" hypothesis.

We begin by formulating a theoretical construct, or *model*, that should describe the behavior of the market, *if the (random) hypothesis were true*. To this end, we translate the term "random movement" into two assumptions:

(a) The chances of the market's rising or falling on a given day are unaffected by its actions on any previous days.
(b) The market is equally likely to go up or down.

Measuring the day-to-day randomness, or lack of randomness, in the market's movements can be accomplished by looking at the lengths of *runs*. We define a *run of length n* (actually, these are runs *up* of length n) to be n consecutive days of market rises, with falls occurring on the day immediately before the first rise and immediately after the nth rise. If the actual behavior of the market's run lengths is markedly different from what assumptions (a) and (b) would predict, we can safely reject the random-movement hypothesis as being unrealistic. In this case, what our model predicts in terms of run length is not too difficult to figure out, even without the benefit of any formal training in probability.

Suppose a fall has occurred, followed by a rise. For a run of length one, the market must next fall. By the two assumptions, this happens half the time, so the probability of a run of length one is $\frac{1}{2}$; our notation for this will be $P(1) = \frac{1}{2}$. The other half the time the market rises, giving the sequence (fall, rise, rise). A run of length *two* occurs if there is now a fall. Again, this happens half the time, making its probability half the half giving the (fall, rise, rise) sequence. Thus, the probability of a run of length two is $\frac{1}{2} \cdot \frac{1}{2} = \frac{1}{4} = P(2)$, and the overall pattern should be clear: the probability of a run of length n is $P(n) = (\frac{1}{2})^n$. Furthermore, if a total of T runs are observed, it seems reasonable to expect $T \cdot (\frac{1}{2})^n$ of them to be of length n.

Table 1.2.1 gives the distribution of runs observed in weekly cash prices at the Chicago Wheat Market between 1883 and 1934 (1). The third column of the table gives the corresponding expected numbers, as calculated from the expression $T \cdot (\frac{1}{2})^n$.

Here the agreement between the actual and the predicted run frequencies seems good enough to lend some credence to assumptions (a) and (b). Before jumping to any rash conclusions, though, we should apply the same two assumptions to a different set of data to see whether a similar level of agree-

TABLE 1.2.1 RUNS IN THE CASH PRICES AT THE CHICAGO WHEAT MARKET

Run Length, n	Observed	Expected $[= 586 \cdot P(n)]$
1	280	293.00
2	147	146.50
3	86	73.25
4	38	36.62
5	15	18.31
6	13	9.16
7+	7	9.16
	586	586.00

ment can be achieved a second time. This is done in Table 1.2.2, where the first three columns list the observed and expected run lengths in Standard and Poor's Composite Stock Price Index during the period from January 1918 to March 1956.

TABLE 1.2.2 RUNS IN STANDARD AND POOR'S COMPOSITE STOCK PRICE INDEX

Run Length, n	Observed	Expected $(p = \frac{1}{2})$	Expected $(p = 0.617)$
1	31	46.00	35.27
2	21	23.00	21.75
3	16	11.50	13.41
4	10	5.75	8.27
5	5	2.87	5.10
6+	9	2.88	8.20
	92	92.00	92.00

This time there is substantial disagreement between columns 2 and 3, a turn of events reflecting unfavorably on the model. We now have two choices: abandon the model or modify it so it fits the data better. One way to do the latter would be to allow for the possibility that the market's rising or falling on any given day might not be a 50–50 proposition. This suggests that assumption (b) should be replaced with

(b)′ The likelihood of a rise in the market is some number p, where $0 \le p \le 1$.

The quantity p is known as a *parameter* of the model. Of course, (b) is a special case of (b′), where $p = \frac{1}{2}$.

Invoking the new assumptions, we can recalculate the probabilities of the various run lengths. For example, following a (fall, rise) sequence, a fall would be expected $100(1 - p)\%$ of the time, so $P(1) = 1 - p$. A fraction p of the time there is another rise. Of this fraction, there is a likelihood $1 - p$ of a fall. Hence the probability of the sequence (fall, rise, rise, fall)—that is, a run of length 2—is $P(2) = p(1 - p)$. In general, $P(n) = p^{n-1}(1 - p)$.

Two questions now arise. Which of the two would be more important for further study depends on the needs and interests of the model maker.

1. Is the initial assumption that $p = \frac{1}{2}$ justified?
2. Given the observed data, what is the best choice (or *estimate*) for p?

To answer question 1 we need to decide whether the discrepancies between the observed and expected run lengths are small enough to be attributed to chance or large enough to render the model invalid. One way to answer question 2 is to seek out that value of p that best "explains" the observations, in terms of maximizing their probability. In the Standard and Poor example, this type of estimate turns out to be $p = 0.617$. The corresponding expected values, based on $P(n) = p^{n-1}(1 - p) = (0.383)(0.617)^{n-1}$, are given in column 4 of Table 1.2.2. Note the dramatic improvement in the agreement between the observed values and the new expected values. Based on what we have seen thus far, we would have to conclude that neither set of data offers any serious refutation of the hypothesis that speculative markets rise and fall randomly.

Comment.

The two questions raised at the conclusion of Case Study 1.2.1 touch on the dual themes of statistical inference: hypothesis testing and estimation. While it is necessary to be vague now about the mathematical techniques these two will require, it should be clear from the example that the notions of probability and expected value will play prominent roles.

Among the useful criteria for classifying experiments into statistical "types" is the nature of the set of possible outcomes. Case Study 1.2.1, for example, described an experiment with a *finite* number of outcomes. A simple extension of that idea leads to experiments with outcomes that are *countably infinite*. Such an outcome set would have been appropriate had we idealized the stock-market model to allow for runs of *any* length. In either case, though, probabilities are calculated as *sums*. The probability that a run will be of either length 1 or length 2, for instance, is

$$P(1) + P(2) = p^0(1 - p)^1 + p^1(1 - p)^1 = 1 - p^2$$

There are many circumstances, however, where the possible outcomes are so numerous and so close together that it becomes convenient to define the outcome set as an interval of real numbers (that is, as an *uncountably infinite* set). Probabilities are then represented as integrals, the continuous analogs of sums. The next case study is based on data that would be considered "continuous."

CASE STUDY 1.2.2

Not all rectangles are created equal. Since antiquity, societies have expressed aesthetic preferences for rectangles having certain width (w) to length (l) ratios. Plato, for example, wrote that rectangles whose sides were in a

$1:\sqrt{3}$ ratio were especially pleasing. (These are the rectangles formed from the two halves of an equilateral triangle.)

Another "standard" calls for the width-to-length ratio to be equal to the ratio of the length to the sum of the width and the length. That is,

$$\frac{w}{l} = \frac{l}{w + l} \qquad (1.2.1)$$

Equation 1.2.1 implies that the width is $\frac{1}{2}(\sqrt{5} - 1)$, or approximately 0.618, times as long as the length. The Greeks called this the golden rectangle and used it often in their architecture (see Figure 1.2.1). Many other cultures were similarly inclined. The Egyptians, for example, built their pyramids out of stones whose faces were golden rectangles. Today, in our society, the golden rectangle remains an architectural and artistic standard, and even items such as drivers' licenses, business cards, and picture frames often have w/l ratios close to 0.618.

Figure 1.2.1

The study described here is an example from a field known as experimental aesthetics. The data are width-to-length ratios of beaded rectangles used by the Shoshoni Indians to decorate their leather goods. The question at issue is whether the golden rectangle can be considered an aesthetic standard for the Shoshonis, just as it was for the Greeks and the Egyptians.

It must be realized, of course, that even if the 0.618 ratio is adopted by a society as an aesthetic standard, there might still be considerable variability in the width-to-length ratios of individual rectangles. What should be close to 0.618, though, is the *average w/l* ratio. This suggests that the hypothesis (H) to be tested might be written

H: The "true" average width-to-length ratio characteristic of Shoshoni artistic work is 0.618.

Table 1.2.3 presents the w/l ratios for 20 rectangles found on Shoshoni handicraft (34).

TABLE 1.2.3 WIDTH-TO-LENGTH RATIOS OF SHOSHONI RECTANGLES

0.693	0.749	0.654	0.670
0.662	0.672	0.615	0.606
0.690	0.628	0.668	0.611
0.606	0.609	0.601	0.553
0.570	0.844	0.576	0.933

The average of these observed ratios is 0.661. The decision to be made, then, is whether H is true and the difference between 0.618 and 0.661 is solely the result of chance fluctuations, or whether H is false and the two societies do have different aesthetic standards. Ultimately, the choice will depend on how likely it is to obtain a *sample* average of 0.661 if, in fact, the *true* average is 0.618. Computing that likelihood, however, is not a simple task, so our final decision will have to be deferred. We return to the problem in Chapter 7.

Case Study 1.2.2 asked whether a given set of observations (coming from a single population) conformed to a theoretically determined standard ($w/l = 0.618$). Another common type of question asks if *several* populations differ among themselves in some particular aspect. For the latter situation, the data will consist of several sets of observations, one set from each of the populations being compared. The next case study is a typical "*k*-sample" problem.

CASE STUDY 1.2.3

In folklore, the full moon is often portrayed as something sinister, a kind of evil force possessing the power to control our behavior. Over the centuries, many prominent writers and philosophers have shared this belief (116). Milton, in *Paradise Lost*, refers to

Demoniac frenzy, moping melancholy
And moon-struck madness.

And Othello, after the murder of Desdemona, laments

It is the very error of the moon,
She comes more near the earth than she was wont
And makes men mad.

On a more scholarly level, Sir William Blackstone, the renowned eighteenth-century English barrister, defined a "lunatic" as

one who hath . . . lost the use of his reason and who hath lucid intervals, sometimes enjoying his senses and sometimes not, and that frequently depending upon changes of the moon.

The possibility of lunar phases' influencing human affairs is a theory not without supporters among the scientific community. Studies by reputable medical researchers have attempted to link the "Transylvania effect," as it has come to be known, with higher suicide rates, pyromania, and even epilepsy. In this example, we look at still another context in which this phenomenon might be expected to occur. Table 1.2.4 shows the admission rates to the emergency room of a Virginia mental health clinic *before*, *during*, and *after* the 12 full moons from August 1971 to July 1972 (11). Notice that for this particular set

TABLE 1.2.4 ADMISSION RATES (PATIENTS/DAY)

Month	Before Full Moon	During Full Moon	After Full Moon
Aug.	6.4	5.0	5.8
Sept.	7.1	13.0	9.2
Oct.	6.5	14.0	7.9
Nov.	8.6	12.0	7.7
Dec.	8.1	6.0	11.0
Jan.	10.4	9.0	12.9
Feb.	11.5	13.0	13.5
Mar.	13.8	16.0	13.1
Apr.	15.4	25.0	15.8
May	15.7	13.0	13.3
June	11.7	14.0	12.8
July	15.8	20.0	14.5
Averages	10.9	13.3	11.4

of data, the average admission rate *is* higher during the full moon than during the rest of the month.

For reasons that we will discuss in Chapter 6, hypothesis tests are always set up so that what is being tested is the absence or negation of any differences from population to population. Following that principle here leads us to state the hypothesis as

H: On the average, there is no difference in mental-hospital admission rates before, during, and after the full moon.

The decision to reject H will be made only if it can be demonstrated that averages as different as 10.9, 13.3, and 11.4 are extremely unlikely to arise by chance alone. An analysis done in Chapter 12 suggests that these averages *are* considerably different, implying that the data could be used as evidence in support of the existence of a Transylvania effect.

Many scientific laws are, from a mathematical viewpoint, discoveries of functional relationships between variable quantities. A familiar example is the formula $s(t) = 16t^2$, where $s(t)$ is the distance in feet an object initially at rest falls in t seconds (neglecting air resistance). The final case study in this section applies an empirical curve-fitting procedure to the problem of estimating the relationship between radiation exposure and cancer mortality rates.

CASE STUDY 1.2.4

The oil embargo of 1973 raised some very serious questions about energy policies in the United States. One of the most controversial is whether nuclear reactors should assume a more central role in the production of electric power. Those in favor point to their efficiency and to the availability of

TABLE 1.2.5 RADIOACTIVE CONTAMINATION AND CANCER MORTALITY IN OREGON

County	Index of Exposure	Cancer Mortality per 100,000
Umatilla	2.49	147.1
Morrow	2.57	130.1
Gilliam	3.41	129.9
Sherman	1.25	113.5
Wasco	1.62	137.5
Hood River	3.83	162.3
Portland	11.64	207.5
Columbia	6.41	177.9
Clatsop	8.34	210.3

nuclear material; those against warn of nuclear "incidents" and emphasize the health hazards posed by low-level radiation.

Since nuclear power is relatively new, there is not an abundance of past experience to draw on. One notable exception, though, is a government reactor that has been in continuous operation for 30 years. What happened there is what environmentalists fear will be a recurrent problem if nuclear reactors are proliferated.

Since World War II, plutonium for use in atomic weapons has been produced at an AEC facility in Hanford, Washington. One of the major safety problems encountered there has been the storage of radioactive wastes. Over the years, significant quantities of these substances—including strontium 90 and cesium 137—have leaked from their open-pit storage areas into the nearby Columbia River, which flows along the Washington–Oregon border, and eventually empties into the Pacific Ocean.

To measure the health consequences of this contamination, an index of exposure was calculated for each of the nine Oregon counties having frontage on either the Columbia River or the Pacific Ocean. This particular index was

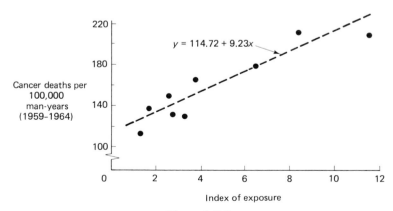

$y = 114.72 + 9.23x$

Cancer deaths per 100,000 man-years (1959-1964)

Index of exposure

Figure 1.2.2

Introduction   Chap. 1

based on several factors, including the county's stream distance from Hanford and the average distance of its population from any water frontage. As a covariate, the cancer mortality rate was determined for each of these same counties.

Table 1.2.5 shows the index of exposure and the cancer mortality rate (deaths per 100,000) for the nine Oregon counties affected (35). Higher index values represent higher levels of contamination.

A simple graph of the data (see Figure 1.2.2) suggests that the cancer mortality rate (y) and the index of exposure (x) vary *linearly*—that is, $y = a + bx$. Finding the numerical values for a and b that position the line so it best fits the data is an important problem in an area of statistics known as regression analysis. Here, the optimal line, based on methods derived in Chapter 10, has the equation $y = 114.72 + 9.23x$.

1.3 A CHAPTER SUMMARY

The concepts of probability lie at the very heart of all statistical problems, the case studies of Section 1.2 being typical examples. Acknowledging that fact, the next two chapters take a close look at some of those concepts. Chapter 2 states the axioms of probability and investigates their consequences. It also covers the basic skills for algebraically manipulating probabilities and gives an introduction to combinatorics, the mathematics of counting. Chapter 3 reformulates much of the material in Chapter 2 in terms of *random variables*, the latter being a concept of great convenience in applying probability to statistics. Over the years, particular measures of probability have emerged as being especially useful: the most prominent of these are profiled in Chapter 4.

Our study of statistics proper begins with Chapter 5, which is a first look at the theory of parameter estimation. Chapter 6 introduces the notion of hypothesis testing, a procedure that, in one form or another, commands a major share of the remainder of the book. From a conceptual standpoint, these are very important chapters: most formal applications of statistical methodology will involve either parameter estimation or hypothesis testing, or both.

Among the probability functions featured in Chapter 4, the *normal distribution*—more familiarly known as the bell-shaped curve—is sufficiently important to merit even further scrutiny. Chapter 7 derives in some detail many of the properties and applications of the normal distribution as well as those of several related probability functions. Much of the theory that supports the methodology appearing in Chapters 8 through 12 comes from Chapter 7.

Chapters 8, 11, and 12 continue the work of Chapter 7, but the emphasis is now on the comparison of several populations, as typified by Case Study 1.2.3. In Case Study 1.2.1 we discussed, in an offhand manner, the extent to which a set of observed values and expected values agreed. Chapter 9 offers a formal approach to that same idea. Linear relationships, such as the radiation exposure index–cancer mortality dependence described in Case Study 1.2.4, are taken up in Chapter 10.

Chapter 13 is an introduction to nonparametric statistics. The objective there is to develop procedures for answering some of the same sorts of questions raised in Chapters 7, 8, 11, and 12, but with fewer initial assumptions.

As a general format, each chapter contains numerous examples and case studies, the latter being actual experimental data taken from recently published journals and periodicals. It is hoped that these examples will make it abundantly clear from the very beginning that while the general orientation of this text is theoretical, the consequences of that theory are never too far from having direct applications in the "real world."

Pierre de Fermat

Blaise Pascal

Probability

Pierre de Fermat (1601–1665)

One of the most influential of seventeenth-century mathematicians, Fermat earned his living as a lawyer and administrator in Toulouse. He shares with Descartes credit for the invention of analytic geometry, but his most important work may have been in number theory. Fermat did not write for publication, preferring instead to send letters and papers to friends. His correspondence with Pascal was the starting point for the development of a mathematical theory of probability.

Blaise Pascal (1623–1662)

Pascal was the son of a nobleman. A prodigy of sorts, he had already published a treatise on conic sections by the age of 16. He also invented one of the early calculating machines to help his father with accounting work. Pascal's contributions to probability were stimulated by his correspondence, in 1654, with Fermat. Later that year he retired to a life of religious meditation.

2.1 INTRODUCTION

The evolution of probability as both a mathematical discipline and a tool of applied science was commented on briefly in Chapter 1. Among the inevitable consequences of probability's widespread adoption by the scientific community was the demand for a rigorous and orderly treatment of its logical foundations, one unfettered by any constraints imposed by specific subjects. In 1933, the great Russian mathematician Andrei Kolmogorov responded to the call for "axiomatization" and published *Grundbegriffe der Wahrscheinlichkeitsrechnung* (*Foundations of the Theory of Probability*).

Grundbegriffe was a landmark piece of work. By extending the approach taken by Richard von Mises, a German mathematician who had a similar objective in mind, Kolmogorov succeeded in reducing the whole of probability to just four simple axions.

The purpose of Chapter 2 is to examine the Kolmogorov–von Mises assumptions and explore their consequences. We begin by addressing some fundamental questions: What does a probability represent? How is it defined? What properties does it have? How is it manipulated mathematically? The chapter concludes with a series of theorems and examples showing how these simple axioms and properties, when properly extended, become enormously useful.

2.2 THE SAMPLE SPACE

We begin our development of probability by defining four key words: *experiment*, *sample outcome*, *sample space*, and *event*. The latter three, all carryovers from classical set theory, give us a familiar mathematical framework within which to work; the former is "new," and provides the conceptual mechanism for casting real-world phenomena into probabilistic terms.

By an *experiment* we will mean any procedure that (1) can be repeated, theoretically, an infinite number of times; and (2) has a well-defined set of possible outcomes. Thus, rolling a pair of dice qualifies as an experiment; so does measuring a hypertensive's blood pressure or doing a spectrographic analysis to determine the carbon content of moon rocks. Each of the potential eventualities of an experiment is referred to as a *sample outcome*, s, and their totality is called the *sample space*, S. To signify the membership of s in S, we write $s \in S$. Any designated collection of sample outcomes, including individual outcomes, the entire sample space, and the null set, constitutes an *event*. The latter is said to *occur* if the outcome of the experiment is one of the members of that event.

Example 2.2.1

Consider the experiment of flipping a coin three times. What is the sample space? Which sample outcomes make up the event A: Majority of coins show heads?

Think of each sample outcome here as an ordered triple, its components representing the outcomes of the first, second, and third tosses, respectively.

Altogether, there are eight different triples, so those eight comprise the sample space:

$$S = \{HHH, HHT, HTH, THH, HTT, THT, TTH, TTT\}$$

By inspection, we see that four of the sample outcomes in S constitute the event A:

$$A = \{HHH, HHT, HTH, THH\}$$

Example 2.2.2

Imagine rolling two dice, the first one red, the second one green. Each sample outcome is an ordered pair (face showing on red die, face showing on green die), and the entire sample space can be represented as a 6×6 matrix (see Figure 2.2.1).

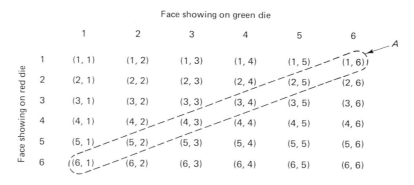

Figure 2.2.1

Gamblers are often interested in the event A that the sum of the faces showing is a 7. Notice in Figure 2.2.1 that the sample outcomes contained in A are the six diagonal entries, (1, 6), (2, 5), (3, 4), (4, 3), (5, 2), and (6, 1).

Example 2.2.3

A local TV station advertises two newscasting positions. If three women (W_1, W_2, W_3) and two men (M_1, M_2) apply, the "experiment" of hiring two co-anchors generates a sample space of 10 outcomes:

$$S = \{(W_1, W_2), (W_1, W_3), (W_2, W_3), (W_1, M_1), (W_1, M_2),$$
$$(W_2, M_1), (W_2, M_2), (W_3, M_1), (W_3, M_2), (M_1, M_2)\}$$

Does it matter here that the two positions being filled are equivalent? Yes. If the station were seeking to hire, say, a sports announcer and a weather forecaster, the number of possible outcomes would be 20: (W_2, M_1), for example, would represent a different staffing assignment than (M_1, W_2).

Example 2.2.4

The number of sample outcomes associated with an experiment need not be finite. Suppose that a coin is tossed until the first tail appears. If the first toss is itself a tail, the outcome of the experiment is T; if the first tail occurs on the second toss, the outcome is HT; and so on. Theoretically, of course, the first tail may *never* occur. By listing just a few sample outcomes, the infinite nature of S is readily apparent:

$$S = \{T, HT, HHT, HHHT, \ldots\}$$

Example 2.2.5

There are three ways to indicate an experiment's sample space. If the number of possible outcomes is small, we can simply list them, as we did in Examples 2.2.1 through 2.2.3. In some cases it may be possible to *characterize* a sample space by showing the structure its outcomes necessarily possess. This is what we did in Example 2.2.4. A third option is to state a mathematical formula that the sample outcomes must satisfy.

A computer programmer is running a subroutine that solves a general quadratic equation, $ax^2 + bx + c = 0$. Her "experiment" consists of choosing values for the three coefficients a, b, and c. Define (1) S and (2) the event A: Equation has two equal roots.

First, the sample space. Since presumably no combinations of finite a, b, and c are inadmissible, we can characterize S by writing a series of inequalities:

$$S = \{(a, b, c): -\infty < a < \infty, -\infty < b < \infty, -\infty < c < \infty\}$$

Defining A requires the well-known result from algebra that a quadratic equation has equal roots if and only if its discriminant, $b^2 - 4ac$, vanishes. Membership in A, then, is contingent on a, b, and c satisfying an equation:

$$A = \{(a, b, c): b^2 - 4ac = 0\}$$

Question 2.2.1 A graduating engineer has signed up for three job interviews. She intends to categorize each one as being either a "success" or a "failure" depending on whether it leads to a plant trip. Write out the appropriate sample space. What outcomes are in the event A: Second success occurs on third interview? In B: First success never occurs? (*Hint:* Notice the similarity between this situation and the coin-tossing experiment described in Example 2.2.1.)

Question 2.2.2 Three dice are tossed, one red, one blue, and one green. What outcomes make up the event A that the sum of the three faces showing equals 5?

Question 2.2.3 An urn contains six chips numbered 1 through 6. Three are drawn out. What outcomes are in the event "Second smallest chip is a 3"?

Question 2.2.4 An ecologist has tranquilized a male bear, a female bear, and a cub. Interested in studying their foraging behavior, she intends to fit them with radio-transmitting collars. To allow each bear to be distinguishable, one from another, each collar will transmit on a different frequency. Write out a sample space showing the ecologist's options in assigning the three different collars to the three different bears.

Question 2.2.5 A woman has her purse snatched by two teenagers. She is subsequently shown a police lineup consisting of five suspects, including the two perpetrators. What is the sample space associated with the experiment "Woman picks two suspects out of lineup"? Which outcomes are in the event, A, that she makes at least one incorrect identification?

Question 2.2.6 Consider the experiment of choosing coefficients for the quadratic equation $ax^2 + bx + c = 0$. Characterize the values of a, b, and c associated with the event A: Equation has complex roots.

Question 2.2.7 In the game of craps, the person rolling the dice (the *shooter*) wins outright if his first toss is a 7 or an 11. If his first toss is a 2, 3, or 12, he loses outright. If his first roll is something else, say, a 9, that number becomes his "point" and he keeps rolling the dice until he either rolls another 9, in which case he wins, or a 7, in which case he loses. Characterize the sample outcomes contained in the event "Shooter wins with a point of 9."

Question 2.2.8 A probability-minded despot offers a convicted murderer a final chance to gain his release. The prisoner is given 20 chips, 10 white and 10 black. All 20 are to be placed into two urns, according to any allocation scheme the prisoner wishes, with the one proviso being that each urn contain at least one chip. The executioner will then pick one of the two urns at random and from that urn, one chip at random. If the chip selected is white, the prisoner will be set free; if it is black, he "buys the farm." Characterize the sample space describing the prisoner's possible allocation options. (Intuitively, which allocation affords the prisoner the greatest chance of survival?)

Associated with events is an *algebra*, a set of rules for manipulating them mathematically. We need to be familiar with these rules because many problems are too complex to be reduced to a single event A defined on a sample space S. Look at the rules governing the game of craps, as outlined in Question 2.2.7. The shooter wins on his initial roll if he throws either a "7" *or* an "11." In the language of Boolean algebra, that event—shooter rolls a "7" or an "11"—is the *union* of two "simple" events—"shooter rolls a 7" and "shooter rolls an 11." If E denotes the union and A and B, the two simple events, we write $E = A \cup B$. The next several definitions and examples review some of the basic tenets of the algebra of sets that we will find useful.

> **Definition 2.2.1.** Let A and B be any two events defined over the same sample space S. Then:
>
> (a) The *intersection* of A and B, written $A \cap B$, is the event whose outcomes belong to both A and B.
>
> (b) The *union* of A and B, written $A \cup B$, is the event whose outcomes belong to either A or B or both.

Example 2.2.6

A single card is drawn from a poker deck. Let A be the event that an ace is selected:

$A = \{$ace of hearts, ace of diamonds, ace of clubs, ace of spades$\}$

Let B be the event "Heart is drawn":

$B = \{2$ of hearts, 3 of hearts, $\ldots$, ace of hearts$\}$

Then

$$A \cap B = \{\text{ace of hearts}\}$$

and

$A \cup B = \{2$ of hearts, 3 of hearts, $\ldots$, ace of hearts,

ace of diamonds, ace of clubs, ace of spades$\}$

(Let C be the event "Club is drawn." Which cards are in $B \cup C$? In $B \cap C$?)

Example 2.2.7

Let A be the set of x's for which $x^2 + 2x = 8$; let B be the set for which $x^2 + x = 6$. Find $A \cap B$ and $A \cup B$.

Since the first equation factors into $(x + 4)(x - 2) = 0$, its solution set is $A = \{-4, 2\}$. Similarly, the second equation can be written $(x + 3)(x - 2) = 0$, making $B = \{-3, 2\}$. Therefore,

$$A \cap B = \{2\}$$

and

$$A \cup B = \{-4, -3, 2\}$$

Example 2.2.8

Consider the electrical circuit pictured in Figure 2.2.2. Let A_i denote the event that switch i fails to close, $i = 1, 2, 3, 4$. Let A be the event "Circuit is not completed." Express A in terms of the A_i's.

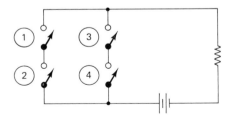

Figure 2.2.2

Call the ① and ② switches, line a; call the ③ and ④ switches, line b. By inspection, the circuit fails only if *both* line a and line b fail. But line a fails only if *either* ① *or* ② (or both) fail. That is, the event that line a fails is the union $A_1 \cup A_2$. Similarly, the failure of line b is the union $A_3 \cup A_4$. The event that the circuit fails, then, is an intersection:

$$A = (A_1 \cup A_2) \cap (A_3 \cup A_4)$$

Definition 2.2.2. Events A and B defined over the same sample space are said to be *mutually exclusive* if they have no outcomes in common—that is, if $A \cap B = \varnothing$, where $\varnothing$ is the null set.

Example 2.2.9

Consider a single throw of two dice. Define A to be the event that the *sum* of the faces showing is odd. Let B be the event that the two faces themselves are odd. Then clearly the intersection is empty, the sum of two odd numbers necessarily being even. In symbols, $A \cap B = \varnothing$. (Recall the event $B \cap C$ asked for in Example 2.2.6.)

Definition 2.2.3. Let A be any event defined on a sample space S. The *complement* of A, written A^C, is the event consisting of all the outcomes in S other than those contained in A.

Example 2.2.10

Let A be the set of (x, y)'s for which $x^2 + y^2 < 1$. Sketch the region in the xy-plane corresponding to A^C.

From analytic geometry, we recognize that $x^2 + y^2 < 1$ describes the interior of a circle of radius 1 centered at the origin. Figure 2.2.3 shows the obvious complement.

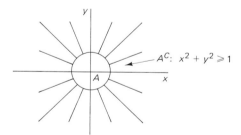

Figure 2.2.3

Question 2.2.9 Sketch the regions in the xy-plane corresponding to $A \cup B$ and $A \cap B$ if

$$A = \{(x, y): 0 < x < 3, 0 < y < 3\}$$

and

$$B = \{(x, y): 2 < x < 4, 2 < y < 4\}$$

Question 2.2.10 Referring to Example 2.2.7, find $A \cap B$ and $A \cup B$ if the two equations were replaced by inequalities: $x^2 + 2x \leq 8$ and $x^2 + x \leq 6$.

Question 2.2.11 An electronic system has four components divided into two pairs. The two components of each pair are wired in parallel; the two pairs are wired in series. Let A_{ij} denote the event "ith component in jth pair fails," $i = 1, 2; j = 1, 2$. Let A be the event "System fails." Write A in terms of the A_{ij}'s.

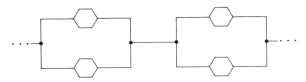

Question 2.2.12 Let $A_i = \{x: 0 \leq x < 1/i\}, i = 1, 2, \ldots, k$. Describe the sets

(a) $\displaystyle\bigcup_{i=1}^{k} A_i$ (b) $\displaystyle\bigcap_{i=1}^{k} A_i$

Question 2.2.13 Define $A = \{x: 0 \leq x \leq 1\}$, $B = \{x: 0 \leq x \leq 3\}$, and $C = \{x: -1 \leq x \leq 2\}$. Draw diagrams showing each of the following sets of points:
(a) $A^C \cap B \cap C$
(b) $A^C \cup (B \cap C)$
(c) $A \cap B \cap C^C$
(d) $((A \cup B) \cap C^C)^C$

Question 2.2.14 In poker, a five-card hand is called a *straight* if its denominations are consecutive—for example, a 4 of hearts, 5 of spades, 6 of spades, 7 of hearts, and 8 of clubs. A hand is called a *flush* if all five cards are in the same suit. If A denotes the set of straights and B, the set of flushes, how many outcomes are in the intersection $A \cap B$? (*Hint:* Picture the deck of 52 cards as a 4×13 matrix where the rows correspond to the four suits and the columns represent the 13 denominations.)

Relationships between events, particularly more than two events, can be difficult to conceptualize using only the formal definitions of union, intersection, and

Venn diagrams

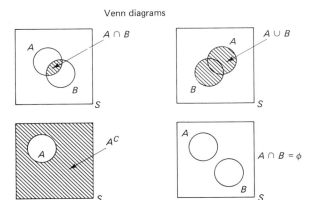

Figure 2.2.4

complement. Fortunately, there is a graphical format that can often greatly simplify the manipulation of complex events. Figure 2.2.4 shows *Venn diagrams* illustrating an intersection, a union, a complement, and two sets that are mutually exclusive. In each case, the shaded interior of a region is intended to represent the outcomes it contains.

Example 2.2.11

When two events A and B are defined on a sample space, we will frequently need to consider either

    (a) the event that *exactly one* (of the two) occurs

or

    (b) the event that *at most one* (of the two) occurs

Getting expressions for each of these is easy if we visualize the corresponding Venn diagrams.

The shaded area in Figure 2.2.5 represents the event E that either A or B, but not both, occurs (that is, exactly one occurs). Just by looking at the

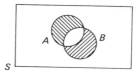

Figure 2.2.5

diagram we can formulate an expression for E. The portion of A, for example, included in E is $A \cap B^C$. Similarly, the portion of B included in E is $B \cap A^C$. It follows, also by inspection, that E can be written as a union:

$$E = (A \cap B^C) \cup (B \cap A^C)$$

[Convince yourself that an equivalent expression for E is $(A \cap B)^C \cap (A \cup B)$.]

Figure 2.2.6 shows the event F that *at most one* (of the two events) occurs. Here a formal expression is obvious:

$$F = (A \cap B)^C$$

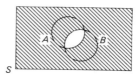

Figure 2.2.6

Example 2.2.12

A fashionable country club has 100 members, 30 of whom are lawyers. Rumor has it that 25 of the club members are liars and that 55 are neither lawyers nor liars. What proportion of the lawyers are liars?

Everyone's initial reaction to problems of this sort is befuddlement: the language itself becomes a barrier. When recast, though, in the terminology of set theory and pictured as a Venn diagram, questions like this are much easier. Let A be the set of lawyers and let B be the set of liars (see Figure 2.2.7).

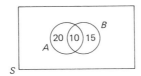

Figure 2.2.7

Also, let the *number* of members in any set Q be denoted $N(Q)$. What we are given reduces to a series of equations:

$$N(S) = 100$$
$$N(A) = 30$$
$$N(B) = 25$$
$$N((A \cup B)^C) = 55$$

Notice that $N((A \cup B)^C) = 55$ implies that

$$N(A \cup B) = 100 - 55 = 45$$

But from Figure 2.2.7,

$$N(A \cup B) = N(A) + N(B) - N(A \cap B) \qquad \text{(why?)}$$

or

$$45 = 30 + 25 - N(A \cap B)$$

Therefore, *10* of the club's members are lawyers *and* liars:

$$N(A \cap B) = 30 + 25 - 45$$
$$= 10$$

(The numbers in Figure 2.2.7 show how the 45 members in $A \cup B$ are distributed among $A \cap B^C$, $A \cap B$, and $A^C \cap B$.) The answer to the question, then, is *one-third*—10 of the club's 30 lawyers are liars.

Question 2.2.15 During orientation week, "Raiders" was shown twice at State University. Among the entering class of 2000 freshmen, 850 went to see it the first time, 690 the second time, while 730 failed to see it either time. How many saw it twice?

Question 2.2.16 Let A and B be any two events. Use Venn diagrams to show that
(a) the complement of their intersection is the union of their complements:

$$(A \cap B)^C = A^C \cup B^C$$

(b) the complement of their union is the intersection of their complements:

$$(A \cup B)^C = A^C \cap B^C$$

(These two results are known as *DeMorgan's Laws*.)

Question 2.2.17   Let A, B, and C be any three events. Use Venn diagrams to show that
(a) $A \cap (B \cup C) = (A \cap B) \cup (A \cap C)$
(b) $A \cup (B \cap C) = (A \cup B) \cap (A \cup C)$

Question 2.2.18   Verify the associative laws for unions and intersections:
(a) $A \cup (B \cup C) = (A \cup B) \cup C$
(b) $A \cap (B \cap C) = (A \cap B) \cap C$

Question 2.2.19   Use a Venn diagram to suggest a formula for $N(A \cup B \cup C)$, the number of outcomes in the union of three events.

Question 2.2.20   At a convention attended by 100 country music buffs, a trade magazine sought to get the fans' opinions of Waylon Jennings, Merle Haggard, and Willie Nelson. Twenty said they liked Jennings but not Haggard or Nelson, 22 were strongly partial to Haggard, and 28 listened only to Willie. Fourteen liked Jennings and Haggard; 18, Haggard and Nelson; 12, Jennings and Nelson; 8 liked all three. It was also reported that 4 did not like any of the three. Comment on the accuracy of the survey.

2.3 THE PROBABILITY FUNCTION

Having introduced in Section 2.2 the twin concepts of experiment and sample space, we are now ready to pursue in a formal way the all-important problem of assigning a "probability" to an experiment's outcome—and, more generally, to an event. The backdrop for our discussion will be the unions, intersections, and complements of set theory.

Over the years, the mathematical definition of probability has undergone a considerable metamorphosis, but the most recent formulations have not invalidated their predecessors. As a consequence, there are, today, four very different ways of defining what is to be meant by the probability of an event.

Historically, the first definition evolved quite naturally out of the gambling context in which so many of the early problems studied by Pascal, Fermat, and others were posed. Imagine an experiment, or game, having n possible outcomes— *and suppose that those outcomes are all equally likely.* If some event A were satisfied by m of those n, it reasonably follows that the "probability of A" [written $P(A)$] should be set equal to m/n. This is called the *classical*, or *a priori*, definition of probability. Recall Example 2.2.2. If two fair dice are tossed, there are $n = 36$ possible outcomes (all equally likely); of those 36, a total of $m = 6$ satisfy the event A: Sum of the faces showing is a 7, so we would write $P(A) = \frac{6}{36}$, or $\frac{1}{6}$.

There are many situations where the classical definition of probability is entirely appropriate. We will see quite a few later in this chapter. Still, it has obvious limitations. What if the outcomes of the experiment are not equally likely? Or the number of outcomes is not finite? Questions such as these provided the impetus for the formation of a more general and more experimentally oriented definition of probability. The first formal statement of what became known as the *empirical*, or *a posteriori*, definition of probability is often credited to the twentieth-century German mathematician Richard von Mises, but the basic notion was implicit in the work of many other probabilists at least a century earlier.

Consider a sample space, S, and any event, A, defined on S. If our experiment were performed *one* time, either A or A^C would be the outcome. If it were performed n times, the resulting set of sample outcomes would be members of A on m occasions, m being some integer between 0 and n, inclusive. Hypothetically, we could continue this process an infinite number of times. As n gets large, the ratio m/n will fluctuate less and less (we will make that statement more precise a little later). The number that m/n converges to is called the *empirical probability* of A: that is, $P(A) = \lim_{n \to \infty} m/n$ (see Figure 2.3.1).

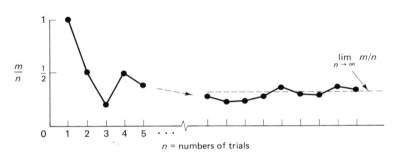

Figure 2.3.1

While the von Mises approach does shore up some of the inadequacies seen in the classical definition of probability, it is not without shortcomings of its own. The limit it focuses on is not the usual one in standard analysis. For example, if a fair coin were to be tossed repeatedly with A being the event "Head appears," it seems reasonable to expect that $P(A) = \lim_{n \to \infty} m/n$ would equal $\frac{1}{2}$. The possibility exists, though, for the coin to come up tails every time, in which case $m/n = 0$, for all n. Also, there is a bit of an inconsistency in extolling $\lim_{n \to \infty} m/n$ as a way of defining a probability *experimentally* when the very act of repeating an experiment under identical conditions an infinite number of times is physically impossible. And left unanswered is the question of how large n must be to give a good approximation for $\lim_{n \to \infty} m/n$.

The next attempt at defining probability was entirely a product of the twentieth century. Modern mathematicians have shown a keen interest in developing subjects axiomatically. It was to be expected, then, that probability would come under such scrutiny, and be defined not as a ratio or as the limit of a ratio, but simply as a function that behaved in accordance with a prescribed set of axioms.

Probability   Chap. 2

The major breakthrough on this front came in 1933 when Andrei Kolmogorov published *Grundbegriffe der Wahrscheinlichkeitsrechnung* (*Foundations of the Theory of Probability*). Kolmogorov's work was a masterpiece of mathematical elegance—it reduced the behavior of the probability function to a set of just three or four simple postulates, three if the sample space is limited to a finite number of outcomes and four if S is infinite.

If S has a finite number of members, Kolmogorov showed that a necessary and sufficient set of axioms for P are the three listed below:

Axiom 1 Let A be any event defined over S. Then $P(A) \geq 0$.

Axiom 2 $P(S) = 1$.

Axiom 3 Let A and B be any two mutually exclusive events defined over S. Then

$$P(A \cup B) = P(A) + P(B)$$

When S has an infinite number of members, a fourth axiom is needed:

Axiom 4 Let $A_1, A_2, \ldots$, be events defined over S. If $A_i \cap A_j = \varnothing$ for each $i \neq j$, then

$$P\left(\bigcup_{i=1}^{\infty} A_i\right) = \sum_{i=1}^{\infty} P(A_i)$$

From these few axioms, we can derive all the other properties of the P function.

Even more recent than the construction of an axiom system have been the efforts to define P in *subjective* terms, as a person's measure of belief that some given event will occur. For example, suppose that we ask "What is the probability that nuclear war will break out in the Mideast sometime in the next five years?" It is impossible to cast such a question meaningfully in a strictly empirical framework. Any number that we might come up with would necessarily be our own personal (subjective) assessment of the situation, based on the various countries' past history, extrapolations of their current policies, and so on.

Regardless of which definition is invoked, the basic properties of the probability function remain the same. The next set of theorems gives six results that we will find particularly useful. All six are almost immediate consequences of Kolmogorov's axioms.

Theorem 2.3.1. $P(A^C) = 1 - P(A)$.

Proof

By Axiom 2 and Definition 2.2.3,

$$P(S) = 1 = P(A \cup A^C)$$

But A and A^C are mutually exclusive, so

$$P(A \cup A^C) = P(A) + P(A^C)$$

and the result follows.

Theorem 2.3.2. $P(\varnothing) = 0$.

Proof
Since $\varnothing = S^C$, $P(\varnothing) = P(S^C) = 1 - P(S) = 0$.

Theorem 2.3.3. If $A \subset B$, then $P(A) \leq P(B)$.

Proof
Note that the event B may be written in the form

$$B = A \cup (B \cap A^C)$$

where A and $(B \cap A^C)$ are mutually exclusive. Therefore,

$$P(B) = P(A) + P(B \cap A^C)$$

which implies that $P(B) \geq P(A)$ since $P(B \cap A^C) \geq 0$.

Theorem 2.3.4. For any event A, $P(A) \leq 1$.

Proof
The proof follows immediately from Theorem 2.3.3 because $A \subset S$ and $P(S) = 1$.

Theorem 2.3.5. Let A_1, A_2, ..., A_n be events defined over S. If $A_i \cap A_j = \varnothing$ for $i \neq j$, then

$$P\left(\bigcup_{i=1}^{n} A_i\right) = \sum_{i=1}^{n} P(A_i).$$

Proof
The proof is a straightforward induction argument with Axiom 3 being the starting point.

Theorem 2.3.6. $P(A \cup B) = P(A) + P(B) - P(A \cap B)$.

Proof
The Venn diagram for $A \cup B$ certainly suggests that the statement of the theorem is true (recall Figure 2.2.4). More formally, we have from Axiom 3 that

$$P(A) = P(A \cap B^C) + P(A \cap B)$$

and

$$P(B) = P(B \cap A^C) + P(A \cap B)$$

Adding these two equations gives

$$P(A) + P(B) = [P(A \cap B^C) + P(B \cap A^C) + P(A \cap B)] + P(A \cap B)$$

By Theorem 2.3.5, the sum in the brackets is $P(A \cup B)$. If we subtract $P(A \cap B)$ from both sides of the equation, the result follows.

Example 2.3.1

A new "no-frills" airline runs a commuter shuttle service using a refurbished World War I bomber. The two-propeller plane will fly if either or both of its engines function. Suppose that P(port engine fails) = 0.10, P(starboard engine fails) = 0.15, and P(both engines fail) = 0.015. What is the probability that the plane will complete its next flight safely?

We can get an answer here very quickly by appealing to Theorem 2.3.1. Since "Plane flies" and "Plane crashes" are complements,

$$P(\text{plane completes next flight safely}) = 1 - P(\text{plane crashes})$$
$$= 1 - P(\text{both engines fail})$$
$$= 1 - 0.015$$
$$= 0.985$$

Example 2.3.2

Consolidated Industries has come under considerable pressure to eliminate its discriminatory hiring practices. Company officials have agreed that during the next five years, 60% of their new employees will be females and 30% will be black. One out of four new employees, though, will be white males. What percentage of black females are they committed to hiring?

There are four basic events relevant to this problem:

B: employee is black

W: employee is white

F: employee is female

M: employee is male

The three probabilities given—$P(F) = 0.60$, $P(B) = 0.30$, and $P(W \cap M) = 0.25$—can be pictured in a 2×2 table (see Figure 2.3.2). Entries in the body of the table are intersection probabilities; the row and column totals represent single-event probabilities.

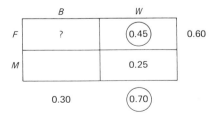

Figure 2.3.2

What we are trying to find, $P(B \cap F)$, can be gotten by repeated applications of Theorem 2.3.1. If $P(B) = 0.30$, then $P(W) = 1 - P(B) = 0.70$. But

$$P(W) = P(W \cap F) + P(W \cap M)$$

or

$$0.70 = P(W \cap F) + 0.25$$

which makes

$$P(W \cap F) = 0.45$$

(see the circled entries in Figure 2.3.2). By a final appeal to Theorem 2.3.1 we see that 15% of newly hired employees will be black females:

$$P(B \cap F) = P(F) - P(W \cap F)$$

$$= 0.60 - 0.45$$

$$= 0.15$$

Example 2.3.3

Winthrop, a premed student, has been summarily rejected by all 126 U.S. medical schools. Desperate, he sends his transcripts and MCATs to the two least selective foreign schools he can think of, the two branch campuses (X and Y) of Swampwater Tech. Based on the success his friends have had there, he estimates that his probability of being accepted at X is 0.7, and at Y, 0.4. He also suspects there is a 75% chance that at least one of his applications will be rejected. What is the probability that at least one of the schools will accept him?

Let A be the event "Branch X accepts him" and B, the event "Branch Y accepts him." We are given that $P(A) = 0.7$, $P(B) = 0.4$, and $P(A^C \cup B^C) = 0.75$; we are trying to find $P(A \cup B)$.

From Theorem 2.3.6,

$$P(A \cup B) = P(A) + P(B) - P(A \cap B)$$

We can get a value for the probability of the intersection by appealing to Question 2.2.16. Since $A^C \cup B^C = (A \cap B)^C$,

$$P(A \cap B) = 1 - P\{(A \cap B)^C\} = 1 - 0.75 = 0.25$$

Substituting, then, into the expression for the union, we find that Winthrop's chances are not all that bad—he has an 85% chance of getting at least one acceptance:

$$P(A \cup B) = 0.7 + 0.4 - 0.25$$

$$= 0.85$$

Question 2.3.1 If $P(A) = \frac{1}{3}$, $P(B) = \frac{1}{2}$, and $P(A \cup B) = \frac{3}{4}$, find
   (a) $P(A \cap B)$
   (b) $P(A^C \cup B^C)$
   (c) $P(A^C \cap B)$

Question 2.3.2 Let A and B be any two events defined on S. Suppose that $P(A) = 0.4$, $P(B) = 0.5$, and $P(A \cap B) = 0.1$. Find the probability that A or B but not both occur.

Question 2.3.3 An experiment has two possible outcomes: the first occurs with probability p; the second, with probability p^2. Find p.

Question 2.3.4 In a soon-to-be released Halloween martial arts movie, a pumpkin has a starring role. Because of the rather violent nature of the film, the producers have also hired a stunt pumpkin. Suppose that the featured pumpkin appears in 40% of all the film's scenes, the stunt pumpkin in 30%, and the two appear simultaneously in 5% of the scenes. What is the probability that in a given scene,
   (a) only the stunt pumpkin appears?
   (b) neither pumpkin appears?

Question 2.3.5 For any events A and B, show that

$$P(A \cap B) \geq 1 - P(A^C) - P(B^C)$$

Question 2.3.6 Express the following probabilities in terms of $P(A)$, $P(B)$, and $P(A \cap B)$.
   (a) $P(A^C \cup B^C)$
   (b) $P(A^C \cap (A \cup B))$

Question 2.3.7 Recall Example 2.3.3. Show that for $P(A)$ and $P(B)$ fixed, $P(A \cup B)$ varies directly with $P(A^C \cup B^C)$. That is, Winthrop's chances of at least one acceptance increase if his chances of at least one rejection increase!

Question 2.3.8 Two cards are drawn successively from a deck of 52. What is the probability that the second card is higher in rank than the first? (*Hint*: Let S be the sample space associated with the second card. Then S is the union of three mutually exclusive events A, B, and C, where A is the set of cards having a higher rank than the first card, B is the set of cards having a lower rank than the first card, and C is the set of cards having the same rank as the first card.)

2.4 DISCRETE PROBABILITY FUNCTIONS

The six theorems proved in Section 2.3 address the problem of "manipulating" probabilities: if $P(A)$ is given, then $P(A^C) = 1 - P(A)$; for any two events A and B, $P(A \cup B) = P(A) + P(B) - P(A \cap B)$; and so on. Before we can put those results to much practical use, though, we need to become more familiar with the probability function itself.

There are two general types of probability functions, *discrete* and *continuous*, the distinction reflecting the nature of the sample space. Some experiments generate sample spaces containing either a finite or a countably infinite number of outcomes (recall Examples 2.2.1 through 2.2.4). Any probability function defined on such a sample space is said to be *discrete*. Any probability function defined on a sample space having an uncountably infinite number of outcomes (recall Example 2.2.5) is said to be *continuous*.

More is at issue here than mere terminology: discrete and continuous probability functions have fundamentally different interpretations and they are treated mathematically in very different ways. Sections 2.4 and 2.5 give the properties associated with each type of function and illustrate each with a few examples. Our involvement with this topic, however, hardly ends with the conclusion of Section 2.5. Much of the first half of this book is concerned explicitly with the definition and application of discrete and continuous probability functions. All that Sections 2.4 and 2.5 provide is background for what lies ahead.

Suppose that the sample space for a given experiment is either finite or countably infinite. Then any P such that

$$\text{(a)} \quad 0 \le P(s) \qquad \text{for each } s \in S$$

$$\text{(b)} \quad \sum_{\text{all } s \in S} P(s) = 1$$

is said to be a *discrete probability function*. An immediate consequence of (a) and (b) is that the probability of any event A is the sum of the probabilities associated with the outcomes in A:

$$P(A) = \sum_{\text{all } s \in A} P(s)$$

Example 2.4.1

In the game of *odd man out*, each player tosses a fair coin. If all the coins turn up the same, *except for one*, the player tossing the "minority" coin is declared the odd man out and is eliminated from the contest. Suppose that three people play odd man out. What is the probability that on the first toss, *someone* will be eliminated?

The experiment here—each of three people tossing a fair coin once—generates a sample space of eight outcomes, all equally likely (see Figure 2.4.1). Six of the eight outcomes in S satisfy the event A: Exactly one coin is different.

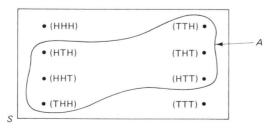

Figure 2.4.1

Therefore, since $P(s) = \frac{1}{8}$ for each $s \in S$,

$$P(A) = P(\text{someone will be eliminated}) = \sum_{\text{all } s \in A} P(s)$$

$$= \frac{1}{8} + \frac{1}{8} + \frac{1}{8} + \frac{1}{8} + \frac{1}{8} + \frac{1}{8}$$

$$= \frac{6}{8}$$

Example 2.4.2

The simplest discrete probability function is the *equally likely* model, where each of an experiment's n possible outcomes is assigned probability $1/n$. Games of chance often have that particular structure, Example 2.4.1 being a case in point. But all of us, whether we are gamblers or not, are already quite familiar with an everyday phenomenon that behaves in accordance with the equally likely model, that phenomenon being your friendly neighborhood photocopy machine.

Suppose that we want to copy 30 pages. When we deposit money for the first page, we are performing a statistical experiment—the outcomes of that experiment are the number of blank pages available in the copier. Our most fervent wish, of course, is that the number of blank pages in the machine's hopper is greater than or equal to the number of pages we need to copy (otherwise frustration ensues, as we are forced to track down the ever-elusive "key operator"). If the copier holds a maximum of 500 blank pages, what is the probability that it runs out of paper?

Assuming that the person preceding us did not leave the machine empty, there will be s blank pages remaining when we put our first coin in, where s is an integer from 1 to 500, inclusive. Furthermore, there is no reason to suppose that the probabilities associated with these 500 outcomes are not all equal.

Figure 2.4.2 is a picture of the sample space and its probability structure.

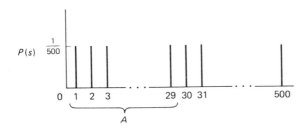

Number of pages remaining, s

Figure 2.4.2

Let A be the event "Machine runs out of paper." Quite clearly, our probability of frustration is 0.058:

$$P(A) = \sum_{\text{all } s \in A} P(s) = P(1) + P(2) + \cdots + P(29)$$

$$= \frac{1}{500} + \frac{1}{500} + \cdots + \frac{1}{500}$$

$$= \frac{29}{500}$$

$$= 0.058$$

Example 2.4.3

A die is loaded in such a way that the probability of any particular face's showing is directly proportional to the number on that face. What is the probability that an even number appears?

To solve this requires that we make use of the fact that the sum of $P(s)$ over S necessarily equals 1. The experiment—tossing a die—generates a sample space containing six outcomes. But the six are not equally likely: by assumption,

$$P(\text{"}i\text{" face appears}) = P(i) = ki, \qquad i = 1, 2, \ldots, 6$$

where k is a constant. From property (b),

$$\sum_{\text{all } s \in S} P(s) = \sum_{i=1}^{6} P(i) = \sum_{i=1}^{6} ki = \frac{6(6+1)}{2} \cdot k = 21k = 1$$

which implies that $k = 1/21$ [and $P(i) = i/21$]. It follows, then, that the probability of A, that the number appearing is even, is 12/21:

$$P(A) = P(\text{even number}) = \sum_{\text{all } s \in A} P(s) = P(2) + P(4) + P(6)$$

$$= \frac{2}{21} + \frac{4}{21} + \frac{6}{21}$$

$$= \frac{12}{21}$$

Example 2.4.4

A fair coin is to be tossed until a head comes up for the first time. What are the chances of that happening on an odd-numbered toss?

Note that the sample space here is countably infinite—and so is the set of outcomes making up the event whose probability we are trying to find. Nevertheless, our approach is basically no different from the steps we followed in Examples 2.4.1 and 2.4.2, except that the analysis requires the evaluation of sums whose indices go from 1 to infinity.

Suppose we let $P(k)$ denote the probability that the first head appears on the kth toss. Since the coin was presumed to be fair, $P(1) = \frac{1}{2}$. Furthermore, we would expect half the coins that showed a tail on the first toss to come up heads on the second, so, intuitively, $\cdot P(2) = \frac{1}{4}$. In general, $P(k) = (\frac{1}{2})^k$, $k = 1, 2, \ldots$ (we will have much more to say about this particular assignment of probabilities in Chapter 4).

Let A be the event "First Head appears on an odd-numbered toss." Then

$$P(A) = P(1) + P(3) + P(5) + \cdots$$

$$= \sum_{i=0}^{\infty} \left(\frac{1}{2}\right)^{2i+1}$$

$$= \frac{1}{2} \sum_{i=0}^{\infty} \left(\frac{1}{4}\right)^{i}$$

Recall the formula for the sum of a geometric series: if $0 < x < 1$,

$$\sum_{k=0}^{\infty} x^k = \frac{1}{1-x}$$

Applying that result to $P(A)$ gives

$$P(A) = \frac{1}{2} \cdot \left(\frac{1}{1 - 1/4}\right)$$

$$= \frac{2}{3}$$

CASE STUDY 2.4.1

For good pedagogical reasons, the principles of probability are always taught by considering events defined on sample spaces generated by the simplest possible experiments. We toss coins, we roll dice, we draw chips from urns. It would be a serious error, though, to infer that the importance of probability extends no further than the nearest gambling table. In its infancy, probability and gambling *were* intimately related. But more than 325 years have passed since Huygens published *De Ratiociniis*. Today, the application of probability to gambling is totally inconsequential compared to the uses it finds in other areas of mathematics, business, medicine, engineering, and science.

Specifically, probability functions—properly chosen—can "model" complex real-world phenomena in much that same way that $P(\text{heads}) = \frac{1}{2}$ describes the behavior of a fair coin. A set of actuarial data provides us with a case in point. Over a period of 3 years ($= 1096$ days) in London, records showed that a total of 903 deaths occurred among males 85 years of age and older (166). Columns 1 and 2 of Table 2.4.1 give the breakdown of those 903 deaths according to the number occurring on a given day. Column 3 gives the *proportion* of days for which exactly s elderly men died.

TABLE 2.4.1

(1) Number of deaths, s	(2) Number of days	(3) Proportion [= Col. (2)/1096]	(4) P(s)
0	484	0.442	0.440
1	391	0.357	0.361
2	164	0.150	0.148
3	45	0.041	0.040
4	11	0.010	0.008
5	1	0.001	0.003
6+	0	0.000	0.000
	1096	1	1

For reasons that will be gone into at length in Chapter 4, the probability function that describes the behavior of this particular phenomenon is

$$P(s) = P(s \text{ elderly men die on a given day})$$

$$= \frac{e^{-0.82}(0.82)^s}{s!}, \qquad s = 0, 1, 2, \ldots \tag{2.4.1}$$

How do we know that the $P(s)$ of Equation 2.4.1 is an appropriate way of assigning probabilities to the "experiment" of elderly men dying? Because it accurately predicts what happened. Column 4 of Table 2.4.1 shows $P(s)$ evaluated for $s = 0, 1, 2, \ldots$. To two decimal places, the agreement between entries in column 3 and entries in column 4 is perfect.

Choosing the probability function best suited to a given phenomenon is one of the most common—and most difficult—tasks faced by a statistician. In gambling situations, the selection of $P(s)$ is usually obvious, as has been the case in the examples seen thus far. In more elaborate contexts, such as the data described in Table 2.4.1, identifying $P(s)$ becomes increasingly difficult. Still, if we know where to look there are clues that can help us make the right choice. How to find probability functions and apply them appropriately are recurring themes of Chapters 3 through 5.

Question 2.4.1 Ace–six flats are a type of crooked dice where the cube is foreshortened in the one–six direction, the effect being that 1's and 6's are more likely than 2's, 3's, 4's, and 5's. Let $P(i) = P(\text{"}i\text{" face appears})$. Suppose that $P(1) = P(6) = \frac{1}{4}$ and $P(2) = P(3) = P(4) = P(5) = \frac{1}{8}$. If two such dice are rolled, what is the probability the sum of the faces will equal 7? [Assume that the probability of rolling an i on the first die and a j on the second is the *product*, $P(i) \cdot P(j)$.] Compare your answer with the probability of rolling a 7 with two fair dice. How would a cheater playing craps use ace–six flats?

Question 2.4.2 Suppose that a rather limited lottery is set up that contains only 12 tickets, numbered 1 through 12. One ticket is to be drawn. Let A be the event "Number on ticket is divisible by 2" and let B be the event "Number on ticket is divisible by 3." Find $P(A \cup B)$.

Question 2.4.3 Recall Example 2.2.3. If the station management did not discriminate when filling the two positions, what is the probability that both coanchors would be women?

Question 2.4.4 Show that

$$P(x) = \frac{1}{1 + \lambda} \left(\frac{\lambda}{1 + \lambda}\right)^x, \qquad x = 0, 1, 2, \ldots; \quad \lambda > 0$$

qualifies as a discrete probability function.

Question 2.4.5 An urn contains five chips, numbered 1 through 5. Three chips are drawn out at random. What is the probability the largest chip in the sample is a 4?

Question 2.4.6 For any outcome in the sample space $S = \{s: s = 2, 3, 4, \ldots\}$, let $P(s) = k \cdot (\frac{2}{3})^s$. Find the value of k that makes $P(s)$ a probability function.

Question 2.4.7 Three fair dice are rolled, one red, one green, and one blue. What is the probability that the three faces are all different and that the blue die equals the sum of the red and the green?

Question 2.4.8 Jean D'Alembert, a renowned eighteenth-century French mathematician, was once asked the following question: What is the probability of getting at least one head in two tosses of a fair coin? D'Alembert answered $\frac{2}{3}$, his argument being that there are three possible outcomes, H, TH, and TT, two of which—H and TH—satisfy the event "at least one head." Discuss his answer and his method of solution.

2.5 CONTINUOUS PROBABILITY FUNCTIONS

Discrete probability functions are *conceptually* easy to define: for experiments whose sample spaces are either finite or countably infinite, $P(s)$ is simply the probability that the outcome is s. Continuous probability functions, in contrast, need to be approached in a more roundabout fashion.

Suppose that the sample space associated with an experiment is an interval of real numbers. That is, S has an uncountably infinite number of outcomes. Let f be a real-valued function defined over S. Then f is said to be a *continuous probability function* if

$$\text{(a)} \quad 0 \leq f(x) \qquad \text{for all } x \in S$$

$$\text{(b)} \quad \int_S f(x)\,dx = 1$$

Furthermore, if A is any event defined on S,

$$P(A) = \int_A f(x)\,dx \tag{2.5.1}$$

(see Figure 2.5.1).

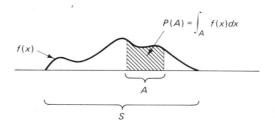

$$P(A) = \int_A f(x)\,dx$$

$f(x)$

A

S

Figure 2.5.1

Equation 2.5.1 is the key to understanding the important distinction between discrete and continuous probability functions. If f is a continuous probability function, $f(x)$ is *not* the probability that the outcome of the experiment is x. Rather, f is that particular function having the property that, for any event A, $P(A)$ is the integral of f over A. Indeed, it follows from Equation 2.5.1 that the probability of any single point, say x', is necessarily 0, even though $f(x')$ may very well be positive:

$$P(\text{outcome is } x') = P(A) = \int_A f(x)\,dx$$

$$= \int_{x'}^{x'} f(x)\,dx$$

$$= 0$$

So, with discrete sample spaces we talk about probabilities associated with *points*; with continuous sample spaces, we talk about probabilities associated with *intervals*, and those probabilities are areas under $f(x)$.

Example 2.5.1

A criminal court judge has heard many cases where the defendant was charged with grand theft auto and the jury returned a guilty verdict. But the final disposition was not always the same. Mitigating circumstances of various kinds led the judge over the years to impose unequal jail terms for what was basically the same offense. Looking back over court transcripts, he sees that x, the imposed sentence length (in years), has a distribution that can be described quite well by a continuous probability function having the form

$$f(x) = \frac{1}{9} x^2 \qquad 0 < x < 3$$

What proportion of those found guilty spent less than a year in jail?

Note, first of all, that $f(x)$ is a legitimate continuous probability function: it satisfies conditions (a) and (b). By inspection, $f(x) \geq 0$ for $0 < x < 3$ and $\int_0^3 \frac{1}{9}x^2\,dx = 1$.

Let A be the event "Prisoner spends less than a year in jail." Then

$$P(A) = \int_A f(x)\, dx$$

$$= \int_0^1 \frac{1}{9} x^2\, dx$$

$$= \frac{1}{27}$$

Figure 2.5.2 shows the area under $f(x)$ corresponding to $P(A)$.

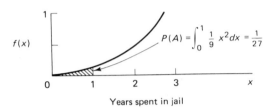

$$P(A) = \int_0^1 \frac{1}{9} x^2 dx = \frac{1}{27}$$

Years spent in jail

Figure 2.5.2

Example 2.5.2

Suppose that an enemy aircraft flies directly over the Alaskan pipeline and fires a single air-to-surface missile. If the missile hits anywhere within 20 feet of the pipeline, major structural damage will be incurred and oil flow will be disrupted. Assume that with a certain sighting device, the (continuous) probability function describing the missile's point of impact is given by

$$f(x) = \begin{cases} \dfrac{60 + x}{3600} & \text{for } -60 < x < 0 \\[2mm] \dfrac{60 - x}{3600} & \text{for } 0 \le x < 60 \\[2mm] 0 & \text{elsewhere} \end{cases}$$

where x is the perpendicular distance from the pipeline to the point of impact. What is the probability that oil flow will be disrupted?

Let A be the event "Flow is disrupted." The probability of A is the area under $f(x)$ above the interval $(-20, +20)$:

$$P(A) = \int_{-20}^{20} f(x)\, dx = \int_{-20}^{0} \frac{60 + x}{3600}\, dx + \int_{0}^{20} \frac{60 - x}{3600}\, dx$$

$$= \left[\frac{60x}{3600} + \frac{x^2}{7200} \right]_{-20}^{0} + \left[\frac{60x}{3600} - \frac{x^2}{7200} \right]_{0}^{20}$$

$$= \frac{1200}{3600} - \frac{400}{7200} + \frac{1200}{3600} - \frac{400}{7200}$$

$$= 0.55$$

Figure 2.5.3 is a graph of $f(x)$ and shows the area representing $P(A)$.

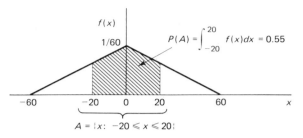

$$P(A) = \int_{-20}^{20} f(x)dx = 0.55$$

$$A = \{x: -20 \leqslant x \leqslant 20\}$$

Figure 2.5.3

Example 2.5.3

In the kinetic theory of gases, the distance, x, that a molecule travels before colliding with another molecule is described probabilistically by an exponential function,

$$f(x) = \left(\frac{1}{\lambda}\right)e^{-x/\lambda} \qquad x > 0$$

where λ is some constant greater than 0 (see Figure 2.5.4). Physicists denote the average distance between collisions as the *mean free path* and define it by the expression

$$\mu = \text{mean free path} = \int_0^\infty x \cdot f(x)\, dx$$

What is the probability that the distance a molecule travels between consecutive collisions is less than half its mean free path?

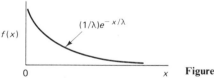

Figure 2.5.4

First, note that the numerical value for the mean free path is just λ:

$$\mu = \int_0^\infty x \cdot \left(\frac{1}{\lambda}\right)e^{-x/\lambda}\, dx$$

$$= \lambda \int_0^\infty y e^{-y}\, dy \qquad (\text{where } y = x/\lambda)$$

$$= \lambda$$

the final step resulting from an integration by parts. Therefore, the probability of the event A, that a molecule's intercollision distance is less than half its mean free path, is given by

$$P(A) = \int_0^{\lambda/2} \left(\frac{1}{\lambda}\right)e^{-x/\lambda}\, dx$$

Making the substitution $y = x/\lambda$, we find that a molecule has a 39% chance of colliding with another molecule before it can traverse half its mean free path:

$$P(A) = \int_0^{1/2} e^{-y} \, dy = -e^{-y}\Big|_0^{1/2} = 1 - e^{-1/2}$$

$$= 0.39$$

10/12 ☆

CASE STUDY   2.5.1

By far the most important of all continuous probability functions is the familiar bell-shaped curve, known more formally as the *normal* (or *Gaussian*) *distribution*. The sample space for the normal distribution is the entire real line; its probability function is given by

$$f(x) = \frac{1}{\sqrt{2\pi}\sigma} \exp\left[-\frac{1}{2}\left(\frac{x-\mu}{\sigma}\right)^2\right]$$

$$-\infty < x < \infty, \quad -\infty < \mu < \infty, \quad \sigma > 0$$

Depending on the values assigned to μ and σ, $f(x)$ takes on a variety of shapes and locations along the x-axis (see Figure 2.5.5).

As an example of a phenomenon that can be modeled by the normal curve, consider the data in Table 2.5.1, which give a recent year's traffic death rates [per 100 million motor vehicle miles (mvm)] for each of our 50 states (102).

TABLE 2.5.1 TRAFFIC DEATHS PER 100 MILLION MVM

Ala.	6.4	La.	7.1	Ohio	4.5
Alaska	8.8	Maine	4.6	Okla.	5.0
Ariz.	6.2	Mass.	3.5	Oreg.	5.3
Ark.	5.6	Md.	3.9	Pa.	4.1
Calif.	4.4	Mich.	4.2	R.I.	3.0
Colo.	5.3	Minn.	4.6	S.C.	6.5
Conn.	2.8	Miss.	5.6	S.Dak.	5.4
Del.	5.2	Mo.	5.6	Tenn.	7.1
Fla.	5.5	Mont.	7.0	Tex.	5.2
Ga.	6.1	N.C.	6.2	Utah	5.5
Hawaii	4.7	N.Dak.	4.8	Va.	4.5
Idaho	7.1	Nebr.	4.4	Vt.	4.7
Ill.	4.3	Nev.	8.0	W.Va.	6.2
Ind.	5.1	N.H.	4.6	Wash.	4.3
Iowa	5.9	N.J.	3.2	Wis.	4.7
Kans.	5.0	N.Mex.	8.0	Wyo.	6.5
Ky.	5.6	N.Y.	4.7		

[handwritten annotations:]
μ = location
σ = scale
1. symmetric wrt μ
2. bell shape
$P(\mu-\sigma \leq x \leq \mu+\sigma) = 68.26\%$
$P(\mu-2\sigma \leq x \leq \mu+2\sigma) = 95.45\%$
standard normal dist
$X \sim N(0,1)$

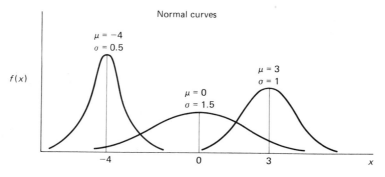

Figure 2.5.5

If we group these data into classes, "2.0–2.9," "3.0–3.9," and so on, we get the histogram shown in Figure 2.5.6. Superimposed over the histogram is the particular normal curve having $\mu = 5.3$ and $\sigma = 1.3$. Quite clearly, $f(x)$ works

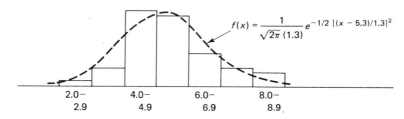

$$f(x) = \frac{1}{\sqrt{2\pi}\,(1.3)}e^{-1/2\,[(x-5.3)/1.3]^2}$$

Traffic deaths per 100 million mvm

Figure 2.5.6

well here—it describes very adequately the state-to-state variability in traffic fatality rates. We will have much more to say about this singularly important distribution—and how the values of μ and σ are estimated—in Chapter 7.

Question 2.5.1 The simplest f that can be defined over a continuous (and bounded) sample space is the *uniform* (or *rectangular*) *distribution*:

$$f(x) = \begin{cases} \dfrac{1}{b-a} & \text{for } a < x < b \\ 0 & \text{elsewhere} \end{cases}$$

Consider a uniform probability function defined over the interval from -1 to 3. Let A be the event "x is less than 2." Find $P(A)$. Graph $f(x)$ and show the area that corresponds to $P(A)$.

Question 2.5.2 In actuarial science, one of the models used for describing mortality is

$$f(t) = k \cdot t^2(100 - t)^2 \qquad 0 \le t \le 100$$

where t denotes the age at which a person dies.
(a) Find k.
(b) Let A be the event " Person lives past 60." Find $P(A)$.

Question 2.5.3 The length of a cotter pin that is part of a wheel assembly is supposed to be 6 cm. The machine that stamps out the parts, though, makes them $6 + x$ cm long, where x varies from pin to pin according to the probability function

$$f(x) = k(x + x^2) \qquad 0 \le x \le 2$$

where k is a constant. If a pin is longer than 7 cm, it is unusable. What proportion of cotter pins produced by this machine will be unusable?

Question 2.5.4 In England from 1875 to 1951 the interval t (in days) between consecutive mining accidents resulting in at least 10 fatalities was well described (96) by the continuous probability function, $f(t) = (1/241)e^{-t/241}$, $t > 0$. Estimate the probability that the gap between consecutive accidents would be somewhere between 50 and 100 days, inclusive.

Question 2.5.5 A batch of small-caliber ammunition is accepted as satisfactory if one shell selected at random is fired and lands within 2 feet of the center of a target. Assume that for a given batch, the probability function describing r, the distance from the target center to a shell's point of impact, is

$$f(r) = \frac{2re^{-r^2}}{1 - e^{-9}} \qquad 0 \le r \le 3$$

Find the probability that the batch will be accepted.

Question 2.5.6 Assume that the reaction time of motorists over the age of 70 to a certain visual stimulus is described by a continuous probability function of the form

$$f(x) = xe^{-x} \qquad x > 0$$

where x is measured in seconds. Let A be the event "Motorist requires longer than 1.5 seconds to react." Find $P(A)$.

Question 2.5.7 Given that

$$f(x) = \frac{1}{\sqrt{2\pi\sigma}} \exp\left[-\frac{1}{2}\left(\frac{x - \mu}{\sigma}\right)^2 \right] \qquad -\infty < x < \infty$$

is a probability function for any number μ and any positive number σ, evaluate

$$\int_0^\infty e^{-4x^2}\, dx$$

2.6 CONDITIONAL PROBABILITY

In this section we will see that the probability of an event A may have to be recomputed if we know for certain that some other event, B, has already occurred. That probabilities *should* change in the light of additional information is certainly not unreasonable. Consider a fair die being tossed, with A defined as the event "6 appears." Clearly, $P(A) = \frac{1}{6}$. But suppose that the die has already been tossed—by someone who refuses to tell us whether or not A occurred but does enlighten us to the point of confirming that B occurred, where B is the event "Even number appears." What are the chances of A now? The answer, of course, is obvious: there are three equally likely even numbers making up the event B—one of them satisfies the event A, so the "updated" probability is $\frac{1}{3}$.

Notice that the effect of additional information, such as the knowledge that B has occurred, is to revise—indeed, to *shrink*—the original sample space S to a new set of outcomes S'. Here, the original S contained six outcomes, the conditional sample space, three (see Figure 2.6.1).

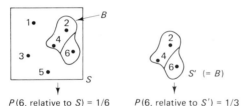

$P(6, \text{relative to } S) = 1/6 \qquad P(6, \text{relative to } S') = 1/3 \qquad$ **Figure 2.6.1**

The symbol $P(A \mid B)$—read "the probability of A given B"—is used to denote a conditional probability. Specifically, $P(A \mid B)$ refers to the probability that A *will occur* given that B *has already occurred*.

It will be convenient to have a formula for $P(A \mid B)$ that can be evaluated in terms of the original S, rather than the revised S'. Motivating such a formula is easy. Suppose that S is a finite sample space with n outcomes and P is the equally likely probability function. Assume that A and B are two events containing a and b outcomes, respectively, and let c denote the number of outcomes in the intersection of A and B (see Figure 2.6.2). Then, from what we have just seen, the conditional

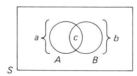

Figure 2.6.2

probability of A given B is the ratio of c to b. But c/b can be written as the ratio of two other ratios:

$$\frac{c}{b} = P(A \mid B) = \frac{c/n}{b/n} = \frac{P(A \cap B)}{P(B)} \qquad (2.6.1)$$

More generally, when the outcomes in S are not equally likely, we make the obvious modification of Equation 2.6.1—we replace the *number* of points in $A \cap B$ and in B with their *probabilities*.

Definition 2.6.1. Let A and B be any two events defined on S such that $P(B) > 0$. The conditional probability of A, assuming that B has already occurred, is written $P(A \mid B)$ and is given by

$$P(A \mid B) = \frac{P(A \cap B)}{P(B)}$$

Notice that Definition 2.6.1 can be reformulated to give an expression for the probability of an intersection: if $P(A \mid B) = P(A \cap B)/P(B)$, then

$$P(A \cap B) = P(A \mid B)P(B) \qquad (2.6.2)$$

Both Equation 2.6.2 and Definition 2.6.1 are extremely important. The next several examples and case studies illustrate some of their applications.

Example 2.6.1

A card is drawn from a poker deck. What is the probability that the card is a club, given that the card is a king?

Intuitively, the answer is $\frac{1}{4}$: the king is equally likely to be a heart, diamond, club, or spade. More formally, let C be the event "Card is a club"; let K be the event "Card is a king." By Definition 2.6.1,

$$P(C \mid K) = \frac{P(C \cap K)}{P(K)}$$

But $P(K) = \frac{4}{52}$ and $P(C \cap K) = P(\text{card is a king of clubs}) = \frac{1}{52}$. Therefore, confirming our suspicion,

$$P(C \mid K) = \frac{1/52}{4/52} = \frac{1}{4}$$

[Notice in this example that the conditional probability $P(C \mid K)$ is numerically the same as the unconditional probability $P(C)$—they both equal $\frac{1}{4}$. This means that our knowledge that K has occurred gives us no additional insight about the chances of C occurring. Two events having this property are said to be *independent*. We will examine the consequences of independence in Section 2.7.]

Example 2.6.2

Home security experts estimate that an untrained house dog has a 70% probability of detecting an intruder—and, given detection, a 50% chance of scaring the intruder away (2). What is the probability that Fido successfully thwarts a burglar?

What is being asked for here is the probability of an intersection. Let A be the event "Fido frightens away burglar" and let B be the event "Fido detects burglar." Then

$$P(\text{Fido thwarts burglar}) = P(\text{Fido detects burglar } and \text{ scares}$$
$$\text{burglar away})$$
$$= P(A \cap B)$$
$$= P(A \mid B) \cdot P(B)$$
$$= (0.50)(0.70)$$
$$= 0.35$$

(The probability of a *trained* watchdog detecting and running off an intruder is estimated to be much higher—on the order of 0.75.)

Example 2.6.3

Recall the probability function describing mortality that was given in Question 2.5.2: if t is the age at which a person dies,

$$f(t) = 3 \times 10^{-9} \cdot t^2(100 - t)^2 \qquad 0 \le t \le 100$$

What is the probability a person will die between the ages of 80 and 85 given that that person has lived to be at least 70?

Let A be the event "$80 \le t \le 85$" and let B be the event "$t > 70$." Then

$$P(\text{person dies between the ages of 80 and 85 given that that}$$
$$\text{person has lived to be at least 70})$$

$$= P(A \mid B) = \frac{P(A \cap B)}{P(B)}$$

$$= \frac{\int_{80}^{85} 3 \times 10^{-9} \cdot t^2(100 - t)^2 \, dt}{\int_{70}^{100} 3 \times 10^{-9} \cdot t^2(100 - t)^2 \, dt}$$

$$= \frac{0.0313}{0.1631}$$

$$= 0.19$$

Example 2.6.4

The possibility of importing liquefied natural gas (LNG) from Algeria has been suggested as one way of coping with a future energy crunch. Complicating matters, though, is the fact that LNG is highly volatile and poses an enormous safety hazard. Any major spill occurring near a U.S. port could result in a fire of catastrophic proportions. The question, therefore, of the *likelihood* of a spill becomes critical input for future policymakers who may have to decide whether or not to implement the proposal.

Two numbers need to be taken into account: (1) the probability that a tanker will have an accident near a port, and (2) the probability that a major spill will develop *given* that an accident has happened. Although no significant spills of LNG have yet occurred anywhere in the world, these probabilities can be approximated from records kept on similar tankers transporting less dangerous cargo. On the basis of such data, it has been estimated (37) that the probability is 8/50,000 that an LNG tanker will have an accident on any one trip. And given that an accident *has* occurred, it is suspected that only three times out of 15,000 will the damage be sufficiently severe that a major spill would be the consequence. Thus, the single-trip probability for a major LNG disaster computes to 3.2×10^{-8}:

$$P(\text{accident occurs and spill develops})$$
$$= P(\text{spill develops} \mid \text{accident occurs}) \cdot P(\text{accident occurs})$$
$$= \left(\frac{3}{15,000}\right)\left(\frac{8}{50,000}\right)$$
$$= 0.000000032$$

[Why do we estimate $P(\text{spill develops} \mid \text{accident occurs})$ and $P(\text{accident occurs})$ *separately*? Why not bypass Definition 2.6.1 entirely and simply estimate $P(\text{accident occurs and spill develops})$?]

CASE STUDY 2.6.1

There once was a brainy baboon
Who always breathed down a bassoon
   For he said, " It appears
   That in billions of years
I shall certainly hit on a tune."
Eddington

The image of a monkey sitting at a typewriter, pecking away at random until he gets lucky and types out a perfect copy of the complete works of William Shakespeare, has long been a favorite model of statisticians and philosophers to illustrate the distinction between something that is theoretically possible but for all practical purposes, impossible. But if that monkey and his typewriter are replaced by a high-technology computer and if we program in the right sorts of conditional probabilities, the prospects for generating *something* intelligible become a little less far-fetched—maybe even disturbingly less far-fetched (8).

Simulating nonnumerical English text requires that 28 characters be dealt with: the 26 letters, the space, and the apostrophe. The simplest approach would be to assign each of those characters a number from 1 to 28. Then a random number in that range would be generated and the character corresponding to that number would be printed. A second random number would be generated, a corresponding second character would be printed; and so on.

Would that be a reasonable model? Of course not. Why should, say, X's have the same chance of being selected as E's when we know that the latter are much more common? At the very least, weights should be assigned to all the characters proportional to their relative probabilities. Table 2.6.1 shows the empirical distribution of the 26 letters, the space, and the apostrophe in the 35,224 characters making up Act III of *Hamlet*. Ranges of random numbers corresponding to each character's frequency are listed in the last column. If two random numbers were generated, say, 27351 and 11616, the computer would print the characters D and O. Doing that, of course, is equivalent to printing a D with probability 0.0312 [= (28425 − 27327)/35244 = 1099/35244] and an O with probability 0.0732 [= (12789 − 10212)/35244 = 2578/35244].

Extending this idea to *sequences* of letters requires an application of Definition 2.6.1. What is the probability, for example, that a T follows an E? By definition,

$$P(T \text{ follows an } E) = P(T \mid E) = \frac{\text{number of ET's}}{\text{number of E's}}$$

The analog of Table 2.6.1, then, would be an array having 28 rows and 28 columns. The entry in the *i*th row and *j*th column would be $P(i \mid j)$, the probability that letter *i* follows letter *j*.

TABLE 2.6.1

Character	Frequency	Probability	Random number range
Space	6934	0.1968	00001–06934
E	3277	0.0930	06935–10211
O	2578	0.0732	10212–12789
T	2557	0.0726	12790–15346
A	2043	0.0580	15347–17389
S	1856	0.0527	17390–19245
H	1773	0.0503	19246–21018
N	1741	0.0494	21019–22759
I	1736	0.0493	22760–24495
R	1593	0.0452	24496–26088
L	1238	0.0351	26089–27326
D	1099	0.0312	27327–28425
U	1014	0.0288	28426–29439
M	889	0.0252	29440–30328
Y	783	0.0222	30329–31111
W	716	0.0203	31112–31827
F	629	0.0178	31828–32456
C	584	0.0166	32457–33040
G	478	0.0136	33041–33518
P	433	0.0123	33519–33951
B	410	0.0116	33952–34361
V	309	0.0088	34362–34670
K	255	0.0072	34671–34925
'	203	0.0058	34926–35128
J	34	0.0010	35129–35162
Q	27	0.0008	35163–35189
X	21	0.0006	35190–35210
Z	14	0.0004	35211–35224

In a similar fashion, conditional probabilities for longer sequences could also be estimated. For example, the probability that an A follows the sequence QU would be the ratio of QUA's to QU's:

$$P(\text{A follows QU}) = P(\text{A} \mid \text{QU}) = \frac{\text{number of QUA's}}{\text{number of QU's}}$$

What does our monkey gain by having a typewriter programmed with probabilities of sequences? Quite a bit. Figure 2.6.3 shows three lines of computer text generated by a program knowing only single-letter frequencies

```
AOOAAORH ONNNDGELC TEFSISO VTALIDMA POESDHEMHIESWON
PJTOMJ FTL FIM TAOFERLMT O NORDEERH HMFIOMRETWOVRCA
OSRIE IEOBOTOGIM NUDSEEWU WHHS AWUA HIDNEVE NL SELTS
```
Figure 2.6.3

(Table 2.6.1). Nowhere does even a single correctly spelled word appear. Contrast that with Figure 2.6.4, showing computer text generated by a program that had been given estimates for conditional probabilities corresponding to all 614,656 ($= 28^4$) *four*-letter sequences. What we get is still garble, but the improvement is astounding—more than 80% of the letter combinations are at least words.

```
A GO THIS BABE AND JUDGEMENT OF TIMEDIOUS RETCH AND NOT LORD
WHAL IF THE EASELVES AND DO AND MAKE AND BASE GATHEM I AY
BEATELLOUS WE PLAY MEANS HOLY FOOL MOUR WORK FROM INMOST
BED BE CONFOULD HAVE MANY JUDGEMENT WAS IT YOU MASSURE'S TO
LADY WOULD HAT PRIME THAT'S OUR THROWN AND DID WIFE FATHER'ST
LIVENGTH SLEEP TITH I AMBITION TO THIN HIM AND FORCE AND LAW'S
MAY BUT SMELL SO AND SPURSELY SIGNOR GENT MUCH CHIEF MIXTURN
```

Figure 2.6.4

One can only wonder how "human" computer-generated text might be if conditional probabilities for, say, seven- or eight-letter sequences were available. Right now they are not, but given the rate our computer technology is developing, they soon will be. When that day comes, our monkey will probably still never come up with text as creative as Hamlet's soliloquy but something ridiculously simpleminded, like a political speech (or a statistics lecture), might not be out of the question.

Question 2.6.1 The table below shows the prediction record of a TV weather forecaster over the past several years. Entries in the table are intersection probabilities.

		Forecast		
		Sunny	Cloudy	Rain
Actual weather	Sunny	0.3	0.05	0.05
	Cloudy	0.04	0.2	0.02
	Rain	0.1	0.04	0.2

(a) What proportion of the days were sunny?

(b) How often was the forecaster wrong?

(c) What was the probability of rain on the days the forecast was "sunny"?

Question 2.6.2 Consider families with two children and assume that the four possible "outcomes"—(younger is a boy, older is a boy), (younger is a boy, older is a girl), and so on—are all equally likely. What is the probability that both children are boys given that at least one is a boy?

Question 2.6.3 Refer to Example 2.6.3. Compare the conditional probability, $P(80 \leq t \leq 85 | t \geq 70)$, with the unconditional probability, $P(80 \leq t \leq 85)$. Which would you expect to be larger?

Question 2.6.4 Suppose that two fair dice are tossed. What is the probability that the sum equals 10 given it exceeds 8?

Question 2.6.5 If $P(A) = a$ and $P(B) = b$, show that

$$P(A|B) \geq \frac{a + b - 1}{b}$$

Question 2.6.6 A sample space S consists of the integers 1 to n, inclusive. Each has an associated probability proportional to its magnitude. One integer is chosen at random. What is the probability that the "1" is chosen given that the number selected is in the first m integers?

Question 2.6.7 If $P(A|B) < P(A)$, show that $P(B|A) < P(B)$.

Question 2.6.8 Suppose that in Question 2.6.2 we ignored the age of the children in two-child families and distinguished only *three* family types: (boy, boy), (girl, boy), and (girl, girl). Would the conditional probability of both children being boys given that at least one is a boy be different from the answer gotten for Question 2.6.2? Explain.

Question 2.6.9 An urn contains 1 red chip and 1 white chip. One is drawn at random. If the chip selected is red, that chip together with two additional red chips are put back into the urn. If it is white, the chip is simply returned to the urn. Then a second chip is drawn. What is the probability of both selections being red?

Question 2.6.10 Part of the support structure for a small pier consists of two oak logs sunk into a riverbank on either side of a wooden platform. It is possible that neither, either, or both of the logs will "settle." On the basis of past experience, a realistic estimate for the chances of each log settling is 0.05. Also, if one settles, the probability of the other doing the same is 0.7. What is the probability that the pier will be level?

Question 2.6.11 A certain kind of avocado is grown in two valleys in southern California. Both areas are sometimes infested with a red aphid that can damage the fruit. Let A be the event that valley X is infested, and B, that valley Y is infested. Suppose that $P(A) = \frac{2}{5}$, $P(B) = \frac{3}{4}$, and $P(A \cup B) = \frac{4}{5}$. If state inspectors find aphids on a shipment of avocadoes coming from valley Y, what is the probability that farmers in valley X are experiencing a similar problem?

Question 2.6.12 Suppose that a probability function is given by

$$f(x) = \left(\frac{1}{\lambda}\right)e^{-x/\lambda} \qquad x > 0$$

Show that the conditional probability of x exceeding some value $s + t$ given that x has already exceeded t is equal to the probability that x exceeds s.

We have seen that conditional probabilities are often useful in evaluating intersection probabilities—that is, $P(A \cap B) = P(A \mid B)P(B) = P(B \mid A)P(A)$. A similar result holds for higher-order intersections. Consider $P(A \cap B \cap C)$. By thinking of $A \cap B$ as a single event—say, D—we can write

$$P(A \cap B \cap C) = P(D \cap C)$$
$$= P(C \mid D)P(D)$$
$$= P(C \mid A \cap B)P(A \cap B)$$
$$= P(C \mid A \cap B)P(B \mid A)P(A)$$

Repeating this same argument for n events, $A_1, A_2, \ldots, A_n$, gives a formula for the general case:

$$P(A_1 \cap A_2 \cap \cdots \cap A_n) = P(A_n \mid A_1 \cap A_2 \cap \cdots \cap A_{n-1})$$
$$\cdot P(A_{n-1} \mid A_1 \cap A_2 \cap \cdots \cap A_{n-2}) \cdot \ldots \cdot P(A_2 \mid A_1) \cdot P(A_1) \qquad (2.6.3)$$

Example 2.6.5

An urn contains 5 white chips, 4 black chips, and 3 red chips. Four chips are drawn sequentially and without replacement. What is the probability of obtaining the sequence (white, red, white, black)?

Figure 2.6.5 shows the evolution of the urn's composition as the desired sequence is assembled. Define the following four events:

A: white chip is drawn on 1st selection

B: red chip is drawn on 2nd selection

C: white chip is drawn on 3rd selection

D: black chip is drawn on 4th selection

Our objective is to find $P(A \cap B \cap C \cap D)$.

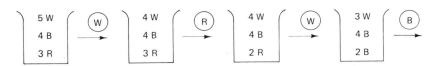

Figure 2.6.5

From Equation 2.6.3,

$$P(A \cap B \cap C \cap D) = P(D \mid A \cap B \cap C) \cdot P(C \mid A \cap B) \cdot P(B \mid A) \cdot P(A)$$

Each of the probabilities on the right-hand side of the equation here can be gotten by just looking at the urns pictured in Figure 2.6.5: $P(D \mid A \cap B \cap C) = \frac{4}{9}$, $P(C \mid A \cap B) = \frac{4}{10}$, $P(B \mid A) = \frac{3}{11}$, and $P(A) = \frac{5}{12}$. Therefore, the probability of drawing a (white, red, white, black) sequence is 0.02:

$$P(A \cap B \cap C \cap D) = \frac{4}{9} \cdot \frac{4}{10} \cdot \frac{3}{11} \cdot \frac{5}{12}$$

$$= \frac{240}{11,880}$$

$$= 0.02$$

CASE STUDY 2.6.2

Since the late 1940s, tens of thousands of eyewitness accounts of strange lights in the skies, unidentified flying objects, even alleged abductions by little green men, have made headlines. None of these incidents, though, has produced any hard evidence, any irrefutable *proof* that Earth has been visited by a race of extraterrestrials. Still, the haunting question remains—are we alone in the universe? Or are there other civilizations, more advanced than ours, waiting for the right moment to make contact?

Until or unless a flying saucer plops down on the White House lawn and a strange-looking creature emerges demanding "Take me to your leader," we may never know whether we have any cosmic neighbors. We can use Equation 2.6.3, though, to speculate on the *probability* of our not being alone.

Although the enormous distances involved have prevented any other planetary systems from being observed, astronomers believe that solar systems like ours are quite common. If so, there are likely to be many planets whose chemical makeups, temperatures, pressures, and so on, are suitable for life. Let those planets be the points in our sample space. Relative to them, we can define three events:

    A: life arises

    B: technical civilization arises (one capable of interstellar communication)

    C: technical civilization is flourishing *now*

In terms of *A*, *B*, and *C*, the probability a habitable planet is presently supporting a technical civilization is the probability of an intersection— specifically, $P(A \cap B \cap C)$. Associating a number with $P(A \cap B \cap C)$ will obviously be difficult; working with its computing formula, $P(C \mid B \cap A) \cdot P(B \mid A) \cdot P(A)$, is also difficult, but a bit more enlightening.

Scientists speculate (143) that life of some kind may arise on one-third of all planets having a suitable environment and that life on maybe 1% of all those planets will evolve into a technical civilization. In our notation, $P(A) = \frac{1}{3}$ and $P(B \mid A) = 1/100$.

More difficult to estimate is $P(C \mid A \cap B)$. On Earth, we have had the capability of interstellar communication (that is, radio astronomy) for only a few decades, so $P(C \mid A \cap B)$, *empirically*, is on the order of 1×10^{-8}. But that may be an overly pessimistic estimate of a technical civilization's ability to endure. It may be true that if a civilization can avoid annihilating itself when it first develops nuclear weapons, its prospects for longevity are fairly good. If that were the case, $P(C \mid A \cap B)$ might be as large as 1×10^{-2}.

Putting these estimates into the computing formula for $P(A \cap B \cap C)$ gives us a range for the probability of a habitable planet currently supporting a technical civilization. The chances may be as small as 3.3×10^{-11} or as "large" as 3.3×10^{-5}:

$$(1 \times 10^{-8})\left(\frac{1}{100}\right)\left(\frac{1}{3}\right) < P(A \cap B \cap C) < (1 \times 10^{-2})\left(\frac{1}{100}\right)\left(\frac{1}{3}\right)$$

or

$$0.000000000033 < P(A \cap B \cap C) < 0.000033$$

A better way to put these figures in some kind of perspective is to think in terms of *numbers* rather than probabilities. Astronomers estimate there are 3×10^{11} habitable planets in our Milky Way galaxy. Multiplying that total by the two limits for $P(A \cap B \cap C)$ gives an indication of *how many* cosmic neighbors we are likely to have. Specifically, $3 \times 10^{11} \cdot 0.000000000033 \doteq 10$, while $3 \times 10^{11} \cdot 0.000033 \doteq 10,000,000$. So, on the one hand, we may be a galactic rarity. At the same time, the probabilities do not preclude the very real possibility that the heavens are abuzz with activity and that our neighbors number in the millions.

Question 2.6.13 An urn contains 6 white chips, 4 black chips, and 5 red chips. Five chips are drawn out, one at a time and without replacement. What is the probability of getting the sequence (black, black, red, white, white)? Suppose that the chips are numbered 1 through 15. What is the probability of getting a specific sequence—say, (2, 6, 4, 9, 13)?

Question 2.6.14 A man has n keys on a key ring, one of which opens the door to his apartment. Having celebrated a bit too much one evening, he returns home only to find himself unable to distinguish one key from another. Resourceful, he works out a fiendishly clever plan: he will choose a key at random and try it. If it fails to open the door, he will discard it and choose at random one of the remaining $n - 1$ keys, and so on. Clearly, the probability that he gains entrance with the first key he selects is $1/n$. Show that the probability the door opens with the *third* key he tries is also $1/n$. (*Hint:* What has to happen before he even gets to the third key?)

Equation 2.6.2 can be extended in still another way. Suppose that S is a sample space partitioned by a set of mutually exclusive and exhaustive events A_1, $A_2, \ldots, A_n$—that is, every sample outcome in S belongs to one and only one A_i (see Figure 2.6.6). Let B be some other event defined over S. What we will derive is a

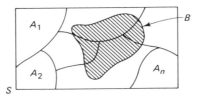

Figure 2.6.6

formula for the unconditional probability for B in terms of the n conditional probabilities, $P(B \mid A_i)$. This will prove to be a fundamentally important result.

Theorem 2.6.1. Let $\{A_i\}_{i=1}^n$ be a set of events defined over S such that $S = \bigcup_{i=1}^{n} A_i$, $A_i \cap A_j = \varnothing$ for $i \neq j$, and $P(A_i) > 0$ for $i = 1, 2, \ldots, n$. For any event B,

$$P(B) = \sum_{i=1}^{n} P(B \mid A_i)P(A_i)$$

Proof.
Because of the first two conditions imposed on the A_i's,

$P(B) = P(B \cap S)$

$$= P\left(B \cap \left(\bigcup_{i=1}^{n} A_i \right) \right) = P\left(\bigcup_{i=1}^{n} (B \cap A_i) \right) \quad \text{(recall Question 2.2.17)}$$

$$= \sum_{i=1}^{n} P(B \cap A_i) \quad \text{(why?)}$$

Apply Equation 2.6.2 to each of the intersection probabilities and the result follows.

Example 2.6.6

In an upstate congressional race, the incumbent Republican (R) is running against a field of three Democrats (D_1, D_2, D_3) seeking the nomination. Political pundits estimate that the probabilities of D_1, D_2, and D_3 winning the primary are 0.35, 0.40, and 0.25, respectively. Furthermore, results from a variety of polls are suggesting that R would have a 40% chance of defeating D_1 in the general election, a 35% chance of defeating D_2, and a 60% chance of defeating D_3. Assuming all these estimates to be accurate, what are the chances that the Republican will retain his seat?

Probability   Chap. 2

Let B denote the event that the Republican wins and let A_i be the event that D_i is nominated, $i = 1, 2, 3$. Then

$$P(B) = \sum_{i=1}^{3} P(B \mid A_i)P(A_i)$$

$$= (0.40)(0.35) + (0.35)(0.40) + (0.60)(0.25)$$

$$= 0.43$$

Example 2.6.7

The Starship *Enterprise* is planning a surprise attack against the Klingons in a neutral quadrant. Possible interference by the Romulans, though, is causing Captain Kirk and Mr. Spock to reassess their strategy. According to Spock's calculations, the probability of the Romulans joining forces with the Klingons is 0.2384. Captain Kirk feels that the probability of the attack's being a success is 0.8 if the *Enterprise* can catch the Klingons alone but only 0.3 if they have to engage both adversaries. Spock claims that the attack would be a tactical misadventure if its probability of success were not at least 0.7306. Should they attack?

No! Let A_1 and A_2 denote the partitioning events " Romulans join in" and "Romulans do not join in," respectively; let B be the event "Attack is a success." Substituting directly into Theorem 2.6.1, we see that the proposed mission's chances of ending favorably are not sufficiently promising:

$$P(B) = P(B \mid A_1)P(A_1) + P(B \mid A_2)P(A_2)$$

$$= (0.3)(0.2384) + (0.80)(0.7616)$$

$$= 0.6808$$

Kirk and Spock would be well advised to obey the Prime Directive for a change!

Example 2.6.8

A toy manufacturer buys ball bearings from three different suppliers—50% of his total order comes from supplier 1, 30% from supplier 2, and the rest from supplier 3. Past experience has shown that the quality control standards of the three suppliers are not all the same. Two percent of the ball bearings produced by supplier 1 are defective, while suppliers 2 and 3 produce defective bearings 3% and 4% of the time, respectively. What proportion of the ball bearings in the toy manufacturer's inventory are defective?

Let A_i be the event "Bearing came from supplier i," $i = 1, 2, 3$. Let B be the event "Bearing in toy manufacturer's inventory is defective." Then

$$P(A_1) = 0.5, \qquad P(A_2) = 0.3, \qquad P(A_3) = 0.2$$

and

$$P(B \mid A_1) = 0.02, \qquad P(B \mid A_2) = 0.03, \qquad P(B \mid A_3) = 0.04$$

Combining these probabilities according to Theorem 2.6.1 gives

$$P(B) = (0.02)(0.5) + (0.03)(0.3) + (0.04)(0.2)$$

$$= 0.027$$

meaning that the manufacturer can expect 2.7% of his ball-bearing stock to be defective.

CASE STUDY   2.6.3

The "drafting" of college basketball star Ralph Sampson by the NBA was one of the more publicized sports stories of 1983. Two teams were competing for Sampson's services—both had "earned" that privilege by having been spectacularly incompetent the year before. In the spring of 1983, a coin-tossing ceremony was held in the office of Larry O'Brien, the league's commissioner. Houston won. With interest naturally focusing on the outcome of the coin toss, no one saw fit to question the rules to which both teams had agreed. Was the procedure by which Houston won the right to negotiate with Sampson "fair"? Yes. Did it make sense, mathematically? No.

The coin-tossing ceremony involved *two* flips of a silver dollar. The purpose of the first was to determine who would call the second. The purpose of the second was to decide which team would have the first pick in the upcoming draft (meaning, in effect, the right to negotiate with Sampson). Houston owner Charlie Thomas won the first toss and then correctly called "heads" on the second.

What is curious about this procedure is that the first toss was entirely superfluous; it had absolutely no effect on the probability of either Houston or Indiana winning the second. If the ceremony had involved only *one* toss, the probability of Houston winning would clearly have been $\frac{1}{2}$. What about the two-toss ceremony? Let B be the event "Houston wins 2nd toss"; let A_1 and A_2 be the events "Houston wins 1st toss" and "Houston loses 1st toss," respectively. By Theorem 2.6.1,

$$P(B) = P(B \mid A_1)P(A_1) + P(B \mid A_2)P(A_2) \tag{2.6.4}$$

But each of the probabilities on the right-hand side of Equation 2.6.4 equals $\frac{1}{2}$ (why?). Therefore,

$$P(B) = \left(\frac{1}{2}\right)\left(\frac{1}{2}\right) + \left(\frac{1}{2}\right)\left(\frac{1}{2}\right)$$

$$= \frac{1}{2}$$

the same winning probability Houston had with the one-toss scenario. [Suppose that unbeknown to any of the participants, the coin was actually biased and $P(\text{heads}) = p \neq \frac{1}{2}$. Would the probability of Houston winning in a one-toss ceremony be something other than $\frac{1}{2}$?]

Question 2.6.15 At the university computing center, 35% of all programs submitted are written in COBOL; the remaining 65% are in FORTRAN. Suppose that 10% of the COBOL programs and 15% of the FORTRAN programs compile on their first run. What is the probability that the next program submitted will compile on its first run?

Question 2.6.16 Recall the "survival" lottery described in Question 2.2.8. What is the probability of release associated with the prisoner's optimal strategy?

Question 2.6.17 Urn I contains two red chips and four white chips; urn II, three red and one white. A chip is drawn at random from urn I and transferred to urn II. Then a chip is drawn from urn II. What is the probability that the chip drawn from urn II is red?

Question 2.6.18 If men constitute 47% of the population and tell the truth 78% of the time, while women tell the truth 63% of the time, what is the probability that a person selected at random will answer a question truthfully?

Question 2.6.19 A card is drawn off the top of a standard deck and dealt face down. Then the next card is turned over. What is the probability that the second card is an ace?

Question 2.6.20 A telephone solicitor is responsible for canvassing three suburbs. In the past, 60% of the completed calls to Belle Meade have resulted in contributions, compared to 55% for Oak Hill and 35% for Antioch. Her list of telephone numbers includes 1000 households from Belle Meade, 1000 from Oak Hill, and 2000 from Antioch. Suppose that she picks a number at random from the list and places the call. What is the probability that she gets a donation?

Question 2.6.21 Urn I contains three red chips and one white chip. Urn II contains two red chips and two white chips. One chip is drawn from each urn and transferred to the other urn. Then a chip is drawn from the first urn. What is the probability that the chip ultimately drawn from urn I is red?

Question 2.6.22 A gambler has three cards: the first is red on both sides; the second, blue on both sides; the third is red on one side and blue on the other. He puts the cards in a hat, shakes the hat thoroughly, draws a card out, and lays it on the table. Suppose that the color showing is red. He then offers to bet you, *even money*, that the side not showing is also red. (In general, he wagers that the side not showing has the same color as the side showing.) Should you bet? [*Hint:* Let D be the event "Red side is down" and let U be the event "Red side is up." Find $P(D \mid U)$.]

The last theorem in this section has an interesting history. Its first explicit statement, coming in 1812, was due to Laplace, but its name derives from the Reverend Thomas Bayes, whose 1763 paper (published posthumously) had already outlined the result. On one level, the theorem is a relatively minor extension of the definition of conditional probability. When viewed from a loftier perspective, though, it takes on some rather profound philosophical implications. These implications, in fact, have precipitated a schism among practicing statisticians: "Bayesians" analyze data one way—"non-Bayesians," another [see (124) or (92)].

Our concern with the result will have nothing to do with its statistical interpretation. We will use it simply as the Reverend Bayes originally intended, as a formula for evaluating a certain kind of "inverse" probability. If we know $P(B \mid A_i)$ for all i, the theorem enables us to compute conditional probabilities "in the other direction"—that is, we can use the $P(B \mid A_i)$'s to find $P(A_j \mid B)$.

Theorem 2.6.2. (*Bayes*) Let $\{A_i\}_{i=1}^n$ be a set of n events, each with positive probability, that partition S in such a way that $\bigcup_{i=1}^n A_i = S$ and $A_i \cap A_j = \emptyset$ for $i \neq j$. For any event B (also defined on S), where $P(B) > 0$,

$$P(A_j \mid B) = \frac{P(B \mid A_j)P(A_j)}{\sum\limits_{i=1}^n P(B \mid A_i)P(A_i)}$$

for any $1 \leq j \leq n$.

Proof

From Definition 2.6.1,

$$P(A_j \mid B) = \frac{P(A_j \cap B)}{P(B)} = \frac{P(B \mid A_j)P(A_j)}{P(B)}$$

But Theorem 2.6.1 allows the denominator to be written as $\sum\limits_{i=1}^n P(B \mid A_i)P(A_i)$ and the result follows.

Example 2.6.9

A weather satellite is sending a binary code of 0's and 1's describing a developing tropical storm. Channel noise, though, can be expected to introduce a certain amount of transmission error. Suppose that the message being relayed is 70% 0's and there is an 80% chance of a given 0 or 1 being received properly. If a "1" is received, what is the probability that a "0" was sent?

Let B_i denote the event "i was sent," $i = 0, 1$, and let A_i denote the event "i is received," $i = 0, 1$. We want to find $P(B_0 \mid A_1)$. According to the given,

$$P(B_0) = 0.7 \qquad \Rightarrow \qquad P(B_1) = 0.3$$

$$P(A_0 \mid B_0) = 0.8 \qquad \Rightarrow \qquad P(A_1 \mid B_0) = 0.2$$

$$P(A_1 \mid B_1) = 0.8 \qquad \Rightarrow \qquad P(A_0 \mid B_1) = 0.2$$

From Theorem 2.6.2, then, the probability of a 0 having been sent given that a 1 was received is 0.37:

$$P(B_0 \mid A_1) = \frac{P(A_1 \mid B_0)P(B_0)}{P(A_1 \mid B_0)P(B_0) + P(A_1 \mid B_1)P(B_1)}$$

$$= \frac{(0.2)(0.7)}{(0.2)(0.7) + (0.8)(0.3)}$$

$$= 0.37$$

Example 2.6.10

A dashboard warning light is supposed to flash red if a car's oil pressure is too low. On a certain model, the probability of the light flashing when it should is 0.99; 2% of the time, though, it flashes for no apparent reason. If there is a 10% chance that the oil pressure really is low, what is the probability that a driver needs to be concerned if the warning light goes on?

Let A_L and A_L^C denote the partitioning events "Oil pressure is low" and "Oil pressure is not low," respectively. Let B denote the event "Red light goes on." Our objective is to find $P(A_L \mid B)$. Written in terms of A_L, A_L^C, and B, what we are given translates into four probability statements:

$$P(B \mid A_L) = 0.99$$

$$P(B \mid A_L^C) = 0.02$$

$$P(A_L) = 0.10 \qquad [\Rightarrow P(A_L^C) = 0.90]$$

By Theorem 2.6.2, then, we see that the driver has an 85% chance for being concerned:

$$P(A_L \mid B) = \frac{P(B \mid A_L)P(A_L)}{P(B \mid A_L)P(A_L) + P(B \mid A_L^C)P(A_L^C)}$$

$$= \frac{(0.99)(0.10)}{(0.99)(0.10) + (0.02)(0.90)}$$

$$= 0.85$$

CASE STUDY 2.6.4

Bayes' theorem has been applied with considerable success to the problem of diagnosing medical conditions—specifically, to estimating the probability that a patient has a certain disease given that a particular diagnostic procedure *says* the patient does. The situation we look at here is one such application and it makes a significant point: that when the disease being checked for is very rare, the number of incorrect diagnoses can be alarmingly high.

Consider the problem of "screening" for cervical cancer (135). Let C be the event of a woman having the disease and B, the event of a positive biopsy—that is, B occurs when the diagnostic procedure indicates that she *does* have cervical cancer. We will assume that $P(C) = 0.0001$, $P(B \mid C) = 0.90$ (the test correctly identifies 90% of all the women who do have the disease), and $P(B \mid C^C) = 0.001$ (the test gives one false positive, on the average, out of every 1000 patients). Find $P(C \mid B)$, the probability a woman actually does have cervical cancer given that the biopsy says she does.

Although the method of solution here is straightforward, the actual numerical answer is not at all what we would expect. From Theorem 2.6.2,

$$P(C \mid B) = \frac{P(B \mid C)P(C)}{P(B \mid C)P(C) + P(B \mid C^C)P(C^C)}$$

$$= \frac{(0.9)(0.0001)}{(0.9)(0.0001) + (0.001)(0.9999)}$$

$$= 0.08$$

That is, only 8% of those women identified as having the disease actually do! Table 2.6.2 shows the strong dependence of $P(C \mid B)$ on $P(C)$ and $P(B \mid C^C)$. In light of these figures, the practicality of large-scale screening programs directed at diseases with low prevalence is open to question, particularly when the diagnostic procedure itself may be a health hazard, as would be the case in using annual chest X-rays to look for tuberculosis.

TABLE 2.6.2

$P(C)$	$P(B \mid C^C)$	$P(C \mid B)$
0.0001	0.001	0.08
	0.0001	0.47
0.001	0.001	0.47
	0.0001	0.90
0.01	0.001	0.90
	0.0001	0.99

Question 2.6.23 Urn I contains 2 white chips and 1 red chip; urn II has 1 white chip and 2 red chips. One chip is drawn at random from urn I and transferred to urn II. Then one chip is drawn from urn II. Suppose that a red chip is selected from urn II. What is the probability that the chip *transferred* was white?

Question 2.6.24 State College is playing Backwater A&M for the conference football championship. If Backwater's first-string quarterback is healthy, A&M has a 75% chance of winning. If they have to start their backup quarterback, their chances of winning drop to 40%. The team physician says that there is a 70% chance that the first-string quarterback will play.
(a) What is the probability that Backwater will win the game?
(b) Suppose that you miss the game but read in the headlines of Sunday's paper that Backwater won. What is the probability that the second-string quarterback started?

Question 2.6.25 Suppose that 0.5% of all the students seeking treatment at a school infirmary are eventually diagnosed as having mononucleosis. Of those who do have mono, 90% complain of a sore throat. But 30% of those not having mono also have sore throats. If a student comes to the infirmary and says that he has a sore throat, what is the probability that he has mono?

Question 2.6.26 During a power blackout, 100 persons are arrested on suspicion of looting. Each is given a polygraph test. From past experience it is known that the polygraph is 90% reliable when administered to a guilty suspect and 98% reliable when given to someone who is innocent. Suppose that of the 100 persons taken into custody, only 12 were actually involved in any wrongdoing. What is the probability that a given suspect is innocent given that the polygraph says he is guilty?

Question 2.6.27 A biased coin, twice as likely to come up heads as tails, is tossed once. If it shows heads, a chip is drawn from urn I, which contains 3 white chips and 4 red chips; if it shows tails, a chip is drawn from urn II, which contains 6 white chips and 3 red chips. Given that a white chip was drawn, what is the probability that the coin came up tails?

Question 2.6.28 Your next-door neighbor has a rather old and temperamental burglar alarm. If someone breaks into his house, the probability of the alarm sounding is 0.95. In the last two years, though, it has gone off on five different nights, each time for no apparent reason. Police records show that the chances of a home being burglarized in your community on any given night are 2 in 10,000. If your neighbor's alarm goes off tomorrow night, what is the probability that his house is being broken into?

Question 2.6.29 A loaded die is tossed for which

$$P(i \text{ appears}) = ki, \qquad i = 1, 2, \ldots, 6$$

If an "i" is rolled, a fair coin is tossed i times. Given that at least one head has appeared, what is the probability that a "2" was rolled?

Question 2.6.30 Bart is planning to murder his rich Uncle Basil in hopes of claiming his inheritance a bit early. Hoping to take advantage of his uncle's predilection for immoderate desserts, Bart has put rat poison in the cherries flambé and cyanide in the chocolate mousse. The probability of the rat poison being fatal is 0.60; the cyanide, 0.90. Based on other dinners he has had with his uncle, Bart estimates that Basil has a 50% chance of asking for the cherries flambé, a 40% chance of ordering the chocolate mousse, and a 10% chance of either requesting something else or skipping dessert altogether. Given that Basil did, indeed, suffer a premature demise, what is the probability that it was the chocolate mousse that did him in?

Question 2.6.31 Josh takes a 20-question multiple-choice exam where each question has five answers. Some of the answers he knows, while others he gets right just by making lucky guesses. Suppose that the conditional probability of his knowing the answer to a randomly selected question given that he got it right is 0.92. How many of the 20 questions was he prepared for?

Question 2.6.32 Recently the U.S. Senate Committee on Labor and Public Welfare investigated the feasibility of setting up a national screening program to detect child abuse. A team of consultants estimated the following probabilities: (1) 1 child in 90 is abused, (2) a physician can detect an abused child 90% of the time, and (3) a screening program would incorrectly label 3% of all nonabused children as abused.

What is the probability that a child is actually abused given that the screening program diagnoses him as such? How does this probability change if the incidence of abuse is 1 in 1000? 1 in 50?

2.7 INDEPENDENCE

Section 2.6 dealt with the problem of reevaluating the probability of a given event in light of the additional information that some other event has already occurred. It sometimes is the case, though, that the probability of the given event remains unchanged, regardless of the outcome of the second event—that is, $P(A \mid B) = P(A) = P(A \mid B^C)$. Events sharing this property are said to be *independent*. Definition 2.7.1 gives a necessary and sufficient condition for two events to be independent.

Definition 2.7.1. Two events A and B are said to be *independent* if $P(A \cap B) = P(A) \cdot P(B)$.

Comment

The fact that the probability of the intersection of two independent events is equal to the product of their individual probabilities follows immediately from our first definition of independence, that $P(A \mid B) = P(A)$. Recall that the definition of conditional probability holds true for *any* two events A and B [provided that $P(B > 0)$]:

$$P(A \mid B) = \frac{P(A \cap B)}{P(B)}$$

But $P(A \mid B)$ can equal $P(A)$ only if $P(A \cap B)$ factors into $P(A)$ times $P(B)$.

Example 2.7.1

Let K be the event of drawing a king from a standard poker deck and D, the event of drawing a diamond. Then, by Definition 2.7.1, K and D are independent because the probability of their intersection—drawing a king of diamonds—is equal to $P(K) \cdot P(D)$:

$$P(K \cap D) = \frac{1}{52} = \frac{1}{13} \cdot \frac{1}{4} = P(K) \cdot P(D)$$

Example 2.7.2

Suppose that A and B are independent events. Does it follow that A^C and B^C are also independent? That is, does $P(A \cap B) = P(A) \cdot P(B)$ guarantee that $P(A^C \cap B^C) = P(A^C) \cdot P(B^C)$?

Yes. The proof is accomplished by equating two different expressions for $P(A^C \cup B^C)$. First, by Theorem 2.3.6,

$$P(A^C \cup B^C) = P(A^C) + P(B^C) - P(A^C \cap B^C) \qquad (2.7.1)$$

But the union of two complements is the complement of their intersection (recall Question 2.2.16). Therefore,

$$P(A^C \cup B^C) = 1 - P(A \cap C) \qquad (2.7.2)$$

Combining Equations 2.7.1 and 2.7.2, we get

$$1 - P(A \cap B) = 1 - P(A) + 1 - P(B) - P(A^C \cap B^C)$$

Since A and B are independent, $P(A \cap B) = P(A) \cdot P(B)$, so

$$P(A^C \cap B^C) = 1 - P(A) + 1 - P(B) - [1 - P(A) \cdot P(B)]$$

$$= [1 - P(A)][1 - P(B)]$$

$$= P(A^C) \cdot P(B^C)$$

the latter factorization implying that A^C and B^C are, themselves, independent. (If A and B are independent, are A and B^C independent?)

It is not immediately obvious how to extend Definition 2.7.1 to, say, *three* events. To call A, B, and C independent, should we require that the probability of the three-way intersection factors into the product of the three original probabilities,

$$P(A \cap B \cap C) = P(A) \cdot P(B) \cdot P(C) \qquad (2.7.3)$$

or should we impose the definition we already have on the three *pairs* of events:

$$P(A \cap B) = P(A) \cdot P(B)$$

$$P(B \cap C) = P(B) \cdot P(C) \qquad (2.7.4)$$

$$P(A \cap C) = P(A) \cdot P(C) \quad ?$$

As the next two examples show, neither condition by itself is sufficient. If three events satisfy Equations 2.7.3 *and* 2.7.4, we will call them independent (or *mutually independent*), but Equation 2.7.3 does not imply Equation 2.7.4, nor does Equation 2.7.4 imply Equation 2.7.3.

Example 2.7.3

Suppose that two fair dice (one red and one green) are thrown, with events A, B, and C defined as follows:

A: a 1 or a 2 shows on the red die
B: a 3, 4, or 5 shows on the green die
C: the dice total is 4, 11, or 12

By direct summation, it is a simple matter to show that $P(A) = \frac{1}{3}$, $P(B) = \frac{1}{2}$, $P(C) = \frac{1}{6}$, $P(A \cap B) = \frac{1}{6}$, $P(A \cap C) = \frac{1}{18}$, $P(B \cap C) = \frac{1}{18}$, and $P(A \cap B \cap$

$C) = \frac{1}{36}$. Note that Equation 2.7.3 is satisfied,

$$P(A \cap B \cap C) = \frac{1}{36} = P(A) \cdot P(B) \cdot P(C) = \left(\frac{1}{3}\right)\left(\frac{1}{2}\right)\left(\frac{1}{6}\right)$$

but Equation 2.7.4 is not:

$$P(B \cap C) = \frac{1}{18} \neq P(B) \cdot P(C) = \left(\frac{1}{2}\right)\left(\frac{1}{6}\right) = \frac{1}{12}$$

Example 2.7.4

A roulette wheel has 36 numbers colored red or black according to the pattern indicated in Figure 2.7.1. Let R be the event "Red number appears"; E, the event "Even number appears"; and T, the event "Total is ≤ 18." Then $P(R) = P(E) = P(T) = \frac{1}{2}$, $P(R \cap E) = \frac{1}{4}$, $P(E \cap T) = \frac{1}{4}$, and $P(R \cap T) = \frac{1}{4}$.

Roulette wheel pattern

1	2	3	4	5	6	7	8	9	10	11	12	13	14	15	16	17	18
R	R	R	R	R	B	B	B	B	R	R	R	R	B	B	B	B	B
36	35	34	33	32	31	30	29	28	27	26	25	24	23	22	21	20	19

Figure 2.7.1

Clearly, R, E, and T are all "pairwise" independent (that is, Equation 2.7.4 holds), yet the probability of the three-way intersection does not factor in accordance with Equation 2.7.3:

$$P(R \cap E \cap T) = \frac{4}{36} = \frac{1}{9} \neq P(R) \cdot P(E) \cdot P(T) = \left(\frac{1}{2}\right)^3 = \frac{1}{8}$$

The upshot of Examples 2.7.3 and 2.7.4 is that for n events to be independent, the probabilities of *all* possible intersections must factor into the product of the probabilities of the component events. Definition 2.7.2 gives the formal statement.

> **Definition 2.7.2.** Events $A_1, A_2, \ldots, A_n$ are said to be *independent* if for every set of indices $i_1, i_2, \ldots, i_k$ between 1 and n, inclusive,
>
> $$P(A_{i_1} \cap A_{i_2} \cap \cdots \cap A_{i_k}) = P(A_{i_1}) \cdot P(A_{i_2}) \cdot \ldots \cdot P(A_{i_k})$$

Example 2.7.5

Suppose that a fair coin is flipped three times. Let H_1 be the event of a head on the first flip; T_2, a tail on the second flip; and H_3, a head on the third flip. Are H_1, T_2, and H_3 independent?

Note, first of all, that

$$P(H_1) = P(T_2) = P(H_3) = \frac{1}{2}$$

Also,

$$P(H_1 \cap T_2) = P(\text{HTH, HTT}) = \frac{2}{8} = \left(\frac{1}{2}\right)\left(\frac{1}{2}\right) = P(H_1) \cdot P(T_2)$$

Similarly,

$$P(H_1 \cap H_3) = \frac{2}{8} = \left(\frac{1}{2}\right)\left(\frac{1}{2}\right) = P(H_1) \cdot P(H_3)$$

and

$$P(T_2 \cap H_3) = \frac{2}{8} = \left(\frac{1}{2}\right)\left(\frac{1}{2}\right) = P(T_2) \cdot P(H_3)$$

Finally,

$$P(H_1 \cap T_2 \cap H_3) = P(\text{HTH}) = \frac{1}{8} = \left(\frac{1}{2}\right)^3 = P(H_1) \cdot P(T_2) \cdot P(H_3)$$

By Definition 2.7.2, then, events H_1, T_2, and H_3 *are* independent.

Question 2.7.1   Suppose that two events A and B, each having nonzero probability, are mutually exclusive. Are they also independent?

Question 2.7.2   A large company is responding to an affirmative-action commitment by setting up hiring quotas by race and sex for office personnel. So far they have agreed to employ the 120 people indicated in the following table. How many black women must they include if they want to be able to claim that the race and sex of the people on their staff are independent?

	White	Black
Male	50	30
Female	40	

Question 2.7.3   Spike is not a terribly bright student. His chances of passing chemistry are 0.35; mathematics, 0.40; and both, 0.12. Are the events "Spike passes chemistry" and "Spike passes mathematics" independent? What is the probability that he fails both subjects?

Question 2.7.4   How many probability equations need to be verified to establish the mutual independence of *four* events?

Question 2.7.5   In a roll of a pair of fair dice (one red and one green), let A be the event the red die shows a 3, 4, or 5; let B be the event the green die shows a 1 or a 2; and let C be the event the dice total is 7. Show that A, B, and C are independent.

Question 2.7.6   In a roll of a pair of fair dice (one red and green), let A be the event of an odd number on the red die, let B be the event of an odd number on the green die, and let C be the event that the sum is odd. Show that any pair of these events are independent but that A, B, and C are not mutually independent.

Question 2.7.7   If $A_1, A_2, \ldots, A_n$ are independent events, show that

$$P(A_1 \cup A_2 \cup \cdots \cup A_n) = 1 - [1 - P(A_1)] \cdot [1 - P(A_2)] \cdot \ldots \cdot [1 - P(A_n)]$$

While the previous examples in this section have focused on the problem of investigating whether or not a set of events are independent (by examining the conditions spelled out in Definition 2.7.2), there are many situations where the independence of $A_1, A_2, \ldots, A_n$ follows immediately from physical considerations. In these cases we can turn the definition around and use it to provide us with an easy method for evaluating probabilities of intersections.

Example 2.7.6

An insurance company has three clients—one in Alaska, one in Missouri, and one in Vermont—whose estimated chances of living to the year 2000 are 0.7, 0.9, and 0.3, respectively. What is the probability that by the end of 1999 the company will have had to pay death benefits to exactly one of the three?

Let A be the event "Alaska client survives through 1999." Define M and V analogously. Then the event E: "Exactly one dies" can be written as the union of three intersections:

$$E = (A \cap M \cap V^C) \cup (A \cap M^C \cap V) \cup (A^C \cap M \cap V)$$

Since each of the intersections is mutually exclusive of the other two,

$$P(E) = P(A \cap M \cap V^C) + P(A \cap M^C \cap V) + P(A^C \cap M \cap V)$$

Furthermore, there is no reason to believe that for all practical purposes the fates of the three are not independent. That being the case, each of the intersection probabilities reduces to a product, and we can write

$$P(E) = P(A) \cdot P(M) \cdot P(V^C) + P(A) \cdot P(M^C) \cdot P(V) + P(A^C) \cdot P(M) \cdot P(V)$$

$$= (0.7)(0.9)(0.7) + (0.7)(0.1)(0.3) + (0.3)(0.9)(0.3)$$

$$= 0.543$$

Comment

"Declaring" events independent for reasons other than those prescribed in Definition 2.7.2 is a necessarily subjective endeavor. Here we might feel fairly certain that a "random" person dying in Alaska will not affect the survival chances of a "random" person residing in Missouri (or Vermont). But there may be special circumstances that invalidate that sort of argument. For example, what if the three individuals in question were mercenaries fighting in an African border war and were all crew members assigned to the same helicopter? In general, all we can do is look at each situation on an individual basis and try to make a reasonable judgment as to whether the occurrence of one event is likely to influence the outcome of another.

Example 2.7.7

Suppose that one of the genes associated with the control of carbohydrate metabolism exhibits two alleles—a dominant W and a recessive w. If the probabilities of the WW, Ww, and ww genotypes in the present generation are p, q, and r, respectively, for both males and females, what are the chances that an individual in the *next* generation will be a ww?

Let A_w denote the event that an offspring receives a w allele from its father; let B_w denote the event that it receives the recessive allele from its mother. What we are being asked to find is $P(A_w \cap B_w)$.

According to the information given,

$$p = P(\text{parent has genotype WW}) = P(WW)$$

$$q = P(\text{parent has genotype Ww}) = P(Ww)$$

$$r = P(\text{parent has genotype ww}) = P(ww)$$

If an offspring is equally likely to receive either of its parent's alleles, the probabilities of A_w and B_w can be easily computed using Theorem 2.6.1:

$$P(A_w) = P(A_w \mid WW)P(WW) + P(A_w \mid Ww)P(Ww) + P(A_w \mid ww)P(ww)$$

$$= 0 \cdot p + \frac{1}{2} \cdot q + 1 \cdot r$$

$$= r + \frac{q}{2} = P(B_w)$$

We are not provided here with any explicit statement regarding the relationship between A_w and B_w but it certainly seems reasonable to presume that the two are independent (why?). Under that hypothesis, the intersection probability we are looking for factors into the two components we have just computed:

$$P(A_w \cap B_w) = P(\text{offspring has genotype ww})$$

$$= P(A_w) \cdot P(B_w)$$

$$= \left(\frac{r + q}{2}\right)^2$$

(The model for allele segregation that we have used here, together with the independence assumption, is called *random Mendelian mating*.)

Example 2.7.8

Diane and Lew are health physicists employed by the Department of Public Health. One of their duties is being "on call" during nonworking hours to handle any nuclear-related incidents that might endanger the public safety. Each carries a pager that can be activated by personnel at Civil Defense. Lew is a conscientious worker and is within earshot of his pager 80% of the time. Not nearly as reliable, Diane is capable of responding to a pager alert only 40% of the time. If Diane and Lew report into Civil Defense independently, what is the probability that at least one of them could be contacted in the event of a nuclear emergency?

Let D and L denote the events "Diane responds to alert" and "Lew responds to alert," respectively. The response capability we are trying to find is the probability of a union, $P(D \cup L)$. From Theorem 2.3.6,

$$P(D \cup L) = P(D) + P(L) - P(D \cap L)$$

By assumption, $P(D) = 0.40$ and $P(L) = 0.80$. Also, since D and L are independent, $P(D \cap L) = P(D) \cdot P(L) = (0.4)(0.8) = 0.32$. There is an 88% chance, then, that *someone* will respond to an alert:

$$P(D \cup L) = 0.4 + 0.8 - 0.32$$
$$= 0.88$$

CASE STUDY 2.7.1

In 1964, a woman shopping in Los Angeles had her purse snatched by a young, blond female wearing a ponytail. The thief fled on foot but was seen shortly thereafter getting into a yellow automobile driven by a black male who had a mustache and a beard. A police investigation subsequently turned up a suspect, one Janet Collins, who was blond, wore a ponytail, and associated with a black male who drove a yellow car and had a mustache. An arrest was made.

Not having any tangible evidence, and no reliable witnesses, the prosecutor sought to build his case on the unlikelihood of Ms. Collins and her companion sharing these characteristics and not being the guilty parties. First, the bits of evidence that were available were assigned probabilities. It was estimated, for example, that the probability of a female wearing a ponytail in Los Angeles was $\frac{1}{10}$. Table 2.7.1 lists the probabilities quoted for the six "facts" agreed on by the victim and the eyewitnesses (38).

TABLE 2.7.1

Characteristic	Probability
Yellow automobile	$\frac{1}{10}$
Man with a mustache	$\frac{1}{4}$
Woman with a ponytail	$\frac{1}{10}$
Woman with blond hair	$\frac{1}{3}$
Black man with beard	$\frac{1}{10}$
Interracial couple in car	$\frac{1}{1000}$

The prosecutor multiplied these six numbers together and claimed that the product, $(\frac{1}{10})(\frac{1}{4}) \cdots (\frac{1}{1000})$, or 1 in 12 million, was the probability of the intersection—that is, the probability that a random couple would fit this description. A probability of 1 in 12 million is so small, he argued, that the only reasonable decision is to find the defendants guilty. The jury agreed, and handed down a verdict of second-degree robbery. Later, though, the Supreme Court of California disagreed. Ruling on an appeal, the higher court reversed the decision, claiming that the probability argument was incorrect and misleading. What do you think? (We will reexamine this evidence in Case Study 2.11.1 and offer a counterargument to the case presented by the prosecution.)

Question 2.7.8 Two fair dice are rolled. What is the probability that the number showing on one will be twice the number appearing on the other?

Question 2.7.9 Urn I has 3 red chips, 2 black chips, and 5 white chips; urn II has 2 red, 4 black, and 3 white. One chip is drawn at random from each urn. What is the probability that both chips are the same color?

Question 2.7.10 Suppose that for both men and women the distribution of blood types in the general population can be summarized by the following figures:

Blood type	Probability
A	0.40
B	0.10
AB	0.05
O	0.45

What is the probability that a man and woman getting married will have different blood types? What assumption are you making? Is it reasonable?

Question 2.7.11 If you were counsel for the defense, how would you counter the prosecution's probability argument given in Case Study 2.7.1?

Question 2.7.12 School board officials are debating whether to require all high school seniors to take a proficiency exam before graduating. A student passing all three parts (mathematics, language skills, and general knowledge) would be awarded a diploma; otherwise, he would receive only a certificate of attendance. A practice test given to this year's 9500 seniors resulted in the following numbers of failures:

Subject area	Number of students failing
Mathematics	3325
Language skills	1900
General knowledge	1425

If "Student fails mathematics," "Student fails language skills," and "Student fails general knowledge" are independent events, what proportion of next year's seniors can be expected to fail to qualify for a diploma? Does independence seem a reasonable assumption in this situation?

Question 2.7.13 Recall Example 2.7.8. Suppose that Mike, with a 60% response probability, is added to the "team" and functions independently of Diane and Lew. If an emergency occurs, what is the probability that at least one of the three will respond?

Question 2.7.14 Consider the following five-switch circuit:

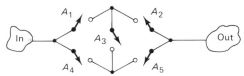

If all switches operate independently and P(switch closes) $= p$, what is the probability the circuit is completed?

Question 2.7.15 Each of m urns contains 3 red chips and 4 white chips. A total of r samples with replacement are taken from each urn. What is the probability that at least one red chip is drawn from at least one urn?

Question 2.7.16 Three points, X_1, X_2, and X_3, are chosen at random in the interval (O, a). A second set of three points, Y_1, Y_2, and Y_3, are chosen at random in the interval (O, b). Let A be the event that X_2 is between X_1 and X_3. Let B be the event that $Y_1 < Y_2 < Y_3$. Find $P(A \cap B)$.

Question 2.7.17 In a certain corporation, protocol for making major decisions follows the flowchart shown below.

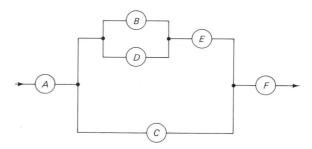

Any proposal is first screened by A. If he approves it, the document is forwarded to B, C, and D. If either B or D concurs, it goes to E. If either E or C say "yes," it moves on to F for a final reading. Only if F is also in agreement does the measure pass. Suppose that A, C, and F each has a 50% chance of saying "yes," whereas B, D, and E will each concur with probability 0.70. If everyone comes to a decision independently, what is the probability that a proposal will pass?

2.8 REPEATED INDEPENDENT TRIALS

It is not uncommon for an experiment to be the composite of a finite or countably infinite number of "subexperiments," each of the latter being performed under essentially the same conditions. We have already seen simple examples of this sort of structure—for instance, tossing a coin twice. In this section we show how useful this model can be by applying it to several more complicated situations.

In general, the subexperiments comprising an experiment are referred to as *trials*. We will restrict our attention here to problems where the trials are independent—that is, for all j, the probability of any given outcome occurring on the jth trial is unaffected by what happened on the preceding $j - 1$ trials. For outcomes $E_1, E_2, \ldots$ defined on a set of such trials,

$$P(E_1 \cap E_2 \cap \cdots) = P(E_1) \cdot P(E_2) \cdot \ldots$$

Example 2.8.1

On her way to work, a commuter encounters four traffic signals. The distance between each of the four is sufficiently great that the probability of getting a green light at any intersection is independent of what happened at any prior intersection. If each light is green for 40 seconds of every minute, what is the probability that the driver has to stop at least three times?

Here the experiment of driving to work can be thought of as the composite of four independent trials, each trial representing a traffic light. The sample space for each of the trials has two possible outcomes, G: "Light is green" and R: "Light is red." By assumption, $P(G) = \frac{40}{60} = \frac{2}{3}$ and $P(R) = \frac{20}{60} = \frac{1}{3}$.

$$
A = \begin{cases}
\dfrac{R}{1} & \dfrac{R}{2} & \dfrac{R}{3} & \dfrac{R}{4} \\[2ex]
\dfrac{R}{1} & \dfrac{R}{2} & \dfrac{R}{3} & \dfrac{G}{4} \\[2ex]
\dfrac{R}{1} & \dfrac{R}{2} & \dfrac{G}{3} & \dfrac{R}{4} \\[2ex]
\dfrac{R}{1} & \dfrac{G}{2} & \dfrac{R}{3} & \dfrac{R}{4} \\[2ex]
\dfrac{G}{1} & \dfrac{R}{2} & \dfrac{R}{3} & \dfrac{R}{4}
\end{cases}
$$

Figure 2.8.1

Figure 2.8.1 shows the five different sequences of outcomes that make up the event A, that the driver has to stop at least three times. Since the sequences are mutually exclusive and the trials are independent,

$$
\begin{aligned}
P(A) = {} & P(R) \cdot P(R) \cdot P(R) \cdot P(R) + P(R) \cdot P(R) \cdot P(R) \cdot P(G) \\
& + P(R) \cdot P(R) \cdot P(G) \cdot P(R) + P(R) \cdot P(G) \cdot P(R) \cdot P(R) \\
& + P(G) \cdot P(R) \cdot P(R) \cdot P(R) \\
= {} & \left(\frac{1}{3}\right)^4 + 4\left(\frac{1}{3}\right)^3 \left(\frac{2}{3}\right)^1 \\
= {} & \frac{1}{9}
\end{aligned}
$$

Example 2.8.2

During the 1978 baseball season, Pete Rose of the Cincinnati Reds set a National League record by hitting safely in 44 consecutive games. Assume that Rose is a .300 hitter and that he comes to bat four times each game. If each at-bat is assumed to be an independent event, what probability might reasonably be associated with a hitting streak of that length?

For this problem we need to invoke the repeated independent trials model *twice*—once for the four at-bats making up a game and a second time for the 44 games making up the streak. Let A_i denote the event "Rose hits safely in ith game," $i = 1, 2, \ldots, 44$. Then

$$
\begin{aligned}
P(\text{Rose hits safely in 44 consecutive games}) \\
= P(A_1 \cap A_2 \cap \cdots \cap A_{44}) \\
= P(A_1) \cdot P(A_2) \cdot \ldots \cdot P(A_{44})
\end{aligned}
\tag{2.8.1}
$$

Since all the $P(A_i)$'s are equal, we can further simplify Equation 2.8.1 by writing

$$P(\text{Rose hits safely in 44 consecutive games}) = [P(A_1)]^{44}$$

To conclude the problem we need a numerical value for $P(A_1)$. It will help to think of the *complement* of A_1. Specifically,

$$P(A_1) = 1 - P(A_1^C)$$

$$= 1 - P(\text{Rose does } not \text{ hit safely in Game 1})$$

$$= 1 - P(\text{Rose makes four outs})$$

$$= 1 - (0.700)^4 \qquad (\text{why?})$$

$$= 0.76$$

Therefore, the probability of a .300 hitter putting together a 44-game streak (during a given set of 44 games) is 0.0000057:

$$P(\text{Rose hits safely in 44 consecutive games})$$

$$= (0.76)^{44}$$

$$= 0.0000057$$

Example 2.8.3

Repeated independent trials problems sometimes involve experiments consisting of a countably infinite number of subexperiments. Conceptually, these problems are approached no differently than the two we have just seen. But algebraically, they require one additional technique, the formula for the sum of a geometric series: if $0 < p < 1$,

$$\sum_{K=0}^{\infty} p^k = \frac{1}{1 - p} \tag{2.8.2}$$

Probabilities associated with games of chance often make use of Equation 2.8.2, a case in point being the familiar game of craps (recall Question 2.2.7). In this example we will translate the rules of craps into equations and compute the probability of the *shooter* (the person rolling the dice) winning.

There are two basic ways the shooter can win: (1) by throwing either a 7 or an 11 on his first roll (this is called a *natural*); or (2) by throwing either a 4, 5, 6, 8, 9, or 10 on his first roll and then throwing that number again *before* he rolls a 7 (this is called *making his point*). Let A_1 be the event that the shooter throws a natural, and let A_4, A_5, A_6, A_8, A_9, and A_{10} be the events that the shooter eventually wins when his point is a 4, 5, 6, 8, 9, or 10, respectively. The A_i's are mutually exclusive, so

$$P(\text{shooter wins}) = P(A_1) + P(A_4) + P(A_5) + P(A_6)$$

$$+ P(A_8) + P(A_9) + P(A_{10})$$

The probability of throwing a natural is easy to compute:

$$P(A_1) = P(7 \text{ or } 11) = P(7) + P(11) = \frac{6}{36} + \frac{2}{36} = \frac{8}{36}$$

To determine the remaining $P(A_i)$'s, we need to think of the game as a series of repeated independent trials. For example, the shooter will win with a point of 4 if he rolls a 4 on the first throw and a 4 on the second *or* a 4 on the first, something other than a 4 or a 7 on the second, and a 4 on the third *or* a 4 on the first, something other than a 4 or a 7 on the second and third, and a 4 on the fourth, and so on. Let B be the event that something other than a 4 or a 7 occurs. Then, appealing again to the fact that these events are all mutually exclusive, we can write

$$P(A_4) = P(4 \text{ on 1st} \cap 4 \text{ on 2nd}) + P(4 \text{ on 1st} \cap B \text{ on 2nd} \cap 4 \text{ on 3rd})$$

$$+ P(4 \text{ on 1st} \cap B \text{ on 2nd} \cap B \text{ on 3rd} \cap 4 \text{ on 4th}) + \cdots$$

By inspection, $P(4) = \frac{3}{36}$ and $P(B) = \frac{27}{36}$. Since each roll is an independent trial,

$$P(A_4) = \left(\frac{3}{36}\right)\left(\frac{3}{36}\right) + \left(\frac{3}{36}\right)\left(\frac{27}{36}\right)\left(\frac{3}{36}\right) + \left(\frac{3}{36}\right)\left(\frac{27}{36}\right)\left(\frac{27}{36}\right)\left(\frac{3}{36}\right) + \cdots$$

$$= \left(\frac{3}{36}\right)^2 \sum_{k=0}^{\infty} \left(\frac{27}{36}\right)^k = \left(\frac{3}{36}\right)^2 \left[\frac{1}{1 - (27/36)}\right]$$

$$= \frac{1}{36}$$

The other $P(A_i)$'s are calculated similarly. Of course, because of symmetry we need only to determine $P(A_4)$, $P(A_5)$, and $P(A_6)$: since the probability of throwing a 4 is the same as the probability of throwing a 10, $P(A_4) = P(A_{10})$—also, $P(A_5) = P(A_9)$ and $P(A_6) = P(A_8)$. Table 2.8.1 summarizes the results. Adding the seven entries in the second column gives the probability that the shooter wins:

$$P(\text{shooter wins}) = \frac{8}{36} + \frac{1}{36} + \cdots + \frac{1}{36}$$

$$= 0.493$$

TABLE 2.8.1

Winning event, A_i	$P(A_i)$
A_1	$\frac{8}{36}$
A_4	$\frac{1}{36}$
A_5	$\frac{16}{360}$
A_6	$\frac{25}{396}$
A_8	$\frac{25}{396}$
A_9	$\frac{16}{360}$
A_{10}	$\frac{1}{36}$

As even-money games of chance go, craps is relatively "fair"—the probability of the shooter winning is not much less than 0.500. On the other hand, the game goes very quickly so a player can still manage to lose a lot of money in a short period of time.

Comment

Money can also be *won* very quickly playing craps. In 1980, a man walked into the Horseshoe Club in downtown Las Vegas carrying two suitcases—one empty, the other stuffed with $777,000 worth of $100 bills. Strolling over to a craps table, he bet the full amount against the shooter. The woman rolling the dice at the table first threw a six, then a nine, *then a seven*. She lost, he won! Moments later, after making a little stop at the cashier's window, he left the casino with *two* suitcases full of money. It was the largest recorded single bet in the history of Las Vegas.

Question 2.8.1 If two fair dice are tossed, what is the smallest number of throws, n, for which the probability of getting at least one double six exceeds 0.5? (*Note*: This was one of the first problems that de Méré communicated to Pascal in 1654.)

Question 2.8.2 A string of eight Christmas tree lights is wired in series. If the probability of any particular bulb failing sometime during the holiday season is 0.05, and if the failures are independent events, what is the probability that the lights will not remain lit?

Question 2.8.3 In a certain Third World nation, statistics show that only 2000 out of 10,000 children born in the early 1960s reached the age of 21. If the same mortality rate is operative over the next generation, how many children should a couple plan to have if they want to be at least 75% certain that at least one of their offspring survives to adulthood?

Question 2.8.4 Players A, B, and C toss a fair coin in order. The first to throw a head wins. What are their respective chances of winning?

Question 2.8.5 A penny may be fair or it may have two heads. We toss it n times and it comes up heads on each occasion. If our initial judgment was that both options for the coin ("fair" or "both sides heads") were equally probable, what is our revised judgment in light of the data?

Question 2.8.6 Suppose that four people (A, B, C, and D) each toss a fair die in order (A first, then B, and so on) until the "6" face appears for the first time. What is the probability that C is the one who rolls the first 6?

Question 2.8.7 A fair die is rolled until a 5 shows. What is the probability that it will take k rolls for that to happen? What is the probability of the first 5 appearing on an even-numbered roll?

Question 2.8.8 Andy, Bob, and Charley have gotten into a disagreement over a female acquaintance, Donna, and decide to settle their dispute with a three-cornered pistol duel. Of the three, Andy is the worst shot, hitting his target only 30% of the time. Charley, a little better, is on-target 50% of the time, while deadeye Bob never misses. The rules they agree to are simple: they are to fire at the target of their choice in succession, and cyclically, in the order Andy, Bob, Charley, Andy, Bob, Charley, and so on, until only one of them is left standing. On each "turn," they get only one shot. Show that Andy's optimal strategy (assuming that he wants to maximize his probability of survival) is to fire his first shot into the ground.

2.9 COMBINATORICS

Combinatorics is a time-honored branch of mathematics concerned with counting, arranging, and ordering. While blessed with a wealth of early contributors (there are references to combinatorial problems in the Old Testament), its emergence as a separate discipline is often credited to the German mathematician and philosopher Gottfried Wilhelm Leibniz (1646–1716), whose 1666 treatise, *Dissertatio de arte combinatoria*, was perhaps the first monograph written on the subject (97).

Applications of combinatorics are rich in both diversity and number. Users range from the molecular biologist trying to determine how many ways genes can be positioned along a chromosome, to a computer scientist studying queueing priorities, to a psychologist modeling the way we learn, to a weekend poker player wondering whether he should draw to a straight or to a flush. Space restrictions dictate that we limit our brief treatment of combinatorics to those results that will reappear later in the book; for a more thorough coverage of the subject, see (177), (41), or (94).

The obligatory place to begin our discussion is with a patently transparent result, yet one so basic to all combinatorial problems that to call it merely a "theorem" would considerably understate its pervasiveness. It will be referred to here as the multiplication rule.

Multiplication Rule. If operation A can be performed in m different ways and operation B in n different ways, the sequence (operation A, operation B) can be performed in $m \cdot n$ different ways.

Proof

At the risk of belaboring the obvious, we can verify the multiplication rule by considering a *tree* diagram (see Figure 2.9.1). Since each version of A can be

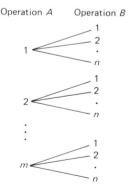

Figure 2.9.1

followed by any of n versions of B, and there are m of the former, the total number of "A, B" sequences that can be pieced together is obviously the product $m \cdot n$.

> **Corollary.** If operation A_i, $i = 1, 2, \ldots, k$, can be performed in n_i ways, $i = 1, 2, \ldots, k$, respectively, then the sequence (operation A_1, operation A_2, ..., operation A_k) can be performed in $n_1 \cdot n_2 \cdot \ldots \cdot n_k$ ways.

Example 2.9.1

In situations where access to a computer must be controlled and monitored, authorized users are assigned a *password*, a sequence of letters and numbers having a prescribed format. Only if the password is keyed in properly can a person log on to the system. How many passwords are possible if each must be in the form

letter   letter   number   number ?

In the terminology of the corollary to the multiplication rule, a password is an ordered sequence of $k = 4$ operations. Clearly, operations A_1 and A_2 can each be "performed" in 26 ways; operations A_3 and A_4, in 10 ways. The total number of distinct passwords, then, is $26 \cdot 26 \cdot 10 \cdot 10$, or *67,600*.

Example 2.9.2

A braille letter is formed by raising at least one dot in a six-dot matrix,

The letter "e," for example, is written

Punctuation marks also have specified dot patterns, as do certain common words, suffixes, and so on. In all, how many different characters can be transcribed using braille?

Think of the dots here, numbered from 1 to 6, as representing six potential operations. In forming a braille letter, we have two options for each dot: we can (1) raise it or (2) *not* raise it. For instance, the letter "e" corresponds to the six-step sequence (raise, don't raise, don't raise, don't raise, raise, don't raise) (see Figure 2.9.2). By the multiplication rule, the six dots generate 2^6 distinct patterns. Figure 2.9.3 shows the entire braille alphabet. [Note that one of the 2^6 patterns has *no* raised dots, a configuration of no use to a blind person. Therefore, the number of *admissible* patterns—that is, the number of letters in the braille alphabet—is *63* ($= 2^6 - 1$).]

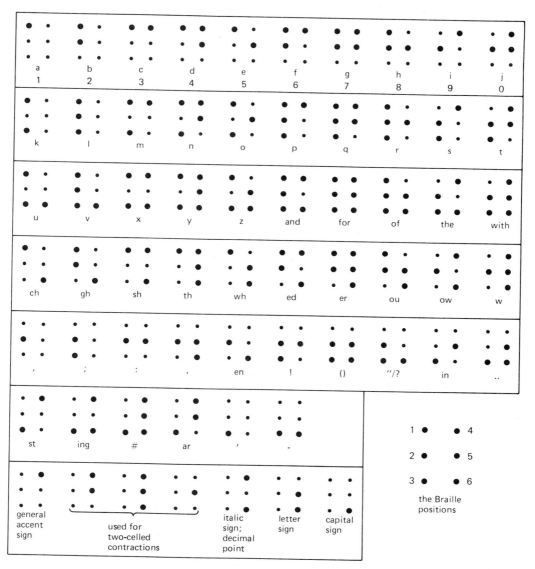

"e" =

Figure 2.9.2

Figure 2.9.3

Jai alai is a three-wall game similar to racquetball except that players wear a curved wicker basket (a *cesta*) over one hand to help them catch and return the ricocheting ball, which travels at speeds in excess of 150 mph. Although popular in many parts of the western hemisphere, jai alai playing in the United States is confined primarily to several cities in Florida, where parimutuel betting is allowed.

In Palm Beach, six jai alai matches, each involving eight teams, are played a night. Anyone picking all six victorious teams wins a special pot; if no one gets all six, that night's money is added to the next night's, and so on, until someone eventually gets lucky. Not long ago, a string of 147 nights went by without anyone picking all six winners (114); accumulated in the pot was more than half a million dollars!

At that point, an enterprising (and anonymous) group of "investors," obviously familiar with the rudiments of combinatorics, entered the picture. Since each match will have one of eight possible winners, the total number of outcomes possible for an entire evening is 8^6, or 262,144. Fortunate enough to have a spare $524,288 lying around the house, the group bought a $2 ticket on each of the 262,144 possible outcomes. Theirs was essentially a sure bet. The only risk involved was the remote possibility that some unsuspecting tourist from Des Moines would be lucky enough to cover the winners by chance, in which case the pot would have to be shared. But that failed to happen; they were the only ones to pick all six. After taxes were deducted, they returned home some $464,038.20 richer—and no doubt with fond memories of the multiplication rule.

Question 2.9.1 A restaurant offers a choice of 4 apetizers, 14 entrees, 6 desserts, and 5 beverages. In how many ways can a diner "design" his evening meal, assuming that he is hungry enough to elect one option from each of the four categories?

Question 2.9.2 A coded message from a CIA operative to his Russian KGB counterpart is to be sent in the form Q4ET, where the first and last entries must be consonants; the second, an integer 1 through 9; and the third, one of the six vowels. How many different ciphers can be transmitted?

Question 2.9.3 How many terms will be included in the expansion of

$$(a + b + c)(d + e + f)(x + y + u + v + w) \quad ?$$

Which of the following will be included in that number: *aeu, cdx, bef, xvw*?

Question 2.9.4 An octave contains 12 distinct notes (on a piano, five black keys and seven white keys). How many different eight-note melodies within a single octave can be written using the white keys only?

Question 2.9.5 Suppose that the format for license plates in a certain state is two letters followed by four numbers.

(a) How many different plates can be made?

(b) How many different plates are there if the letters can be repeated but no two numbers can be the same?

(c) How many different plates can be made if repetitions of numbers and letters is allowed except that no plate can have four zeros?

Question 2.9.6 How many integers between 100 and 999 have distinct digits, and how many of those are odd numbers?

Question 2.9.7 A fast-food restaurant offers customers a choice of eight toppings that can be added to a hamburger. How many different hamburgers can be ordered?

Question 2.9.8 In baseball there are 24 different "base-out" configurations (runner on first—two outs, bases loaded—none out, and so on). Suppose that a new game, sleazeball, is played where there are seven bases (excluding home plate) and each team gets five outs an inning. How many base-out configurations would be possible in sleazeball?

Question 2.9.9 In international Morse code, each letter in the alphabet is symbolized by a series of dots and dashes: the letter "a," for example, is encoded as "· –". What is the maximum number of dots and/or dashes needed to represent any letter in the English alphabet?

The problem of counting ordered sequences has a number of variations. One of these is sufficiently important for our purposes that it needs to be singled out. Imagine a finite set of n distinct elements. We will call any ordered arrangement of k of those elements a *permutation* (of length k). For example, given three elements—A, B, and C—there are six different permutations of length 2: AB, AC, BC, BA, CA, and CB.

> **Theorem 2.9.1.** The number of permutations of length k that can be formed from a set of n distinct elements, repetitions not allowed, is
> $$n(n-1)(n-2) \cdots (n-k+1) = \frac{n!}{(n-k)!}$$

Proof. Any of the n objects may occupy the first position in the arrangement, any of $n-1$ the second, and so on—the number of choices available for filling the kth position will be $n-k+1$ (see Figure 2.9.4). The theorem follows, then, from the multiplication rule: there will be $n(n-1) \cdots (n-k+1)$ ordered arrangements.

Choices: $\dfrac{(n)}{1}$ $\quad \dfrac{(n-1)}{2}$ $\quad \cdots \quad$ $\dfrac{(n-k+2)}{k-1}$ $\quad \dfrac{(n-k+1)}{k}$

Permutations of length k

Figure 2.9.4

Corollary. The number of ways to permute an entire set of n distinct objects is $n!$.

Example 2.9.3

How many permutations of length $k = 3$ can be formed from the set of $n = 4$ distinct elements, A, B, C, and D?

According to Theorem 2.9.1, the number should be 24:

$$\frac{n!}{(n-k)!} = \frac{4!}{(4-3)!} = \frac{4 \cdot 3 \cdot 2 \cdot 1}{1} = 24$$

Confirming that figure, Table 2.9.1 lists the entire set of 24 permutations and illustrates the argument used in the proof of the theorem.

TABLE 2.9.1

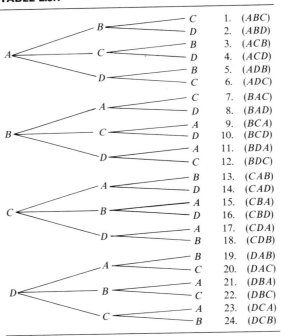

Example 2.9.4

Years ago—long before Rubik cube fever had become epidemic—puzzles were much simpler. One of the more popular combinatorial-related diversions was a 4×4 grid consisting of 15 movable squares and one empty space. The object was to maneuver as quickly as possible an arbitrary configuration (Figure 2.9.5a) into a specific pattern (Figure 2.9.5b). How many different ways could the puzzle be arranged?

Probability   Chap. 2

13	1	8	7
6	9	3	11
1	10	■	4
5	12	15	14

(a)

1	2	3	4
5	6	7	8
9	10	11	12
13	14	15	

(b)

Figure 2.9.5

More than you might think. Take the empty space to be square number 16 and imagine the four rows of the grid laid end to end to make a 16-digit sequence. Each permutation of that sequence corresponds to a different pattern for the grid. By the corollary to Theorem 2.9.1, the number of ways to arrange the puzzle is 16!, more than 20 trillion (20,922,789,888,000, to be exact). *That total is more than 50 times the number of stars in the entire Milky Way galaxy.* (*Note*: Not all of the 16! permutations can be generated without physically removing some of the tiles. Think of the 2 × 2 version of Figure 2.9.5 with tiles numbered 1 through 3. How many of the 4! theoretical configurations can actually be formed?)

Comment

Computing $n!$ can be quite cumbersome, even for n's that are fairly small—16! is already in the trillions. Calculators can help for small n, but even their capacity is quickly exceeded. Fortunately, there is a fairly easy to use approximation: according to *Stirling's formula*,

$$n! \doteq \sqrt{2\pi}\, n^{n+1/2} e^{-n}$$

In practice, we apply Stirling's formula by writing

$$\log(n!) \doteq \log(\sqrt{2\pi}) + \left(n + \frac{1}{2}\right) \log(n) - n \log(e)$$

and then exponentiating the right-hand side.

Example 2.9.5

A deck of 52 cards is shuffled and dealt face up in a row. In how many ways can the four aces be adjacent?

Like many combinatorial problems, this one cannot be solved by simply substituting into a formula. We first need to break the problem down into simpler components. Figure 2.9.6 shows there are 49 basic positionings of the *set* of four adjacent aces relative to the other 48 cards. (Conceptually, we can put the four aces in front of the first nonace, behind the last nonace, or in any of the 47 spaces *between* the nonaces.)

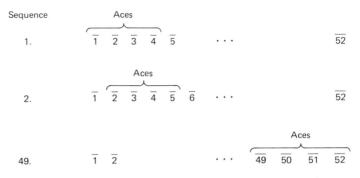

Figure 2.9.6

Consider the first sequence pictured in Figure 2.9.6. Given that the aces are to be located in the first four places, they can still be permuted in 4! ways (by the corollary to Theorem 2.9.1). Similarly, the 48 nonaces can be permuted in 48! ways. Furthermore, since each of the 4! permutations of the aces can be matched with any of the 48! permutations of the nonaces, the total number of arrangements generated by that first sequence is the product, $4! \cdot 48!$ (by the multiplication rule).

The argument we just gave, of course, can be applied to *any* of the 49 sequences indicated in Figure 2.9.6. It follows, then, that the total number of ways the four aces can be adjacent is $49 \cdot 4! \cdot 48!$, a number on the order of 10^{64}.

Question 2.9.10 The board of a large corporation has six members willing to be nominated for office. How many different "president/vice-president/treasurer" slates could be submitted to the stockholders?

Question 2.9.11 How many ways can a set of four tires be put on a car if all the tires are interchangeable? How many ways are possible if two of the four are snow tires?

Question 2.9.12 Use Stirling's formula to approximate 30!. (*Note*: The exact answer is 265,252,859,812,268,935,315,188,480,000,000.)

Question 2.9.13 In her sonnet with the famous first line, "How do I love thee? Let me count the ways," Elizabeth Barrett Browning listed eight. Suppose that Ms. Browning had gone into the greeting-card business and composed verses using all possible choices and orderings of four of those ways. For how many years could she have corresponded with her favorite beau on a daily basis and never sent the same card twice?

Question 2.9.14 How many ways can a 12-member cheerleading squad (6 men and 6 women) pair up to form 6 male–female teams?

Question 2.9.15 Four men and four women are to be seated in a row of chairs numbered 1 through 8.

(a) How many total arrangements are possible?

(b) How many arrangements are possible if the men are required to sit in alternate chairs?

Question 2.9.16 Four Nigerians (*A*, *B*, *C*, *D*), three Chinese (#, *, &), and three Greeks (α, β, γ) are lined up at the box office, waiting to buy tickets for the World's Fair.

(a) How many ways can they position themselves if the Nigerians are to hold the first four places in line; the Chinese, the next three; and the Greeks, the last three?

(b) How many arrangements are possible if members of the same nationality must stay together?

(c) How many different queues can be formed?

Question 2.9.17 Suppose that 10 people, including you and a friend, line up for a group picture. How many ways can the photographer rearrange the line if she wants to keep exactly three people between you and your friend?

Question 2.9.18 A new horror movie, "Friday the 13th Part VIII," stars Jason's great-grandson as a psychotic trying to dismember 8 camp counselors, 4 men and 4 women. In how many ways can he dispatch all the men before any of the women?

Question 2.9.19 Theorem 2.9.1 was the first mathematical result known to have been proved by induction, that feat being accomplished in 1321 by Levi ben Gerson. Assume that we do not know the multiplication rule. Prove the theorem the way Levi ben Gerson did.

Question 2.9.20 In how many ways can a pack of 52 cards be dealt to 13 players, 4 to each, so that every player has one card of each suit?

Question 2.9.21 If the definition of $n!$ is to hold for all nonnegative integers n, show that it follows that $0!$ must equal 1.

Order is not always a meaningful characteristic of a collection of elements, our efforts earlier in this section notwithstanding. Consider a poker player being dealt a five-card hand. Whether he receives a 2 of hearts, 4 of clubs, 9 of clubs, jack of hearts, and ace of diamonds *in that order*, or in any one of the other $5! - 1$ permutations of those particular five cards is irrelevant—the hand is still the same. As the last set of examples in this section bear out, there are many such situations—problems where our only legitimate concern is with the composition of a set of elements, not their precise order.

We call a collection of *k unordered* elements a *combination of size k*. For example, given a set of $n = 4$ distinct elements—*A*, *B*, *C*, and *D*—there are *six* ways to form combinations of size 2:

$$A \text{ and } B \qquad B \text{ and } C$$

$$A \text{ and } C \qquad B \text{ and } D$$

$$A \text{ and } D \qquad C \text{ and } D$$

A general formula for counting combinations can be derived quite easily from what we already know about counting permutations.

Theorem 2.9.2. The number of ways to form combinations of size k from a set of n distinct objects, repetitions not allowed, is denoted by the symbol $\binom{n}{k}$, where

$$\binom{n}{k} = \frac{n!}{k!(n-k)!}$$

Proof. Let the symbol $\binom{n}{k}$ denote the number of combinations satisfying the conditions of the theorem. Since each of those combinations can be ordered in $k!$ ways, the product $k!\binom{n}{k}$ must equal the number of *permutations* of length k that can be formed from n distinct elements. But n distinct elements can be formed into permutations of length k in $n(n-1)\cdots(n-k+1) = n!/(n-k)!$ ways. Therefore,

$$k!\binom{n}{k} = \frac{n!}{(n-k)!}$$

Solving for $\binom{n}{k}$ gives the result.

Comment

It often helps to think of combinations in the context of drawing objects out of an urn. If an urn contains n chips labeled 1 through n, the number of ways we can reach in and draw out different samples of size k is $\binom{n}{k}$. In deference to this sampling interpretation for the formation of combinations, $\binom{n}{k}$ is usually read "n things taken k at a time" or "n choose k."

Comment

The symbol $\binom{n}{k}$ appears in the statement of a familiar theorem from algebra,

$$(x + y)^n = \sum_{k=0}^{n} \binom{n}{k} x^k y^{n-k}$$

Since the expression being raised to a power involves *two* terms, x and y, the constants $\binom{n}{k}$, $k = 0, 1, \ldots, n$, are commonly referred to as *binomial coefficients*.

Example 2.9.6

Eight politicians meet at a fund-raising dinner. How many greetings can be exchanged if each politician shakes hands with every other politician exactly once?

Imagine the politicians to be eight chips—1 through 8—in an urn. A handshake corresponds to an unordered sample of size 2 chosen from that urn. Since repetitions are not allowed (even the most obsequious and overzealous of campaigners would not shake hands with himself!), Theorem 2.9.2 applies, and the total number of handshakes is

$$\binom{8}{2} = \frac{8!}{2!6!}$$

or *28*.

Example 2.9.7

The basketball recruiter for Swampwater Tech has scouted 16 former NBA starters that he thinks he can pass off as JUCO transfers—six are guards, seven are forwards, and three are centers. Unfortunately, his slush fund of illegal alumni donations is at an all-time low and he can afford to buy new Trans-Ams for only nine of the players. If he wants to keep 3 guards, 4 forwards, and 2 centers, how many ways can he parcel out the cars?

This is a combination problem that also requires an application of the multiplication rule. First, note there are $\binom{6}{3}$ *sets* of three guards that could be chosen to receive Trans-Ams (think of drawing a set of three names out of an urn containing six names). Similarly, the forwards and centers can be bribed in $\binom{7}{4}$ and $\binom{3}{2}$ ways, respectively. It follows from the multiplication rule, then, that the total number of ways to divvy up the cars is the product

$$\binom{6}{3} \cdot \binom{7}{4} \cdot \binom{3}{2}$$

or *2100* $(= 20 \cdot 35 \cdot 3)$.

Example 2.9.8

Binomial coefficients have many interesting properties. One of the most famous involves *Pascal's triangle*, a numerical array where each entry is equal to the sum of the two figures appearing diagonally above it (see Figure 2.9.7). If we number the rows (beginning with 0 for the top if the triangle) and the columns (going from left to right), each entry in Figure 2.9.7 can be replaced by a binomial coefficient (see Figure 2.9.8). From the latter array we see that

$$\begin{array}{ccccccccc} & & & & 1 & & & & \\ & & & 1 & & 1 & & & \\ & & 1 & & 2 & & 1 & & \\ & 1 & & 3 & & 3 & & 1 & \\ 1 & & 4 & & 6 & & 4 & & 1 \\ & \vdots & & & & & & \vdots & \end{array}$$

Figure 2.9.7

$$\begin{array}{ccccccccc} & & & & \binom{0}{0} & & & & \\ & & & \binom{1}{0} & & \binom{1}{1} & & & \\ & & \binom{2}{0} & & \binom{2}{1} & & \binom{2}{2} & & \\ & \binom{3}{0} & & \binom{3}{1} & & \binom{3}{2} & & \binom{3}{3} & \\ \binom{4}{0} & & \binom{4}{1} & & \binom{4}{2} & & \binom{4}{3} & & \binom{4}{4} \\ & \vdots & & & & & & \vdots & \end{array}$$

Figure 2.9.8

the formation of Pascal's triangle can be summarized with a single equation:

$$\binom{n+1}{k} = \binom{n}{k} + \binom{n}{k-1}$$

(2.9.1)

Equation 2.9.1 is called a *combinatorial identity*—it holds for any values of n and k for which the symbols are defined.

Question 2.9.22 The crew of Apollo 17 consisted of two pilots and one geologist. Suppose that NASA had actually trained a total of nine pilots and four geologists. How many possible Apollo 17 crews could have been formed?
(a) Assume that the two pilot positions have identical duties.
(b) Assume that the two pilot positions are really a pilot and a copilot.

Question 2.9.23 How many straight lines can be drawn between five points (A, B, C, D, and E), no three of which are collinear?

Question 2.9.24 Vanessa is a cabaret singer who always opens her act by telling four jokes. Her current engagement is scheduled to run for four months. If she gives one performance a night and never wants to repeat the same set of jokes on any two nights, what is the minimum number of jokes she must have in her repertoire?

Question 2.9.25 The Alpha Beta Zeta sorority is trying to fill a pledge class of nine new members during fall rush. Among the 25 available candidates, 15 have been judged marginally acceptable and 10, highly desirable. How many ways can the pledge class be chosen to give a two-to-one ratio of highly desirable to marginally acceptable candidates?

Question 2.9.26 Prove the identity cited in Example 2.9.8, that

$$\binom{n+1}{k} = \binom{n}{k} + \binom{n}{k-1}$$

Question 2.9.27 A man wants to walk from X to Y, passing through intersection O in the process. How many different paths can he take, assuming that he never wants to go out of his way?

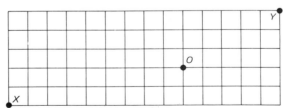

Question 2.9.28 Prove that $\sum_{k=0}^{n} \binom{n}{k} = 2^n$.

Question 2.9.29 Compare the coefficients of t^k in $(1 + t)^d(1 + t)^e = (1 + t)^{d+e}$ to prove that

$$\sum_{j=0}^{k} \binom{d}{j}\binom{e}{k-j} = \binom{d+e}{k}$$

2.10 COMBINATORIAL PROBABILITY

In Section 2.9 our concern focused on counting the number of ways a given operation, or sequence of operations, could be performed. In Section 2.10 we want to couple those enumeration results with the notion of probability. Putting the two together makes a lot of sense—there are many combinatorial problems where an enumeration, by itself, does not provide information particularly relevant. A poker player, for example, is not interested in knowing the total *number* of ways he can draw to an inside straight; he *is* interested, though, in his *probability* of drawing to an inside straight.

In a combinatorial setting, making the transition from an enumeration to a probability is easy. If there are n ways to perform a certain operation and a total of m of those satisfy some stated condition—call it A—then $P(A)$ is defined to be the ratio, m/n. This assumes, of course, that all possible outcomes are equally likely.

Historically, the "m over n" idea is what motivated the early work of Pascal, Fermat, and Huygens (recall Section 1.1). Today we recognize that not all probabilities are so easily characterized. Nevertheless, the m/n model—the so-called *classical* definition of probability—is entirely appropriate for describing a wide variety of phenomena.

Example 2.10.1

An urn contains eight chips, numbered 1 through 8. A sample of three is drawn without replacement. What is the probability that the largest chip in the sample is a "5"?

Let A be the event "Largest chip in sample is a 5." Figure 2.10.1 shows what must happen in order for A to occur: (1) the "5" chip must be selected,

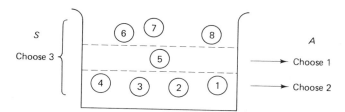

Figure 2.10.1

and (2) two chips must be drawn from the subpopulation of chips numbered 1 through 4. By the multiplication rule, the number of samples satisfying event A is the product $\binom{1}{1} \cdot \binom{4}{2}$.

The sample space S for the experiment of drawing three chips from the urn contains $\binom{8}{3}$ outcomes, all equally likely. In this situation, then, $m = \binom{1}{1} \cdot \binom{4}{2}$, $n = \binom{8}{3}$, and

$$P(A) = \frac{\binom{1}{1} \cdot \binom{4}{2}}{\binom{8}{3}}$$

$$= \frac{1 \cdot 6}{56}$$

$$= 0.11$$

Example 2.10.2

One of the more instructive—and to some, one of the more useful—applications of combinatorics is the calculation of probabilities associated with various poker hands. It will be assumed in what follows that five cards are dealt from a poker deck and that no other cards are showing, although some may already have been dealt. The sample space is the set of $\binom{52}{5} = 2{,}598{,}960$ different hands, each having probability 1/2,598,960. We will derive the probabilities of being dealt (a) a *full house*, (b) *one pair*, and (c) a *straight*. Probabilities for the various other kinds of poker hands (two pairs, three-of-a-kind, flush, and so on) are gotten in much the same way.

  (a) *Full house.* A full house consists of three cards of one denomination and two of another. Denominations for the three-of-a-kind can be chosen in $\binom{13}{1}$ ways. Then, given that a denomination has been decided on, the three requisite suits can be selected in $\binom{4}{3}$ ways. Applying the same reasoning to the pair gives $\binom{12}{1}$ available denominations, each having $\binom{4}{2}$ possible choices of suits. Thus, by the multiplication rule,

$$P(\text{full house}) = \frac{\binom{13}{1}\binom{4}{3}\binom{12}{1}\binom{4}{2}}{\binom{52}{5}} = 0.00144$$

  (b) *One pair.* To qualify as a one-pair hand, the five cards must include two of the same denomination and three "single" cards—cards whose denominations match neither the pair nor each other. For the pair, there are $\binom{13}{1}$ possible denominations and, once selected, $\binom{4}{2}$ possible suits. Denominations for the three single cards can be chosen $\binom{12}{3}$ ways (see Question 2.10.8), and each card can have any of $\binom{4}{1}$ suits. Multiplying all these factors together and dividing by $\binom{52}{5}$ gives a probability of 0.42:

$$P(\text{one pair}) = \frac{\binom{13}{1}\binom{4}{2}\binom{12}{3}\binom{4}{1}\binom{4}{1}\binom{4}{1}}{\binom{52}{5}} = 0.42$$

(c) *Straight.* A straight is five cards having consecutive denominations *but not all in the same suit*—for example, a 4 of diamonds, 5 of hearts, 6 of hearts, 7 of clubs, and 8 of diamonds. An ace may be counted "high" or "low," which means that (10, jack, queen, king, ace) is a straight and so is (ace, 2, 3, 4, 5). (If five consecutive cards are all in the same suit, the hand is called a *straight flush.* The latter is considered a fundamentally different type of hand in the sense that a straight flush "beats" a straight.) To get the numerator for P(straight) we will first ignore the condition that all five cards not be in the same suit and simply count the number of hands having consecutive denominations. Note there are 10 sets of consecutive denominations of length 5: (ace, 2, 3, 4, 5), (2, 3, 4, 5, 6), ..., (10, jack, queen, king, ace). With no restrictions on the suits, each card can be either a diamond, heart, club, or spade. It follows, then, that the number of five-card hands having consecutive denominations is $10 \cdot \binom{4}{1}^5$. But 40 ($= 10 \cdot 4$) of those hands are straight flushes. Therefore,

$$P(\text{straight}) = \frac{10 \cdot \binom{4}{1}^5 - 40}{\binom{52}{5}} = 0.00392$$

Table 2.10.1 shows the probabilities associated with all the different poker hands. In general, hand i beats hand j if $P(\text{hand } i) < P(\text{hand } j)$.

TABLE 2.10.1

Hand	Probability
One pair	0.42
Two pairs	0.048
Three-of-a-kind	0.021
Straight	0.0039
Flush	0.0020
Full house	0.0014
Four-of-a-kind	0.00024
Straight flush	0.000014
Royal flush	0.0000015

Example 2.10.3

There are a number of problems in probability whose solutions defy our intuition. Perhaps the most famous of these is the *birthday problem*. On hearing it for the first time, everyone's reaction is the same: the answer is absurd. The solution, though, is easily derived using only the basic ideas from Section 2.9, and the mathematics is unassailably correct.

Suppose that a group of k randomly selected individuals is assembled. What is the probability that at least two will have the same birthday? Most people would guess that for k relatively small—say, less than 50—a match would be very unlikely. Not so. The odds of at least one match are better than 50–50 if as few as 23 people are present. And when the group does number 50, there is a 97% chance of at least one match!

The counting results we need to appeal to are the multiplication rule and Theorem 2.9.1. Imagine a group of k people, lined up in a row. Omitting leap year, each of those k individuals might have any one of 365 birthdays. By the multiplication rule, the group generates a total of 365^k different sets of birthdays, the collection of which comprises our sample space.

Define the event A to be "At least two people have the same birthday." If each person is assumed to have an equal chance of being born on any given day, the 365^k sequences in the sample space will be equally likely, and

$$P(A) = \frac{\text{number of sequences in } A}{365^k}$$

Notice that each birthday sequence in the sample space belongs to exactly one of the following two categories:

1. At least two people have the same birthday.
2. All k people have different birthdays.

It follows that

$$\text{number of outcomes in } A = 365^k - \text{number of outcomes in } A^C \quad (2.10.1)$$

Equation 2.10.1 is a key step. Why? Because we can easily find the number of sequences in A^C using Theorem 2.9.1. A simple subtraction, then, gives us the number of sequences in A.

Clearly, the number of ways to form birthday sequences for k people subject to the restriction that all k birthdays must be different (the number of outcomes in A^C) is

$$365(364) \cdots (365 - k + 1) = \frac{(365)!}{(365 - k)!}$$

Therefore,

$$P(A) = P(\text{at least two people have the same birthday})$$
$$= \frac{(365)^k - 365(364) \cdots (365 - k + 1)}{(365)^k}$$

Table 2.10.2 gives the value of $P(A)$ for k values of 15, 22, 23, 40, 50, and 70. Notice how much larger the $P(A)$ values are than our intuition would suggest.

TABLE 2.10.2

k	$P(A) = P$(at least two have same birthday)
15	0.253
22	0.475
23	0.507
40	0.891
50	0.970
70	0.999

Comment

To facilitate the computation of $P(A)$, it was assumed that a person had the same chance of being born on any given day of the year. Birth certificates, however, show that that assumption is not entirely valid, births being somewhat more common during the summer than during the winter. It has been proved, though, that any such nonuniformity serves only to *increase* the value of $P(A)$ (110). Thus, with 23 people, the smallest possible probability of at least one match is 0.507, and that occurs when each birthday is equally likely.

Comment

Presidential biographies offer one opportunity to "confirm" the unexpectedly large values Table 2.10.2 gives for $P(A)$. And they do. Among our $k = 40$ presidents, two did have the same birthday—Harding and Polk were both born on November 2. More surprising, though, are the death dates of the presidents, where there were *four* matches: Adams, Jefferson, and Monroe all died on July 4 and Fillmore and Taft both died on March 8.

Question 2.10.1 To determine an "odd man out," n players each toss a fair coin. If one player's coin turns up differently from all the others, that person is declared the odd man out. Let A be the event that someone is declared an odd man out. Find $P(A)$.

Question 2.10.2 Recall Example 2.9.5. What is the *probability* that the four aces in a well-shuffled deck are all adjacent?

Question 2.10.3 A committee of 50 politicians is to be chosen from among our 100 U.S. Senators. If the selection is done at random, what is the probability that each state will be represented?

Question 2.10.4 If three fair dice are tossed—one red, one white, and one green—what is the probability that the sum of the faces showing will be less than or equal to 5? greater than or equal to 5?

Question 2.10.5 Bordering a large estate is a stand of 15 European white birches, all in a row. Three of the trees have leaves damaged by some sort of parasite, and those three are adjacent to one another. What probability would you associate with that event?

Question 2.10.6 Thirteen tombstones in a country churchyard are arranged in three rows—four in the first row, five in the second, and four in the third. Suppose that two women are buried in each row. Assuming each arrangement to be equally likely, what is the probability that in each row the women occupy the two leftmost positions? Is it reasonable to assume that the arrangements will be equally likely? Explain.

Question 2.10.7 Recall Question 2.9.16. If the 10 people line up at random, what is the probability that the Nigerians will occupy the first four positions?

Question 2.10.8 For one-pair poker hands, why is the number of denominations for the three single cards $\binom{12}{3}$ rather than $\binom{12}{1}\binom{11}{1}\binom{10}{1}$?

Question 2.10.9 An apartment building has eight floors. If seven people get on the elevator on the first floor, what is the probability they all want to get off on different floors? On the same floor? What assumption are you making? Does it seem reasonable? Explain.

Question 2.10.10 Six dice are rolled one time. What is the probability that each of the six faces appears?

Question 2.10.11 Five cards are dealt from a poker deck. Find the probabilities associated with each of the following hands.
(a) two pairs
(b) three-of-a-kind
(c) four-of-a-kind
(d) flush (five cards in the same suit, but not having consecutive denominations)
(e) royal flush (10, jack, queen, king, and ace in the same suit)

Question 2.10.12 A poker player is dealt a 3 of diamonds, an 8 of clubs, a 9 of clubs, a 10 of spades, and an ace of spades. If he discards the 3 and the ace, what are the chances he draws to a straight?

Question 2.10.13 Consider a set of 10 urns, nine of which each contains 3 white chips and 3 red chips, while the tenth contains 5 white chips and 1 red chip. An urn is picked at random. Then a sample of size 3 is drawn without replacement from that urn. If all three chips drawn are white, what is the probability that the urn being sampled is the one with 5 white chips?

Question 2.10.14 Suppose that a randomly selected group of k people are brought together. What is the probability that exactly one pair has the same birthday?

Question 2.10.15 A pinochle deck has 48 cards, two of each of six denominations (9, J, Q, K, 10, A) and the usual four suits. Among the many hands that count for meld is a *roundhouse*, which occurs when a player has a king and queen of each suit. In a hand of 12 cards, what is the probability of getting a "bare" roundhouse (a king and queen of each suit and no other kings or queens)?

Question 2.10.16 A coke hand in bridge is one where none of the 13 cards is an ace or is higher than a 9. What is the probability of being dealt such a hand?

Question 2.10.17 A somewhat inebriated conventioneer finds himself in the embarrassing predicament of being unable to predetermine whether his next step will be forward or backward. What is the probability that after hazarding n such maneuvers he will have stumbled forward a distance of r steps? (*Hint*: Let x denote the number of steps he takes forward and y, the number backward. Then $x + y = n$ and $x - y = r$.)

2.11 COMBINATORIAL PROBABILITY: TWO SPECIAL CASES—THE HYPERGEOMETRIC AND BINOMIAL DISTRIBUTIONS

The problems discussed in Section 2.10 were all combinatorial in nature but beyond that they shared no special similarity—the particular enumeration technique needed to solve the birthday problem, for example, was entirely different from what was used for the poker hands. Moreover, none of those examples qualifies as a bona fide problem "type" in the sense that its structure serves as a model for a wide range of other applications. In this section we look at two formulas that *do* represent entire categories of combinatorial problems. We begin by deriving the *hypergeometric distribution*.

Imagine an urn containing N chips, of which r are red and w are white. Suppose that we reach in and draw out—all at one time—a sample of size n. A certain number of chips in that sample will be red. Call that number Y. What we want to derive is a formula giving the probability that Y is equal to any specified value, say, k.

> **Theorem 2.11.1.** Suppose that an urn contains r red chips and w white chips ($r + w = N$). If n chips are drawn out at random, without replacement, and Y denotes the total number of red chips selected, Y is said to have a *hypergeometric distribution*, and
>
> $$P(Y = k) = \frac{\binom{r}{k}\binom{w}{n-k}}{\binom{N}{n}} \qquad \max\{0, n - w\} \le k \le \min(n, r)$$

Proof. The proof follows the same rationale as that used in finding the probabilities of poker hands. Consider the chips to be distinguishable. From Theorem 2.9.2, the total number of ways to select a sample of size n is $\binom{N}{n}$.

Similarly, there are $\binom{r}{k}$ ways to select a sample of k red chips, and, for each of those, $\binom{w}{n-k}$ ways to select enough white chips $(n-k)$ to fill out the sample.

By the multiplication rule, then, the number of ways to form samples having exactly k red chips is $\binom{r}{k}\binom{w}{n-k}$. Since each of the $\binom{N}{n}$ possible samples is assumed to be equally likely, it follows that the probability of getting exactly k red chips is the ratio $\binom{r}{k}\binom{w}{n-k}\Big/\binom{N}{n}$.

Comment

The name "hypergeometric" derives from a series introduced by the Swiss mathematician and physicist, Leonhard Euler, in 1769:

$$1 + \frac{ab}{c}x + \frac{a(a+1)b(b+1)}{2!\,c(c+1)}x^2 + \frac{a(a+1)(a+2)b(b+1)(b+2)}{3!\,c(c+1)(c+2)}x^3 + \cdots$$

This is an expansion of considerable flexibility: given appropriate values for a, b, and c, it reduces to many of the standard infinite series used in analysis. In particular, if a is set equal to 1, and b and c are set equal to each other, it reduces to the familiar *geometric* series,

$$1 + x + x^2 + x^3 + \cdots$$

hence the name *hypergeometric*. The relationship of the probability function in Theorem 2.11.1 to Euler's series becomes apparent if we set $a = -n$, $b = -r$, $c = w - n + 1$, and multiply the series by $\binom{w}{n}\Big/\binom{N}{n}$. Then the coefficient of x^k will be

$$\frac{\binom{r}{k}\binom{w}{n-k}}{\binom{N}{n}}$$

the value the theorem gives for $P(Y = k)$.

Example 2.11.1

Keno is among the most popular games played in Las Vegas even though it ranks as one of the least "fair" in the sense that the odds are overwhelmingly in favor of the house. (Betting on keno is only a little less foolish than playing a slot machine!) A keno card has 80 numbers, 1 through 80, from which the player selects a sample of size k, where k can be anything from 1 to 15. The "caller" then announces 20 winning numbers, chosen at random from the 80. If—and how much—the player wins depends on how many of his numbers match the 20 identified by the caller. Suppose that a player bets on a 10-spot ticket. What is his probability of "catching" five numbers?

Consider an urn containing 80 numbers, 20 of which are winners, and 60 losers (see Figure 2.11.1). By betting on a 10-spot ticket, the player, in effect, is drawing a sample of size 10 from that urn. Let Y denote the number of

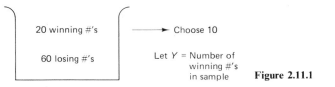

20 winning #'s

60 losing #'s

Choose 10

Let Y = Number of winning #'s in sample

Figure 2.11.1

winning numbers included among the player's 10 selections. What we are trying to find is $P(Y = 5)$.

By Theorem 2.11.1 (with $r = 20$, $w = 60$, $n = 10$, $N = 80$, and $k = 5$), the player has approximately a 5% chance of guessing exactly five winning numbers:

$$P(Y = 5) = \frac{\binom{20}{5}\binom{60}{5}}{\binom{80}{10}} = 0.05$$

(What are the player's chances of getting *none* right?)

Example 2.11.2

Urn I contains five red chips and four white chips; urn II contains four red and five white. Two chips are to be transferred from urn I to urn II. Then a single chip is to be drawn from urn II (see Figure 2.11.2). What is the probability that the chip drawn from the second urn will be white?

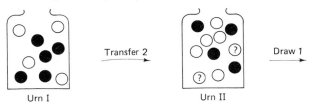

Transfer 2

Draw 1

Urn I

Urn II

Figure 2.11.2

Let W be the event "White chip is drawn from urn II." Let A_i, $i = 0, 1, 2$, denote the event "i white chips are transferred from urn I to urn II." Then, by Theorem 2.6.1,

$$P(W) = P(W \mid A_0)P(A_0) + P(W \mid A_1)P(A_1) + P(W \mid A_2)P(A_2)$$

Note that $P(W \mid A_i) = (5 + i)/11$ and that $P(A_i)$ is gotten directly from Theorem 2.11.1. Therefore,

$$P(W) = \left(\frac{5}{11}\right)\frac{\binom{4}{0}\binom{5}{2}}{\binom{9}{2}} + \left(\frac{6}{11}\right)\frac{\binom{4}{1}\binom{5}{1}}{\binom{9}{2}} + \left(\frac{7}{11}\right)\frac{\binom{4}{2}\binom{5}{0}}{\binom{9}{2}}$$

$$= \left(\frac{5}{11}\right)\left(\frac{10}{36}\right) + \left(\frac{6}{11}\right)\left(\frac{20}{36}\right) + \left(\frac{7}{11}\right)\left(\frac{6}{36}\right)$$

$$= \frac{53}{99}$$

Example 2.11.3

Muffy is studying for a history exam covering the French Revolution that will consist of five essay questions selected at random from a list of 10 the professor has handed out to the class in advance. Not exactly a Napoleon buff, Muffy would like to avoid researching all ten questions but still be reasonably assured of getting a fairly good grade. Specifically, she wants to have at least an 85% chance of getting at least four of the five questions right. Will it be sufficient if she studies eight of the 10 questions?

No. Think of the questions as being two kinds of chips in an urn—there are the eight whose answers Muffy will know and the two she will be unprepared for (see Figure 2.11.3). In making out the test, the professor is drawing a

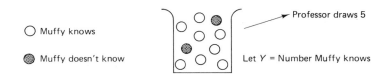

Figure 2.11.3

random sample of size 5. Let Y denote the number in the sample coming from the eight questions Muffy will have prepared. Unfortunately,

$$P(Y \geq 4) = P(Y = 4) + P(Y = 5)$$

$$= \frac{\binom{8}{4}\binom{2}{1}}{\binom{10}{5}} + \frac{\binom{8}{5}\binom{2}{0}}{\binom{10}{5}} = \frac{196}{252}$$

$$= 0.78$$

so it's back to the books! By studying eight questions she has only a 78% chance (rather than 85%) of getting at least four correct. (Would Muffy satisfy her "85%" requirement if she prepared for *nine* of the questions?)

Question 2.11.1 A Scrabble set consists of 54 consonants and 44 vowels. What is the probability that your initial draw (of seven letters) will be all consonants? six consonants and one vowel? five consonants and two vowels?

Question 2.11.2 Urn I contains 4 red chips, 3 white chips, and 2 blue chips. Urn II has 3 red, 4 white, and 5 blue. Two chips are drawn at random and without replacement from each urn. What is the probability that all four chips are the same color?

Question 2.11.3 X-rays show that 6 of Muffy's 32 teeth have cavities. Unfortunately, her dentist is a bit myopic and drills 6 of her teeth *at random*. What is the probability that fewer than half of the cavities are properly drilled?

Question 2.11.4 Six terminals, numbered 1 through 6, are on-line to a mainframe computer; all are ready to execute their programs. You and a friend are working on terminals 2 and 5. At random, the computer selects three terminals and advances them in the access priority queue. What is the probability that both your terminal and your friend's terminal were among the three selected to be advanced?

Question 2.11.5 The Admissions Committee of a medical school has ten applicants, eight white and two who are members of ethnic minorities, from which the remaining n positions in the first-year class are to be filled. If the committee chooses at random from the 10, what is the smallest value of n that will ensure a probability greater than $\frac{1}{2}$ of admitting at least one minority applicant?

Question 2.11.6 Show directly that the set of probabilities associated with the hypergeometric distribution sum to 1. *Hint*: Expand the identity

$$(1 + \mu)^N = (1 + \mu)^r(1 + \mu)^{N-r}$$

and equate coefficients.

Tradition dictated that the hypergeometric distribution be introduced in the context of drawing chips from an urn. It is similarly *de rigueur* to motivate the second of our special cases, the *binomial distribution*, by tossing coins.

Imagine three distinguishable coins being tossed, each having a probability p of coming up heads. The set of possible outcomes are the eight listed in Table 2.11.1. If the probability of any of the coins coming up heads is p, then the probability of the *sequence* (H, H, H) is p^3, since the coin tosses qualify as independent trials. Similarly, the probability of (T, H, H) is $(1 - p)p^2$. The fourth column of Table 2.11.1 shows the probabilities associated with each of the three-coin sequences.

TABLE 2.11.1

1st coin	2nd coin	3rd coin	Probability	Y = number of heads
H	H	H	p^3	3
H	H	T	$p^2(1 - p)$	2
H	T	H	$p^2(1 - p)$	2
T	H	H	$p^2(1 - p)$	2
H	T	T	$p(1 - p)^2$	1
T	H	T	$p(1 - p)^2$	1
T	T	H	$p(1 - p)^2$	1
T	T	T	$(1 - p)^3$	0

Suppose that our main interest in the coin tosses, though, is only in the *number* of heads that occurred: whether the actual sequence was, say, (H, H, T) or (H, T, H) is immaterial. That being the case, it makes sense to define a variable Y equal to the number of heads observed in a sequence of three tosses. The last column of Table 2.11.1 lists the value of Y associated with each outcome in the sample space. Note that *three* outcomes, each having an individual probability of $p^2(1 - p)$, yield the value $Y = 2$. It follows, then, that the probability associated with a Y value of 2 is $3p^2(1 - p)$. Table 2.11.2 displays the entire probability distribution for Y.

TABLE 2.11.2

Y = number of heads	$P(Y = k)$
0	$(1 - p)^3$
1	$3p(1 - p)^2$
2	$3p^2(1 - p)$
3	p^3

More generally, suppose that n coins are tossed, so Y can equal any integer from 0 through n. What can we write down for $P(Y = k)$? If we argue by analogy, $P(Y = k)$ will equal the probability associated with any particular sequence having k heads [and $(n - k)$ tails] multiplied by the number of ways to position the k heads and the $n - k$ tails. The latter factor derives from a familiar combinatorial problem. Think of an urn with n chips, numbered 1 through n. Every distinct sample of size k chosen from that urn could represent the position numbers of the k heads in an ordered sequence of length n. But how many samples of size k are there? By, Theorem 2.9.2, $\binom{n}{k}$. Of course, the probability associated with any particular sequence having exactly k heads is $p^k(1 - p)^{n-k}$. Therefore, $P(Y = k) = \binom{n}{k} p^k(1 - p)^{n-k}$, for $k = 0, 1, \ldots, n$.

What we have just derived is the *binomial distribution*. Theorem 2.11.2 summarizes the model and formally states its assumptions.

> **Theorem 2.11.2.** Consider a series of n independent trials, each resulting in one of two possible outcomes, "success" or "failure." Let $p = P(\text{success occurs at any given trial})$ and assume that p remains constant from trial to trial. Let the variable Y denote the total number of successes in the n trials. Then Y is said to have a *binomial distribution* and
>
> $$P(Y = k) = \binom{n}{k} p^k(1 - p)^{n-k} \qquad k = 0, 1, \ldots, n$$

Comment

Recall that every discrete probability function must satisfy the equation

$$\sum_{\text{all } s} P(s) = 1 \qquad (2.11.1)$$

For the binomial distribution, Equation 2.11.1 follows immediately from a well-known theorem in algebra: for any a and b,

$$(a + b)^n = \sum_{k=0}^{n} \binom{n}{k} a^k b^{n-k}$$

Let $a = p$ and $b = 1 - p$. Then

$$\sum_{\text{all } s} P(s) = \sum_{k=0}^{n} \binom{n}{k} p^k (1 - p)^{n-k} = (p + 1 - p)^n = 1^n = 1$$

Example 2.11.4

Each day the price of a new computer stock moves up one point or down one point with probabilities $\frac{3}{4}$ and $\frac{1}{4}$, respectively. What is the probability that after six days the stock will have returned to its original quotation? Assume that the daily price fluctuations are independent events.

In order for the stock price to be the same on day 7 that it was on day 1, it must have gone up three times and down three times during the six days in between. Let Y denote the number of price *increases* during a six-day period. Then

$$P(\text{stock returns to original price after 6 days})$$

$$= P(Y = 3) = \binom{6}{3}\left(\frac{3}{4}\right)^3 \left(1 - \frac{3}{4}\right)^{6-3}$$

$$= 0.13$$

Example 2.11.5

In a nuclear reactor, the fission process is controlled by inserting into the radioactive core a number of special rods whose purpose is to absorb the neutrons emitted by the critical mass and thereby slow down the nuclear chain reaction. When functioning properly, these rods serve as the first-line defense against a core meltdown.

Suppose that a particular reactor has 10 of these control rods (in "real life" there would be more than 100), each operating independently, and each having a 0.80 probability of being properly inserted in the event of an "incident." Furthermore, suppose that a meltdown will be prevented if at least half the rods perform satisfactorily. What is the probability that, upon demand, the system will fail?

If Y denotes the number of control rods that function as they should, a system failure occurs if $Y \leq 4$. By Theorem 2.11.2, the probability of that happening is 0.007:

$$P(\text{system will fail}) = P(Y \leq 4) = \sum_{k=0}^{4} \binom{10}{k}(0.80)^k (1 - 0.80)^{10-k}$$

$$= \binom{10}{0}(0.80)^0 (0.20)^{10} + \cdots$$

$$+ \binom{10}{4}(0.80)^4 (0.20)^6$$

$$= 0.000 + 0.000 + 0.000 + 0.001 + 0.006$$

$$= 0.007$$

Example 2.11.6

For reasons not entirely clear, Doomsday Airlines books a daily shuttle service from Altoona to Hoboken. They offer two round-trip flights, one on a two-engine prop plane, the other on a four-engine prop plane. Suppose that each engine on each plane will fail independently with the same probability p and that each plane will arrive safely at its destination only if at least half its engines remain in working order. Assuming that you want to continue living, for what values of p would you prefer to fly in the two-engine plane?

Let Y denote the number of engines on each plane that remain operable. For the two-engine plane,

$$P(\text{flight lands safely}) = P(Y \geq 1) = \sum_{k=1}^{2} \binom{2}{k}(1-p)^k p^{2-k} \qquad (2.11.2)$$

For the four-engine plane,

$$P(\text{flight lands safely}) = P(Y \geq 2) = \sum_{k=2}^{4} \binom{4}{k}(1-p)^k p^{4-k} \qquad (2.11.3)$$

When to opt for the two-engine plane, then, reduces to an algebra problem: we look for the values of p for which

$$\sum_{k=1}^{2} \binom{2}{k}(1-p)^k p^{2-k} > \sum_{k=2}^{4} \binom{4}{k}(1-p)^k p^{4-k} \qquad (2.11.4)$$

To minimize the number of terms that need to be dealt with, it will prove expedient to rephrase the problem using the complements of Equations 2.11.2 and 2.11.3. The set of p values for which Inequality 2.11.4 is true is equivalent to the set for which

$$\sum_{k=0}^{1} \binom{4}{k}(1-p)^k p^{4-k} > \binom{2}{0}(1-p)^0 p^2$$

Simplifying the inequality

$$\binom{4}{0}(1-p)^0 p^4 + \binom{4}{1}(1-p)^1 p^3 > \binom{2}{0}(1-p)^0 p^2$$

gives

$$(3p-1)(p-1) < 0 \qquad (2.11.5)$$

But $(p-1)$ is never positive, so Inequality 2.11.5 will be true only when $(3p-1) > 0$, which gives $p > \frac{1}{3}$ as the desired solution set. Figure 2.11.4 is a graph of the two "safe return" probabilities, $P(Y \geq 1 \mid n = 2)$ and $P(Y \geq 2 \mid n = 4)$, as a function of p.

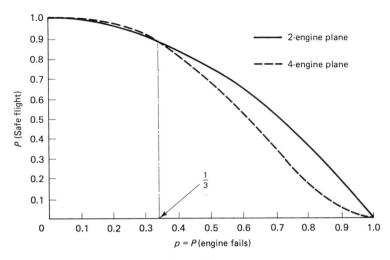

Figure 2.11.4

Consider again the evidence presented in *People* v. *Collins*, as outlined in Case Study 2.7.1. The prosecution's case rested on the unlikelihood of a given couple's matching up with the six characteristics reported by the several eyewitnesses of the crime. It was estimated that the joint occurrence of a white female with blond hair combed in a ponytail riding in a yellow car with a black male having a beard and a mustache was on the order of 1 in 12 million—a number so small, the prosecution contended, that Ms. Collins and her male friend were clearly guilty, a classic open-and-shut case. Not so, argued the counsel for the defense. By approaching the same data from a slightly different perspective, they were able to show that "reasonable doubt" had not really been eliminated, despite the apparent 12-million-to-1 odds.

Suppose that n is taken to be the total number of couples who could conceivably have been in the area and perpetrated the crime, and p, the probability that any such couple would share the six characteristics introduced by the prosecution as evidence. Define Y to be the number of couples matching up with the eyewitness accounts. It is not unreasonable to assume that Y is binomial, in which case

$$P(Y = k) = \binom{n}{k} p^k (1 - p)^{n-k}, \qquad k = 0, 1, \ldots, n$$

Therefore,

$$P(Y = 1) = np(1 - p)^{n-1}$$

and

$$P(Y \geq 1) = 1 - P(Y = 0) = 1 - (1 - p)^n$$

from which it follows that

$$P(Y > 1) = 1 - (1 - p)^n - np(1 - p)^{n-1}$$

Now, consider the ratio

$$\frac{P(Y > 1)}{P(Y \geq 1)} = \frac{1 - (1 - p)^n - np(1 - p)^{n-1}}{1 - (1 - p)^n} \qquad (2.11.6)$$

= P(more than one of the n couples fit the description given that at least one does)

= P(there is at least one other couple who could have committed the crime)

If $P(Y > 1)/P(Y \geq 1)$ is anything other than a very small number, we would have to accept the possibility that Ms. Collins and her friend have a pair of lookalikes and that perhaps *they* were the culprits.

Table 2.11.3 shows the value of the probability ratio (Equation 2.11.6) for various values of n befitting a large metropolitan area and for the prosecutor's estimate of p (= 1/12,000,000). What the last column makes clear is that $P(Y > 1)/P(Y \geq 1)$ is *not* a particularly small number.

TABLE 2.11.3

p	n	$P(Y > 1)/P(Y \geq 1)$
1/12,000,000	1,000,000	0.0402
1/12,000,000	2,000,000	0.0786
1/12,000,000	5,000,000	0.1875
1/12,000,000	10,000,000	0.3479

Looked at in this way, the evidence is certainly not as incriminating as the "1-in-12,000,000" argument would have us believe. At least that was the opinion of the California Supreme Court: based on the probability argument just presented, they overturned the initial verdict of "guilty" that had been handed down by the Superior Court of Los Angeles County.

Question 2.11.7 If a family has four children, is it more likely they will have two boys and two girls or three of one sex and one of the other? Assume that the probability of any child being a boy is $\frac{1}{2}$.

Question 2.11.8 Suppose that since the early 1950s some 10,000 independent UFO sightings have been reported to civil authorities. If the probability of any particular sighting's being genuine is on the order of 1 in 100,000, what is the probability that at least one sighting was genuine?

Question 2.11.9 The captain of a Navy gunboat orders a volley of 25 missiles to be fired at random along a 500-foot stretch of shoreline that he hopes to establish as a beachhead. Dug into the beach is a 30-foot-long bunker serving as the enemy's first line of defense. What is the probability that exactly three shells will hit the bunker?

Question 2.11.10 Let $b(k; n, p)$ denote the probability that the variable in a binomial model equals k—that is,

$$b(k; n, p) = \binom{n}{k} p^k (1 - p)^{n-k}$$

Prove the recursion formula,

$$b(k; n, p) = \frac{(n - k + 1)p}{k(1 - p)} \cdot b(k - 1; n, p)$$

Question 2.11.11 In his 1889 publication, *Natural Inheritance*, the renowned British scientist Sir Francis Galton described a pinball-type board that he called a *quincunx*. As pictured below, the quincunx has five rows of pegs, the pegs in each row being the same distance apart.

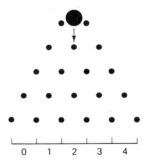

| | | | | |
|0|1|2|3|4|

Figure Q2.11.11

At the bottom of the board are five cells, numbered 0 through 4. A ball is introduced at the top of the board between the pegs in the first row. After wedging past those two pegs, it will hit the middle peg in the second row and veer either to the right or to the left, then strike a peg in the third row—again veer to the right or to the left—and so on. If the ball has a 50–50 chance of going either direction each time it hits a peg, what is the probability that it ends up in cell 3?

Question 2.11.12 Two baseball teams are negotiating a format for deciding how to determine the league champion. Two possibilities have been mentioned: a "best two of three" playoff series or a "best three of five" series. Suppose the members of one of the teams estimate that they have a 55% chance of defeating their opponent on any given day. Which of the two playoff schemes should they support? Compute the two relevant probabilities. Assume that each game can be considered an independent trial. Does your answer make sense intuitively?

Question 2.11.13 Recall Example 2.5.1. Suppose that the records of three former prisoners, all sentenced for grand theft auto, are examined. What is the probability that exactly two of the three were released in less than a year?

Question 2.11.14 Samuel Pepys was the greatest diarist of the English language. He was also a friend of Sir Isaac Newton and, in 1693, sought the latter's advice in a matter related to gambling. Phrased in modern terminology, Pepys' question was a binomial problem: Is it more likely to get at least one 6 when six dice are rolled, at least two 6's when 12 dice are rolled, or at least three 6's when 18 dice are rolled? After considerable correspondence [see (147)], Newton was able to convince a skeptical Pepys that the former has the greatest likelihood. Compute these three probabilities.

Question 2.11.15 Experience has shown that only $\frac{1}{3}$ of all patients having a certain disease will recover if given the standard treatment. A new drug is to be tested on a group of 12 volunteers. If the FDA requires that at least seven of these patients recover before it will license the new drug, what is the probability that the treatment will be discredited even if it has the potential to increase an individual's recovery rate to $\frac{1}{2}$?

Question 2.11.16 Let $B(k; n, p)$ denote the lower tail of the binomial distribution:

$$B(k; n, p) = \sum_{r=0}^{k} \binom{n}{r} p^r (1 - p)^{n-r}$$

Show that

$$B(k; n, p) = (n - k)\binom{n}{k} \int_{0}^{1-p} t^{n-k-1}(1 - t)^k \, dt$$

(*Hint*: Differentiate both sides of the equation with respect to p.)

3

Random Variables

Jakob (Jacques) Bernoulli (1654–1705)

One of a Swiss family producing eight distinguished scientists, Jakob was forced by his father to pursue theological studies, but his love of mathematics eventually led him to a university career. He and his brother, Johann, were the most prominent champions of Leibniz' calculus on continental Europe, the two using the new theory to solve numerous problems in physics and mathematics. Bernoulli's main work in probability, *Ars Conjectandi*, was published after his death by his nephew, Nikolaus, in 1713.

3.1 INTRODUCTION

Throughout most of Chapter 2, probability functions were defined in terms of the elementary outcomes making up an experiment's sample space. Thus, if two fair dice were tossed, a P value was assigned to each of the 36 possible pairs of upturned faces: $P((3, 2)) = \frac{1}{36}$, $P((2, 3)) = \frac{1}{36}$, $P((4, 6)) = \frac{1}{36}$, and so on. We have already seen, though, that in certain situations some attribute of an outcome may hold more interest for the experimenter than the outcome itself. A craps player, for example, may be concerned only that he throws a 7, not whether the 7 was the result of a 5 and a 2, a 4 and a 3, or a 6 and 1. That being the case, it makes sense to replace the 36-member sample space of (x, y) pairs with the more relevant (and simpler) 11-member set of all possible two-dice *sums*, $S = \{x + y : x + y = 2, 3, \ldots, 12\}$.

In this chapter we investigate the consequences of redefining an experiment's sample space. More is at stake than simply the number of outcomes in S—the probability structure also changes. Consider, again, the craps player. The original sample space contains 36 outcomes, *all equally likely*. The revised sample space contains 11 outcomes, but the latter are *not* equally likely. The probability of getting a sum equal to 2, for example, is $1/36[= P((1, 1))]$; but the probability of getting a sum equal to 3 is $2/36[= P((1, 2)) + P((2, 1))]$. Other values for the "sum" probabilities are obtained similarly (see Figure 3.1.1).

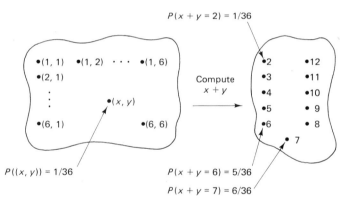

Figure 3.1.1

In general, rules for redefining sample spaces—like going from (x, y)'s to $(x + y)$'s—are called *random variables*. As a conceptual framework, random variables are of fundamental importance: they provide a single rubric under which *all* probability problems may be brought. Even in cases where the original sample space needs no redefinition—that is, where the measurement recorded is the measurement of interest—the concept still applies: we simply take the random variable to be the identity mapping.

On the whole, this is a nuts-and-bolts chapter. Its primary purpose is to introduce the bits and pieces of random variable terminology and technique that make up the mathematics of probability and statistics; the really important applications of these notions will come later.

3.2 DENSITIES AND DISTRIBUTIONS

The place to begin a discussion of random variables is by formalizing the general notion introduced in Section 3.1. We start with a definition.

Definition 3.2.1. A real-valued function whose domain is the sample space S is called a *random variable*. We denote random variables by uppercase letters, often X, Y, or Z.

Figure 3.2.1 is a "picture" of Definition 3.2.1. Any random variable, Y, is simply a mapping from the sample space S to the real line. If the range of the mapping contains either a finite or a countably infinite number of values, the random variable is said to be *discrete*; if the range includes an interval of real numbers, bounded or unbounded, the random variable is said to be *continuous*.

Figure 3.2.1

Associated with each discrete random variable Y is a *probability density function* (or *pdf*), $f_Y(y)$. By definition, $f_Y(y)$ is the sum of all the probabilities associated with outcomes in S that get mapped into y by the random variable Y. That is,

$$f_Y(y) = P(\{s \in S \mid Y(s) = y\})$$

Conceptually, $f_Y(y)$ describes the probability structure induced on the real line by the random variable Y.

For instance, in the dice example described in Section 3.1, $Y(s) = Y((a, b)) = a + b$, from which it follows that

$$f_Y(3) = P(\text{dice sum equals } 3)$$
$$= P(\{s \in S \mid Y(s) = 3\})$$
$$= P((1, 2), (2, 1))$$
$$= \frac{1}{36} + \frac{1}{36} = \frac{2}{36}$$

Figure 3.2.2 shows $f_Y(y)$ for all y.

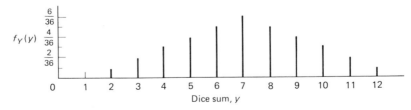

Figure 3.2.2

For notational simplicity, we will delete all references to s and S and write

$$f_Y(y) = P(Y = y)$$

In words, $f_Y(y)$ is the "probability that the random variable Y takes on the value y."

Associated with each continuous random variable Y is also a probability density function, $f_Y(y)$, but $f_Y(y)$ in this case is *not* the probability that the random variable Y takes on the value y. Rather, $f_Y(y)$ is a continuous curve having the property that for all a and b,

$$P(a \leq Y \leq b) = P(\{s \in S \mid a \leq Y(s) \leq b\}) = \int_a^b f_Y(y)\, dy$$

(see Figure 3.2.3).

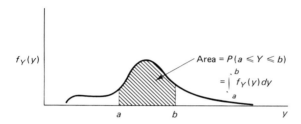

Figure 3.2.3

Example 3.2.1

Random variables are being developed formally for the first time in this chapter, but the concept and the terminology have already been introduced. Both the binomial and the hypergeometric distributions derived in Chapter 2 are examples of discrete random variables. In the binomial, for instance, the original sample space is the set of all possible 2^n sequences of successes or failures occurring at the n independent trials:

$$S = \{(s, s, s, \ldots, s), \quad (f, s, s, \ldots, s), \ldots, \quad (f, f, f, \ldots, f)\}$$

If we let the random variable Y denote the total number of successes in the n trials, then

$$Y(f, f, f, \ldots, f) = 0$$
$$Y(s, f, f, \ldots, f) = 1$$
$$Y(f, s, f, \ldots, f) = 1$$
$$\vdots$$
$$Y(s, s, s, \ldots, s) = n$$

Furthermore, the probability distribution for Y—that is, the probability structure induced by Y on the real line—is given by Theorem 2.11.2: if $P(\text{success}) = P(s) = p$, then

$$P(Y = k) = \binom{n}{k} p^k (1 - p)^{n-k} \qquad k = 0, 1, \ldots, n$$

Similarly, the hypergeometric random variable Y is defined to be the number of red chips contained in a sample of size n drawn without replacement from an urn holding r red chips and w white chips. Here the original sample space has $\binom{r + w}{n}$ outcomes; the smaller set of possible Y values is restricted to $0, 1, \ldots$, min (r, n). In the original sample space, each of the $\binom{r + w}{n}$ outcomes is equally likely; the probability distribution of the random variable Y is given by Theorem 2.11.1:

$$P(Y = k) = \frac{\binom{r}{k}\binom{w}{n - k}}{\binom{r + w}{n}} \qquad k = 0, 1, \ldots, \text{min } (r, n)$$

Example 3.2.2

A not very skillful dart player throws two darts at the board pictured in Figure 3.2.4. Assume that his throws are independent and that the probability of a dart landing in any particular region is proportional to the area of that

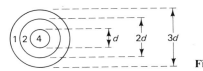

Figure 3.2.4

region. Furthermore, assume that his final score is defined to be the *product* of the points assigned to the regions where the two darts land. What is the probability that he earns a score of 4?

If (a, b) denotes the event of getting a points with the first dart and b with the second, then S contains a total of 9 outcomes:

$$S = \{(1, 1), (1, 2), (1, 4), (2, 1), (2, 2), (2, 4), (1, 4), (2, 4), (4, 4)\}$$

Let Y be the random variable that computes the player's score:

$$Y(s) = Y((a, b)) = ab$$

By inspection, the outcomes in S that get mapped by Y into a score of 4 are $(1, 4), (4, 1)$, and $(2, 2)$. Therefore,

$$P(Y = 4) = P(\{s \in S \mid Y(s) = 4\})$$
$$= P((1, 4), (4, 1), (2, 2))$$
$$= P((1, 4)) + P((4, 1)) + P((2, 2))$$

Since the throws are assumed to be independent, $P((a, b)) = P(a) \cdot P(b)$, so

$$P(Y = 4) = P(1) \cdot P(4) + P(4) \cdot P(1) + P(2) \cdot P(2) \qquad (3.2.1)$$

To find $P(1)$, $P(2)$, and $P(4)$ we need to compute the areas of the three regions shown in Figure 3.2.4. By simple geometry,

$$\text{area of "4" region} = \frac{\pi d^2}{4}$$

$$\text{area of "2" region} = \pi d^2 - \frac{\pi d^2}{4} = \frac{3\pi d^2}{4}$$

$$\text{area of "1" region} = \frac{9\pi d^2}{4} - \pi d^2 = \frac{5\pi d^2}{4}$$

Using the proportionality assumption, then, we get that

$$P(4) = \frac{\pi d^2/4}{9\pi d^2/4} = \frac{1}{9}$$

$$P(2) = \frac{3\pi d^2/4}{9\pi d^2/4} = \frac{3}{9}$$

$$P(1) = \frac{5\pi d^2/4}{9\pi d^2/4} = \frac{5}{9}$$

A final substitution into Equation 3.2.1 gives the probability the player scores a total of 4:

$$P(Y = 4) = \left(\frac{5}{9}\right)\left(\frac{1}{9}\right) + \left(\frac{1}{9}\right)\left(\frac{5}{9}\right) + \left(\frac{3}{9}\right)\left(\frac{3}{9}\right)$$

$$= \frac{19}{81}$$

In general, the random variable defined here can take on any of five values: 1, 2, 4, 8, and 16. Table 3.2.1 shows the entire probability distribution of the random variable Y. The same information is shown graphically in Figure 3.2.5.

TABLE 3.2.1

y	$P(Y = y)$
1	$\frac{25}{81}$
2	$\frac{30}{81}$
4	$\frac{19}{81}$
8	$\frac{6}{81}$
16	$\frac{1}{81}$

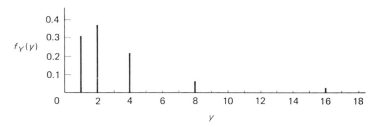

Figure 3.2.5

Example 3.2.3

Continuous random variables often involve the identity mapping, $Y(s) = s$. When they do, the questions they seek to answer are identical to the sorts of problems we discussed in Section 2.5.

Consider a manufacturer of electrical equipment who markets a light bulb that has an average life expectancy of 3000 hours. He offers a money-back guarantee on bulbs that fail to last at least 300 hours. For what proportion of his sales will he need to make a refund?

Experience has shown that the lifetime, s, of mechanical and electrical equipment is often described by an exponential probability function,

$$f(s) = \left(\frac{1}{\lambda}\right)e^{-s/\lambda} \qquad s \geq 0$$

where λ is a constant equal to the equipment's average life expectancy. [In the terminology of Section 2.5, $f(s)$ is a *continuous* probability function: it models the probabilistic behavior of a sample space having an uncountably infinite number of outcomes, $S = \{s : s \geq 0\}$.]

Now, to the consumer, the relevant information contained in the "experiment" of operating a light bulb is simply how long it lasts—that is, s. Here, then, the original sample space needs no redefinition: the measurement recorded—s—is the measurement the consumer wants to know. Therefore, to rephrase this problem in random variable terminology we let Y denote the light bulb's lifetime and we set $Y(s) = s$. It follows, of course, with an identity mapping that the probability structure induced by the random variable on the real line is identical to what was present in the original sample space. Specifically,

$$f_Y(y) = \left(\frac{1}{3000}\right)e^{-y/3000} \qquad y \geq 0 \tag{3.2.2}$$

Integrating Equation 3.2.2 over the appropriate region shows that *10%* of the light bulbs are likely to be returned:

P(light bulb lasts less than 300 hours)

$$= P(Y < 300) = P(\{s \in S \mid Y(s) < 300\})$$

$$= P(s < 300)$$

$$= \int_0^{300} \left(\frac{1}{3000}\right) e^{-y/3000} \, dy$$

$$= \int_0^{0.1} e^{-u} \, du = -e^{-u} \Big|_0^{0.1} = -0.90 + 1$$

$$= 0.10$$

Figure 3.2.6 shows the area under $f_Y(y)$ corresponding to the probability that a light bulb burns for fewer than 300 hours.

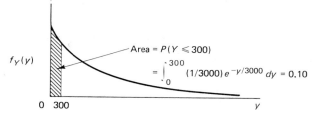

$f_Y(y)$

Area $= P(Y \leqslant 300)$

$$= \int_0^{300} (1/3000) e^{-y/3000} \, dy = 0.10$$

0   300 y **Figure 3.2.6**

Example 3.2.4

By definition, the *median*, y^*, of a continuous probability distribution is that value of y for which $P(Y < y^*) = P(Y > y^*)$. Suppose that a random variable is described by the probability density function

$$f_Y(y) = 3y^2 \qquad 0 \leq y \leq 1$$

Find y^*.

First, note that the probability that Y *equals* y^* is 0 [since $\int_{y^*}^{y^*} f_Y(y) \, dy = 0$]. It follows that the probability of Y being strictly less than y^* is $\frac{1}{2}$, and we can write

$$P(Y < y^*) = \frac{1}{2} = \int_0^{y^*} 3y^2 \, dy$$

$$= \frac{3y^3}{3} \Big|_0^{y^*}$$

$$= (y^*)^3$$

Therefore,

$$y^* = \sqrt[3]{\frac{1}{2}}$$

$$= 0.79$$

(see Figure 3.2.7).

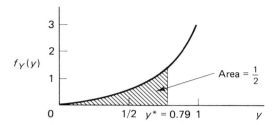

Figure 3.2.7

Comment

The difference between a product's *average* life expectancy and its *median* life expectancy is often exploited in advertising. For example, the median $y*$ for the light bulb *pdf* described in Example 3.2.3 is 2079 hours $[\int_0^{2079} (1/3000) e^{-y/3000} \, dy = 0.50]$. What would be written on cartons containing those bulbs, though, would be their average life expectancy, 3000 hours. Thus, while 3000 hours may be an impressive-sounding figure, far fewer than half the bulbs will last that long. We will return to the very important notion of an "average" later in this chapter.

Question 3.2.1 The sample space for an experiment has three outcomes, all equally likely:

$$S = \{(1, 2, 0), (2, 1, 3), (4, 1, 1)\}$$

Define the random variable

$$Y = Y(s) = Y((a, b, c)) = a + b + c$$

Find the probability distribution of Y.

Question 3.2.2 In the first football game of the season, Vanderbilt's quarterback throws 30 passes. Let Y denote the number caught by the opposing team's middle linebacker. If the probability of the middle linebacker catching any given pass is 0.10, find $P(Y = 4)$. Assume that the passes represent independent events.

Question 3.2.3 Urn I contains 3 red chips and 4 white chips; urn II has 6 red and 4 white. One chip is drawn from urn I, two are drawn from urn II. Let a denote the number of red chips in the sample coming from urn I; let b denote the number of red chips in the sample coming from urn II.
(a) List the possible outcomes (a, b) and compute the probability associated with each.
(b) Define the random variable $Y((a, b)) = ab$. Find $f_Y(y)$ for all y.

Question 3.2.4 Suppose that five people, including you and a friend, line up at random. Let Y denote the number of people standing between you and your friend. Find the probability density function for the random variable Y.

Question 3.2.5 A basketball player has a 70% foul-shooting percentage. Let Y denote the shot on which she makes her first free throw. Write down a general formula for $f_Y(y)$, $y = 1, 2, \ldots$.

Question 3.2.6 Let Y be a continuous random variable described by the pdf

$$f_Y(y) = 3(1 - y)^2 \qquad 0 < y < 1$$

Find $P(|Y - \frac{1}{2}| > \frac{1}{4})$. Draw a graph of $f_Y(y)$ and show the area representing the probability in question.

Question 3.2.7 Suppose the pdf for Y is given by

$$f_Y(y) = \frac{1}{y^2} \qquad y > 1$$

Find y^*, the median of the Y distribution.

Question 3.2.8 Scores on a freshman chemistry test are described by the pdf

$$f_Y(y) = 6y(1 - y) \qquad 0 \le y \le 1$$

where Y represents the proportion of questions a student answers correctly. Any score below 0.40 is a failing mark.
(a) What is the probability that a student fails?
(b) If six students take the exam, what is the probability that exactly 2 fail?

Associated with every random variable, discrete or continuous, is a *cumulative distribution function* (or *cdf*), $F_Y(y)$. The relationship between pdf's and cdf's is important, as we will see later in this chapter.

> **Definition 3.2.2.** Let Y be a random variable defined on a sample space S with probability function P. For any real number y, the *cumulative distribution function of Y* [abbreviated *cdf* and written $F_Y(y)$] is the probability associated with the set of sample points in S that get mapped by Y into values on the real line *less than or equal to y*. Formally,
>
> $$F_Y(y) = P(\{s \in S \mid Y(s) \le y\})$$
>
> As we did with pdf's, references to s and S will usually be deleted and we write
>
> $$F_Y(y) = P(Y \le y)$$

If Y is a discrete random variable, $F_Y(y)$ is a step function, with jumps occurring at the values of y for which $f_Y(y) > 0$. If Y is a continuous random variable, $F_Y(y)$ is itself continuous and monotonically nondecreasing.

Example 3.2.5

A fair die is rolled four times. Let the random variable Y denote the number of sixes that appear. Find $F_Y(y)$, the cdf for Y.

Note, first, that Y has a binomial distribution with $n = 4$ and $p = \frac{1}{6}$. From Theorem 2.11.2,

$$f_Y(y) = P(Y = y) = \binom{4}{y}\left(\frac{1}{6}\right)^y\left(\frac{5}{6}\right)^{4-y} \qquad y = 0, 1, 2, 3, 4$$

(see Figure 3.2.8).

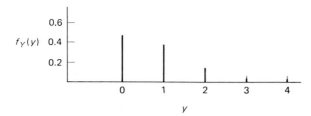

Figure 3.2.8

To find $F_Y(y)$, we have to look at Figure 3.2.8 and see which ranges of y values give the same value for $P(Y \leq y)$. For example, if y is any number less than 0, $F_Y(y) = P(Y \leq y) = 0$; for any y in the semiopen interval $[0, 1)$,

$$F_Y(y) = P(Y \leq y) = P(Y = 0) = \binom{4}{0}\left(\frac{1}{6}\right)^0\left(\frac{5}{6}\right)^4 = \left(\frac{5}{6}\right)^4 = 0.48;$$

for $1 \leq y < 2$,

$$F_Y(y) = P(Y \leq y) = P(Y = 0) + P(Y = 1)$$

$$= \binom{4}{0}\left(\frac{1}{6}\right)^0\left(\frac{5}{6}\right)^4 + \binom{4}{1}\left(\frac{1}{6}\right)^1\left(\frac{5}{6}\right)^3 = \left(\frac{5}{6}\right)^4 + 4\left(\frac{1}{6}\right)\left(\frac{5}{6}\right)^3 = 0.86.$$

Continuing this argument establishes $F_Y(y)$ to be a step function with jumps at the points $y = 0, 1, 2, 3,$ and 4:

$$F_Y(y) = \begin{cases} 0 & y < 0 \\ \left(\dfrac{5}{6}\right)^4 = 0.48 & 0 \leq y < 1 \\ \left(\dfrac{5}{6}\right)^4 + 4\left(\dfrac{1}{6}\right)\left(\dfrac{5}{6}\right)^3 = 0.86 & 1 \leq y < 2 \\ \left(\dfrac{5}{6}\right)^4 + 4\left(\dfrac{1}{6}\right)\left(\dfrac{5}{6}\right)^3 + 6\left(\dfrac{1}{6}\right)^2\left(\dfrac{5}{6}\right)^2 = 0.98 & 2 \leq y < 3 \\ \left(\dfrac{5}{6}\right)^4 + 4\left(\dfrac{1}{6}\right)\left(\dfrac{5}{6}\right)^3 + 6\left(\dfrac{1}{6}\right)^2\left(\dfrac{5}{6}\right)^2 + 4\left(\dfrac{1}{6}\right)^3\left(\dfrac{5}{6}\right) = 0.99 & 3 \leq y < 4 \\ 1 & 4 \leq y \end{cases}$$

Figure 3.2.9 is a graph of $F_Y(y)$.

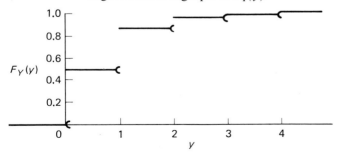

Figure 3.2.9

Example 3.2.6

The distance, Y, a molecule of gas travels before colliding with another molecule is described by the pdf

$$f_Y(y) = \left(\frac{1}{\lambda}\right)e^{-y/\lambda} \qquad y > 0$$

where λ is a constant characteristic of the particular gas being considered (recall Example 2.5.3). Find $F_Y(y)$.

Here, Y is continuous, so its cdf will be continuous. For $y < 0$, $F_Y(y) = P(Y \leq y)$ is clearly 0. A short integration gives $F_Y(y)$ for $y \geq 0$:

$$F_Y(y) = P(Y \leq y) = \int_{-\infty}^{y} f_Y(t)\, dt$$
$$= \int_{-\infty}^{0} 0\, dt + \int_{0}^{y} \left(\frac{1}{\lambda}\right)e^{-t/\lambda}\, dt$$
$$= 1 - e^{-y/\lambda}$$

(see Figure 3.2.10).

Intercollision distance, y **Figure 3.2.10**

Example 3.2.7

If a pdf is piecewise continuous and has discontinuities at a finite number of points, its cdf will still be continuous. Figure 3.2.11 shows the pdf for a piecewise continuous random variable Y, where

$$f_Y(y) = \begin{cases} 0 & y < 0 \\ 2y & 0 \leq y \leq \frac{1}{2} \\ 6 - 6y & \frac{1}{2} < y \leq 1 \\ 0 & y > 1 \end{cases}$$

Find the corresponding cdf.

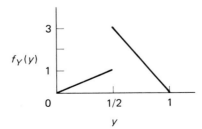

Figure 3.2.11

Looking at the form of $f_Y(y)$, we see that $F_Y(y)$ will have four different functional expressions, depending on whether $y < 0$, $0 \le y \le \frac{1}{2}$, $\frac{1}{2} < y \le 1$, or $y > 1$. For $y < 0$, $F_Y(y) = P(Y \le y) = 0$. For $0 \le y \le \frac{1}{2}$,

$$F_Y(y) = P(Y \le y) = \int_{-\infty}^{0} 0 \, dx + \int_{0}^{y} 2t \, dt$$

$$= y^2$$

(see Figure 3.2.12). For $\frac{1}{2} < y \le 1$,

$$F_Y(y) = P(Y \le y) = \int_{-\infty}^{0} 0 \, dt + \int_{0}^{1/2} 2t \, dt + \int_{1/2}^{y} (6 - 6t) \, dt$$

$$= \frac{1}{4} + (6t - 3t^2)\Big|_{1/2}^{y}$$

$$= 6y - 3y^2 - 2$$

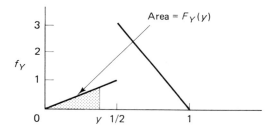

Figure 3.2.12

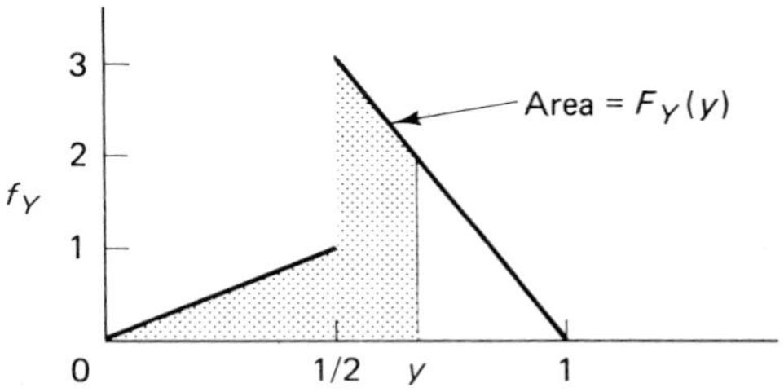

Figure 3.2.13

(see Figure 3.2.13). Of course, for $y > 1$, $F_Y(y) = 1$. Putting all these cases together gives

$$F_Y(y) = \begin{cases} 0 & y < 0 \\ y^2 & 0 \le y \le \frac{1}{2} \\ 6y - 3y^2 - 2 & \frac{1}{2} < y \le 1 \\ 1 & y > 1 \end{cases}$$

Figure 3.2.14 is a graph of $F_Y(y)$.

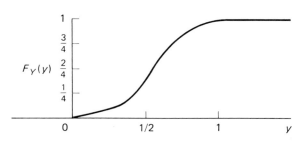

Figure 3.2.14

Question 3.2.9 An insurance company runs an actuarial trainee program that has a 20% attrition rate. Suppose that three people enter the program. Find and graph the cdf of the number who finish.

Question 3.2.10 Graph the cdf corresponding to a random variable whose pdf is given by

$$f_Y(y) = \frac{1}{3} \qquad y = 0, 1, 2$$

Question 3.2.11 Let Y denote the number of aces a poker player receives in a five-card hand. Graph the cdf for Y.

Question 3.2.12 An urn contains n chips numbered 1 through n. Suppose that the probability of any chip, k, being drawn is proportional to k. A single chip is selected. Let the random variable Y denote the number on the chip drawn. Find $F_Y(y)$.

Question 3.2.13 Find and graph the cdf associated with a random variable distributed *uniformly* over the interval (a, b) (see Question 2.5.1).

Question 3.2.14 Let Y be a random variable with pdf

$$f_Y(y) = \begin{cases} \dfrac{3}{4} & 0 \le y \le 1 \\[2mm] \dfrac{1}{4} & 2 \le y \le 3 \\[2mm] 0 & \text{elsewhere} \end{cases}$$

(a) Graph $f_Y(y)$
(b) Find and graph $F_Y(y)$.

Question 3.2.15 If the pdf for Y is

$$f_Y(y) = \begin{cases} 0 & |y| > 1 \\ 1 - |y| & |y| \le 1 \end{cases}$$

find and graph $F_Y(y)$.

Question 3.2.16 Recall the cotter-pin example of Question 2.5.3. If

$$f_Y(y) = k(y + y^2) \qquad 0 \le y \le 2$$

find $F_Y(y)$.

There are certain properties of cdf's that follow immediately from the way they are defined. These give us an alternative way of computing probabilities. Since $F_Y(y) = P(Y \le y)$, then

(a) $P(Y > y) = 1 - F_Y(y)$

(b) $P(a < Y \le b) = F_Y(b) - F_Y(a)$

(c) $P(Y = t) = F_Y(t) - \lim_{y \to t^-} F_Y(y)$

Properties (a) and (b) are obvious; property (c) deserves a word of explanation. If Y is a continuous random variable, $F_Y(y)$ is continuous and $\lim_{y \to t^-} F_Y(y) = F_Y(t)$, implying that $P(Y = t) = 0$ for all t. If Y is discrete, $F_Y(t) - \lim_{y \to t^-} F_Y(y)$ will be 0 everywhere except at the points for which $f_Y(t) > 0$. Figure 3.2.15 is a graph of the

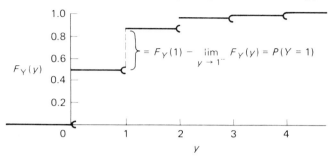

Figure 3.2.15

cdf found for the binomial distribution described in Example 3.2.5. For a point such as $t = 1.5$,

$$F_Y(t) - \lim_{y \to t^-} F_Y(y) = P(Y \leq 1.5) - \lim_{y \to 1.5^-} P(Y \leq y)$$

$$= P(Y = 0) + P(Y = 1) - \{P(Y = 0) + P(Y = 1)\}$$

$$= 0$$

and the latter, of course, *is* equal to $P(Y = 1.5)$. On the other hand, for a point like $t = 1.0$,

$$F_Y(t) - \lim_{y \to t^-} F_Y(y) = P(Y \leq 1.0) - \lim_{y \to 1^-} P(Y \leq y)$$

$$= P(Y = 0) + P(Y = 1) - P(Y = 0)$$

$$= P(Y = 1)$$

(see Figure 3.2.15). Given the graph, then, of any cdf—discrete or continuous—we can reconstruct the probability associated with any point y by simply subtracting from $F_Y(y)$ its left-hand limit.

The most important property of cdf's, though, for our purposes will be the relationship between a continuous pdf and the *derivative* of its cdf. We will first use this property in Section 3.5 but it will reappear in a number of derivations later in the book.

Theorem 3.2.1. If Y is a continuous random variable with pdf $f_Y(y)$ and cdf $F_Y(y)$, then

$$f_Y(y) = F_Y'(y)$$

provided $F_Y'(y)$ exists at all but a finite number of points.

Proof. The statement of the theorem follows immediately from the Fundamental Theorem of calculus and the Intermediate Value Theorem.

Example 3.2.8

Suppose that a random variable has the cdf shown in Figure 3.2.16. In functional form,

$$F_Y(y) = \begin{cases} 0 & y < 0 \\ \dfrac{3y}{4} & 0 \le y \le 1 \\ \dfrac{3}{4} & 1 < y \le 2 \\ \dfrac{y}{4} + \dfrac{1}{4} & 2 < y \le 3 \\ 1 & y > 3 \end{cases}$$

Find the corresponding pdf.

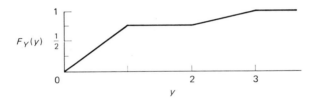

Figure 3.2.16

Since $F_Y(y)$ is everywhere continuous and $F'_Y(y)$ exists at all but four points, we can use Theorem 3.2.1 to find $f_Y(y)$. Differentiating $F_Y(y)$ gives

$$f_Y(y) = F'_Y(y) = \begin{cases} 0 & y < 0 \\ \dfrac{3}{4} & 0 \le y \le 1 \\ 0 & 1 < y \le 2 \\ \dfrac{1}{4} & 2 < y \le 3 \\ 0 & y > 3 \end{cases}$$

which shows that $f_Y(y)$ is a simple step-type function (see Figure 3.2.17).

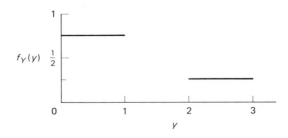

Figure 3.2.17

Random Variables   Chap. 3

Question 3.2.17 A continuous random variable Y has a cdf given by

$$F_Y(y) = \begin{cases} 0 & y < 0 \\ y^2 & 0 \le y < 1 \\ 1 & y \ge 1 \end{cases}$$

Find $P(\frac{1}{2} < Y \le \frac{3}{4})$ two ways—first, by using the cdf and second, by using the pdf.

Question 3.2.18 Suppose that $f_Y(y)$ is a symmetric and continuous pdf: $f_Y(y) = f_Y(-y)$, for $y > 0$. Show that

$$P(-a < Y < a) = 2 \cdot F_Y(a) - 1$$

for $a > 0$.

Question 3.2.19 A random variable Y has cdf

$$F_Y(y) = \begin{cases} 0 & y < 1 \\ \ln y & 1 \le y \le e \\ 1 & e < y \end{cases}$$

Find
(a) $P(Y < 2)$
(b) $P(2 < Y \le 2\frac{1}{2})$
(c) $P(2 < Y < 2\frac{1}{2})$
(d) $f_Y(y)$

Question 3.2.20 Suppose that Y is a continuous random variable defined over the interval A. Describe an event, B, having the property that the probability of B is 1, but B is not certain to occur. Describe an event, C, for which $P(C) = 0$, yet C is not impossible.

Question 3.2.21 Among the most famous of all meteor showers are the Perseids, which occur each year in early August. In some areas the frequency of visible Perseids averages six per quarter hour. The probability model describing Y, the number of meteors a person sees, has the pdf

$$f_Y(y) = \frac{e^{-6}6^y}{y!} \qquad y = 0, 1, 2, \ldots$$

Find the probability a person sees—in a given quarter hour—at least half as many meteors as he might expect to see.

Question 3.2.22 The cdf for a random variable Y is defined by $F_Y(y) = 0$ for $y < 0$; $F_Y(y) = 4y^3 - 3y^4$ for $0 \le y \le 1$; and $F_Y(y) = 1$ for $y > 1$. Find $P(\frac{1}{4} \le Y \le \frac{3}{4})$ by integrating $f_Y(y)$.

Question 3.2.23 Let Y be a random variable denoting the age at which a piece of equipment fails. In reliability theory, the probability that an item fails at time y

given that it has survived until time y is called the *hazard rate*, $h(y)$. In terms of pdf's and cdf's,

$$h(y) = \frac{f_Y(y)}{1 - F_Y(y)}$$

Find $h(y)$ if Y is described by an exponential pdf, $f_Y(y) = (1/\lambda)e^{-y/\lambda}$, $y > 0$. Interpret your answer.

3.3 JOINT DENSITIES

Section 3.2 introduced the basic terminology for describing the probabilistic behavior of a *single* random variable, whether that variable is discrete or continuous. Such information, while adequate for many problems, is insufficient in situations where more than one random variable affects the outcome of an experiment. For example, consider an electronic system containing two components, one for backup, but both under load. Suppose that the only way the system will fail is if both components fail. The distribution, then, of Z, the *system's* life, depends "jointly" on the distributions of X and Y, the component lives. Knowing only $f_X(x)$ and $f_Y(y)$, though, will not necessarily provide us with enough information to determine $f_Z(z)$. What we need is a probability function describing the "simultaneous" behavior of X and Y.

More generally, we may have to deal with the joint behavior of n random variables. Management of a fast-food chain, for example, may want to predict their total sales volume coming from five local franchises, each competing to some extent for the same pool of customers. Or consider the problem faced by medical researchers in trying to sort out which of several factors—serum cholesterol, triglyceride level, blood pressure, genetic predisposition, and so on—contribute to a patient's risk of a coronary, and to what extent. As might be expected, when the number of random variables increases, the mathematical task of describing their joint behavior becomes that much more difficult. To keep technical details from obscuring underlying concepts, we will concentrate first on the two-variable case. At the end of the section, analogous results for the n-variable generalization will be outlined.

Definition 3.3.1.

(a) Suppose that X and Y are two discrete random variables defined on the same sample space S. The *joint probability density function of X and Y* (or *joint pdf*) is denoted $f_{X,Y}(x, y)$, where

$$f_{X,Y}(x, y) = P(\{s \in S \mid X(s) = x, Y(s) = y\})$$

$$= P(X = x, Y = y)$$

(b) Suppose that X and Y are two continuous random variables defined over the same sample space S. The joint pdf of X and Y, $f_{X,Y}(x, y)$, is the surface having the property that for any region R in the xy-plane,

$$P((X, Y) \in R) = P(\{s \in S \mid (X(s), Y(s)) \in R\})$$

$$= \int_R \int f_{X,Y}(x, y) \, dx \, dy$$

Example 3.3.1

A supermarket has two express lines. Let X and Y denote the number of customers in the first and in the second, respectively, at any given time. During nonrush hours, the joint pdf of X and Y is summarized by the following table:

		X		
	0	1	2	3
0	0.1	0.2	0	0
1	0.2	0.25	0.05	0
2	0	0.05	0.05	0.025
3	0	0	0.025	0.05

(Y labels the rows 0, 1, 2, 3)

Find $P(|X - Y| = 1)$, the probability that X and Y differ by exactly 1.
By definition,

$$P(|X - Y| = 1) = \sum_{|x-y|=1} \sum f_{X,Y}(x, y)$$

$$= f_{X,Y}(0, 1) + f_{X,Y}(1, 0) + f_{X,Y}(1, 2)$$

$$+ f_{X,Y}(2, 1) + f_{X,Y}(2, 3) + f_{X,Y}(3, 2)$$

$$= 0.2 + 0.2 + 0.05 + 0.05 + 0.025 + 0.025$$

$$= 0.55$$

[Would you expect $f_{X,Y}(x, y)$ to be symmetric? Would you expect the event $|X - Y| \geq 2$ to have zero probability?]

Example 3.3.2

A study shows the daily number of hours, X, a teenager watches television and the daily number of hours, Y, he works on his homework are approximated by the joint pdf,

$$f_{X,Y}(x, y) = xye^{-(x+y)} \qquad x > 0, \quad y > 0$$

What is the probability a teenager chosen at random spends at least twice as much time watching television as he does working on his homework?

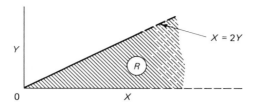

Figure 3.3.1

The region, R, in the xy-plane corresponding to the event "$X \geq 2Y$" is shown in Figure 3.3.1. It follows that $P(X \geq 2Y)$ is the volume under $f_{X,Y}(x, y)$ above the region R:

$$P(X \geq 2Y) = \int_0^\infty \int_0^{x/2} xye^{-(x+y)} \, dy \, dx$$

Separating variables, we can write

$$P(X \geq 2Y) = \int_0^\infty xe^{-x} \left[\int_0^{x/2} ye^{-y} \, dy \right] dx$$

What the double integral eventually reduces to is $\frac{7}{27}$:

$$P(X \geq 2Y) = \int_0^\infty xe^{-x} \left[1 - \left(\frac{x}{2} + 1 \right) e^{-x/2} \right] dx$$

$$= \int_0^\infty xe^{-x} \, dx - \int_0^\infty \frac{x^2}{2} e^{-3x/2} \, dx - \int_0^\infty xe^{-3x/2} \, dx$$

$$= 1 - \frac{16}{54} - \frac{4}{9}$$

$$= \frac{7}{27}$$

Example 3.3.3

Consider the electronic system mentioned in the beginning of this section. Suppose that the two components operate independently and have identical performance characteristics. Let X and Y be random variables denoting their life spans. Experience with wear-out times suggests that in certain cases a good choice for $f_{X,Y}(x, y)$ would be

$$f_{X,Y}(x, y) = \begin{cases} \lambda^2 e^{-\lambda(x+y)} & x \geq 0, \quad y \geq 0 \\ 0 & \text{otherwise} \end{cases}$$

where λ is some constant greater than 0.

Suppose that the manufacturer advertises a money-back guarantee if the system fails to last for more than 1000 hours. What are the chances of a given system's being returned for a refund?

Since the system fails only if both components fail,

$$P(\text{refund}) = P(X \le 1000, \quad Y \le 1000)$$

$$= \int_0^{1000} \int_0^{1000} \lambda^2 e^{-\lambda(x+y)} \, dy \, dx$$

$$= \int_0^{1000} \left(\int_0^{1000} \lambda^2 e^{-\lambda x} e^{-\lambda y} \, dy \right) dx$$

The integration in parentheses is done with respect to y, with x being treated as a constant. The expression $\lambda e^{-\lambda x}$ can therefore be factored out of the y integration, so

$$\int_0^{1000} \left(\int_0^{1000} \lambda^2 e^{-\lambda x} e^{-\lambda y} \, dy \right) dx = \int_0^{1000} \lambda e^{-\lambda x} \left(\int_0^{1000} \lambda e^{-\lambda x} \, dy \right) dx$$

$$= \int_0^{1000} \lambda e^{-\lambda x} (1 - e^{-\lambda \cdot 1000}) \, dx$$

$$= (1 - e^{-\lambda \cdot 1000}) \int_0^{1000} \lambda e^{-\lambda x} \, dx$$

Evaluating the final integral and multiplying by the factor that precedes it, we find that

$$P(\text{refund}) = (1 - e^{-1000\lambda})^2$$

Example 3.3.4

Let X, Y, and $f_{X, Y}(x, y)$ be defined as they were in Example 3.3.3. However, suppose the system itself is modified so that one component is kept on reserve—and activated only when the other needs replacing. As before, the system fails only when both components burn out, but now the component lives are cumulative. Again we seek the probability that the system fails in 1000 hours or less. Translated into a statement about X and Y, the probability of a refund reduces to

$$P(\text{refund}) = P(X + Y \le 1000) = \int_R \int \lambda^2 e^{-\lambda(x+y)} \, dy \, dx$$

where R is the shaded region shown in Figure 3.3.2.

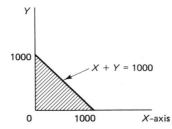

Figure 3.3.2

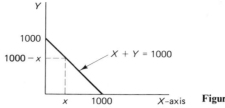

Figure 3.3.3

As in the preceding example, the double integral is evaluated by first holding x constant. Here, though, the upper limit of the integration for y depends on x: for x fixed, y varies between 0 and $1000 - x$ (see Figure 3.3.3). Thus

$$\int_R \int f_{X, Y}(x, y)\, dy\, dx = \int_0^{1000} \left(\int_0^{1000-x} \lambda^2 e^{-\lambda x} e^{-\lambda y}\, dy \right) dx$$

$$= \int_0^{1000} \lambda e^{-\lambda x} \left(\int_0^{1000-x} \lambda e^{-\lambda y}\, dy \right) dx$$

$$= \int_0^{1000} \lambda e^{-\lambda x} (1 - e^{-\lambda(1000-x)})\, dx$$

$$= \int_0^{1000} (\lambda e^{-\lambda x} - \lambda e^{-\lambda \cdot 1000})\, dx$$

$$= 1 - e^{-\lambda \cdot 1000} - 1000\lambda e^{-\lambda \cdot 1000}$$

Collecting terms gives

$$P(\text{system fails in 1000 hours or less}) = 1 - (1 + 1000\lambda)e^{-1000\lambda}$$

Comment

Note that with a bit more effort we can get the pdf for $X + Y$. Let $Z = X + Y$. The argument we used to find $P(Z \leq 1000)$ obviously holds for any $z > 0$. Therefore,

$$P(\text{system fails in } z \text{ hours or less}) = P(Z \leq z) = F_Z(z) = 1 - (1 + \lambda z)e^{-\lambda z}$$

Differentiating gives $f_Z(z)$:

$$f_Z(z) = F_Z'(z) = \lambda^2 z e^{-\lambda z} \qquad z > 0$$

Question 3.3.1 If $f_{X, Y}(x, y) = cxy$ at the points $(1, 1)$, $(2, 1)$, $(2, 2)$, and $(3,1)$, and equals 0 elsewhere, find c.

Question 3.3.2 An urn contains 4 red chips, 3 white chips, and 2 blue chips. A random sample of size 3 is drawn without replacement. Let X denote the number of white chips in the sample and Y, the number of blue. Write down a formula for the joint pdf of X and Y.

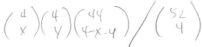

$$\left(\begin{array}{c} 4 \\ x \end{array} \right)\left(\begin{array}{c} 4 \\ y \end{array} \right)\left(\begin{array}{c} 44 \\ 4-x-y \end{array} \right) \Big/ \left(\begin{array}{c} 52 \\ 4 \end{array} \right)$$

Question 3.3.3 Four cards are drawn from a standard poker deck. Let X be the number of kings drawn and Y, the number of queens. Find $f_{X,Y}(x, y)$.

Question 3.3.4 An advisor looks over the schedules of his 50 students to see how many math and science courses each has registered for in the coming semester. He summarizes his results in a table.

		Number of math courses, X		
		0	1	2
Number	0	11	6	4
of science	1	9	10	3
courses, Y	2	5	0	2

What is the probability that a student selected at random will have signed up for more math courses than science courses?

Question 3.3.5 Suppose that two fair dice are tossed one time. Let X denote the number of 2's that appear, and Y the number of 3's. Write out the matrix giving the joint probability density function for X and Y. Suppose a third random variable, Z, is defined, where $Z = X + Y$. Use $f_{X,Y}(x, y)$ to find $f_Z(z)$.

Question 3.3.6 Suppose that X and Y have a bivariate uniform density over the unit square:

$$f_{X,Y}(x, y) = \begin{cases} c & 0 < x < 1, \quad 0 < y < 1 \\ 0 & \text{elsewhere} \end{cases}$$

(a) Find c.
(b) Find $P(0 < X < \frac{1}{2}, 0 < Y < \frac{1}{4})$.

Question 3.3.7 Let X and Y have the joint pdf

$$\int_0^\infty \int_{Y/3}^Y 2e^{-(x+y)}\, dx\, dy$$

$$f_{X,Y}(x, y) = 2e^{-(x+y)} \qquad 0 < x < y, \quad 0 < y$$

Find $P(Y < 3X)$.

Question 3.3.8 A point is chosen at random from the interior of a circle whose equation is $x^2 + y^2 \le 4$. Let the random variables X and Y denote the x- and y-coordinates of the sampled point. Find $f_{X,Y}(x, y)$.

Not surprisingly, the consideration of joint *pdf*'s leads to the consideration of joint *cdf*'s. As was true with their analogs in the univariate case, the two share some important properties.

> **Definition 3.3.2.** Let X and Y be two random variables defined on the same sample space S. The *joint cumulative distribution function* (or *joint cdf*) of X *and* Y is denoted $F_{X,Y}(x, y)$, where
>
> $$F_{X,Y}(x, y) = P(\{s \in S \mid X(s) \le x \quad \text{and} \quad Y(s) \le y\})$$
> $$= P(X \le x, \quad Y \le y)$$
>
> The domain of $F_{X,Y}(x, y)$ is the set of all pairs of real numbers.

Given the joint pdf $f_{X,Y}(x, y)$ of two random variables X and Y, we can easily find the variables' joint cdf. If X and Y are discrete,

$$F_{X,Y}(x, y) = \sum_{X \leq x, Y \leq y} f_{X,Y}(x, y)$$

If X and Y are continuous,

$$F_{X,Y}(x, y) = \int_{-\infty}^{y} \int_{-\infty}^{x} f_{X,Y}(u, v) \, du \, dv$$

We will also find it useful later to go in the other direction—that is, to find $f_{X,Y}(x, y)$ given $F_{X,Y}(x, y)$. Recall in the univariate case that when Y is continuous $f_Y(y)$ is the derivative of $F_Y(y)$. Here we have to differentiate *twice*.

> **Theorem 3.3.1.** Let X and Y be continuous random variables defined over the same sample space S and let $F_{X,Y}(x, y)$ be their joint cdf. Then
>
> $$f_{X,Y}(x, y) = \frac{\partial^2}{\partial x \, \partial y} F_{X,Y}(x, y)$$

Proof. See (46).

Question 3.3.9 Consider the experiment of simultaneously tossing a fair coin and rolling a fair die. Let X denote the number of heads showing on the coin and Y, the number of spots showing on the die.
(a) List the outcomes in S.
(b) Find $F_{X,Y}(1, 2)$.
(c) What would the graph of $F_{X,Y}(x, y)$ look like?

Question 3.3.10 An urn contains 12 chips—4 red, 3 black, and 5 white. A sample of size 4 is to be drawn without replacement. Let X denote the number of white chips in the sample; Y, the number of red. Find $F_{X,Y}(1, 2)$.

Question 3.3.11 Let the random variables X and Y denote the number of letters and the number of vowels, respectively, in each word of the following quotation:

Saepe creat molles aspera spina rosas.

Tabulate $F_{X,Y}(x, y)$.

Question 3.3.12 Prove that

$$P(a < X \leq b, c < Y \leq d) = F_{X,Y}(b, d) - F_{X,Y}(a, d)$$
$$- F_{X,Y}(b, c) + F_{X,Y}(a, c)$$

Question 3.3.13 Find and graph $f_{X,Y}(x, y)$ if the joint cdf for random variables X and Y is

$$F_{X,Y}(x, y) = xy \qquad 0 < x < 1, \quad 0 < y < 1$$

Question 3.3.14 Find the joint pdf associated with two random variables X and Y whose joint cdf is

$$F_{X,Y}(x, y) = (1 - e^{-\lambda y})(1 - e^{-\lambda x}) \qquad x > 0, \quad y > 0$$

Joint pdf's and single-variable pdf's are, themselves, intimately related: given an $f_{X,Y}(x, y)$, we can "recover" the individual pdf's for X and Y by integrating out (or summing over) the unwanted variable. The necessity to do this arises quite often in some of the proofs we will see in later chapters.

> **Theorem 3.3.2.** Let X and Y be discrete random variables with joint pdf $f_{X,Y}(x, y)$. The *individual* pdf's for X and Y—$f_X(x)$ and $f_Y(y)$, respectively—can be gotten from the joint pdf by an appropriate summation:
>
> $$\text{(a)} \quad f_X(x) = \sum_{\text{all } y} f_{X,Y}(x, y) \qquad \text{(b)} \quad f_Y(y) = \sum_{\text{all } x} f_{X,Y}(x, y)$$
>
> Similarly, if X and Y are both continuous, $f_X(x)$ and $f_Y(y)$ can be gotten from $f_{X,Y}(x, y)$ by an appropriate integration:
>
> $$\text{(a)} \quad f_X(x) = \int_{-\infty}^{\infty} f_{X,Y}(x, y) \, dy \qquad \text{(b)} \quad f_Y(y) = \int_{-\infty}^{\infty} f_{X,Y}(x, y) \, dx$$

Proof. The proof will be given for the discrete case only. The analogous result for continuous X and Y is derived in much the same way.

By definition,

$$f_X(x) = P(X = x) = P\left\{ \bigcup_{\text{all } y} (X = x, Y = y) \right\}$$

But the events $(X = x, Y = y_i)$ and $(X = x, Y = y_j)$ are mutually exclusive for $i \neq j$, implying that

$$P\left\{ \bigcup_{\text{all } y} (X = x, Y = y) \right\} = \sum_{\text{all } y} P(X = x, Y = y)$$

Therefore,

$$f_X(x) = \sum_{\text{all } y} P(X = x, Y = y) = \sum_{\text{all } y} f_{X,Y}(x, y)$$

Example 3.3.5

Consider an experiment consisting of three flips of a fair coin. Let X denote the number of heads on the last flip; Y, the total number of heads for the three tosses. Table 3.3.1 shows the (x, y) value associated with each of the eight possible outcomes. Find $f_X(x)$ and $f_Y(y)$.

TABLE 3.3.1

Outcome	(x, y)	$P(X = x, Y = y)$
(H, H, H)	(1, 3)	$\frac{1}{8}$
(T, H, H)	(1, 2)	$\frac{1}{8}$
(H, T, H)	(1, 2)	$\frac{1}{8}$
(H, H, T)	(0, 2)	$\frac{1}{8}$
(T, T, H)	(1, 1)	$\frac{1}{8}$
(T, H, T)	(0, 1)	$\frac{1}{8}$
(H, T, T)	(0, 1)	$\frac{1}{8}$
(T, T, T)	(0, 0)	$\frac{1}{8}$

By appropriately combining the second and third columns of Table 3.3.1, we can write the joint pdf of X and Y as the 2×4 matrix in Table 3.3.2. Note that summing across the rows gives $f_X(x)$, while summing down the columns yields $f_y(y)$.

TABLE 3.3.2

		\multicolumn{4}{c}{Y}	
		0	1	2	3	$f_X(x)$
X	0	$\frac{1}{8}$	$\frac{1}{4}$	$\frac{1}{8}$	0	$\frac{1}{2}$
	1	0	$\frac{1}{8}$	$\frac{1}{4}$	$\frac{1}{8}$	$\frac{1}{2}$
	$f_Y(y)$	$\frac{1}{8}$	$\frac{3}{8}$	$\frac{3}{8}$	$\frac{1}{8}$	

Comment

When $f_{X, Y}(x, y)$ is written as a matrix, the individual densities will appear, as they do in Table 3.3.2, as "margins." For that reason, $f_X(x)$ and $f_Y(y)$, in the context of a joint pdf, are often referred to as *marginal densities*. The word "marginal," though, is nothing but a reminder that a second variable was originally involved—there is absolutely no difference between a pdf and a marginal pdf.

Example 3.3.6

Suppose that X and Y are two random variables jointly distributed over the first quadrant of the xy-plane according to the pdf,

$$f_{X, Y}(x, y) = \begin{cases} y^2 e^{-y(x + 1)} & x \geq 0, \quad y \geq 0 \\ 0 & \text{elsewhere} \end{cases}$$

Find the two marginal pdf's.

First, consider $f_X(x)$. By Theorem 3.3.2,

$$f_X(x) = \int_{-\infty}^{\infty} f_{X,Y}(x, y)\, dy = \int_0^{\infty} y^2 e^{-y(x+1)}\, dy$$

In the integrand, substitute

$$u = y(x + 1)$$

making $du = (x + 1)\, dy$. This gives

$$f_X(x) = \frac{1}{x+1} \int_0^{\infty} \frac{u^2}{(x+1)^2}\, e^{-u}\, du = \frac{1}{(x+1)^3} \int_0^{\infty} u^2 e^{-u}\, du$$

After applying integration by parts (twice) to $\int_0^{\infty} u^2 e^{-u}\, du$, we get

$$f_X(x) = \frac{1}{(x+1)^3} \left[-u^2 e^{-u} - 2u e^{-u} - 2e^{-u} \right]\Big|_0^{\infty}$$

$$= \frac{1}{(x+1)^3} \left[2 - \lim_{u \to \infty} \left(\frac{u^2}{e^u} + \frac{2u}{e^u} + \frac{2}{e^u} \right) \right]$$

$$= \frac{2}{(x+1)^3}$$

Finding $f_Y(y)$ is a bit easier:

$$f_Y(y) = \int_{-\infty}^{\infty} f_{X,Y}(x, y)\, dx = \int_0^{\infty} y^2 e^{-y(x+1)}\, dx$$

$$= y^2 e^{-y} \int_0^{\infty} e^{-yx}\, dx = y^2 e^{-y} \left(\frac{1}{y} \right) \left(-e^{-yx} \right)\Big|_0^{\infty}$$

$$= y e^{-y}$$

Question 3.3.15 Find $f_X(x)$ and $f_Y(y)$ if the joint pdf for X and Y is

$$f_{X,Y}(x, y) = \frac{1}{x} \qquad 0 < y < x, \quad 0 < x < 1$$

Question 3.3.16 Suppose X and Y have a joint pdf given by

$$f_{X,Y}(x, y) = 6x \qquad 0 < x < 1, \quad 0 < y < 1 - x$$

Find $f_X(x)$ and $f_Y(y)$. Also, sketch the joint pdf and the two marginals.

Question 3.3.17 The campus recruiter for an international conglomerate classifies the large number of students she interviews into three categories—the lower quarter, the middle half, and the upper quarter. If she meets six students on a given morning, what is the probability that they will be evenly divided among the three categories? What is the marginal probability that exactly two will belong to the middle half?

Question 3.3.18 For each of the following joint pdf's, find $f_X(x)$ and $f_Y(y)$.

(a) $f_{X,Y}(x, y) = \frac{1}{2}$ $0 < x < 2, \quad 0 < y < 1$

(b) $f_{X,Y}(x, y) = xye^{-(x+y)}$ $x > 0, \quad y > 0$

(c) $f_{X,Y}(x, y) = 2e^{-(x+y)}$ $0 < x < y, \quad 0 < y$

(d) $f_{X,Y}(x, y) = \frac{3}{2} \cdot y^2$ $0 \le x \le 2, \quad 0 \le y \le 1$

(e) $f_{X,Y}(x, y) = c(x + y)$ $0 < x < 1, \quad 0 < y < 1$

(f) $f_{X,Y}(x, y) = 2$ $0 < x < y < 1$

(g) $f_{X,Y}(x, y) = ye^{-xy - y}$ $x > 0, \quad y > 0$

(h) $f_{X,Y}(x, y) = 4xy$ $0 < x < 1, \quad 0 < y < 1$

(i) $f_{X,Y}(x, y) = \frac{2}{3} \cdot (x + 2y)$ $0 < x < 1, \quad 0 < y < 1$

The definitions and theorems in this section extend in a very straightforward way to situations involving more than two variables. The joint pdf for n discrete random variables, for example, is denoted $f_{X_1, \ldots, X_n}(x_1, \ldots, x_n)$, where

$$f_{X_1, \ldots, X_n}(x_1, \ldots, x_n) = P(X_1 = x_1, \ldots, X_n = x_n)$$

For n continuous random variables, the joint pdf is that function $f_{X_1, \ldots, X_n}(x_1, \ldots, x_n)$ having the property that for any region R in n-space,

$$P((X_1, \ldots, X_n) \in R) = \underset{R}{\iint \cdots \int} f_{X_1, \ldots, X_n}(x_1, \ldots, x_n) \, dx_1 \cdots dx_n$$

And if $F_{X_1, \ldots, X_n}(x_1, \ldots, x_n)$ is the joint *cdf* of continuous random variables $X_1, \ldots, X_n$—that is, $F_{X_1, \ldots, X_n}(x_1, \ldots, x_n) = P(X_1 \le x_1, \ldots, X_n \le x_n)$—then

$$f_{X_1, \ldots, X_n}(x_1, \ldots, x_n) = \frac{\partial^n}{\partial X_1 \cdots \partial X_n} F_{X_1, \ldots, X_n}(x_1, \ldots, x_n)$$

The notion of a marginal pdf also extends readily, although in the n-variate case, a marginal pdf can, itself, be a joint pdf. Given $X_1, \ldots, X_n$, the marginal pdf of any subset of r of those variables $(X_{i_1}, X_{i_2}, \ldots, X_{i_r})$ is derived by integrating (or summing) the joint pdf with respect to the remaining $n - r$ variables $(X_{j_1}, X_{j_2}, \ldots, X_{j_{n-r}})$. Formally, if the X_i's are all continuous,

$$f_{X_{i_1}, \ldots, X_{i_r}}(x_{i_1}, \ldots, x_{i_r}) = \int_{-\infty}^{\infty} \int_{-\infty}^{\infty} \cdots \int_{-\infty}^{\infty} f_{X_1, \ldots, X_n}(x_1, \ldots, x_n) \, dX_{j_1} \cdots dX_{j_{n-r}}$$

Question 3.3.19 A certain brand of fluorescent bulbs will last, on the average, 1000 hours. Suppose that four of these bulbs are installed in an office. What is the probability that all four are still functioning after 1050 hours? If X_i denotes the ith bulb's life, assume that

$$f_{X_1, X_2, X_3, X_4}(x_1, x_2, x_3, x_4) = \prod_{i=1}^{4} \left(\frac{1}{1000}\right) e^{-x_i/1000}$$

for $x_i > 0$, $i = 1, 2, 3, 4$.

Question 3.3.20 A hand of six cards is dealt from a standard poker deck. Let X denote the number of aces, Y the number of kings, and Z the number of queens.

(a) Write down a formula for $f_{X, Y, Z}(x, y, z)$.

(b) Find $f_{X, Y}(x, y)$ and $f_{X, Z}(x, z)$.

Question 3.3.21 Suppose that the random variables X, Y, and Z have the multivariate pdf

$$f_{X, Y, Z}(x, y, z) = (x + y)e^{-z}$$

for $0 < x < 1, 0 < y < 1$, and $z > 0$. Find (a) $f_{X, Y}(x, y)$, (b) $f_{Y, Z}(y, z)$, and (c) $f_Z(z)$.

Question 3.3.22 The four random variables W, X, Y, and Z have the multivariate pdf

$$f_{W, X, Y, Z}(w, x, y, z) = 16wxyz$$

for $0 < w < 1$, $0 < x < 1$, $0 < y < 1$, and $0 < z < 1$. Find the marginal pdf, $f_{W, X}(w, x)$, and use it to compute $P(0 < W < \frac{1}{2}, \frac{1}{2} < X < 1)$.

3.4 INDEPENDENT RANDOM VARIABLES

The concept of independent events that was introduced in Section 2.7 leads quite naturally to a similar definition for independent random variables.

> **Definition 3.4.1.** Random variables X and Y are said to be *independent* if for any intervals A and B,
>
> $$P(X \in A, Y \in B) = P((X, Y) \in A \otimes B) = P(X \in A) \cdot P(Y \in B)$$

Although it *defines* independence, Definition 3.4.1 is of little value in *establishing* independence. Given a continuous X and a continuous Y, it would be impossible to check all intervals A and B to verify the relationship between $P(X \in A, Y \in B)$ and $P(X \in A) \cdot P(Y \in B)$. A more workable characterization of this property is provided in Theorem 3.4.1.

> **Theorem 3.4.1.** Two random variables X and Y are independent if and only if
>
> $$f_{X, Y}(x, y) = f_X(x) \cdot f_Y(y)$$
>
> for all x and y.

Proof. We will show that independence implies that the joint pdf factors into a product of the marginals. The converse is left as an exercise.

The proof for discrete X and Y is trivial. Suppose that X and Y are discrete and independent; let a and b be any two numbers. Then

$$f_{X, Y}(a, b) = P(X = a, Y = b) = P(X = a) \cdot P(Y = b) = f_X(a) \cdot f_Y(b)$$

Now, let both X and Y be continuous (and independent), and take A and B to be two arbitrary intervals. We can write

$$\int_A \left(\int_B f_{X,Y}(x, y) \, dy \right) dx = \int_A \int_B f_{X,Y}(x, y) \, dy \, dx = P(X \in A, Y \in B)$$

$$= P(X \in A) \cdot P(Y \in B)$$

$$= \int_A f_X(x) \, dx \cdot \int_B f_Y(y) \, dy$$

$$= \int_A \left(\int_B f_X(x) \cdot f_Y(y) \, dy \right) dx$$

Note that the functions

$$\int_B f_{X,Y}(x, y) \, dy \quad \text{and} \quad \int_B f_X(x) f_Y(y) \, dy$$

have equal integrals over every interval A; therefore, by a theorem of calculus, the two are equal. But then it follows that $f_{X,Y}(x, y)$ and $f_X(x) \cdot f_Y(y)$ have equal integrals over every interval B, so by the same theorem, *they* are equal.

In Chapter 2, extending the notion of independence from *two* events to n events proved to be something of a problem: the independence of each subset of the n events had to be checked separately (recall Definition 2.7.2). This is not necessary in the case of random variables: to generalize Theorem 3.4.1, we need only establish that the joint pdf factors into a product of the n marginals.

Definition 3.4.2. The n random variables $X_1, X_2, \ldots, X_n$ are said to be *independent* if, for all $x_1, x_2, \ldots, x_n$,

$$f_{X_1, X_2, \ldots, X_n}(x_1, x_2, \ldots, x_n) = f_{X_1}(x_1) \cdot f_{X_2}(x_2) \cdots f_{X_n}(x_n)$$

Definition 3.4.3. Let $X_1, X_2, \ldots, X_n$ be a set of n independent random variables, all having the same pdf. Then $X_1, X_2, \ldots, X_n$ are said to be a *random sample of size n.*

Example 3.4.1

Consider k urns, each holding n chips, numbered 1 through n. A chip is to be drawn at random from each urn. What is the probability that all k chips will bear the same number?

If $X_1, X_2, \ldots, X_k$ denote the numbers on the 1st, 2nd, ..., and kth chips, respectively, we are looking for the probability that $X_1 = X_2 = \cdots = X_k$. In terms of the joint pdf,

$$P(X_1 = X_2 = \cdots = X_k) = \sum_{x_1 = x_2 = \cdots = x_k} f_{X_1, X_2, \ldots, X_k}(x_1, x_2, \ldots, x_k)$$

Each of the selections here is obviously independent of all the others so the joint pdf factors according to Definition 3.4.2, and we can write

$$P(X_1 = X_2 = \cdots = X_k) = \sum_{i=1}^{n} f_{X_1}(x_i) \cdot f_{X_2}(x_i) \cdots f_{X_k}(x_i)$$

$$= n \cdot \left(\frac{1}{n} \cdot \frac{1}{n} \cdot \cdots \cdot \frac{1}{n} \right)$$

$$= \frac{1}{n^{k-1}}$$

Question 3.4.1   Write down the joint probability density function for a random sample of size n drawn from the exponential pdf, $f_X(x) = (1/\lambda)e^{-x/\lambda}$, $x > 0$.

Question 3.4.2   Two fair dice are tossed. Let X denote the number appearing on the first die and Y, the number on the second. Show that X and Y are independent.

Question 3.4.3   A joint pdf is given by

$$f_{X_1, X_2}(x_1, x_2) = 12x_1 x_2(1 - x_2)$$

for $0 < x_1 < 1$ and $0 < x_2 < 1$. Are X_1 and X_2 independent?

Question 3.4.4   Let X_1 and X_2 be independent random variables each having the pdf

$$f_X(x) = \frac{x}{2} \qquad 0 < x < 2$$

Find the joint *cdf* of X_1 and X_2.

Question 3.4.5   Let X and Y be random variables with joint pdf

$$f_{X, Y}(x, y) = k \qquad 0 \le x \le 1, \quad 0 \le y \le 1, \quad 0 \le x + y \le 1$$

Give a geometric argument to show that X and Y are not independent.

Question 3.4.6   Recall the joint pdf used in Example 3.3.3—$f_{X, Y}(x, y) = \lambda^2 e^{-\lambda(x+y)}$, $x \ge 0$, $y \ge 0$. Are the random variables X and Y independent? Does your answer agree with the expression we found for P(refund)? Explain.

3.5 COMBINING AND TRANSFORMING RANDOM VARIABLES

Frequently, the random variables being measured in an experiment are not, themselves, the researcher's ultimate objective. What *is* of primary interest is some function of those variables. Example 3.3.4 is a case in point: there the life, Z, of an electronic system depended on the *sum*, $X + Y$, of two component lives. At issue, of course, was $f_Z(z)$. Other situations may require a change in the *scale* of a random variable. For example, suppose that X is originally calibrated in degrees Fahrenheit but the forces of metric conversion demand its reexpression in degrees Celsius. How does the pdf of Y, the Celsius random variable, compare to that of X?

In cases such as these it is inefficient to compute the pdf's of the "new" random variables by returning to elementary principles. Easier methods are available. If $Y = u(X)$ and $f_X(x)$ is known, we can easily find $f_Y(y)$ by writing the cdf for X in terms of y and then differentiating (recall Theorem 3.2.1). A similar technique works well in bivariate problems where $Z = u(X, Y)$ and we seek $f_Z(z)$.

Example 3.5.1

Suppose that a random variable X has pdf

$$f_X(x) = \begin{cases} 6x(1-x) & 0 < x < 1 \\ 0 & \text{elsewhere} \end{cases}$$

Let Y be a second random variable functionally related to X according to Equation 3.5.1:

$$Y = 2X + 1 \tag{3.5.1}$$

What is the pdf for Y?

Note, first, that the random variable Y maps the range of X-values, $(0, 1)$, onto the interval $(1, 3)$ (see Figure 3.5.1).

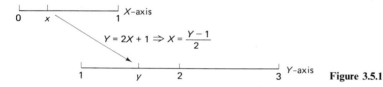

Figure 3.5.1

Let $1 < y < 3$. Then

$$F_Y(y) = P(Y \le y) = P(2X + 1 \le y) = P\left(X \le \frac{y-1}{2}\right)$$

$$= F_X\left(\frac{y-1}{2}\right)$$

But

$$F_X(t) = \begin{cases} 0 & t \le 0 \\ \displaystyle\int_0^t 6x(1-x)\,dx = 3t^2 - 2t^3 & 0 < t < 1 \\ 1 & t \ge 1 \end{cases}$$

Therefore,

$$F_Y(y) = F_X\left(\frac{y-1}{2}\right) = \begin{cases} 0 & y \le 1 \\ 3\left(\dfrac{y-1}{2}\right)^2 - 2\left(\dfrac{y-1}{2}\right)^3 \\ \quad = -\dfrac{y^3}{4} + \dfrac{3y^2}{2} - \dfrac{9y}{4} + 1 & 1 < y < 3 \\ 1 & y \ge 3 \end{cases}$$

Differentiating $F_Y(y)$ gives $f_Y(y)$:

$$f_Y(y) = \begin{cases} 0 & y \leq 1 \\ -\dfrac{3y^2}{4} + 3y - \dfrac{9}{4} & 1 < y < 3 \\ 0 & y \geq 3 \end{cases}$$

Example 3.5.2

In doing Monte Carlo studies, it is sometimes necessary to generate a series of exponential random variables, $Y_1, Y_2, \ldots$, where $f_{Y_i}(y) = (1/\lambda)e^{-y/\lambda}$, $y > 0$. Many computers do not have readily available software for generating such variables, but they do have subroutines for generating *uniform* random variables. By using an appropriate transformation, we can use the latter to construct the former.

Let X denote a random variable uniformly distributed over the unit interval. Define the transformation

$$Y = -\lambda \ln X$$

We will show that the pdf for Y is exponential with parameter λ.

By definition,

$$F_Y(y) = P(Y \leq y) = P(-\lambda \ln X \leq y)$$

$$= P\left(\ln X > -\frac{y}{\lambda}\right) = P(X > e^{-y/\lambda})$$

Thus the probability that Y is less than or equal to y is equal to the probability associated with that portion of the X-axis lying to the *right* of $e^{-y/\lambda}$ (see Figure 3.5.2). But since X is uniform over $(0, 1)$,

$$F_Y(y) = P(X > e^{-y/\lambda}) = \int_{e^{-y/\lambda}}^{1} 1 \cdot dx = 1 - e^{-y/\lambda}$$

Therefore,

$$f_Y(y) = F_Y'(y) = \left(\frac{1}{\lambda}\right)e^{-y/\lambda}$$

the result we set out to prove.

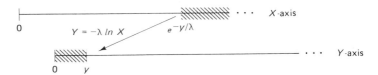

Figure 3.5.2

Example 3.5.3

Assume that the velocity of a gas molecule of mass m is a random variable X with pdf $f_X(x) = ax^2 e^{-bx^2}$, $x > 0$, where a and b are constants depending on the gas. Find the pdf for the kinetic energy, Y, of such a molecule, where $Y = (m/2)X^2$.

Let $F_Y(y)$ be the cdf for Y. Then

$$F_Y(y) = P(Y \le y) = P\left(\frac{m}{2} \cdot X^2 \le y\right)$$

$$= P\left(X^2 \le \frac{2y}{m}\right)$$

Figure 3.5.3 is a diagram of the X-axis and the Y-axis. Note the interval along the X-axis that gets mapped into the event " $Y \le y$." It follows that

$$F_Y(y) = P\left(X^2 \le \frac{2y}{m}\right) = P\left(X \le \sqrt{\frac{2y}{m}}\right)$$

$$= \int_0^{\sqrt{2y/m}} ax^2 e^{-bx^2}\, dx$$

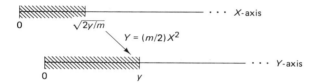

$Y = (m/2)X^2$

Figure 3.5.3

Finally, from a theorem in calculus,

$$f_Y(y) = F_Y'(y) = a\left(\sqrt{\frac{2y}{m}}\right)^2 e^{-b(\sqrt{2y/m})^2} \cdot \frac{d}{dy}\left(\sqrt{\frac{2y}{m}}\right)$$

$$= a \cdot \frac{2y}{m} \cdot e^{-b(2y/m)} \cdot \sqrt{\frac{2}{m}} \cdot \frac{1}{2} \cdot y^{-1/2}$$

$$= \frac{a\sqrt{2}}{m^{3/2}} \cdot \sqrt{y} e^{-2by/m} \qquad y > 0$$

Question 3.5.1 An urn contains 4 red chips and 3 white chips. Two are drawn out at random without replacement. Let X denote the number of red chips in the sample. Find the pdf for Y, where $Y = 2X - 1$.

Question 3.5.2 Suppose that the random variable X takes on the values -1, 1, and 4, each with probability $\frac{1}{3}$, and is 0 elsewhere. Find the pdf of $Y = 3X + 2$.

Question 3.5.3 Let X have the uniform pdf over the unit interval. Find $f_Y(y)$, where $Y = -3X - 4$.

Question 3.5.4 If X has the exponential pdf,

$$f_X(x) = 2e^{-2x} \qquad x > 0$$

find the *cdf* for $Y = 2X + 4$.

Question 3.5.5 If $f_X(x) = 6x(1 - x)$, $0 < x < 1$, find $f_Y(y)$, where $Y = 2X - 3$.

Question 3.5.6 Suppose that X has the uniform pdf over the interval (a, b). What linear transformation of X represents a random variable having the uniform pdf over $(0, 1)$?

Question 3.5.7 Suppose that X has pdf $f_X(x) = e^{-x}$, $x > 0$.
(a) Find the pdf for $Y = 1/X$.
(b) Find the pdf for $Y = \ln X$.

Question 3.5.8 A random variable X has density function $f_X(x) = \frac{3}{8}x^2$ over the interval $0 \le x \le 2$ (and 0, elsewhere). Suppose that a circle is generated by a radius whose length is the value of X. Find the density function for Y, the *area* of the circle.

Question 3.5.9 The temperature, X, achieved in a certain chemical reaction varies considerably from experiment to experiment but appears to be described quite well by a pdf of the form

$$f_X(x) = xe^{-x^2/2} \qquad x > 0$$

where x is measured in degrees Fahrenheit. The conversion formula for going from degrees Fahrenheit (X) to degrees Celsius (Y) is

$$Y = \frac{5}{9} \cdot (X - 32)$$

Describe the distribution of temperatures in terms of degrees Celsius. That is, find $f_Y(y)$.

Question 3.5.10 Find the pdf of $R = A \cdot \sin \theta$, where A is a constant and θ is uniformly distributed on $(-\pi/2, \pi/2)$. Variables like R arise quite often in the theory of ballistics. If a projectile is fired at an angle α with a velocity v, the distance R that it travels can be expressed as $R = (v^2/g) \cdot \sin 2\alpha$, where g is the gravitational constant.

In doing transformation problems, extreme care must be exercised in identifying the proper set of x's that get mapped into the event $Y \le y$. The best approach is always to sketch a diagram like the ones in Figures 3.5.1 and 3.5.2; the next two examples show why.

Example 3.5.4

Let X have the uniform density over the interval $(-1, 2)$:

$$f_X(x) = \begin{cases} \dfrac{1}{3} & -1 < x < 2 \\ 0 & \text{elsewhere} \end{cases}$$

Find the density function for Y, where $Y = X^2$.

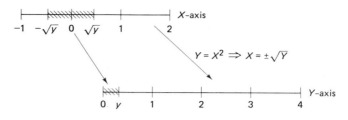

Figure 3.5.4

Here, the set of the x's that get mapped into the interval $Y \leq y$ has a different form depending on whether $0 \leq y < 1$ or $1 \leq y < 4$. To see this, suppose, first of all, that $0 \leq y < 1$. Then the set of x's that get mapped into $0 \leq Y \leq y$ are those from $-\sqrt{y}$ to $+\sqrt{y}$ (see Figure 3.5.4). Therefore,

$$F_Y(y) = P(Y \leq y) = P(-\sqrt{y} \leq X \leq +\sqrt{y})$$

$$= \int_{-\sqrt{y}}^{\sqrt{y}} \left(\frac{1}{3}\right) dx = \frac{2\sqrt{y}}{3} \qquad (3.5.2)$$

On the other hand, suppose $1 \leq y < 4$. Now, from Figure 3.5.4, the corresponding x's range from -1 to $\sqrt{y}$, and

$$F_Y(y) = P(Y \leq y) = P(-1 < X \leq \sqrt{y})$$

$$= \int_{-1}^{\sqrt{y}} \left(\frac{1}{3}\right) dx = \frac{\sqrt{y} + 1}{3} \qquad (3.5.3)$$

Equations 3.5.2 and 3.5.3, together with the obvious "boundary" cases (when $y < 0$ and $y \geq 4$), define $F_Y(y)$:

$$F_Y(y) = \begin{cases} 0 & y < 0 \\ \dfrac{2\sqrt{y}}{3} & 0 \leq y < 1 \\ \dfrac{\sqrt{y} + 1}{3} & 1 \leq y < 4 \\ 1 & y \geq 4 \end{cases}$$

Then, by differentiating $F_Y(y)$, we get

$$f_Y(y) = \begin{cases} \dfrac{1}{3\sqrt{y}} & 0 \leq y < 1 \\ \dfrac{1}{6\sqrt{y}} & 1 \leq y < 4 \\ 0 & \text{elsewhere} \end{cases}$$

Example 3.5.5

Suppose that X and Y have a joint uniform density over the unit square:

$$f_{X,Y}(x, y) = \begin{cases} 1 & 0 < x < 1, 0 < y < 1 \\ 0 & \text{elsewhere} \end{cases}$$

Find the pdf for their product—that is, find $f_Z(z)$, where $Z = XY$.

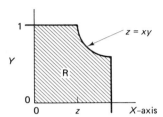

Figure 3.5.5

For $0 < z < 1$, $F_Z(z)$ is the volume above the shaded region in Figure 3.5.5. Specifically,

$$F_Z(z) = P(Z \le z) = P(XY \le z) = \iint\limits_R f_{X,Y}(x, y)\, dy\, dx$$

By inspection, we see that the double integral over R can be split up into *two* double integrals—one letting x range from 0 to z, the other having x-values extend from z to 1:

$$F_Z(z) = \int_0^z \left(\int_0^1 1\, dy \right) dx + \int_z^1 \left(\int_0^{z/x} 1\, dy \right) dx$$

But

$$\int_0^z \left(\int_0^1 1\, dy \right) dx = \int_0^z 1\, dx = z$$

and

$$\int_z^1 \left(\int_0^{z/x} 1\, dy \right) dx = \int_z^1 \left(\frac{z}{x} \right) dx = z \ln x \Big|_z^1 = -z \ln z$$

It follows that

$$F_Z(z) = \begin{cases} 0 & z \le 0 \\ z - z \ln z & 0 < z < 1 \\ 1 & z \ge 1 \end{cases}$$

in which case

$$f_Z(z) = \begin{cases} -\ln z & 0 < z < 1 \\ 0 & \text{elsewhere} \end{cases}$$

Question 3.5.11 Let X and Y have the bivariate uniform pdf over the unit square. Define $Z = X + Y$. Find $F_Z(z)$. (*Hint*: Consider two cases, $0 < z \leq 1$ and $1 < z < 2$. For each case, draw a diagram showing the appropriate region of integration.)

Question 3.5.12 Let X and Y be continuous random variables for which $f_{X,Y}(x, y) = x + y$, $0 < x < 1, 0 < y < 1$. Find $f_Z(z)$, where $Z = XY$.

Question 3.5.13 Suppose that the random variables X and Y have the joint uniform density over the unit square. Find (a) the cdf and (b) the pdf for $Z = X/Y$.

Question 3.5.14 A number is chosen at random from the interval $(0, 3)$. A second number is chosen independently and at random from the interval $(0, 4)$. What is the eightieth percentile of the sum of the two numbers? [By definition, the 80th percentile, z_{80}, is that number for which $P(X + Y \leq z_{80}) = 0.80$.]

For reasons that will become apparent a little later, we frequently have to deal with the problem of finding the pdf of the *sum* of two or more identically distributed and independent random variables. Examples 3.5.6 and 3.5.7 illustrate the simplest of such problems, situations where only two variables are to be added. Notice that in the discrete case we work directly with the pdf, $P(X + Y = z)$. When X and Y are continuous we first find the *cdf* for the sum—then differentiate.

Example 3.5.6

Suppose that X and Y are two independent binomial random variables, each with the same success probability but defined on m and n trials, respectively. Specifically,

$$f_X(k) = \binom{m}{k} p^k q^{m-k} \qquad k = 0, 1, \ldots, m$$

and

$$f_Y(k) = \binom{n}{k} p^k q^{n-k} \qquad k = 0, 1, \ldots, n$$

Let $Z = X + Y$. Find $f_Z(z)$.

Being a sum of x- and y-values, any given z can come about in a number of different ways, all mutually exclusive: $z = 0 + z = 1 + (z - 1)$, and so on. Therefore,

$$P(Z = z) = P\{(X = 0, Y = z) \cup (X = 1, Y = z - 1) \cup \cdots \cup (X = z, Y = 0)\}$$

$$= \sum_{k=0}^{z} P(X = k, Y = z - k)$$

and, since X and Y are independent,

$$P(Z = z) = \sum_{k=0}^{z} P(X = k) \cdot P(Y = z - k)$$

$$= \sum_{k=0}^{z} \binom{m}{k} p^k q^{m-k} \binom{n}{z-k} p^{z-k} q^{n-z+k}$$

$$= \sum_{k=0}^{z} \binom{m}{k} \binom{n}{z-k} p^z q^{m+n-z}$$

Recall the statement of Question 2.9.29,

$$\sum_{k=0}^{z} \binom{m}{k} \binom{n}{z-k} = \binom{m+n}{z}$$

from which it follows that

$$P(Z = z) = f_Z(z) = \binom{m+n}{z} p^z q^{m+n-z} \qquad z = 0, 1, \ldots, m+n$$

Notice that the binomial distribution "reproduces" itself—that is, X and Y were binomial and their sum, $Z = X + Y$, also proves to be binomial. Not all random variables share this property. The sum of two independent *uniform* random variables, for example, is not itself uniform (recall Question 3.5.11).

Example 3.5.7

Let X and Y be two independent random variables with pdf's

$$f_X(x) = e^{-x} \qquad x > 0$$

and

$$f_Y(y) = e^{-y} \qquad y > 0$$

Define $Z = X + Y$. Then

$$F_Z(z) = P(Z \le z) = P(X + Y \le z)$$

Here $F_Z(z)$ is the volume above the shaded triangular region, R, shown in Figure 3.5.6.

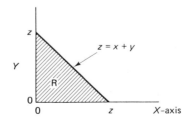

Figure 3.5.6

For $z > 0$,

$$F_Z(z) = \iint_R f_{X,Y}(x, y) \, dy \, dx = \int_0^z e^{-x} \left(\int_0^{z-x} e^{-y} \, dy \right) dx$$

$$= \int_0^z e^{-x}(1 - e^{x-z}) \, dx = \int_0^z (e^{-x} - e^{-z}) \, dx$$

$$= -e^{-x} \Big|_0^z - xe^{-z} \Big|_0^z$$

$$= 1 - e^{-z} - ze^{-z}$$

and the pdf for the sum readily follows:

$$f_Z(z) = F'_Z(z) = \begin{cases} ze^{-z} & z > 0 \\ 0 & \text{elsewhere} \end{cases}$$

Question 3.5.15 Let X and Y be two independent random variables with probability density functions

$$f_X(x) = \frac{e^{-r}r^x}{x!} \qquad x = 0, 1, 2, \ldots$$

and

$$f_Y(y) = \frac{e^{-s}s^y}{y!} \qquad y = 0, 1, 2, \ldots$$

Let $Z = X + Y$. Find $f_Z(z)$. Do X and Y reproduce themselves?

Question 3.5.16 If X and Y have the joint pdf

$$f_{X,Y}(x, y) = 2(x + y) \qquad 0 \le x \le y \le 1$$

find $f_Z(z)$, where $Z = X + Y$.

Question 3.5.17 Let X and Y have the joint pdf

$$f_{X,Y}(x, y) = xe^{-x} \cdot e^{-y} \qquad x > 0, \quad y > 0$$

Find $f_Z(z)$ for $Z = X + Y$.

Question 3.5.18 If a random variable X is independent of two random variables Y and Z, prove that X is also independent of the *sum* of Y and Z.

The cdf method we have used in this section for finding the pdf's of functions of random variables is extremely powerful and its generality makes it very useful. For certain kinds of problems, though, a more direct application of the technique is available. Theorems 3.5.1 and 3.5.2 give formulas for $f_Y(y)$ if (1) $Y = aX + b$ or (2) $Y = X^2$. Theorem 3.5.3 gives a general formula for $f_Z(z)$ when $Z = X + Y$, and X and Y are independent. The proofs are all straightforward and will be omitted.

Theorem 3.5.1. Let X be a random variable with probability density function $f_X(x)$. Let $a \neq 0$ and b be constants, and define $Y = aX + b$.

(a) If X is discrete,

$$f_Y(y) = f_X\left(\frac{y-b}{a}\right)$$

(b) If X is continuous,

$$f_Y(y) = \frac{1}{|a|} f_X\left(\frac{y-b}{a}\right)$$

Theorem 3.5.2. Let X be a continuous random variable with pdf $f_X(x)$. Let $Y = X^2$. For $y > 0$,

$$f_Y(y) = \frac{1}{2\sqrt{y}} \left(f_X(\sqrt{y}) + f_X(-\sqrt{y})\right)$$

Theorem 3.5.3. Let X and Y be independent random variables with pdf's $f_X(x)$ and $f_Y(y)$, respectively. Let $Z = X + Y$.

(a) If X and Y are discrete,

$$f_Z(z) = \sum_{\text{all } x} f_X(x) f_Y(z - x)$$

(b) If X and Y are continuous,

$$f_Z(z) = \int_{-\infty}^{\infty} f_X(x) f_Y(z - x)\, dx$$

Comment

An integral of the form

$$\int_{-\infty}^{\infty} f_X(x) f_Y(z - x)\, dx$$

is referred to as the *convolution* of the functions f_X and f_Y. Besides their frequent appearances in random-variable problems, convolutions turn up in many areas of mathematics and engineering.

Question 3.5.19 Use Theorems 3.5.1 and 3.5.3 to rederive the pdf's worked out in Examples 3.5.6 and 3.5.7.

Question 3.5.20 Find $f_Y(y)$ if $Y = X^2$ and $f_X(x) = 6x(1 - x), 0 < x < 1$.

Question 3.5.21 Let X be a random variable with pdf

$$f_X(x) = 2x \qquad 0 < x < 1$$

Let $Y = 3X - 1$. Find $f_Y(y)$.

Question 3.5.22 Let X and Y have the joint pdf

$$f_{X,Y}(x, y) = 4e^{-x-4y}$$

for $x > 0$ and $y > 0$. Define $Z = X + Y$. Find $f_Z(z)$.

Question 3.5.23 Let X_1, X_2, and X_3 be independent random variables with pdf

$$f_{X_i}(x) = e^{-x} \qquad x > 0, \quad i = 1, 2, 3$$

Let $Z = X_1 + X_2 + X_3$. Find $f_Z(x)$.

3.6 ORDER STATISTICS

The single-variable transformations taken up in Section 3.5 all involved some standard mathematical operation, such as $Y = aX + b$ or $Y = X^2$. The bivariate transformations were similarly arithmetic, typically being concerned with either sums or products. In this section we will consider a different sort of transformation, one involving the *ordering* of an entire *set* of random variables. This particular transformation has wide applicability in many areas of statistics, and we will see some of its consequences in later chapters. Here, though, we will limit our discussion to one basic result: a derivation of the marginal pdf for the *i*th largest observation, $i = 1, 2, \ldots, n$, in a random sample of size n.

> **Definition 3.6.1.** Let X be a continuous random variable for which x_1, $x_2, \ldots, x_n$ are the values of a random sample of size n. Reorder the x_i's from smallest to largest:
>
> $$x_1' < x_2' < \cdots < x_n'$$
>
> (No two of the x_i's are equal, except with probability zero, since X is continuous.) Define the random variable X_i' to have the value x_i', $1 \le i \le n$. Then X_i' is called the *i*th *order statistic*. Sometimes X_n' and X_1' are denoted $X_{\max}$ and $X_{\min}$, respectively.

Example 3.6.1

Suppose that four measurements are made on the random variable X: $x_1 = 3.4$, $x_2 = 4.6$, $x_3 = 2.6$, and $x_4 = 3.2$. The corresponding ordered sample would be

$$2.6 < 3.2 < 3.4 < 4.6$$

The random variable representing the smallest observation would be denoted X_1', with its value for this particular sample being 2.6. Similarly, the value for the second order statistic, X_2', is 3.2, and so on.

Theorem 3.6.1. Let X be a continuous random variable with probability density function $f_X(x)$. If a random sample of size n is drawn from $f_X(x)$, the marginal pdf for the ith order statistic is given by

$$f_{X_i}(y) = \frac{n!}{(i-1)!(n-i)!}[F_X(y)]^{i-1}[1 - F_X(y)]^{n-i}f_X(y)$$

for $1 \leq i \leq n$.

Proof. A standard way to verify a theorem of this sort is to take advantage of the fact that we already know the correct answer and give an induction argument on i; here, it should be noted, the induction is slightly unusual in that it proceeds from n to 1 rather than from 1 to n.

The first step, then, is to find $f_{X_n}(y)$. This can be done easily by differentiating the corresponding cdf. Note that

$$F_{X_n}(y) = P(X'_n \leq y) = P(X_1, X_2, \ldots, X_n \leq y)$$
$$= P(X_1 \leq y) \cdot P(X_2 \leq y) \cdot \cdots \cdot P(X_n \leq y)$$
$$= [F_X(y)]^n$$

Therefore, $f_{X_n}(y) = F'_{X_n}(y) = n[F_X(y)]^{n-1}f_X(y)$, which, according to Theorem 3.6.1, is the appropriate form for $i = n$.

Next comes the induction step: we assume the theorem to be correct for $f_{X_{i'+1}}(y)$ and seek an expression for $f_{X_i}(y)$. By definition,

$$F_{X_i}(y) = P(X'_i \leq y) = P(X'_i \leq y, X'_{i+1} \leq y) + P(X'_i \leq y, X'_{i+1} > y)$$
$$= P(X'_{i+1} \leq y) + P(X'_i \leq y, X'_{i+1} > y)$$

The first summand on the right is just $F_{X_{i'+1}}(y)$; the second is obtained by noticing that exactly i of the X_i's must be less than or equal to y, and this can occur in $\binom{n}{i}$ ways. Thus, appealing to the binomial distribution,

$$F_{X_i}(y) = F_{X_{i'+1}}(y) + \binom{n}{i}[F_X(y)]^i[1 - F_X(y)]^{n-i} \tag{3.6.1}$$

We then take the derivative of Equation 3.6.1,

$$f_{X_i}(y) = \frac{n!}{i!(n-i-1)!}[F_X(y)]^i[1 - F_X(y)]^{n-i-1}f_X(y) + \frac{n!}{i!(n-i)!}$$
$$\cdot \{i[F_X(y)]^{i-1}[1 - F_X(y)]^{n-i} - (n-i)[F_X(y)]^i[1 - F_X(y)]^{n-i-1}\}f_X(y)$$
$$= \frac{n!}{i!(n-i)!}[F_X(y)]^{i-1}[1 - F_X(y)]^{n-i-1}f_X(y)$$
$$\cdot \{(n-i)F_X(y) + i[1 - F_X(y)] - (n-i)F_X(y)\}$$
$$= \frac{n!}{i!(n-i)!}[F_X(y)]^{i-1}[1 - F_X(y)]^{n-i-1}f_X(y)[i(1 - F_X(y))]$$
$$= \frac{n!}{(i-1)!(n-i)!}[F_X(y)]^{i-1}[1 - F_X(y)]^{n-i}f_X(y)$$

Comparing the latter expression with the statement of Theorem 3.6.1 completes the induction.

> **Corollary.** Let $X_1, X_2, \ldots, X_n$ be a random sample of size n from the continuous pdf $f_X(x)$. Let $X_{\min}$ and $X_{\max}$ denote the smallest and largest order statistics, respectively. Then
>
> (a) $f_{X_{\min}}(y) = n \cdot f_X(y) \cdot [1 - F_X(y)]^{n-1}$
> (b) $f_{X_{\max}}(y) = n \cdot f_X(y) \cdot [F_X(y)]^{n-1}$

Example 3.6.2

Suppose that many years of observation have confirmed that the annual maximum flood tide X (in feet) for a certain river has pdf

$$f_X(x) = \frac{1}{20} \qquad 20 < x < 40$$

(*Note*: It is unlikely that flood tides would be described by anything as simple as a uniform pdf. We are making that choice here solely to facilitate the mathematics.) The Army Corps of Engineers are planning to build a levee along a certain portion of the river, and they want to build it high enough so that there is only a 30% chance that the second worst flood in the next 33 years will overflow the embankment. How high should the levee be? (We assume that there will be only one potential flood per year.)

Let h be the desired height. If $X_1, X_2, \ldots, X_{33}$ denote the flood tides for the next $n = 33$ years, what we require of h is that

$$P(X'_{32} > h) = 0.30$$

As a starting point, notice that for $20 < y < 40$,

$$F_X(y) = \int_{20}^{y} \frac{1}{20}\, dx = \frac{y}{20} - 1$$

Therefore,

$$f_{X_{32}'}(y) = \frac{33!}{31!1!}\left(\frac{y}{20} - 1\right)^{31}\left(2 - \frac{y}{20}\right)^{1} \cdot \frac{1}{20}$$

and h is the solution of the integral equation

$$\int_{h}^{40} (33)(32)\left(\frac{y}{20} - 1\right)^{31}\left(2 - \frac{y}{20}\right)^{1} \cdot \frac{dy}{20} = 0.30 \qquad (3.6.2)$$

If we make the substitution

$$u = \frac{y}{20} - 1$$

Equation 3.6.2 simplifies to

$$P(X'_{32} > h) = 33(32) \int_{(h/20)-1}^{1} u^{31}(1 - u)\, du$$

$$= 1 - 33\left(\frac{h}{20} - 1\right)^{32} + 32\left(\frac{h}{20} - 1\right)^{33} \tag{3.6.3}$$

Setting the right-hand side of Equation 3.6.3 equal to 0.30 and solving for h by trial and error gives

$$h = 39.3 \text{ feet}$$

Question 3.6.1 Four measurements are taken on a random variable having pdf $f_X(x) = (\frac{1}{3})e^{-x/3}$, $x > 0$. Find $P(X_4 \le 5.1)$ and $P(X'_4 \le 5.1)$. (*Hint*: Keep in mind that X'_4 will be less than or equal to 5.1 only if *all* the observations are less than or equal to 5.1.)

Question 3.6.2 Let $X_1, X_2, \ldots, X_n$ be a random sample from $f_X(x) = e^{-x}$, $x > 0$. Define $Y = X'_1$. Find $f_Y(y)$, the probability density function for the smallest exponential order statistic.

Question 3.6.3 What is the probability that the larger of two random variables from any continuous pdf will exceed the $(100p)$th percentile? (See Question 3.5.14.)

Question 3.6.4 Suppose the length of time, in minutes, that you have to wait at a bank teller's window is uniformly distributed over the interval (0, 10). If you go to the bank four times during the next month, what is the probability that your second longest wait will be less than 5 minutes?

3.7 CONDITIONAL DENSITIES

We have already seen that many of the concepts defined in Chapter 2 relating to the probabilities of *events*—for example, independence—have their random-variable counterparts. Another of these carryovers is the notion of a conditional probability, or, in what will be our present terminology, a *conditional probability density function*. Applications of conditional pdf's are not uncommon. The height and girth of a tree, for instance, can be considered a pair of random variables. While it is easy to measure girth, it can be difficult to determine height; thus it might be of interest to a lumberman to know the probabilities of a Ponderosa pine's attaining certain heights given a known value for its girth. Or consider the plight of a school board member agonizing over which way to vote on a proposed budget increase. Her task would be that much easier if she knew the conditional probability that x additional tax dollars would stimulate an average increase of y points among twelfth-graders taking a standardized proficiency exam.

In the case of discrete random variables, a conditional pdf can be treated in the same way as a conditional probability. Note the similarity between Definitions 3.7.1 and 2.6.1.

Definition 3.7.1. Let X and Y be discrete random variables. The *conditional probability density function of Y given x*—that is, the probability that Y takes on the value y given that X is equal to x—is denoted $f_{Y|x}(y)$ and given by

$$f_{Y|x}(y) = P(Y = y \mid X = x) = \frac{f_{X,Y}(x, y)}{f_X(x)}$$

for $f_X(x) \neq 0$.

Example 3.7.1

Recall the coin-tossing experiment in Example 3.3.5; Table 3.7.1 reproduces the joint pdf, $f_{X,Y}(x, y)$. A simple application of Definition 3.7.1 gives

$$f_{Y|0}(0) = \frac{1/8}{1/2} = \frac{1}{4} \qquad f_{Y|1}(0) = \frac{0}{1/2} = 0$$

$$f_{Y|0}(1) = \frac{1/4}{1/2} = \frac{1}{2} \qquad f_{Y|1}(1) = \frac{1/8}{1/2} = \frac{1}{4}$$

and so on.

TABLE 3.7.1

		Y				
		0	1	2	3	$f_X(x)$
X	0	$\frac{1}{8}$	$\frac{1}{4}$	$\frac{1}{8}$	0	$\frac{1}{2}$
	1	0	$\frac{1}{8}$	$\frac{1}{4}$	$\frac{1}{8}$	$\frac{1}{2}$
	$f_Y(y)$	$\frac{1}{8}$	$\frac{3}{8}$	$\frac{3}{8}$	$\frac{1}{8}$	

Comment

The notion of a conditional pdf generalizes easily to situations involving more than two random variables. For example, if X, Y, and Z have the joint pdf $f_{X,Y,Z}(x, y, z)$, the *joint conditional pdf* of, say, X and Y given that $Z = z$ is the ratio

$$f_{X,Y|z}(x, y) = \frac{f_{X,Y,Z}(x, y, z)}{f_Z(z)}$$

Question 3.7.1 Five cards are dealt from a standard poker deck. Let X be the number of aces received, and Y, the number of kings. Compute $P(X = 2 \mid Y = 2)$.

Question 3.7.2 Suppose that X and Y have the joint pdf

$$f_{X,Y}(x, y) = \frac{xy^2}{39}$$

for the points $(1, 2)$, $(1, 3)$, $(2, 2)$, and $(2, 3)$, and is 0, otherwise. Find the conditional probability that X is 1 given that Y is 2.

Question 3.7.3 An urn contains 8 red chips, 6 white chips, and 4 blue chips. A sample of size 3 is drawn without replacement. Let X denote the number of red chips in the sample and Y, the number of white chips. Find an expression for $f_{Y|x}(y)$.

Question 3.7.4 A fair coin is tossed five times. Let Y denote the total number of heads occurring in the five tosses and let X denote the number of heads occurring in the last two tosses. Find the conditional pdf $f_{Y|x}(y)$.

Question 3.7.5 Suppose that the random variables X, Y, and Z have a trivariate distribution described by the joint pdf

$$f_{X, Y, Z}(x, y, z) = \frac{xy}{9z}$$

defined for the points $(1, 1, 1)$, $(2, 1, 2)$, $(1, 2, 2)$, $(2, 2, 2)$, and $(2, 2, 1)$. Tabulate the joint conditional pdf of X and Y given each of the possible values for z.

Question 3.7.6 Let X have the pdf

$$f_X(x) = \binom{n}{x} p^x(1 - p)^{n - x} \qquad x = 0, 1, \ldots, n$$

and let Y have the pdf

$$f_Y(y) = \binom{n}{y} p^y(1 - p)^{n - y} \qquad y = 0, 1, \ldots, n$$

Define $Z = X + Y$. Show that the conditional pdf $f_{X|z}(x)$ is hypergeometric.

If the variables X and Y are continuous, we can still appeal to the quotient $f_{X, Y}(x, y)/f_X(x)$ as the definition of $f_{Y|x}(y)$ and argue its propriety by analogy. A more satisfying approach, though, is to arrive at the same conclusion by taking the limit of Y's "conditional" *cdf*.

If X is continuous, a direct evaluation of $F_{Y|x}(y) = P(Y \leq y \,|\, X = x)$, via Definition 2.6.1, is impossible, since the denominator would be 0. Alternatively, we can think of $P(Y \leq y \,|\, X = x)$ as a limit:

$$P(Y \leq y \,|\, X = x) = \lim_{h \to 0} P(Y \leq y \,|\, x \leq X \leq x + h)$$

$$= \lim_{h \to 0} \frac{\displaystyle\int_x^{x+h} \int_{-\infty}^y f_{X, Y}(t, u) \, du \, dt}{\displaystyle\int_x^{x+h} f_X(t) \, dt}$$

Evaluating the quotient of the limits gives $0/0$, so l'Hospital's rule is indicated:

$$P(Y \leq y \,|\, X = x) = \lim_{h \to 0} \frac{\dfrac{d}{dh} \displaystyle\int_x^{x+h} \int_{-\infty}^y f_{X, Y}(t, u) \, du \, dt}{\dfrac{d}{dh} \displaystyle\int_x^{x+h} f_X(t) \, dt} \qquad (3.7.1)$$

By the Fundamental Theorem of calculus,

$$\frac{d}{dh} \int_x^{x+h} g(t)\, dt = g(x + h)$$

which simplifies Equation 3.7.1 to

$$P(Y \le y \mid X = x) = \lim_{h \to 0} \frac{\displaystyle\int_{-\infty}^y f_{X,Y}[(x + h), u]\, du}{f_X(x + h)}$$

$$= \frac{\displaystyle\int_{-\infty}^y \lim_{h \to 0} f_{X,Y}(x + h, u)\, du}{\displaystyle\lim_{h \to 0} f_X(x + h)} = \int_{-\infty}^y \frac{f_{X,Y}(x, u)}{f_X(x)}\, du$$

provided that the limit operation and the integration can be interchanged [see (6) for a discussion of when such an interchange is valid]. It follows from this last expression that $f_{X,Y}(x, u)/f_X(x)$ behaves as a conditional probability density function should, and we are justified in extending Definition 3.7.1 to the continuous case.

Example 3.7.2

Let X and Y be continuous random variables with joint pdf

$$f_{X,Y}(x, y) = \begin{cases} \left(\dfrac{1}{8}\right)(6 - x - y) & 0 < x < 2, \quad 2 < y < 4 \\ 0 & \text{elsewhere} \end{cases}$$

Find (a) $f_X(x)$, (b) $f_{Y|x}(y)$, and (c) $P(2 < Y < 3 \mid x = 1)$.
(a) From Theorem 3.3.2,

$$f_X(x) = \int_{-\infty}^{\infty} f_{X,Y}(x, y)\, dy = \int_2^4 \left(\frac{1}{8}\right)(6 - x - y)\, dy$$

$$= \left(\frac{1}{8}\right)(6 - 2x) \qquad 0 < x < 2$$

(b) Substituting into the " continuous " statement of Definition 3.7.1,

$$f_{Y|x}(y) = \frac{f_{X,Y}(x, y)}{f_X(x)} = \frac{(1/8)(6 - x - y)}{(1/8)(6 - 2x)}$$

$$= \frac{6 - x - y}{6 - 2x} \qquad 0 < x < 2, \quad 2 < y < 4$$

(c) To find $P(2 < Y < 3 \mid x = 1)$ we simply integrate $f_{Y \mid 1}(y)$ over the interval $2 < Y < 3$:

$$P(2 < Y < 3 \mid x = 1) = \int_2^3 f_{Y \mid 1}(y)\, dy$$

$$= \int_2^3 \frac{5 - y}{4}\, dy$$

$$= \frac{5}{8}$$

[A partial check that the derivation of a conditional pdf is correct can be gotten by integrating $f_{Y \mid x}(y)$ over the entire range of Y. That integral should be 1. Here, $\int_{-\infty}^{\infty} f_{Y \mid x}(y)\, dy = \int_2^4 [(5 - y)/4]\, dy$ *does* equal 1.]

Question 3.7.7 Let X be a nonnegative random variable. We say that X is *memoryless* if

$$P(X > s + t \mid X > t) = P(X > s) \qquad \text{for all } s, t \geq 0$$

Show that a random variable with pdf $f_X(x) = (1/\lambda)e^{-x/\lambda}$, $x > 0$, is memoryless.

Question 3.7.8 Given the joint pdf

$$f_{X, Y}(x, y) = 2e^{-(x + y)} \qquad 0 < x < y, \quad y > 0$$

find
(a) $P(Y < 1 \mid X < 1)$
(b) $P(Y < 1 \mid X = 1)$
(c) $f_{Y \mid x}(y)$

Question 3.7.9 Find the conditional pdf of Y given x if

$$f_{X, Y}(x, y) = x + y$$

for $0 \leq x \leq 1$ and $0 \leq y \leq 1$.

Question 3.7.10 If

$$f_{X, Y}(x, y) = 2 \qquad x \geq 0, \quad y \geq 0, \quad x + y \leq 1$$

show that the conditional pdf of Y given x is uniform.

Question 3.7.11 Suppose that

$$f_{Y \mid x}(y) = \frac{2y + 4x}{1 + 4x} \qquad \text{and} \qquad f_X(x) = \frac{1}{3} \cdot (1 + 4x)$$

for $0 < x < 1$ and $0 < y < 1$. Find the marginal pdf for Y.

Question 3.7.12 Suppose that X and Y are jointly distributed according to the joint pdf

$$f_{X, Y}(x, y) = \frac{2}{5} \cdot (2x + 3y) \qquad 0 \leq x \leq 1, \quad 0 \leq y \leq 1$$

Find (a) $f_X(x)$, (b) $f_{Y \mid x}(y)$, and (c) $P(\tfrac{1}{4} \leq Y \leq \tfrac{3}{4} \mid X = \tfrac{1}{2})$.

Question 3.7.13 If X and Y have the joint pdf

$$f_{X,Y}(x, y) = 2 \qquad 0 < x < y < 1$$

find $P(0 < X < \frac{1}{2} \mid Y = \frac{3}{4})$.

Question 3.7.14 Suppose that X_1, X_2, X_3, X_4, and X_5 have the joint pdf

$$f_{X_1, X_2, X_3, X_4, X_5}(x_1, x_2, x_3, x_4, x_5) = 32 x_1 x_2 x_3 x_4 x_5$$

for $0 < x_i < 1$, $i = 1, 2, \ldots, 5$. Find the joint conditional pdf of X_1, X_2, and X_3 given that $X_4 = x_4$ and $X_5 = x_5$.

3.8 EXPECTED VALUES

Probability density functions, as we have already seen, provide a global overview of a random variable's behavior. If X is discrete, $f_X(x)$ gives $P(X = x)$ for all x; if X is continuous, and A is any interval, or countable union of intervals, $P(X \in A) = \int_A f_X(x)\, dx$. Detail that explicit, though, is not always necessary—or even helpful. There are times when a more prudent strategy is to focus the information contained in $f_X(x)$ by summarizing certain of its features with single numbers. The search for these numbers—and the investigation, interpretation, and application of their properties—is what the remainder of this chapter is primarily about.

The first feature of a pdf that we will examine is *central tendency*, a term referring to the "average" value of a random variable. Consider the pdf's $f_X(x)$ and $g_Y(y)$ pictured in Figure 3.8.1. Although we obviously cannot predict with certainty what values any future X's and Y's will take on, it seems clear that X values will tend to lie somewhere near m_X, and Y values, somewhere near m_Y. In some sense, then, we can characterize $f_X(x)$ by m_X and $g_Y(y)$ by m_Y.

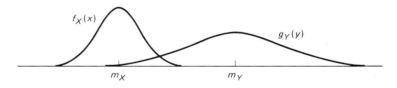

Figure 3.8.1

The most frequently used measure for describing central tendency—that is, for quantifying m_X and m_Y—is the *expected value*. Discussed at some length in this section and in Section 3.9, the expected value of a random variable is a slightly more abstract formulation of what we are already familiar with in simple discrete settings as the arithmetic average.

Gambling affords a familiar illustration of the notion of an expected value. Consider the game of roulette. After bets are placed, the croupier spins the wheel, and declares one of 38 numbers, 00, 0, 1, 2, ..., 36, to be the winner. Disregarding what seems to be a perverse tendency of many roulette wheels to land on numbers for which no money has been wagered, we will assume that each of these 38 numbers is equally likely. Suppose that our particular bet is $1 on "odds." If X denotes our winnings, then X takes on the value 1 if an odd number occurs, and -1 otherwise. By the uniformity assumption,

$$f_X(1) = P(X = 1) = \frac{18}{38} = \frac{9}{19}$$

and

$$f_X(-1) = P(X = -1) = \frac{20}{38} = \frac{10}{19}$$

It follows that we will win $1 $\frac{9}{19}$ of the time and lose $1 $\frac{10}{19}$ of the time. Intuitively, then, if we persist in this foolishness, we stand to *lose*, on the average, a little more than 5 cents each time we play the game:

$$\text{"expected" winnings} = \$1 \cdot \frac{9}{19} + (-\$1) \cdot \frac{10}{19}$$

$$= -\$0.053 \doteq -5\cancel{c}$$

The number -0.053 will be called the expected value of X.

Physically, an expected value can be thought of as the center of gravity for $f_X(x)$. Imagine two bars of height $\frac{10}{19}$ and $\frac{9}{19}$ positioned along a weightless X-axis at the points -1 and $+1$, respectively (see Figure 3.8.2). If a fulcrum were placed at the point -0.053, the system would be in balance. It is in this sense that we can interpret the expected value of a random variable as being a measure of a pdf's "location."

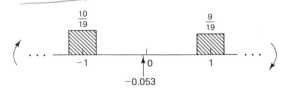

Figure 3.8.2

If X is a discrete random variable taking on each of its values, $x_1, x_2, \ldots, x_n$, with the same probability, the expected value of X is simply the everyday notion of an arithmetic average or mean:

$$\text{expected value of } X = \sum_{i=1}^{n} x_i \cdot \frac{1}{n} = \frac{1}{n} \sum_{i=1}^{n} x_i$$

Extending this idea to a discrete X described by an arbitrary pdf, $f_X(x)$, gives

$$\text{expected value of } X = \sum_{i=1}^{\infty} x_i \cdot f_X(x_i) \qquad (3.8.1)$$

Thus, Equation 3.8.1 can be viewed as a sort of generalized mean, a number marking off, in some sense, the "center" of $f_X(x)$. Although the idea of an average predates Pythagoras (the circle of Pythagoreans knew of more than nine different kinds of means), the probabilistic concept of an expected value made its debut in 1657 in Huygens' *De Ratiociniis in Aleae Ludo*.

> **Definition 3.8.1.** Let X be a random variable with probability density function $f_X(x)$. The *expected value of X* is denoted $E(X)$, or μ, and is given by
>
> (a) $E(X) = \sum\limits_{\text{all } x} x f_X(x)$ if X is discrete
>
> (b) $E(X) = \int_{-\infty}^{\infty} x f_X(x)\, dx$ if X is continuous

Comment

It is assumed that both the sum and the integral in Definition 3.8.1 converge absolutely:

$$\sum_{\text{all } x} |x| f_X(x) < \infty \qquad \int_{-\infty}^{\infty} |x|\, f_X(x)\, dx < \infty$$

If not, we say that X has no finite expected value. One immediate reason for requiring *absolute* convergence is that a convergent sum that is not absolutely convergent depends on the order in which the terms are added, and order should not be a consideration when defining a mean.

Example 3.8.1

An urn contains nine chips, five red and four white. Three are drawn out at random without replacement. Let X denote the number of red chips in the sample. Find $E(X)$.

From Section 2.11, we recognize X to be a hypergeometric random variable, where

$$P(X = x) = f_X(x) = \frac{\binom{5}{x}\binom{4}{3-x}}{\binom{9}{3}} \qquad x = 0, 1, 2, 3$$

Therefore,

$$E(X) = \sum_{x=0}^{3} x \cdot \frac{\binom{5}{x}\binom{4}{3-x}}{\binom{9}{3}}$$

$$= (0)\left(\frac{4}{84}\right) + (1)\left(\frac{30}{84}\right) + (2)\left(\frac{40}{84}\right) + (3)\left(\frac{10}{84}\right)$$

$$= \frac{5}{3}$$

Example 3.8.2

Recall Example 2.5.1, where the prison sentence X (in years) of persons convicted of grand theft auto had the pdf,

$$f_X(x) = \frac{1}{9} \cdot x^2 \qquad 0 < x < 3 \tag{3.8.2}$$

What is the *average* length of time these people spend in jail?

Here X is continuous, so by part (b) of Definition 3.8.1,

$$E(X) = \int_{-\infty}^{\infty} x \cdot f_X(x)\, dx = \int_{0}^{3} x \cdot \left(\frac{1}{9}\right) x^2\, dx$$

$$= \frac{x^4}{36}\Big|_{0}^{3}$$

$$= 2.25 \text{ years}$$

Comment

Note that the *median* of a distribution is not necessarily the same as its expected value. Let m denote the median for the pdf of Equation 3.8.2. Then

$$P(X < m) = \int_{0}^{m} \frac{1}{9} \cdot x^2\, dx$$

$$= \frac{m^3}{27}$$

which *must* equal $\frac{1}{2}$, implying that

$$m = \sqrt[3]{\frac{27}{2}}$$

$$= 2.38 \text{ years}$$

In general, $E(X)$ and m are equal only if $f_X(x)$ is symmetric.

Example 3.8.3

Among the more common versions of the "numbers" racket is a game called D.J., its name deriving from the fact that the winning ticket is determined from Dow Jones averages. Three sets of stocks are used, Industrials, Transportations, and Utilities. Traditionally, the three are quoted at two different times, 11 A.M. and noon. The last digits of the earlier quotation are arranged to form a three-digit number; the noon quotation generates a second three-digit number, formed the same way. Those two numbers are then added together and the last three digits of that sum become the winning pick. Figure 3.8.3 shows a set of quotations for which *906* would be declared the winner.

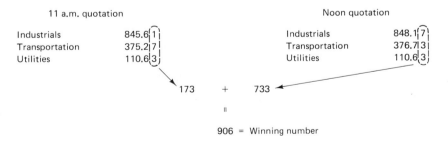

Figure 3.8.3

The payoff in D.J. is 700 to 1. Suppose that we bet $5. How much do we stand to win, or lose, *on the average*?

Let p denote the probability of our number being the winner and let X denote our earnings. Then

$$X = \begin{cases} \$3500 & \text{with probability } p \\ -\$5 & \text{with probability } 1 - p \end{cases}$$

and

$$E(X) = \$3500 \cdot p - \$5 \cdot (1 - p)$$

Our intuition would suggest (and this time it would be correct!) that each of the possible winning numbers, 000 through 999, is equally likely. That being the case, $p = 1/1000$ and

$$E(X) = \$3500 \cdot \left(\frac{1}{1000}\right) - \$5 \cdot \left(\frac{999}{1000}\right) = -\$1.50$$

On the average, then, we lose $1.50 on a $5 bet. (There are actually two types of bets that can be placed in D.J.—a *straight bet* is what we have just described; a *box bet* is a wager on a specific number—say, 906—as well as on all of its permutations. The payoff on the latter is 700/6 to 1.)

Example 3.8.4

Let X be a binomial random variable defined on n trials, where $p = P(\text{success})$. Find $E(X)$.

Applying Definition 3.8.1 to the pdf for a binomial, we can write

$$E(X) = \sum_{k=0}^{n} k \cdot f_X(k) = \sum_{k=0}^{n} k \binom{n}{k} p^k (1 - p)^{n-k}$$

$$= \sum_{k=0}^{n} \frac{k \cdot n!}{k!(n-k)!} p^k (1 - p)^{n-k}$$

$$= \sum_{k=1}^{n} \frac{n!}{(k-1)!(n-k)!} p^k (1 - p)^{n-k} \tag{3.8.3}$$

At this point, a "trick" is called for. If $E(X) = \sum_{\text{all } k} g(k)$ can be factored in such a way that $E(X) = h \sum_{\text{all } k} f_Y(k)$, where $f_Y(k)$ is the pdf for some random variable Y, then $E(X) = h$, since the sum of a pdf over its entire range is 1. Here, suppose that np is factored out of Equation 3.8.3. Then

$$E(X) = np \sum_{k=1}^{n} \frac{(n-1)!}{(k-1)!(n-k)!} p^{k-1}(1 - p)^{n-k}$$

$$= np \sum_{k=1}^{n} \binom{n-1}{k-1} p^{k-1}(1 - p)^{n-k}$$

Now, let $j = k - 1$. It follows that

$$E(X) = np \sum_{j=0}^{n-1} \binom{n-1}{j} p^j (1-p)^{n-j-1}$$

Finally, letting $m = n - 1$ gives

$$E(X) = np \sum_{j=0}^{m} \binom{m}{j} p^j (1-p)^{m-j}$$

and, since the value of the sum is 1 (why?),

$$E(X) = np \qquad (3.8.4)$$

Equation 3.8.4 should come as no surprise. If a multiple-choice test, for example, has 100 questions, each with five possible answers, we would "expect" to get 20 correct if we were just guessing. But $20 = E(X) = 100(\frac{1}{5}) = np$.

Example 3.8.5

Consider the following game. A fair coin is flipped until the first tail appears; we win \$2 if it appears on the first toss, \$4 if it appears on the second toss, and, in general, $\$2^k$ if it first occurs on the kth toss. Let the random variable X denote our winnings. How much should we have to pay in order for this to be a fair game? [*Note*: A fair game is one where the difference between the ante and $E(X)$ is 0.]

Known as the St. Petersburg paradox, this problem has a rather unusual answer. First, note that

$$f_X(2^k) = P(X = 2^k) = \frac{1}{2^k} \qquad k = 1, 2, \ldots$$

Therefore,

$$E(X) = \sum_{\text{all } x} x f_X(x) = \sum_{k=1}^{\infty} 2^k \cdot \frac{1}{2^k} = 1 + 1 + 1 + \cdots$$

which is a divergent sum. That is, X does not have a finite expected value, so in order for this game to be fair, our ante would have to be an infinite amount of money!

Comment

Mathematicians have been trying to "explain" the St. Petersburg paradox for almost 200 years. The answer seems clearly absurd—no gambler would consider paying even \$25 to play such a game, much less an infinite amount—yet the computations involved in showing that X has no finite expected value are unassailably correct. Where the difficulty lies, according to one common theory, is with our inability to put in perspective the very small probabilities of winning very large payoffs. Furthermore, the problem assumes that our opponent has infinite capital, which is an impossible state of affairs. We get a much more reasonable answer for $E(X)$ if the stipulation is added that our winnings can be at most, say, \$1000 (see Question 3.8.8) or if the payoffs are assigned according to some formula other than 2^k (see Question 3.8.9).

Comment

There are two important lessons to be learned from the St. Petersburg paradox. First is the realization that $E(X)$ is not necessarily a meaningful characterization of the "location" of a distribution. Question 3.8.15 shows another situation where the formal computation of $E(X)$ gives a similarly inappropriate answer. Second, we need to be aware that the notion of expected value is not necessarily synonymous with the concept of "worth." Just because a game, for example, has a positive expected value—even a very *large* positive expected value—does not imply that someone would want to play it. Suppose, for example, that you had the opportunity to spend your last $10,000 on a sweepstakes ticket where the prize was a billion dollars but the probability of winning was only 1 in 10,000. The expected value of such a bet would be over $90,000,

$$E(X) = \$1,000,000,000 \left(\frac{1}{10,000}\right) + (-\$10,000)\left(\frac{9,999}{10,000}\right)$$

$$= \$90,001$$

but it is doubtful that many people would rush out to buy a ticket. (Economists have long recognized the distinction between a payoff's numerical value and its perceived desirability. They refer to the latter as *utility*.)

Example 3.8.6

Suppose that N people are to be given a blood test to see which of them, if any, has a certain rare disease. The obvious approach is to examine each person's blood individually, meaning that a total of N tests will be run. But that may be inefficient. An alternative strategy is to pool the N blood samples into groups of k—and run the test on each group. If a given test proves negative, all k in that group are free from the disease. If the result is positive, each person's blood in that group must be rerun, ultimately resulting in $k + 1$ tests being done on those particular k individuals. Suppose that the probability of any one person showing a positive result is p. What will be the expected number of tests under the "pooling" plan—and how does it compare to N?

Let X be the random variable counting the number of tests done on a pooled sample. Then X takes on only two values, either 1 or $k + 1$, where

$$f_X(1) = P(\text{none of the } k \text{ gives a positive test})$$
$$= (1 - p)^k$$

and

$$f_X(k + 1) = 1 - (1 - p)^k$$

By Definition 3.8.1,

$$E(X) = 1 \cdot (1 - p)^k + (k + 1)[1 - (1 - p)^k]$$
$$= (k + 1) - k(1 - p)^k$$

Since there are approximately N/k groups of size k, the total expected number of tests is roughly

$$\left(\frac{N}{k}\right) \cdot E(X) = \left(\frac{N}{k}\right)[(k+1) - k(1-p)^k] = N\left[1 - (1-p)^k + \left(\frac{1}{k}\right)\right]$$

Table 3.8.1 gives $(N/k) \cdot E(X)$ for p values of $1/100$ and $1/1000$ and for $k = 2, 5, 10, 20, 40$, and 100. Notice that pooling can reduce considerably the average number of tests that need to be performed, especially when p is very small.

TABLE 3.8.1

p	k	Expected number of tests
1/100	2	0.52N
	5	0.25N
	10	0.20N
	20	0.23N
	40	0.36N
	100	0.64N
1/1000	2	0.50N
	5	0.20N
	10	0.11N
	20	0.07N
	40	0.06N
	100	0.10N

Example 3.8.7

As part of a reliability study, a total of n items are put "on test." Suppose that each has the same exponential failure time distribution—that is, if T_i denotes the time when the ith item fails,

$$f_{T_i}(t) = \left(\frac{1}{\lambda}\right)e^{-t/\lambda} \qquad t > 0$$

where λ is a constant greater than 0. On the average, how long will we have to wait for the *first* failure?

Let T denote the first failure time. Note that T exceeds some particular value t only if all the T_i's exceed t, an event that occurs with probability $e^{-nt/\lambda}$:

$$P(T > t) = P(T_1 > t, T_2 > t, \ldots, T_n > t)$$
$$= P(T_1 > t) \cdot P(T_2 > t) \cdot \ldots \cdot P(T_n > t)$$
$$= \int_t^\infty \left(\frac{1}{\lambda}\right)e^{-t_1/\lambda}\, dt_1 \cdot \ldots \cdot \int_t^\infty \left(\frac{1}{\lambda}\right)e^{-t_n/\lambda}\, dt_n$$
$$= e^{-t/\lambda} \cdot e^{-t/\lambda} \cdot \ldots \cdot e^{-t/\lambda}$$
$$= e^{-nt/\lambda}$$

It follows that the *cdf* for T is

$$F_T(t) = P(T \le t) = 1 - e^{-nt/\lambda}$$

and its *pdf* can be gotten via Theorem 3.2.1:

$$f_T(t) = F'_T(t) = \left(\frac{n}{\lambda}\right)e^{-nt/\lambda} \qquad t > 0$$

Having found $f_T(t)$, we can easily determine $E(T)$:

$$E(T) = \int_0^\infty t \cdot \left(\frac{n}{\lambda}\right)e^{-nt/\lambda}\, dt$$

Let $u = nt/\lambda$. Then

$$E(T) = \frac{\lambda}{n}\int_0^\infty ue^{-u}\, du$$

$$= \frac{\lambda}{n}$$

Thus, while the average waiting time for any particular item to fail is λ [$= \int_0^\infty t(1/\lambda)e^{-t/\lambda}\, dt$], that figure reduces to λ/n if we are simply waiting for whichever of the n items fails first.

Comment

The relationship between $E(T_i)$ and $E(T)$—specifically, the latter's dependence on n—reflects an everyday phenomenon with which we all are familiar. As our necessities of life become more and more complicated, whether that means owning a car with all the latest options or relying on a comprehensive word-processing system in the office, the waiting time T for *something* to malfunction gets shorter and shorter. In a sense, we find ourselves in a no-win situation: as technology advances, the expected failure times (λ) of individual manufactured goods will increase, but those same improvements encourage the multiplicity of components and complexity of objectives that can have the opposite effect of *decreasing* the expected lifetime (λ/n) for the overall system.

Question 3.8.1 Find $E(X)$ for the urn problem described in Example 3.8.1 if the three chips are drawn *with* replacement.

Question 3.8.2 Let X be uniformly distributed over the interval (a, b). Find $E(X)$.

Question 3.8.3 In Case Study 2.4.1, the *pdf*

$$f_S(s) = \frac{e^{-0.82}(0.82)^s}{s!} \qquad s = 0, 1, 2, \ldots$$

was shown to provide a good model for the daily number of fatalities among London senior citizens. Show that $E(S) = 0.82$. (*Hint*: Follow the same approach used in Example 3.8.4.)

Question 3.8.4 Below are the last five lines of Shelley's poem, *Ozymandias*:

> " My name is Ozymandias, king of kings:
> Look on my works, ye Mighty, and despair! "
> Nothing beside remains. Round the decay
> Of that colossal wreck, boundless and bare
> The lone and level sands stretch far away.

Suppose that a word is selected at random from those lines. What is its expected length?

Question 3.8.5 Stan Musial retired from baseball with a lifetime batting average of .331. On the average, how many hits did he get in a game where he had five official at-bats?

Question 3.8.6 Find $E(X)$ for the following *pdf*'s:
(a) $f_X(x) = \frac{1}{3}$ $x = 6, 15,$ and $22; 0,$ elsewhere
(b) $f_X(x) = 3(1 - x)^2$ $0 \le x \le 1$
(c) $f_X(x) = 4xe^{-2x}$ $x > 0$
(d) $f_X(x) = \begin{cases} \frac{3}{4} & 0 \le x \le 1 \\ \frac{1}{4} & 2 \le x \le 3 \\ 0 & \text{elsewhere} \end{cases}$

(e) $f_X(x) = \sin x$ $0 \le x \le \pi/2$

Question 3.8.7 Show that

$$f_X(x) = \frac{1}{x^2}   x \ge 1$$

is a valid pdf but does not have a finite expected value.

Question 3.8.8 How much would you have to ante to make the St. Petersburg game "fair" (recall Example 3.8.5) if the most you could win was $1000? That is, the payoffs are 2^k for $1 \le k \le 9$, and $1000 for $k \ge 10$.

Question 3.8.9 For the St. Petersburg problem (Example 3.8.5), find the expected payoff if
(a) the amounts won are c^k instead of 2^k, where $0 < c < 2$.
(b) the amounts won are $\log 2^k$. [This was a modification suggested by D. Bernoulli (a nephew of James Bernoulli) to take into account the decreasing marginal utility of money—the more you have, the less useful a bit more is.]

Question 3.8.10 A fair die is rolled 3 times. Let X denote the number of different faces showing, $X = 1, 2, 3$. Find $E(X)$.

Question 3.8.11 A local tavern has six bar stools. The bartender predicts that if two strangers come into the bar they will sit in such a way as to leave at least two stools between them.
(a) If two strangers do come in, but choose their seats at random, what is the probability of the bartender's prediction coming true?
(b) Compute the expected value of the number of stools between the two customers, under the assumption that they seat themselves randomly.

Question 3.8.12 Two distinct integers are chosen at random from the first five positive integers. Compute the expected value of the absolute value of the difference of the two numbers.

Question 3.8.13 A single elimination tennis tournament has eight players. Assume that the players can be ranked from 1 to 8, with player 1 always being able to defeat player 2, player 2 always superior to player 3, and so on. If the initial "draw" for the tournament is done at random, what is the expected rank of the runner-up? [See (181) for a proof showing that as the number of players increases, the expected rank of the runner-up converges to 3.]

Question 3.8.14 Suppose that two evenly matched teams are playing in the World Series. On the average, how many games will be played? (The winner is the first team to get four victories.) Assume that each game is an independent event.

Question 3.8.15 An urn contains 1 white chip and 1 black chip. A chip is drawn at random. If it is white, the "game" is over; if it is black, that chip and another black one are put into the urn. Then another chip is drawn at random from the "new" urn and the same rules for ending or continuing the game are followed (if the chip is white, the game is over; if the chip is black, it is replaced in the urn, together with another chip of the same color). The drawings continue until a white chip is selected. Show that the expected number of drawings necessary to get a white chip is not finite.

3.9 PROPERTIES OF EXPECTED VALUES

We defined $E(X)$ in Section 3.8—the obvious next step is to investigate some of its mathematical properties. Three of the most important will be taken up in this section. We will show that (1) the expected value of a linear combination is equal to the linear combination of expected values

$$E(a_1 X_1 + \cdots + a_n X_n) = a_1 E(X_1) + \cdots + a_n E(X_n)$$

(2) the expected value of a *function* of X, $Y = g(X)$, can be gotten by simply summing (or integrating) $g(x) \cdot f_X(x)$—we do not have to determine $f_Y(y)$ and evaluate the sum (or integral) of $y \cdot f_Y(y)$; and (3) if X and Y are independent, the expected value of their product is equal to the product of their expected values.

We begin by considering the problem of finding the expected value of a sum. What we derive proves to be a very useful result: the examples that follow Theorem 3.9.1 and its corollary show very clearly that if a random variable X can be written in the form $X = \sum_{i=1}^{n} a_i X_i$, it may be much easier to evaluate $\sum_{i=1}^{n} a_i E(X_i)$ than it is to work with $E(X)$ directly.

> **Theorem 3.9.1.** For any random variables X and Y and any numbers a and b,
>
> $$E(aX + bY) = aE(X) + bE(Y)$$
>
> provided both $E(X)$ and $E(Y)$ exist.

Proof. We will prove the theorem by establishing that $E(aX) = aE(X)$ and $E(X + Y) = E(X) + E(Y)$. Only the continuous case will be considered.

Let $f_{aX}(t)$ be the probability density function for aX. Then

$$E(aX) = \int_{-\infty}^{\infty} t \cdot f_{aX}(t) \, dt = \int_{-\infty}^{\infty} t \cdot \frac{1}{|a|} \cdot f_X\left(\frac{t}{a}\right) dt$$

the last equality being a consequence of Theorem 3.5.1. Assume that $a > 0$, so $a = |a|$. Making the substitutions $x = t/a$ and $dx = (1/a)\, dt$, we can write

$$E(aX) = \int_{-\infty}^{\infty} ax\left(\frac{1}{a}\right) f_X(x) a \, dx = a \int_{-\infty}^{\infty} x \cdot f_X(x) \, dx = aE(X)$$

The derivation for $a < 0$ follows similarly.

The crux of the second part of the proof is to find an expression for $E(X + Y)$. One approach would be to derive a general formula for $f_{X+Y}(t)$ and then apply Definition 3.8.1. A simpler and more illuminating mode of attack, though, is to return to the basic concept of an expected value. The random variable $X + Y$ takes on values of the form $x + y$, with the "likelihood" of such a value being $f_{X,Y}(x, y)$. It follows that $E(X + Y)$ should equal

$$\sum_{\text{all } x} \sum_{\text{all } y} (x + y) f_{X,Y}(x, y) \qquad \text{or} \qquad \int_{-\infty}^{\infty} \int_{-\infty}^{\infty} (x + y) f_{X,Y}(x, y) \, dy \, dx$$

depending on the nature of X and Y.

Once we have written $E(X + Y)$ in terms of $f_{X,Y}(x, y)$, verifying that $E(X + Y) = E(X) + E(Y)$ becomes a simple exercise in the definition of marginal distributions. Consider the case where X and Y are continuous. By splitting the integrand into two parts, we find that

$$E(X + Y) = \int_{-\infty}^{\infty} \int_{-\infty}^{\infty} (x + y) f_{X,Y}(x, y) \, dy \, dx$$

$$= \int_{-\infty}^{\infty} \int_{-\infty}^{\infty} x \cdot f_{X,Y}(x, y) \, dy \, dx + \int_{-\infty}^{\infty} \int_{-\infty}^{\infty} y \cdot f_{X,Y}(x, y) \, dy \, dx$$

$$= \int_{-\infty}^{\infty} x\left(\int_{-\infty}^{\infty} f_{X,Y}(x, y) \, dy\right) dx + \int_{-\infty}^{\infty} y\left(\int_{-\infty}^{\infty} f_{X,Y}(x, y) \, dx\right) dy$$

$$= \int_{-\infty}^{\infty} x \cdot f_X(x) \, dx + \int_{-\infty}^{\infty} y \cdot f_Y(y) \, dy$$

But the latter two integrals are simply $E(X)$ and $E(Y)$, respectively, so the result is proved.

An easy induction argument extends Theorem 3.9.1 to the case of n variables. We will omit the details.

Corollary. Given n random variables $X_1, X_2, \ldots, X_n$ and a set of n constants $a_1, a_2, \ldots, a_n$,

$$E(a_1 X_1 + a_2 X_2 + \cdots + a_n X_n) = a_1 E(X_1) + a_2 E(X_2) + \cdots + a_n E(X_n)$$

Comment

An equivalent way of writing the statement of the corollary is to say that "E is a linear transformation."

A good example for demonstrating the usefulness of the corollary is the problem of finding the expected value of a binomial random variable. Contrast the simplicity of the approach given in Example 3.9.1 with the intricacies of the direct summation described in Example 3.8.4. Two even more difficult problems that are much simplified by the corollary are taken up in Examples 3.9.2 and 3.9.3.

Example 3.9.1

Let X be a binomial random variable defined on n independent trials, each trial resulting in success with probability p. Find $E(X)$.

Note, first, that X can be thought of as a sum, $X = X_1 + X_2 + \cdots + X_n$, where X_i represents the number of successes occurring at the ith trial:

$$X_i = \begin{cases} 1 & \text{if the } i\text{th trial produces a success} \\ 0 & \text{if the } i\text{th trial produces a failure} \end{cases}$$

(Any X_i defined in this way on an individual trial is called a *Bernoulli* random variable. Every binomial, then, can be thought of as the sum of n independent Bernoulli's.) By assumption, $f_{X_i}(1) = p$ and $f_{X_i}(0) = 1 - p$, $i = 1, 2, \ldots, n$. Using the corollary,

$$E(X) = E(X_1) + E(X_2) + \cdots + E(X_n)$$
$$= n \cdot E(X_1)$$

the last step a consequence of the X_i's having identical distributions. But

$$E(X_1) = 1 \cdot p + 0 \cdot (1 - p) = p$$

so $E(X) = np$, which is what we found before (recall Equation 3.8.4).

Example 3.9.2

A disgruntled secretary is upset about having to stuff envelopes. Handed a box of n letters and n envelopes, she vents her frustration by putting the letters into the envelopes *at random*. How many people, on the average, will receive their correct mail?

If X denotes the number of envelopes properly stuffed, what we want is $E(X)$. However, applying Definition 3.8.1 here would prove formidable because of the difficulty in getting a workable expression for $f_X(x)$ [see (90)]. By using the corollary to Theorem 3.9.1, though, we can solve the problem easily.

Let X_i denote a random variable equal to the number of correct letters put into the ith envelope, $i = 1, 2, \ldots, n$. Then X_i equals 0 or 1, and

$$f_{X_i}(k) = P(X_i = k) = \begin{cases} \dfrac{1}{n} & \text{for } k = 1 \\ \dfrac{n-1}{n} & \text{for } k = 0 \end{cases}$$

But $X = X_1 + X_2 + \cdots + X_n$ and $E(X) = E(X_1) + E(X_2) + \cdots + E(X_n)$. Furthermore, each of the X_i's has the same expected value, $1/n$:

$$E(X_i) = \sum_{k=0}^{1} k \cdot P(X_i = k) = 0 \cdot \frac{n-1}{n} + 1 \cdot \frac{1}{n} = \frac{1}{n}$$

It follows that

$$E(X) = \sum_{i=1}^{n} E(X_i) = n \cdot \left(\frac{1}{n}\right)$$

$$= 1$$

showing that, *regardless of n*, the expected number of properly stuffed envelopes is 1. (Are the X_i's independent? Does it matter?)

Example 3.9.3

The honor count in a (13-card) bridge hand can vary from 0 to 37 according to the formula

honor count $= 4 \cdot$ (number of aces) $+ 3 \cdot$ (number of kings)
$+ 2 \cdot$ (number of queens) $+ 1 \cdot$ (number of jacks)

What is the expected honor count of North's hand?

The solution here is a bit unusual in that we use the corollary *backwards*. If X_i, $i = 1, 2, 3, 4$, denotes the honor count for North, South, East, and West, respectively, and if X denotes the analogous sum for the entire deck, we can write

$$X = X_1 + X_2 + X_3 + X_4$$

But

$$X = E(X) = 4 \cdot 4 + 3 \cdot 4 + 2 \cdot 4 + 1 \cdot 4 = 40$$

By symmetry, $E(X_i) = E(X_j)$, $i \neq j$, so it follows that $40 = 4 \cdot E(X_1)$, which implies that *10* is the expected honor count of North's hand. (Try doing this problem directly, without making use of the fact that the deck's honor count is 40.)

Question 3.9.1 Ten fair dice are thrown. What is the expected value of the sum of the faces showing?

Question 3.9.2 Suppose that X_i is a random variable for which $E(X_i) = \mu$, $i = 1, 2, \ldots, n$. Under what conditions will

$$E\left(\sum_{i=1}^{n} a_i X_i\right) = \mu \quad ?$$

Question 3.9.3 Let $X_1, X_2, \ldots, X_n$ be a random sample of size n drawn from a population with expected value $E(X)$. The *sample mean*, denoted $\bar{X}$, is the arithmetic average of the X_i's:

$$\bar{X} = \frac{1}{n} \sum_{i=1}^{n} X_i$$

Find $E(\bar{X})$.

Question 3.9.4 Marksmanship competition at a certain level requires each contestant to take 10 shots with each of two different hand guns. Final scores are computed by taking a weighted average of four times the number of bull's-eyes made with the first gun plus six times the number gotten with the second. If Cathie has a 30% chance of hitting the bull's-eye with each shot from the first gun and a 40% chance with each shot from the second gun, what is her expected score?

Question 3.9.5 If $X = 3X_1 + 4X_2 + 2X_3$, where

$$f_{X_1}(x) = \binom{6}{x}\left(\frac{1}{3}\right)^x\left(\frac{2}{3}\right)^{6-x} \qquad x = 0, 1, \ldots, 6$$

$$f_{X_2}(x) = \frac{e^{-4}4^x}{x!} \qquad x = 0, 1, 2, \ldots$$

and

$$f_{X_3}(x) = 5e^{-5x} \qquad x > 0$$

find $E(5X)$.

Question 3.9.6 Eight cards are drawn from a poker deck. What is the expected number of clubs?

Question 3.9.7 Cards are dealt one by one off the top of a shuffled poker deck until the first ace appears. On the average, how many cards are "above" that first ace?

Question 3.9.8 Suppose that the daily closing price of a stock goes up an eighth of a point with probability p and down an eighth of a point with probability q, where $p > q$. After n days how much gain can we expect the stock to have achieved? Assume that the daily price fluctuations are independent events.

Question 3.9.9 An urn contains n chips numbered 1 through n. A sample of size r ($< n$) is drawn. What is the expected value of the sum of the chips in the sample? Does it matter if the selections are made with replacement or without replacement?

The utility of linear combinations notwithstanding, there are many situations that call for the expected value of a *nonlinear* function of one or more random variables. Extending what we have already done to cover this more general case is not particularly difficult: the technique introduced in the proof of Theorem 3.9.1 for finding the expected value of $X + Y$ also works for arbitrary functions.

Theorem 3.9.2. If X is a discrete random variable with pdf $f_X(x)$ and if $g(x)$ is any function of X, then

$$E[g(X)] = \sum_{\text{all } x} g(x) \cdot f_X(x)$$

provided that

$$\sum_{\text{all } x} |g(x)| f_X(x) < \infty$$

If X is a continuous random variable,

$$E[g(X)] = \int_{-\infty}^{\infty} g(x) \cdot f_X(x)\, dx$$

provided that

$$\int_{-\infty}^{\infty} |g(x)| f_X(x)\, dx < \infty$$

Proof. We will prove the result for the discrete case. See (136) for details showing how the argument is modified when $f_X(x)$ is continuous.

Let $Y = g(X)$. The set of all possible x-values, $x_1, x_2, \ldots$, will give rise to a set of y-values, $y_1, y_2, \ldots$, where, in general, more than one x may be associated with a given y. Let S_j be the set of x's for which $g(x) = y_j$ [so $\bigcup_j S_j$ is the entire set of x-values for which $f_X(x)$ is defined]. We obviously have that $P(Y = y_j) = P(X \in S_j)$, and we can write

$$E(Y) = \sum_j y_j \cdot P(Y = y_j) = \sum_j y_j \cdot P(X \in S_j)$$

$$= \sum_j y_j \sum_{x \in S_j} f_X(x)$$

$$= \sum_j \sum_{x \in S_j} y_j \cdot f_X(x)$$

$$= \sum_j \sum_{x \in S_j} g(x) f_X(x) \qquad \text{(why?)}$$

$$= \sum_{\text{all } x} g(x) f_X(x)$$

Since it is being assumed that $\sum_{\text{all } x} |g(x)| f_X(x) < \infty$, the statement of the theorem holds.

Example 3.9.4

Suppose that X is a random variable whose pdf is nonzero only for the three values -2, 1, and $+2$:

x	$f_X(x)$
-2	$\frac{5}{8}$
1	$\frac{1}{8}$
2	$\frac{2}{8}$
	$\overline{1}$

Let $Y = g(X) = X^2$. Verify the statement of Theorem 3.9.2 by computing $E(Y)$ two ways—first, by finding $f_Y(y)$ and summing $y \cdot f_Y(y)$ over y and, second, by summing $g(x) \cdot f_X(x)$ over x.

By inspection, the pdf for Y is defined for only two values, 1 and 4:

$y (= x^2)$	$f_Y(y)$
1	$\frac{1}{8}$
4	$\frac{7}{8}$
	$\overline{1}$

Following the first approach we just mentioned gives

$$E(Y) = \sum_y y \cdot f_Y(y) = 1 \cdot \left(\frac{1}{8}\right) + 4 \cdot \left(\frac{7}{8}\right)$$

$$= \frac{29}{8}$$

To find the expected value via Theorem 3.9.2, we take

$$E[g(X)] = \sum_x x^2 \cdot f_X(x) = (-2)^2 \cdot \frac{5}{8} + (1)^2 \cdot \frac{1}{8} + (2)^2 \cdot \frac{2}{8}$$

with the sum here reducing to the answer we already found, 29/8.

For this particular situation, neither approach was easier than the other. In general, that will not be the case. Finding $f_Y(y)$ is often quite difficult, and on those occasions Theorem 3.9.2 can be of great benefit.

Example 3.9.5

Consolidated Industries is planning to market a new product and they are trying to decide how many to manufacture. They estimate that each item sold will return a profit of m dollars; each one not sold represents an n dollar loss. Furthermore, they suspect the demand for the product, V, will have an exponential distribution,

$$f_V(v) = \left(\frac{1}{\lambda}\right)e^{-v/\lambda} \qquad v > 0$$

How many items should the company produce if they want to maximize their expected profit? (Assume that n, m, and λ are known.)

If a total of x items are made, the company's profit function, Q, is given by

$$Q = Q(v) = \begin{cases} mv - n(x - v) & \text{if} \quad v < x \\ mx & \text{if} \quad v \geq x \end{cases}$$

It follows that their *expected* profit is

$$E(Q) = \int_0^\infty Q \cdot f_V(v) \, dv$$

$$= \int_0^x [(m + n)v - nx]\left(\frac{1}{\lambda}\right)e^{-v/\lambda} \, dv + \int_x^\infty mx \cdot \left(\frac{1}{\lambda}\right)e^{-v/\lambda} \, dv \quad (3.9.1)$$

The integration here is straightforward, though a bit tedious. What Equation 3.9.1 eventually simplifies to is

$$E(Q) = \lambda \cdot (m + n) - \lambda \cdot (m + n)e^{-x/\lambda} - nx$$

To find the optimal production level, we need to solve $dE(Q)/dx = 0$ for x. But

$$\frac{dE(Q)}{dx} = (m + n)e^{-x/\lambda} - n$$

and the latter equals 0 when

$$x = -\lambda \cdot \ln\left(\frac{n}{m + n}\right)$$

(How many items should the company manufacture if the demand function is

$$f_V(v) = 0.0001e^{-0.0001v} \qquad v > 0$$

and m and n are fixed at 50 cents and \$2, respectively?)

Question 3.9.10 Let X have the probability density function

$$f_X(x) = \begin{cases} 2(1 - x) & 0 < x < 1 \\ 0 & \text{elsewhere} \end{cases}$$

Suppose that $Y = g(X) = X^3$. Find $E(Y)$ two different ways.

Question 3.9.11 A tool and die company makes castings for steel stress-monitoring gauges. Their annual profit, Q, in hundreds of thousands of dollars, can be expressed as a function of product demand, x:

$$Q = Q(x) = 2(1 - e^{-2x})$$

Suppose that the demand (in thousands) for their castings follows an exponential pdf, $f_X(x) = 6e^{-6x}$, $x > 0$. Find the company's expected profit.

Theorem 3.9.2 can be extended to include functions of any number of random variables. We formally state the generalization only for the bivariate case, but the examples that follow involve functions of two, three, and four random variables. For a proof of Theorem 3.9.3, see (118).

Theorem 3.9.3. Suppose that X and Y are discrete random variables with joint density function $f_{X,Y}(x, y)$. Let $g(x, y)$ be any function of X and Y. Then

$$E[g(X, Y)] = \sum_{\text{all } x} \sum_{\text{all } y} g(x, y) \cdot f_{X,Y}(x, y)$$

provided that

$$\sum_{\text{all } x} \sum_{\text{all } y} |g(x, y)| \cdot f_{X,Y}(x, y) < \infty$$

If X and Y are continuous random variables,

$$E[g(X, Y)] = \int_{-\infty}^{\infty} \int_{-\infty}^{\infty} g(x, y) \cdot f_{X,Y}(x, y) \, dx \, dy$$

provided that

$$\int_{-\infty}^{\infty} \int_{-\infty}^{\infty} |g(x, y)| \cdot f_{X,Y}(x, y) \, dx \, dy < \infty$$

Example 3.9.6

In Example 3.3.5 we considered two random variables whose joint pdf was given by the 2×4 matrix shown in Table 3.9.1.

TABLE 3.9.1

		\multicolumn{4}{c}{Y}
		0	1	2	3
X	0	$\frac{1}{8}$	$\frac{1}{4}$	$\frac{1}{8}$	0
	1	0	$\frac{1}{8}$	$\frac{1}{4}$	$\frac{1}{8}$

Define

$$g(x, y) = 3x - 2xy + y$$

Find $E[g(X, Y)]$ directly, and then by using Theorem 3.9.3.

Let $Z = 3X - 2XY + Y$. By inspection, Z can take on the values 0, 1, 2, and 3 with pdf $f_Z(z)$ as shown in Table 3.9.2.

TABLE 3.9.2

z	0	1	2	3
$f_Z(z)$	$\frac{1}{4}$	$\frac{1}{2}$	$\frac{1}{4}$	0

From Definition 3.8.1, then, $E[g(X, Y)]$ is equal to 1:

$$E[g(X, Y)] = E(Z) = \sum_{\text{all } z} z \cdot f_Z(z)$$

$$= 0 \cdot \frac{1}{4} + 1 \cdot \frac{1}{2} + 2 \cdot \frac{1}{4} + 3 \cdot 0$$

$$= 1$$

We get the same answer for $E[g(X, Y)]$ by applying Theorem 3.9.3 to the joint pdf given in Table 3.9.1:

$$E[g(X, Y)] = 0 \cdot \frac{1}{8} + 1 \cdot \frac{1}{4} + 2 \cdot \frac{1}{8} + 3 \cdot 0 + 3 \cdot 0 + 2 \cdot \frac{1}{8} + 1 \cdot \frac{1}{4} + 0 \cdot \frac{1}{8}$$

$$= 1$$

The advantage, of course, enjoyed by the latter solution is that we avoid the intermediate step of having to determine $f_Z(z)$.

Example 3.9.7

An electrical circuit has three resistors, R_X, R_Y, and R_Z, wired in parallel (see Figure 3.9.1). The nominal resistance of each is 15 ohms, but their *actual* resistances, X, Y, and Z, vary between 10 and 20 according to the joint pdf,

$$f_{X, Y, Z}(x, y, z) = \frac{1}{675,000} (xy + xz + yz) \qquad \begin{array}{l} 10 \leq x \leq 20 \\ 10 \leq y \leq 20 \\ 10 \leq z \leq 20 \end{array}$$

What is the expected resistance for the circuit?

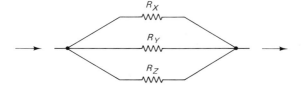

Figure 3.9.1

Let R denote the circuit's resistance. A well-known result in physics holds that

$$\frac{1}{R} = \frac{1}{X} + \frac{1}{Y} + \frac{1}{Z}$$

or, equivalently,

$$R = \frac{XYZ}{XY + XZ + YZ} = R(X, Y, Z)$$

Integrating $R(x, y, z) \cdot f_{X, Y, Z}(x, y, z)$ shows that the expected resistance is 5.0:

$$E(R) = \int_{10}^{20} \int_{10}^{20} \int_{10}^{20} \frac{xyz}{xy + xz + yz} \cdot \frac{1}{675{,}000} (xy + xz + yz) \, dx \, dy \, dz$$

$$= \frac{1}{675{,}000} \int_{10}^{20} \int_{10}^{20} \int_{10}^{20} xyz \, dx \, dy \, dz$$

$$= 5.0$$

Example 3.9.8

Two points, $Q = (x, y)$ and $Q' = (x', y')$, are chosen at random inside a square having sides of length s. What is the expected value of D^2, the square of the distance between Q and Q'? (See Figure 3.9.2.)

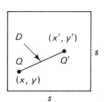

Figure 3.9.2

Written in terms of the coordinates of Q and Q', D^2 is a function of *four* random variables,

$$D^2 = g(X, X', Y, Y') = (X' - X)^2 + (Y' - Y)^2$$

Therefore,

$$E(D^2) = \int_0^s \int_0^s \int_0^s \int_0^s [(x' - x)^2 + (y' - y)^2]$$

$$\cdot f_{X, X', Y, Y'}(x, x', y, y') \, dx \, dx' \, dy \, dy'$$

Since X, X', Y, and Y' are all to be chosen at random over the interval $(0, s)$,

$$f_{X, X', Y, Y'}(x, x', y, y') = f_X(x) \cdot f_X(x') \cdot f_Y(y) \cdot f_Y(y')$$

$$= \frac{1}{s^4}$$

so the expression for $E(D^2)$ simplifies to

$$E(D^2) = \frac{1}{s^4} \int_0^s \int_0^s \int_0^s \int_0^s [(x' - x)^2 + (y' - y)^2] \, dx \, dx' \, dy \, dy'$$

The integrations here are lengthy, but straightforward; what they finally simplify to is

$$E(D^2) = \frac{s^2}{3}$$

Question 3.9.12   Two fair dice are tossed one time. Let X denote the number of 2's that appear and Y, the number of 3's. Let $Z = XY^2$. Find $E[g(X, Y)]$ two ways.

Question 3.9.13   Suppose that $f_{X, Y}(x, y) = e^{-x-y}$, $x > 0$, $y > 0$. Find $E(X + Y)$.

Question 3.9.14   Find $E(R)$ for a two-resistor circuit similar to the one described in Example 3.9.7 where $f_{X, Y}(x, y) = k(x + y)$, $10 \le x \le 20$, $10 \le y \le 20$.

Question 3.9.15   Two points, X and Y, are chosen at random along perpendicular sides of the unit square. What is the expected value of the area of the rectangle whose sides are of length X and Y?

Question 3.9.16   An urn contains n chips numbered 1 through n. A sample of size 2 is drawn without replacement. Show that the expected value of the product of the two drawn is

$$\frac{1}{12}(n + 1)(3n + 2)$$

The final result in this section is a special case of Theorem 3.9.3 dealing with the expected value of a *product* of random variables. Note that unlike the opening theorem in this section that addressed the expected values of sums, the statement we are about to see for products is true *only if the variables are independent*.

Theorem 3.9.4.   If X and Y are independent random variables,

$$E(XY) = E(X) \cdot E(Y)$$

provided both $E(X)$ and $E(Y)$ exist.

Proof. Let X and Y both be discrete. Using Theorems 3.4.1 and 3.9.3, we can easily establish the desired factorization:

$$E(XY) = \sum_{\text{all } x} \sum_{\text{all } y} xy \cdot f_{X, Y}(x, y)$$

$$= \sum_{\text{all } x} \sum_{\text{all } y} xy \cdot f_X(x) \cdot f_Y(y)$$

$$= \sum_{\text{all } x} x \cdot f_X(x) \cdot \left[\sum_{\text{all } y} y \cdot f_Y(y) \right]$$

$$= E(X) \cdot E(Y)$$

The proof for continuous X and Y is left as an exercise.

Comment

Theorem 3.9.4 is not an if and only if statement: just because $E(XY)$ equals $E(X) \cdot E(Y)$, it does not follow that X and Y are independent. See (90) for a counterexample.

Question 3.9.17 Two fair dice are tossed. What is the expected value of the product of the faces showing?

Question 3.9.18 Let $X_1, X_2, \ldots, X_n$ be a set of mutually independent and continuous random variables. Show that

$$E(X_1 X_2 \cdots X_n) = E(X_1) \cdot E(X_2) \cdots E(X_n)$$

3.10 THE VARIANCE

The expected value is a good enough measure of central tendency, but it still leaves out some critical information about a random variable's behavior. Unless we are also provided with some indication of how *spread out* a random variable's probability density function is, the expected value by itself can be misleading, and we are prey to absurdities such as, "A person with his head in the freezer and feet in the oven is *on the average* quite comfortable." More quantitatively, a football team may have an interior line averaging 200 pounds if it has five players all weighing close to that figure, or if it has four 150-pounders and a 400-pound behemoth at left tackle. One would expect the opposition to recognize the tactical significance of such a difference in weight dispersion rather quickly. Finally, a random variable representing a game where $1 is won or lost with equal probability has expected value zero, as does one where $100,000 is the stake; yet, surely, these are different games (or, at the very least, played by different people).

Given, then, that the assessment of a random variable's dispersion, or variability, is a worthwhile pursuit, how should we proceed? One seemingly reasonable approach would be to average, in the generalized sense, the deviations of the values of X from their expected value. Unfortunately, this proves to be futile, since the numerical value of such an average will always be 0:

$$E(X - \mu) = E(X) - \mu = \mu - \mu = 0 \qquad (3.10.1)$$

Another possibility would be to modify Equation 3.10.1 by making all the deviations positive—that is, replace $E(X - \mu)$ with $E(|X - \mu|)$. This does work, and it *is* sometimes used to measure dispersion, but the absolute value is somewhat troublesome mathematically: it does not have a simple arithmetic formula, nor is it a differentiable function. *Squaring* the deviations proves to be a much more attractive solution.

> **Definition 3.10.1.** The *variance* of a random variable X, denoted Var (X), is the expected value of its squared deviations from μ
>
> $$\text{Var}(X) = E[(X - \mu)^2]$$
>
> where $\mu = E(X)$. [If $E(X^2)$ is not finite, the variance is not defined.] Often the symbol σ^2 is used in place of Var (X).

Comment

One unfortunate consequence of Definition 3.10.1 is that the units for the variance are the square of the units for X: if X is measured in inches, the units for Var (X) are inches^2. This causes obvious problems in relating the variance back to the sample values. In applied statistics, the variance is often replaced as a measure of dispersion by the *standard deviation*, where

$$\text{standard deviation of } X = \sqrt{\text{Var } (X)}$$

For some random variables, Definition 3.10.1 does not provide a particularly convenient computing formula for Var (X). Theorem 3.10.1 gives an equivalent expression that is often much easier to use.

Theorem 3.10.1. Let X be a random variable having mean μ and for which $E(X^2)$ is finite. Then

$$\text{Var } (X) = E(X^2) - \mu^2$$

Proof. The proof is a simple application of the distributive property of expected values:

$$\text{Var } (X) = E((X - \mu)^2)$$

$$= E(X^2 - 2\mu X + \mu^2)$$

$$= E(X^2) - 2\mu E(X) + \mu^2$$

$$= E(X^2) - 2\mu^2 + \mu^2$$

$$= E(X^2) - \mu^2$$

Example 3.10.1

An urn contains 5 chips, 2 red and 3 white. Suppose that two are drawn out at random, *without replacement*. Let X denote the number of red chips in the sample. Find Var (X).

Regardless of which formula we elect to use, Definition 3.10.1 or Theorem 3.10.1, we first need to find μ. Here, since X is hypergeometric,

$$\mu = E(X) = \sum_{x=0}^{2} x \cdot \frac{\binom{2}{x}\binom{3}{2-x}}{\binom{5}{2}} = 0.8$$

The variance via Definition 3.10.1, then, becomes

$$\text{Var}(X) = E[(X - \mu)^2] = \sum_{\text{all } x} (x - \mu)^2 \cdot f_X(x)$$

$$= (0 - 0.8)^2 \cdot \frac{\binom{2}{0}\binom{3}{2}}{\binom{5}{2}} + (1 - 0.8)^2 \cdot \frac{\binom{2}{1}\binom{3}{1}}{\binom{5}{2}} + (2 - 0.8)^2 \cdot \frac{\binom{2}{2}\binom{3}{0}}{\binom{5}{2}}$$

$$= 0.36$$

To find Var (X) using Theorem 3.10.1, we begin by computing $E(X^2)$. From Theorem 3.9.2,

$$E(X^2) = \sum_{\text{all } x} x^2 \cdot f_X(x) = 0^2 \cdot \frac{\binom{2}{0}\binom{3}{2}}{\binom{5}{2}} + 1^2 \cdot \frac{\binom{2}{1}\binom{3}{1}}{\binom{5}{2}} + 2^2 \cdot \frac{\binom{2}{2}\binom{3}{0}}{\binom{5}{2}}$$

$$= 1.00$$

Therefore, according to our second formula,

$$\text{Var}(X) = E(X^2) - \mu^2 = 1.00 - (0.8)^2$$

$$= 0.36$$

confirming what we calculated earlier.

Question 3.10.1 Find Var (X) for the urn problem of Example 3.10.1 if the sampling is done *with* replacement.

Question 3.10.2 Find the variance of X if

$$f_X(x) = \begin{cases} \dfrac{3}{4} & 0 \le x \le 1 \\[2mm] \dfrac{1}{4} & 2 \le x \le 3 \\[2mm] 0 & \text{elsewhere} \end{cases}$$

Question 3.10.3 Ten equally qualified applicants, six men and four women, apply for three lab technician positions. Unable to justify choosing any of the applicants over all the others, the personnel director decides to select the three at random. Let Y denote the number of men hired. Compute the standard deviation of Y.

Question 3.10.4 Compute the variance for a uniform random variable defined on the unit interval.

Question 3.10.5 Use Theorem 3.10.1 to find the variance of the random variable X, where

$$f_X(x) = 3(1 - x)^2 \qquad 0 < x < 1$$

Question 3.10.6 Consider the pdf defined by

$$f_X(x) = \frac{2}{x^3} \qquad x \geq 1$$

Show that (a) $\int_1^\infty f_X(x)\, dx = 1$, (b) $E(X) = 2$, and (c) Var (X) is not finite.

Question 3.10.7 Frankie and Johnny play the following game. Frankie selects a number at random from the interval $[a, b]$. Johnny, not knowing Frankie's number, is to pick a second number from that same interval and pay Frankie an amount, W, equal to the squared difference between the two [so $0 \leq W \leq (b-a)^2$]. What should be Johnny's strategy if he wants to minimize his expected loss?

Examined in Section 3.9 were some of the properties associated with expected values. Analogous results for the variance are motivated by asking two simple questions: (1) how is the variance affected by a linear transformation, and (2) if $Y = X_1 + X_2 + \cdots + X_n$, how is the variance of Y related to the variance of the X_i's?

Theorem 3.10.2. Let X be a random variable and let a and b be constants. Define $Y = aX + b$. Then

$$\text{Var } (Y) = a^2 \text{ Var } (X)$$

Proof. Since $E(aX + b) = a\mu + b$,

$$\begin{aligned}
\text{Var } (Y) &= E([(aX + b) - (a\mu + b)]^2) \\
&= E(a^2(X - \mu)^2) \\
&= a^2 E((X - \mu)^2) \\
&= a^2 \text{ Var } (X)
\end{aligned}$$

Example 3.10.2

A random variable X is described by the pdf

$$f_X(x) = 2x \qquad 0 < x < 1$$

What is the standard deviation of Y, where $Y = 3X + 2$? First, we need to find the variance of X. But

$$E(X) = \int_0^1 x \cdot 2x\, dx = \frac{2}{3}$$

and

$$E(X^2) = \int_0^1 x^2 \cdot 2x\, dx = \frac{1}{2}$$

so

$$\text{Var}(X) = E(X^2) - \mu^2 = \frac{1}{2} - \left(\frac{2}{3}\right)^2$$

$$= \frac{1}{18}$$

Then, by Theorem 3.10.2,

$$\text{Var}(Y) = (3)^2 \cdot \text{Var}(X) = 9 \cdot \frac{1}{18}$$

$$= \frac{1}{2}$$

which makes the standard deviation of Y equal to $\sqrt{\frac{1}{2}}$, or 0.71.

Question 3.10.8 If $E(X) = \mu$ and Var $(X) = \sigma^2$, show that

$$E\left(\frac{X - \mu}{\sigma}\right) = 0 \quad \text{and} \quad \text{Var}\left(\frac{X - \mu}{\sigma}\right) = 1$$

Question 3.10.9 Over a recent 90-year period, temperatures recorded in Bismarck, North Dakota, in December ranged from $-36°F$ to $+72°F$ (174). Assume that the standard deviation of the distribution of daily extremes is approximately $18°F$. What is the corresponding standard deviation in degrees *Celsius*? [*Note*: If X is a temperature recorded in degrees Fahrenheit and Y is the same temperature expressed in degrees Celsius, then $Y = \frac{5}{9} \cdot (X - 32)$.]

Question 3.10.10 Let Y be a uniform random variable defined on the interval $(1000, 3000)$. Use Theorem 3.10.2 to find the variance of Y by considering the transformation $X = (Y - 2000)/1000$.

If $Y = X_1 + X_2 + \cdots + X_n$, the expected value of Y is the sum of the corresponding $E(X_i)$'s *whether or not the X_i's are independent*. What would be the parallel result for variances is not true: the variance of Y is not necessarily equal to the sum of the Var (X_i)'s. Equality *does* hold, though, if the X_i's are independent.

Theorem 3.10.3. Let $X_1, X_2, \ldots, X_n$ be independent random variables and let $Y = X_1 + X_2 + \cdots + X_n$. Then

$$\text{Var}(Y) = \text{Var}(X_1) + \text{Var}(X_2) + \cdots + \text{Var}(X_n)$$

Proof. We give a proof for $Y = X_1 + X_2$; an easy induction completes the argument for general n. From Theorems 3.9.1 and 3.10.1,

$$\text{Var}(Y) = E((X_1 + X_2)^2) - [E(X_1) + E(X_2)]^2$$

Writing out the squares gives

$$\text{Var } (Y) = E(X_1^2 + 2X_1X_2 + X_2^2) - [E(X_1)]^2 - 2E(X_1)E(X_2) - [E(X_2)]^2$$

$$= E(X_1^2) - [E(X_1)]^2 + E(X_2^2) - [E(X_2)]^2$$

$$+ 2[E(X_1X_2) - E(X_1)E(X_2)] \qquad (3.10.2)$$

By the independence of X_1 and X_2, $E(X_1X_2) = E(X_1)E(X_2)$, making the last term in Equation 3.10.2 vanish. The remaining terms combine to give the desired result: $\text{Var } (Y) = \text{Var } (X_1) + \text{Var } (X_2)$.

Example 3.10.3

The binomial random variable, being a sum of n independent Bernoulli's is an obvious candidate for Theorem 3.10.3. Let X_i denote the number of successes occurring on the ith trial. Then

$$X_i = \begin{cases} 1 & \text{with probability } p \\ 0 & \text{with probability } 1 - p \end{cases}$$

Write $Y = X_1 + X_2 + \cdots + X_n$. We want to find the variance of Y. But

$$E(X_i) = 1 \cdot p + 0 \cdot q$$

and

$$E(X_i^2) = (1)^2 \cdot p + (0)^2 \cdot (1 - p) = p$$

so

$$\text{Var } (X_i) = EX_i^2 - [E(X_i)]^2 = p - p^2$$

$$= p(1 - p)$$

It follows, then, that the *variance of a binomial random variable is* $np(1 - p)$:

$$\text{Var } (Y) = \sum_{i=1}^{n} \text{Var } (X_i) = np(1 - p)$$

Example 3.10.4

In statistics, we often have to draw inferences based on $\bar{X}$, the average computed from a random sample of n observations. Two important properties of $\bar{X}$ need to be remembered. First, if the X_i's come from a population whose mean is μ, then $E(\bar{X}) = \mu$ (recall Question 3.9.3). Second, if the X_i's come from a population whose variance is σ^2, then $\text{Var } (\bar{X}) = \sigma^2/n$.

To see the latter, we can appeal to Theorems 3.10.2 and 3.10.3. Let

$$Y = \bar{X} = \frac{1}{n} \sum_{i=1}^{n} X_i = \frac{1}{n} \cdot X_1 + \frac{1}{n} \cdot X_2 + \cdots + \frac{1}{n} \cdot X_n$$

Then

$$\text{Var}(Y) = \text{Var}(\bar{X}) = \left(\frac{1}{n}\right)^2 \cdot \text{Var}(X_1) + \left(\frac{1}{n}\right)^2$$

$$\cdot \text{Var}(X_2) + \cdots + \left(\frac{1}{n}\right)^2 \cdot \text{Var}(X_n)$$

$$= \left(\frac{1}{n}\right)^2 \sigma^2 + \left(\frac{1}{n}\right)^2 \sigma^2 + \cdots + \left(\frac{1}{n}\right)^2 \sigma^2$$

$$= \frac{\sigma^2}{n}$$

Both these results—$E(\bar{X}) = \mu$ and $\text{Var}(\bar{X}) = \sigma^2/n$—will be used repeatedly in the chapters ahead.

Question 3.10.11 A mason is contracted to build a patio retaining wall. Plans call for the base of the wall to be a row of 50 10-inch bricks, each separated by $\frac{1}{2}$-inch-thick mortar. Suppose that the bricks used are randomly chosen from a population of bricks whose mean length is 10 inches and whose standard deviation is $\frac{1}{32}$ inch. Also, suppose that the mason, on the average, will make the mortar $\frac{1}{2}$ inch thick, but the actual dimension varies from brick to brick, the standard deviation of the thicknesses being $\frac{1}{16}$ inch. What is the standard deviation of L, the length of the first row of the wall? What assumption are you making?

Question 3.10.12 Let X be a binomial random variable based on n trials and a success probability of p_X; let Y be an independent binomial random variable based on m trials and a success probability of p_Y. Find $E(W)$ and $\text{Var}(W)$, where $W = 4X + 6Y$.

Question 3.10.13 An electric circuit has six resistors wired in series, each nominally being 5 ohms. What is the maximum standard deviation that can be allowed in the manufacture of these resistors if the combined circuit resistance is to have a standard deviation no greater than 0.4 ohm?

Question 3.10.14 Carry out the induction argument to complete the proof of Theorem 3.10.3.

Question 3.10.15 A gambler plays n hands of poker. If he wins the kth hand, he collects k dollars; if he loses the kth hand, he collects nothing. Let T denote his total winnings in n hands. Assuming that his chances of winning each hand are constant and are independent of his success or failure at any other hand, find $E(T)$ and $\text{Var}(T)$.

3.11 HIGHER MOMENTS

The quantities we have identified as the mean and the variance are actually special cases of what are referred to more generally as a random variable's *moments*. Specifically, $E(X)$ is the *first moment about the origin* and σ^2, the *second moment about the mean.* As the terminology suggests, we will have occasion to define even "higher" moments of X.

It is not our intention to pursue the computation and interpretation of higher moments to any great extent. Suffice it to say that just as μ and σ^2 reflect a random variable's location and dispersion, respectively, so is it possible to characterize other aspects of a distribution in terms of other moments. We will see why we want to do that in Chapter 5. Here our limited objective is simply to define what a higher moment is and give a useful existence criterion for knowing whether or not that moment is finite.

> **Definition 3.11.1.** Let X be any random variable with pdf $f_X(x)$. For any positive integer r,
>
> (a) The rth *moment of X about the origin*, μ_r, is given by
>
> $$\mu_r = E(X^r)$$
>
> provided $\int_{-\infty}^{\infty} |x|^r \cdot f_X(x)\, dx < \infty$ (or provided that the analogous condition on the *summation* of $|x|^r$ holds, if X is discrete). When $r = 1$, we usually delete the subscript and write $E(X)$ as μ, rather than μ_1.
>
> (b) The rth *moment of X about the mean*, μ_r', is given by
>
> $$\mu_r' = E[(X - \mu)^r]$$
>
> provided that the same finiteness conditions hold that were cited in part (a).

Comment

We can express μ_r' in terms of μ_j, $j = 1, 2, \ldots, r$, by simply writing out the binomial expansion of $(X - \mu)^r$:

$$\mu_r' = E[(X - \mu)^r] = \sum_{j=0}^{r} \binom{r}{j} E(X^j)(-\mu)^{r-j}$$

Thus

$$\mu_2' = E[(X - \mu)^2] = \sigma^2 = \mu_2 - \mu_1^2$$

$$\mu_3' = E[(X - \mu)^3] = \mu_3 - 3\mu_1\mu_2 + 2\mu_1^3$$

$$\mu_4' = E[(X - \mu)^4] = \mu_4 - 4\mu_1\mu_3 + 6\mu_1^2\mu_2 - 3\mu_1^4$$

and so on.

Example 3.11.1

The *skewness* of a pdf can be measured in terms of its third moment about the mean. If a pdf is symmetric, $E[(X - \mu)^3]$ will obviously be 0; for pdf's not symmetric, $E[(X - \mu)^3]$ will not be zero. In practice, the symmetry (or lack of symmetry) of a pdf is often measured by the *coefficient of skewness*, γ_1, where

$$\gamma_1 = \frac{E[(X - \mu)^3]}{\sigma^3}$$

Dividing μ_3' by σ^3 makes γ_1 dimensionless.

A second "shape" parameter in common use is the *coefficient of kurtosis*, γ_2, which involves the *fourth* moment about the mean. Specifically,

$$\gamma_2 = \frac{E[(X - \mu)^4]}{\sigma^4} - 3$$

For certain pdf's, γ_2 is a useful measure of peakedness: relatively "flat" pdf's are said to be *platykurtic*; more peaked pdf's are called *leptokurtic* [see (90)].

Question 3.11.1 Let X be a uniform random variable defined over the interval $(0, 2)$. Find an expression for the rth moment of X about the origin. Also, use the binomial expansion as described in the comment to find $E[(X - \mu)^6]$.

Question 3.11.2 Find the coefficient of skewness for an exponential random variable having the pdf

$$f_X(x) = e^{-x} \qquad x > 0$$

Use the fact that if k is a positive integer,

$$\int_0^\infty x^k e^{-x}\, dx = k!$$

Question 3.11.3 If $Y = aX + b$, show that Y has the same coefficients of skewness and kurtosis as X.

Earlier in this chapter we encountered random variables whose means did not exist—recall, for example, the St. Petersburg paradox. More generally, there are random variables having certain of their higher moments finite and certain others, not finite. Addressing the question of whether or not a given $E(X^j)$ is finite is an easy to use existence theorem.

> **Theorem 3.11.1.** If the kth moment of a random variable exists, all moments of order less than k exist.

Proof. Let $f_X(x)$ be the pdf of X. By Definition 3.11.1, $E(X^k)$ exists if and only if

$$\int_{-\infty}^\infty |x|^k \cdot f_X(x)\, dx < \infty \tag{3.11.1}$$

Let $1 \leq j < k$. To prove the theorem we must show that

$$\int_{-\infty}^\infty |x|^j \cdot f_X(x)\, dx < \infty$$

is implied by Inequality 3.11.1. But

$$\int_{-\infty}^{\infty} |x|^j \cdot f_X(x)\, dx = \int_{|x| \le 1} |x|^j \cdot f_X(x)\, dx + \int_{|x| > 1} |x|^j \cdot f_X(x)\, dx$$

$$\le \int_{|x| \le 1} f_X(x)\, dx + \int_{|x| > 1} |x|^j \cdot f_X(x)\, dx$$

$$\le 1 + \int_{|x| > 1} |x|^j \cdot f_X(x)\, dx$$

$$\le 1 + \int_{|x| > 1} |x|^k \cdot f_X(x)\, dx < \infty$$

Therefore, $E(X^j)$ exists, $j = 1, 2, \ldots, k - 1$.

Example 3.11.2

Many of the random variables that play a major role in statistics have moments existing for *all k*, as does, for instance, the normal distribution introduced in Case Study 2.5.1. Still, it is not difficult to find well-known models for which this is *not* true. A case in point is the *Student t distribution*, a probability function widely used in inference procedures.

The pdf for a Student *t* random variable is given by

$$f_X(x) = \frac{c(n)}{\left(1 + \dfrac{x^2}{n}\right)^{(n+1)/2}} \qquad -\infty < x < \infty, \quad n \ge 1$$

where *n* is referred to as the distribution's "degrees of freedom" and *c(n)* is a constant. By definition, the (2*k*)th moment is the integral

$$E(X^{2k}) = c(n) \cdot \int_{-\infty}^{\infty} \frac{x^{2k}}{\left(1 + \dfrac{x^2}{n}\right)^{(n+1)/2}}\, dx$$

Is $E(X^{2k})$ finite?

Not necessarily. Recall from calculus that an integral of the form

$$\int_{-\infty}^{\infty} \frac{1}{x^{\alpha}}\, dx$$

will converge only if $\alpha > 1$. Also, the convergence properties for integrals of

$$\frac{x^{2k}}{\left(1 + \dfrac{x^2}{n}\right)^{(n+1)/2}}$$

are the same as those for

$$\frac{x^{2k}}{(x^2)^{(n+1)/2}} = \frac{1}{x^{n+1-2k}}$$

Therefore, if $E(X^{2k})$ is to be finite, we must have

$$n + 1 - 2k > 1$$

or, equivalently, $2k < n$. Thus a Student t random variable with, say, $n = 9$ degrees of freedom has $E(X^8) < \infty$, but no moment of order higher than eight exists.

Question 3.11.4 Suppose that the random variable X is described by the pdf

$$f_X(x) = c \cdot x^{-6} \qquad x > 1$$

(a) Find c.
(b) What is the highest moment of X that exists?

Question 3.11.5 Make the substitution $z = 1/(1 + x^2/n)$ and find an expression for the $(2k)$th moment of a Student t random variable (where $2k < n$).

3.12 MOMENT-GENERATING FUNCTIONS

Finding moments of random variables, particularly the higher moments defined in Section 3.11, is conceptually straightforward but can be difficult to accomplish in practice: depending on the nature of $f_X(x)$, integrals of the form $\int_{-\infty}^{\infty} x^r f_X(x)\, dx$ are not always easy to evaluate. Fortunately, an alternative method is available. For some densities, we can find a *moment-generating function* (or *mgf*), $M_X(t)$, one of whose properties is that its rth derivative evaluated at 0 is equal to $E(X^r)$. If $M_X(t)$ can be found, $M_X^{(r)}(t)$ will often be easier to evaluate than $\int_{-\infty}^{\infty} x^r f_X(x)\, dx$.

Moment-generating functions can also be extremely useful in deriving the distribution of a *sum* of independent random variables. Such problems are important in statistics and typically difficult: recall, for instance, Example 3.5.6, where it took considerable effort to prove that the sum of two binomials is also binomial. Using moment-generating functions, that particular problem is trivial.

> **Definition 3.12.1.** Let X be a random variable. The *moment-generating function for X*, denoted $M_X(t)$, is given by
>
> $$M_X(t) = E(e^{tX})$$
>
> at all values of t for which the expected value exists.

Before investigating the properties of moment-generating functions, we will first simply *find* $M_X(t)$ for two of the densities we have already encountered, the binomial and the exponential. The computations here are a straightforward application of Theorem 3.9.2.

Example 3.12.1

Let X be a binomial random variable with pdf

$$f_X(x) = \binom{n}{x} p^x (1-p)^{n-x} \qquad x = 0, 1, \ldots, n$$

Derive the moment-generating function for X.

From Definition 3.12.1, and with the help of the formula for a binomial expansion, we find that $M_X(t) = (1 - p + pe^t)^n$:

$$M_X(t) = E(e^{tX}) = \sum_{x=0}^{n} e^{tx} \cdot \binom{n}{x} p^x (1-p)^{n-x} = \sum_{x=0}^{n} \binom{n}{x} (pe^t)^x (1-p)^{n-x}$$

$$= (1 - p + pe^t)^n$$

Here $M_X(t)$ is defined for all values of t.

Example 3.12.2

Suppose that X has an exponential pdf, $f_X(x) = (1/\lambda)e^{-x/\lambda}$, $x > 0$. Find $M_X(t)$.

Since the exponential is a continuous random variable, $M_X(t)$ is an integral:

$$M_X(t) = E(e^{tX}) = \int_0^\infty e^{tx} \cdot \left(\frac{1}{\lambda}\right) e^{-x/\lambda} \, dx$$

$$= \int_0^\infty \left(\frac{1}{\lambda}\right) e^{(t - 1/\lambda)x} \, dx$$

$$= \frac{1}{\lambda t - 1} \cdot e^{(t - 1/\lambda)x} \Big|_0^\infty$$

$$= \frac{1}{\lambda t - 1} \left[\lim_{x \to 0} e^{(t - 1/\lambda)x} - 1 \right]$$

Notice that the limit here exists (and equals 0) only if $t - 1/\lambda < 0$. Thus, for $t < 1/\lambda$, the moment-generating function of X is given by

$$M_X(t) = \frac{1}{1 - \lambda t}$$

For $t \geq 1/\lambda$, $M_X(t)$ fails to exist.

Question 3.12.1 Let X be a discrete random variable with

$$f_X(x) = \begin{cases} p(1-p)^{x-1} & x = 1, 2, \ldots \\ 0 & \text{elsewhere} \end{cases}$$

Show that $M_X(t) = (pe^t)/(1 - qe^t)$, where $q = 1 - p$. (*Hint*: Recall the formula for the sum of a geometric series,

$$\sum_{t=0}^{\infty} r^t = \frac{1}{1 - r} \qquad \text{for} \quad 0 < r < 1.)$$

Question 3.12.2 Show that the moment-generating function of the random variable X having pdf $f_X(x) = \frac{1}{3}$, $-1 < x < 2$, is

$$M_X(t) = \begin{cases} \dfrac{e^{2t} - e^{-t}}{3t} & t \neq 0 \\ 1 & t = 0 \end{cases}$$

Question 3.12.3 Let X have pdf

$$f_X(x) = \begin{cases} x & 0 \le x \le 1 \\ 2 - x & 1 \le x \le 2 \\ 0 & \text{elsewhere} \end{cases}$$

Find $M_X(t)$.

Theorem 3.12.1 shows that $M_X(t)$ does, indeed, generate moments. We give a sketch of the proof, leaving its details as an exercise.

Theorem 3.12.1. Let X be a random variable with probability density function $f_X(x)$. [If X is continuous, $f_X(x)$ must be sufficiently smooth to allow the order of differentiation and integration to be interchanged.] Let $M_X(t)$ be the moment-generating function for X. Then

$$M_X^{(r)}(0) = E(X^r)$$

provided that the rth moment exists.

Proof. We will verify the theorem for the continuous case where r is either 1 or 2. The extensions to discrete random variables and to an arbitrary positive integer r are straightforward.

For $r = 1$,

$$M_X^{(1)}(0) = \frac{d}{dt} \int_{-\infty}^{\infty} e^{tx} f_X(x)\, dx \bigg|_{t=0} = \int_{-\infty}^{\infty} \frac{d}{dt} e^{tx} f_X(x)\, dx \bigg|_{t=0}$$

$$= \int_{-\infty}^{\infty} x e^{tx} f_X(x)\, dx \bigg|_{t=0} = \int_{-\infty}^{\infty} x e^{0 \cdot x} f_X(x)\, dx$$

$$= \int_{-\infty}^{\infty} x f_X(x)\, dx = E(X)$$

For $r = 2$,

$$M_X^{(2)}(0) = \frac{d^2}{dt^2} \int_{-\infty}^{\infty} e^{tx} f_X(x)\, dx \bigg|_{t=0} = \int_{-\infty}^{\infty} \frac{d^2}{dt^2} e^{tx} f_X(x)\, dx \bigg|_{t=0}$$

$$= \int_{-\infty}^{\infty} x^2 e^{tx} f_X(x)\, dx \bigg|_{t=0} = \int_{-\infty}^{\infty} x^2 e^{0 \cdot x} f_X(x)\, dx \bigg|_{t=0}$$

$$= \int_{-\infty}^{\infty} x^2 f_X(x)\, dx = E(X^2)$$

Example 3.12.3

Let X be a binomial random variable. Use Theorem 3.12.1 to find $E(X)$ and Var (X).

For notational simplicity, write $q = 1 - p$, so $f_X(x) = \binom{n}{x} p^x q^{n-x}$, $x = 0, 1, \ldots, n$. From Example 3.12.1, $M_X(t) = (q + pe^t)^n$. Taking the first two derivatives of $M_X(t)$, we get

$$M_X^{(1)}(t) = n(q + pe^t)^{n-1} \cdot pe^t$$

and

$$M_X^{(2)}(t) = n(q + pe^t)^{n-1} \cdot pe^t + pe^t \cdot n(n-1)(q + pe^t)^{n-2} pe^t$$

Set $t = 0$ in both expressions. Then

$$M_X^{(1)}(0) = n(q + p)^{n-1} p$$
$$= np = E(X)$$

and

$$M_X^{(2)}(0) = n(q + p)^{n-1} \cdot p + pn(n-1)(q + p)^{n-2} p$$
$$= np + p^2 n(n-1) = E(X^2)$$

Note that the expression for $E(X)$—np—checks with what we derived earlier in Examples 3.8.4 and 3.9.1.

To find the variance of X, recall that Var $(X) = E(X^2) - (E(X))^2$. Therefore,

$$\text{Var }(X) = np + p^2 n(n-1) - (np)^2$$
$$= np - np^2$$
$$= np(1 - p)$$

the same expression we obtained in Example 3.10.3.

Question 3.12.4 Show that the moment generating function for a *Poisson* random variable having pdf

$$f_X(x) = \frac{e^{-\lambda}\lambda^x}{x!} \qquad x = 0, 1, \ldots$$

is $M_X(t) = e^{-\lambda + \lambda e^t}$. Use $M_X(t)$ to verify that $E(X) = \lambda$ and Var $(X) = \lambda$.

Question 3.12.5 Find $E(X)$ and Var (X) for the *geometric* random variable described in Question 3.12.1.

Question 3.12.6 Find $E(X^3)$ if X is an exponential random variable with pdf $f_X(x) = (1/\lambda)e^{-x/\lambda}$, $x > 0$.

Question 3.12.7 Let $C_X(t) = \ln M_X(t)$. Show that $C_X'(0) = \mu$ and $C_X''(0) = \sigma^2$.

As mentioned earlier, moment-generating functions do more than generate moments: they also characterize probability density functions and facilitate the derivation of $f_Z(z)$, where $Z = X_1 + X_2 + \cdots + X_n$. The next two theorems state the basic results we need. The first gives the uniqueness property of moment-generating functions: if X and Y have the same mgf's, then they must necessarily have the same pdf's. The second gives two results that are of particular help in manipulating moment-generating functions. The proof of Theorem 3.12.2 is beyond the scope of this text. The proof of Theorem 3.12.3 will be left as an exercise.

Theorem 3.12.2. Suppose that X and Y are random variables for which $M_X(t) = M_Y(t)$ for some interval of t's containing 0. Then $f_X = f_Y$.

Theorem 3.12.3
(a) Let X be a random variable with moment-generating function $M_X(t)$. Let $Y = aX + b$. Then

$$M_Y(t) = e^{bt} M_X(at)$$

(b) Let $X_1, X_2, \ldots, X_n$ be independent random variables with moment-generating functions $M_{X_1}(t)$, $M_{X_2}(t), \ldots$, and $M_{X_n}(t)$, respectively. Let $Y = X_1 + X_2 + \cdots + X_n$. Then

$$M_Y(t) = M_{X_1}(t) \cdot M_{X_2}(t) \cdot \cdots \cdot M_{X_n}(t)$$

For the final example in this section, we apply Theorems 3.12.2 and 3.12.3 to the transformation problem described in Example 3.5.6. Notice how much the solution is simplified when we use moment-generating functions.

Example 3.12.4

Suppose that X and Y are independent binomial random variables defined on m and n trials, respectively. Let the success probability, p, be the same for both sets of trials. Find the pdf for Z, where $Z = X + Y$.

From part (b) of Theorem 3.12.3,

$$M_Z(t) = M_X(t) \cdot M_Y(t) = (q + pe^t)^m \cdot (q + pe^t)^n$$
$$= (q + pe^t)^{m+n} \qquad (3.12.1)$$

Does the right-hand side of Equation 3.12.1 look familiar? It should. The moment-generating function for a binomial random variable defined on $m + n$ trials would be $(q + pe^t)^{m+n}$. By Theorem 3.12.2, then, it follows that the pdf for Z is binomial—specifically,

$$f_Z(z) = \binom{m+n}{z} p^z q^{m+n-z} \qquad z = 0, 1, \ldots, m+n$$

Question 3.12.8 Let X and Y be two independent exponential random variables with pdf's

$$f_X(x) = \left(\frac{1}{\lambda}\right)e^{-x/\lambda} \qquad x > 0$$

and

$$f_Y(y) = \left(\frac{1}{\lambda}\right)e^{-y/\lambda} \qquad y > 0$$

Let $Z = X + Y$. Does Z have an exponential distribution?

Question 3.12.9 Let X and Y be two independent Poisson random variables, each with parameter λ (see Question 3.12.4). Find the pdf of their sum, $X + Y$.

Question 3.12.10 The moment-generating function for a *normal* random variable having pdf

$$f_X(x) = \frac{1}{\sqrt{2\pi}\sigma}\exp\left[-\frac{1}{2}\left(\frac{x-\mu}{\sigma}\right)^2\right] \qquad -\infty < x < \infty$$

can be shown to equal

$$M_X(t) = e^{\mu t + (\sigma^2 t^2/2)}$$

Suppose that $X_1, X_2, \ldots, X_n$ are n independent normal random variables each with mean μ and variance σ^2.

(a) Find the pdf of Y, where $Y = X_1 + X_2 + \cdots + X_n$.

(b) Find the pdf of $\bar{Y}$, where $\bar{Y} = (1/n)\sum_{i=1}^{n} X_i$.

Question 3.12.11 Prove Theorem 3.12.3.

3.13 CHEBYSHEV'S INEQUALITY

If X is a continuous random variable with pdf $f_X(x)$—and if we *know* $f_X(x)$—then

$$P(a \le X \le b) = \int_a^b f_X(x)\, dx \qquad (3.13.1)$$

for any interval (a, b). If, on the other hand, X is a random variable and we know nothing about its pdf, then obviously the only probability statement we can make about X lying between a and b is the trivial one,

$$0 \le P(a \le X \le b) \le 1 \qquad (3.13.2)$$

Equation 3.13.1 and Inequality 3.13.2 represent the extremes in making pronouncements about a random variable's behavior; in this section we look at a famous result that strikes a compromise.

Suppose that $f_X(x)$ is unknown *but we do know the variance of X*. In light of Theorem 3.12.2, which says that knowing *all* the moments of X uniquely characterizes $f_X(x)$, it seems reasonable to assume that knowing *one* of those moments—namely Var (X)—should "buy" us something. And, indeed, it does. What we can derive is an upper bound for the probability that X lies outside an ϵ-neighborhood of μ. That is, we can find a nontrivial c, where $c = c(\sigma^2, \epsilon)$, such that

$$P(|X - \mu| \geq \epsilon) \leq c(\sigma^2, \epsilon)$$

for *any* random variable X having mean μ and variance σ^2.

Theorem 3.13.1 (Chebyshev's inequality). Let X be any random variable with mean μ and variance σ^2. For any $\epsilon > 0$,

$$P(|X - \mu| < \epsilon) > 1 - \frac{\sigma^2}{\epsilon^2}$$

or, equivalently,

$$P(|X - \mu| \geq \epsilon) \leq \frac{\sigma^2}{\epsilon^2}$$

Proof. In the continuous case,

$$\text{Var } (X) = \int_{-\infty}^{\infty} (x - \mu)^2 f_X(x) \, dx$$

$$= \int_{-\infty}^{\mu - \epsilon} (x - \mu)^2 f_X(x) \, dx + \int_{\mu - \epsilon}^{\mu + \epsilon} (x - \mu)^2 f_X(x) \, dx$$

$$+ \int_{\mu + \epsilon}^{\infty} (x - \mu)^2 f_X(x) \, dx$$

Omitting the nonnegative middle integral gives an inequality:

$$\text{Var } (X) \geq \int_{-\infty}^{\mu - \epsilon} (x - \mu)^2 f_X(x) \, dx + \int_{\mu + \epsilon}^{\infty} (x - \mu)^2 f_X(x) \, dx$$

$$\geq \int_{|x - \mu| \geq \epsilon} (x - \mu)^2 f_X(x) \, dx$$

$$\geq \int_{|x - \mu| \geq \epsilon} \epsilon^2 f_X(x) \, dx$$

$$= \epsilon^2 P(|X - \mu| \geq \epsilon)$$

Division by ϵ^2 completes the proof. (If X is discrete, replace the integrals with summations.)

Example 3.13.1

A student makes 100 check transactions between receiving his January and February bank statements. Rather than subtract the amounts he spends exactly, he rounds off each checkbook entry to the nearest dollar. Use Chebyshev's inequality to get an upper bound for the probability that the student's accumulated error (either positive or negative) after his 100 transactions is $5 or more.

Let X_i denote the round-off error associated with the ith transaction, $i = 1, 2, \ldots, 100$. It can be assumed (why?) that the X_i's are independent random variables and follow a uniform pdf over the interval $(-\$\frac{1}{2}, +\$\frac{1}{2})$. Therefore,

$$E(X_i) = 0$$

and

$$\text{Var } (X_i) = \frac{1}{12}$$

(see Question 3.10.4).

Let $X = X_1 + X_2 + \cdots + X_{100}$ denote the student's total accumulated error. There is no simple way to find the pdf for X but its expected value and variance can be derived easily: by the corollary to Theorem 3.9.1,

$$E(X) = \mu = 0$$

and by Theorem 3.10.3,

$$\text{Var } (X) = \frac{100}{12} = 8.3$$

Substituting $E(X)$ and Var (X) into Theorem 3.13.1, we can write

$$P(|X| \geq \$5) = P(|X - \mu| \geq \epsilon) \leq \frac{8.3}{25} = 0.33$$

There is *at most* a 33% chance, then, that the student's total will be off by as much as $5.

Example 3.13.2

The relationship between the *sample* mean of a random sample of n observations, $\bar{X} = (1/n) \sum_{i=1}^{n} X_i$, and the (unknown) *true* mean, μ, of the pdf from which the X_i's are taken is a problem of pivotal importance in statistics. Typically, our only source of information about μ is what we can glean from $\bar{X}$. Any attempt, then, to draw an inference about μ will necessarily be predicated on the probabilistic behavior of $\bar{X}$.

Historically, one of the earliest results relating the behavior of the random variable $\bar{X}$ to the constant μ was the *weak law of large numbers*, which says that $\bar{X}$ "converges" to μ as n gets large. More formally, if $X_1, X_2, \ldots, X_n$ are continuous random variables with the same mean μ and the same variance σ^2, then for every $\epsilon > 0$ and $\delta > 0$ there is an N such that

$$P(|\bar{X} - \mu| > \epsilon) < \delta$$

if $n > N$. Equivalently,

$$\lim_{n \to \infty} P(|\bar{X} - \mu| > \epsilon) = 0 \qquad (3.13.3)$$

That Equation 3.13.3 is true follows almost immediately from Theorem 3.13.1. Since $\bar{X} = (1/n)(X_1 + X_2 + \cdots + X_n)$,

$$\text{Var } (\bar{X}) = \frac{1}{n^2} \sum_{i=1}^{n} \text{Var } (X_i) = \frac{1}{n^2} n\sigma^2$$

$$= \frac{\sigma^2}{n}$$

and, of course, $E(\bar{X}) = \mu$. By Chebyshev's inequality,

$$P(|\bar{X} - \mu| > \epsilon) \leq \frac{\text{Var } (\bar{X})}{\epsilon^2}$$

But $\text{Var } (\bar{X})/\epsilon^2 = \sigma^2/n\epsilon^2$ and the latter goes to 0 as $n \to \infty$.

Question 3.13.1 Suppose that X is an exponential random variable with pdf $f_X(x) = e^{-x}$, $x > 0$.
(a) Compute the *exact* probability that X takes on a value more than two standard deviations away from its mean.
(b) Use Chebyshev's inequality to get an upper bound for the probability asked for in part (a).

Question 3.13.2 A fair die is tossed 100 times. Let X_k denote the outcome on the kth roll. Use Theorem 3.13.1 to get a lower bound for the probability that $X = X_1 + X_2 + \cdots + X_{100}$ is between 300 and 400.

Question 3.13.3 Suppose that the distribution of scores on an IQ test has mean 100 and standard deviation 16. Show that the probability of a student having an IQ above 148 or below 52 is at most $\frac{1}{9}$.

Question 3.13.4 Use Chebyshev's inequality to get a lower bound for the number of times a fair coin must be tossed in order for the probability to be at least 0.90 that the ratio of the observed number of heads to the total number of tosses be between 0.4 and 0.6?

4

Special Distributions

Lambert Adolphe Jacques Quetelet (1796–1874)

Although he maintained lifelong literary and artistic interests, Quetelet's mathematical talents led him to a doctorate from the University of Ghent and from there to a college teaching position in Brussels. In 1833 he was appointed astronomer at the Brussels Royal Observatory, after having been largely responsible for its founding. His work with the Belgian census marked the beginning of his pioneering efforts in what today would be called mathematical sociology. Quetelet was well known throughout Europe in scientific and literary circles: at the time of his death he was a member of more than 100 learned societies.

4.1 INTRODUCTION

Certain probability functions occur sufficiently often, in theoretical as well as applied contexts, to make their study both useful and rewarding. Almost invariably these functions occur in *families*, where they are indexed by one or more unknown parameters. To single out a particular member of such a family, numerical values must be assigned to each of the function's parameters. We have already encountered the one-parameter exponential family, $f_X(x) = (1/\lambda)e^{-x/\lambda}$, in this regard: recall that $f_X(x) = (1/3000)e^{-x/3000}$ was used as a model for the life distribution of light bulbs (see Example 3.2.3). Another example is the binomial distribution, $P(X = k) = \binom{n}{k}p^k(1 - p)^{n-k}$, $k = 0, 1, \ldots, n$: this is a family indexed by *two* parameters, n and p.

For the statistician, many families of probability functions merit special attention—far more than can be discussed in a text of this length. Here, only those finding particularly extensive application will be dealt with. See (67) or the Johnson and Kotz series (84, 85, 86) for a more extensive survey.

Altogether, a total of five families of probability functions—the Poisson, normal, geometric, negative binomial, and gamma—will be introduced in the next several sections. By varying the numerical values of their parameters, we will see that these five families can accommodate—that is, adequately describe—a surprisingly large number of real-world phenomenon. In certain cases it will prove enlightening to recount briefly the history of a distribution—how it came to be discovered and what its first applications were. Some of the results presented here date back more than 150 years and are now part of the bedrock of modern statistical theory.

4.2 THE POISSON DISTRIBUTION

Simeon Denis Poisson (1781–1840) was an eminent French mathematician and physicist, an academic administrator of some note, and, according to an 1826 letter from the mathematician Abel to a friend, a man who knew "how to behave with a great deal of dignity." One of Poisson's many interests was the application of probability to the law, and in 1837 he wrote *Recherches sur la Probabilité de Jugements*. This text contained a good deal of mathematics, including a limit theorem for the binomial distribution.[1] Although initially viewed as little more than a welcome approximation for hard-to-compute binomial probabilities, this particular result was destined for bigger things: it was the analytical seed out of which grew what is now one of the most important of all probability models, the Poisson distribution.

---

[1] Although credit for this theorem is given to Poisson, there is some evidence that DeMoivre may have discovered it almost a century earlier (31).

> **Theorem 4.2.1.** Let λ be a fixed number and n an arbitrary positive integer. For each nonnegative integer x,
>
> $$\lim_{n \to \infty} \binom{n}{x} p^x (1-p)^{n-x} = \frac{e^{-\lambda} \lambda^x}{x!}$$
>
> Poisson
>
> $\lambda = np$
>
> where $p = \lambda/n$.

Proof. We begin by rewriting the binomial probability in terms of λ:

$$\lim_{n \to \infty} \binom{n}{x} p^x (1-p)^{n-x} = \lim_{n \to \infty} \binom{n}{x} \left(\frac{\lambda}{n}\right)^x \left(1 - \frac{\lambda}{n}\right)^{n-x}$$

$$= \lim_{n \to \infty} \frac{n!}{x!(n-x)!} \lambda^x \left(\frac{1}{n^x}\right) \left(1 - \frac{\lambda}{n}\right)^{-x} \left(1 - \frac{\lambda}{n}\right)^n$$

$$= \frac{\lambda^x}{x!} \lim_{n \to \infty} \frac{n!}{(n-x)!} \frac{1}{(n-\lambda)^x} \left(1 - \frac{\lambda}{n}\right)^n$$

But since $[1 - (\lambda/n)]^n \to e^{-\lambda}$ as $n \to \infty$, we need only show that

$$\frac{n!}{(n-x)!(n-\lambda)^x} \to 1$$

to prove the theorem. However, note that

$$\frac{n!}{(n-x)!(n-\lambda)^x} = \frac{n(n-1) \cdots (n-x+1)}{(n-\lambda)(n-\lambda) \cdots (n-\lambda)}$$

a quantity which, indeed, tends to 1 as $n \to \infty$.

Example 4.2.1

Tables 4.2.1 and 4.2.2 give an indication of the *rate* at which

$$\binom{n}{x} p^x (1-p)^{n-x}$$

converges to

$$\frac{e^{-np}(np)^x}{x!}$$

In both cases $\lambda = np$ is equal to 1, but in the former, n is set equal to 5—in the latter, to 100. We see in Table 4.2.1 ($n = 5$) that for some x the agreement between the binomial probability and Poisson's limit is not very good. If n is as large as 100, though (Table 4.2.2), the agreement is remarkably good for all x.

TABLE 4.2.1 BINOMINAL PROBABILITIES AND POISSON LIMITS; $n = 5$ AND $p = \frac{1}{5}(\lambda = 1)$

x	$\binom{5}{x}(0.2)^x(0.8)^{5-x}$	$\dfrac{e^{-1}(1)^x}{x!}$
0	0.328	0.368
1	0.410	0.368
2	0.205	0.184
3	0.051	0.061
4	0.006	0.015
5	0.000	0.003
6+	0	0.001
	$\overline{1.000}$	$\overline{1.000}$

TABLE 4.2.2 BINOMIAL PROBABILITIES AND POISSON LIMITS; $n = 100$ AND $p = \frac{1}{100}(\lambda = 1)$

x	$\binom{100}{x}(0.01)^x(0.99)^{100-x}$	$\dfrac{e^{-1}(1)^x}{x!}$
0	0.366032	0.367879
1	0.369730	0.367879
2	0.184865	0.183940
3	0.060999	0.061313
4	0.014942	0.015328
5	0.002898	0.003066
6	0.000463	0.000511
7	0.000063	0.000073
8	0.000007	0.000009
9	0.000001	0.000001
10	0.000000	0.000000
	$\overline{1.000000}$	$\overline{0.999999}$

The next example is typical of how convenient the Poisson limit can be in approximating binomial probabilities. The data concern the incidence of childhood leukemia and whether or not the disease is contagious, a question that has been around for quite a few years but still has medical researchers baffled.

CASE STUDY 4.2.1

Leukemia is a rare form of cancer whose cause and mode of transmission remain largely unknown. While evidence abounds that excessive exposure to radiation can increase a person's risk of contracting the disease, it is at the same time true that most cases occur among persons whose history contains no such overexposure. A related issue, one maybe even more basic than the causality question, concerns the *spread* of the disease. It is safe to say that the prevailing medical opinion is that most forms of leukemia are not contagious—still, the hypothesis persists that *some* forms of the disease, particularly the childhood variety, may be. What continues to fuel this speculation are the discoveries of so-called "leukemia clusters," aggregations in time and space of unusually large numbers of cases.

To date, one of the most frequently cited leukemia clusters in the medical literature occurred during the late 1950s and early 1960s in Niles, Illinois, a suburb of Chicago (69). In the $5\frac{1}{3}$-year period from 1956 to the first four months of 1961, physicians in Niles reported a total of eight cases of leukemia among children less than 15 years of age. The number at risk (that is, the number of residents in that age range) was 7076. To assess the likelihood of that many cases occurring in such a small population, it is necessary to look first at the leukemia incidence in neighboring towns. For all of Cook county, excluding Niles, there were 1,152,695 children less than 15—and among those, 286 diagnosed cases of leukemia. That gives an average $5\frac{1}{3}$-year leukemia rate of 24.8 cases per 100,000:

$$\frac{286 \text{ cases for } 5\frac{1}{3} \text{ years}}{1,152,695 \text{ children}} \times \frac{100,000}{100,000} = 24.8 \text{ cases}/100,000 \text{ children}/5\frac{1}{3} \text{ years}$$

Now, imagine the 7076 children in Niles to be a series of $n = 7076$ (independent) Bernoulli trials, each having a probability of $p = 24.8/100,000 = 0.000248$ of contracting leukemia. The question then becomes, given an n of 7076 and a p of 0.000248, how likely is it that eight "successes" would occur? (The expected number, of course, would be $7076 \times 0.000248 = 1.75$.) Actually, for reasons that will be elaborated on in Chapter 6, it will prove more meaningful to consider the related event, eight *or more* cases occurring in a $5\frac{1}{3}$-year span. If the probability associated with the latter is very small, it could be argued that leukemia did not occur randomly in Niles and that, perhaps, contagion was a factor.

Using the binomial distribution, we can express the probability of eight or more cases as

$$P(8 \text{ or more cases}) = \sum_{k=8}^{7076} \binom{7076}{k}(0.000248)^k(0.999752)^{7076-k}$$

a computation not entirely pleasant to contemplate. With the aid of Poisson's theorem, though, the problem is trivial. By letting $\lambda = np = 7076 \times 0.000248 = 1.75$, we easily obtain 0.00049 as the Poisson limit:

$$P(8 \text{ or more cases}) \doteq \sum_{k=8}^{\infty} \frac{e^{-1.75}(1.75)^k}{k!}$$

$$= 1 - \sum_{k=0}^{7} \frac{e^{-1.75}(1.75)^k}{k!}$$

$$= 1 - 0.99951$$

$$= 0.00049$$

How close can we expect 0.00049 to be to the "true" binomial sum? Very close. Considering the accuracy of the Poisson limit when n is as small as 100 (recall Table 4.2.2), we should feel very confident here, where n is 7076.

Interpreting the 0.00049 probability is not nearly as easy as assessing its accuracy. The fact that the probability is so very small tends to denigrate the hypothesis that leukemia in Niles occurred at random. On the other hand, rare events, such as clusters, *do* happen by chance. The basic difficulty in putting the probability associated with a given cluster in any meaningful perspective is not knowing in how many similar communities leukemia did *not* exhibit a tendency to cluster. That there is no obvious way to do this is one reason the leukemia controversy is still with us.

Question 4.2.1 A chromosome mutation believed to be linked with colorblindness is known to occur, on the average, once in every 10,000 births. If 20,000 babies are born this year in a certain city, what is the probability that at least one will develop color-blindness? What is the exact probability model that applies here?

Question 4.2.2 Given 400 people, estimate the probability that 3 or more will have a birthday on July 4. Assume there are 365 days in a year and each is equally likely to be the birthday of a randomly chosen person.

Question 4.2.3 Suppose that 1% of all items in a supermarket are unmarked. A customer buys 10 items and proceeds to check out through the express lane. Estimate the probability that the customer will be delayed because one or more of the items require a price check.

Question 4.2.4 Use the Poisson approximation to calculate the probability that at most 1 person in 500 will have a birthday on Christmas. Assume there are 365 days in a year.

Question 4.2.5 Astronomers estimate that as many as 100 billion stars in the Milky Way galaxy may be encircled by planets. Let p denote the probability that any such solar system contains intelligent life. How small can p be and still give a 50–50 chance that there is intelligent life in at least one other solar system in our galaxy?

Poisson's theorem certainly sent no shock waves through the scientific community: on the contrary, it languished in mathematical limbo for over 50 years, attracting little interest and finding no real application. Its resurrection finally came in 1898 when Ladislaus von Bortkiewicz, a German professor born in Russia of Polish ancestry, wrote a monograph entitled *Das Gesetz der Kleinen Zahlen* (*The Law of Small Numbers*). Included in this work was Poisson's theorem (with due credit), possibly the first consideration of the Poisson limit as a probability distribution in its own right, and, most important, the use of the "Poisson" to model real-world phenomena. (Bortkiewicz was one of several nineteenth-century scholars—we shall read of others in Section 4.3—who pioneered the use of probability distributions as data models. Before the early nineteenth century, probability and statistics were viewed as being distinctly different subjects.)

Theorem 4.2.2. Let $p(x, \lambda)$ denote Poisson's limit,

$$p(x, \lambda) = \frac{e^{-\lambda}\lambda^x}{x!} \qquad \text{for} \quad x = 0, 1, \ldots$$

Then $p(x, \lambda)$ is a probability function. Also, if X is a random variable with pdf $p(x, \lambda)$, then $E(X) = \lambda$ and Var $(X) = \lambda$. We will call X a *Poisson random variable*.

Proof. To show that $p(x, \lambda)$ qualifies as a probability function, note, first of all, that $p(x, \lambda) \geq 0$ for all nonnegative integers x. Also, it sums to 1:

$$\sum_{x=0}^{\infty} p(x, \lambda) = \sum_{x=0}^{\infty} \frac{e^{-\lambda}\lambda^x}{x!} = e^{-\lambda} \sum_{x=0}^{\infty} \frac{\lambda^x}{x!} = e^{-\lambda} \cdot e^{\lambda} = 1$$

since $\sum_{x=0}^{\infty} \frac{\lambda^x}{x!}$ is the Taylor series expansion of e^{λ}.

The expected value for X can be gotten by direct summation:

$$E(X) = \sum_{x=0}^{\infty} x p(x, \lambda)$$

$$= \sum_{x=0}^{\infty} x \frac{e^{-\lambda}\lambda^x}{x!}$$

$$= \lambda \sum_{x=1}^{\infty} \frac{e^{-\lambda}\lambda^{x-1}}{(x-1)!}$$

$$= \lambda \sum_{y=0}^{\infty} \frac{e^{-\lambda}\lambda^y}{y!} = \lambda \cdot 1 = \lambda$$

where $y = x - 1$. The proof that Var $(X) = \lambda$ is similar and will be left as an exercise.

Question 4.2.6 Derive the variance for a Poisson random variable. [*Hint:* Consider the factorial moment, $E(X(X-1))$.]

Question 4.2.7 Show that the moment-generating function for a Poisson random variable is

$$M_X(t) = e^{-\lambda + \lambda e^t}$$

and use $M_X(t)$ to verify the formulas for $E(X)$ and Var (X).

Question 4.2.8 If the random variable X has a Poisson distribution such that $P(X = 1) = P(X = 2)$, find $P(X = 4)$.

Question 4.2.9 Suppose that X and Y are two independent random variables with moment-generating functions

$$M_X(t) = e^{-4+4e^t} \qquad \text{and} \qquad M_Y(t) = e^{-6+6e^t}$$

respectively. Let $Z = X + Y$. Find $f_Z(z)$. (*Hint*: See Question 4.2.7.)

Question 4.2.10 If X and Y are two independent random variables with moment-generating functions

$$e^{-4+4e^t} \qquad \text{and} \qquad \left(\frac{1}{3} + \frac{2}{3} e^t\right)^2$$

respectively, find $P(X = 2Y)$. (*Hint*: See Question 4.2.7.)

Among the several phenomena that Bortkiewicz successfully "fit" with the Poisson model, the one best remembered is the Prussian cavalry data described in Case Study 4.2.2. Historically, the greatest contribution of this particular application was its implicit suggestion that an incredibly wide variety of phenomena might, indeed, fall under the purview of the Poisson distribution.

CASE STUDY 4.2.2

During the latter part of the nineteenth century, Prussian officials gathered information on the hazards that horses posed to cavalry soldiers. A total of 10 cavalry corps were monitored over a period of 20 years (13). Recorded for each year and each corps was X, the number of fatalities due to kicks. Table 4.2.3 shows the empirical distribution of X for these 200 "corps-years."

TABLE 4.2.3 OBSERVED HORSE-KICK FATALITIES

x = Number of Deaths	Observed Number of Corps-Years in Which x Fatalities Occurred
0	109
1	65
2	22
3	3
4	1
	$\overline{200}$

Altogether there were 122 fatalities $[109(0) + 65(1) + 22(2) + 3(3) + 1(4)]$, meaning that the observed fatality *rate* was 122/200, or 0.61 fatalities per corps-year. For reasons that will be discussed in Chapter 5, Bortkiewicz set λ equal to 0.61 and proposed as a model for X,

$$P(X = x) = p(x, 0.61) = \frac{e^{-0.61}(0.61)^x}{x!}$$

Multiplying $p(x, 0.61)$ by 200 would then yield the *expected* number of years in which x fatalities occurred. Clearly, the agreement between the second and third columns of Table 4.2.4 is excellent. We would infer from this that the phenomenon of Prussian soldiers' being kicked to death by their horses was modeled very well by the Poisson distribution.

TABLE 4.2.4 POISSON-DISTRIBUTION
PREDICTION OF HORSE-KICK FATALITIES

x	Observed Number of Corps-Years	Expected Number of Corps-Years
0	109	108.7
1	65	66.3
2	22	20.2
3	3	4.1
4	1	0.6
	200	199.9

With Bortkiewicz's work pointing the way, not many years passed before the Poisson distribution was firmly entrenched in the repertoire of mathematicians, statisticians, and scientists of all types. What began as a simple limiting form of the binomial has since been shown to be an excellent model for the enumeration of such disparate phenomena as radioactive emission, outbreaks of wars, the positioning of stars in space, telephone calls originating at pay telephones, traffic accidents along given stretches of road, mine disasters, and even misprints in books.

We might well ask what causes this distribution to appear so often. Or, turning the question around, what can all these various phenomena, so apparently different, possibly have in common? To answer that, we need to reexamine the implications of Poisson's theorem.

Consider one of the applications just mentioned: calls dialed from a pay telephone. Suppose that the calls are placed at an average rate of λ per unit time. We imagine a time interval of length T partitioned into n subintervals, each having length T/n. We suppose that these subintervals are so small that the probability of more than one call originating during any particular subinterval is negligible. Finally, we assume that events defined on these intervals are independent—that is, a call made during the ith subinterval has no effect on the probability of a call made during the jth subinterval, $i \neq j$.

If p_n denotes the probability of a call being placed during any subinterval, the probability of exactly k calls originating during the entire interval of length T is given by the binomial

$$P(X = k) = b(k, n, p_n) = \binom{n}{k} p_n^k (1 - p_n)^{n-k}$$

It follows that the total number of calls placed during an interval of length T will be approximately np_n (the mean for a binomial random variable); on the other hand, since λ is the call rate per unit time, the number made should be approximately λT. We may assume, therefore, that $np_n \to \lambda T$ as $n \to \infty$. Equivalently,

$$b(k, n, p_n) \doteq b\left(k, n, \frac{\lambda T}{n}\right)$$

but, by Poisson's theorem,

$$b\left(k, n, \frac{\lambda T}{n}\right) \to p(k, \lambda T) \qquad (4.2.1)$$

Equation 4.2.1, then, is what all these phenomena have in common. They are all basically a series of independent, Bernoulli trials, where n is large, p is small, and the Poisson limit to the binomial is operative.

CASE STUDY 4.2.3

Among the early research efforts in radioactivity was a 1910 study of alpha emission done by the now-famous Rutherford and Geiger (142). Their experimental setup consisted of a polonium source placed a short distance from a small screen. For each of 2608 eighth-minute intervals, the two physicists recorded the number of α particles impinging on the screen (columns 1 and 2 of Table 4.2.5).

TABLE 4.2.5 RUTHERFORD–GEIGER STUDY

Number of α Particles Recorded, x	Observed Frequency	Expected Frequency
0	57	54
1	203	211
2	383	407
3	525	526
4	532	508
5	408	394
6	273	254
7	139	140
8	45	68
9	27	29
10	10	11
11+	6	6
	$\overline{2608}$	$\overline{2608}$

The particular Poisson being fitted here (column 3) is

$$P(X = x) = \frac{e^{-3.87}(3.87)^x}{x!}$$

Note that the large number of observations tends to magnify the discrepancies between the observed and expected columns. The *relative* errors, though, as measured by

$$\frac{\text{observed frequency} - \text{expected frequency}}{\text{expected frequency}}$$

are really quite small—this would be considered a very good fit.

CASE STUDY 4.2.4

In the 432 years from 1500 to 1931, war broke out somewhere in the world a total of 299 times. (By definition, a military action was a war if it either was legally declared, involved over 50,000 troops, or resulted in significant boundary realignments. To achieve greater uniformity from war to war, major confrontations were split into smaller "subwars": World War I, for example, was treated as five separate wars.) Table 4.2.6 gives the distribution of the number of years in which x wars broke out (132). The last column gives the expected frequencies based on the Poisson model, $P(X = x)$ $= e^{-0.69}(0.69)^x/x!$.

TABLE 4.2.6 OUTBREAKS OF WAR

Number of Wars, x, Beginning in a Given Year	Observed Frequency	Expected Frequency
0	223	217
1	142	149
2	48	52
3	15	12
4+	4	2
	$\overline{432}$	$\overline{432}$

One explanation for the excellent fit evident in Table 4.2.6 is that human society has a "hostility level"—as measured by the number of new wars initiated per year—that is constant. However, a similar breakdown showing the distribution of the number of wars *ending* in a given year also proved to be well described by a Poisson [see (132)]. Putting these two facts together, it could be speculated that what is constant in society is not a desire for war or a desire for peace, but a desire for change.

Question 4.2.11 The Brown's Ferry incident of 1975 focused national attention on the ever-present danger of fires breaking out in nuclear power plants. The Nuclear Regulatory Commission has estimated that with present technology there will be, on the average, one fire for every 10 nuclear-reactor years. Suppose that a certain state put two reactors on line in 1985. Assuming the incidence of fires can be described by a Poisson distribution, what is the probability that by 1990 at least two fires will have occurred? Does the Poisson assumption seem reasonable here?

Question 4.2.12   Flaws in a particular kind of metal sheeting occur at an average rate of one per 10 ft^2. What is the probability of two or more flaws in a 5-by-8-foot sheet?

Question 4.2.13   A radioactive source is metered for 2 hours, during which time the total number of alpha particles counted is 482. What is the probability that exactly three particles will be counted during the next minute? No more than three?

Question 4.2.14   A certain young lady is quite popular with her male classmates. She receives, on the average, four phone calls a night. What is the probability that tomorrow night the number of calls she receives will exceed her average by more than one standard deviation?

Question 4.2.15   Records were kept during World War II of the number of bombs falling in south London and their precise points of impact. The particular part of the city studied was divided into 576 areas, each of area $\frac{1}{4}$ km^2. The numbers of areas experiencing x hits, $x = 0, 1, 2, 3, 4, 5$, are listed (21).

Number of hits, x	Frequency
0	229
1	211
2	93
3	35
4	7
5	1
	$\overline{576}$

Compute the expected frequencies to see how well the Poisson model applies. Use

$$P(X = x) = \frac{e^{-0.93}(0.93)^x}{x!} \qquad x = 0, 1, 2, \ldots$$

Question 4.2.16   In a certain published book of 520 pages, 390 typographical errors occur. What is the probability that one page, selected randomly by the printer as a sample of her work, will be free from errors?

Question 4.2.17   Let X be a Poisson random variable with parameter λ. Show that the probability X is even is $\frac{1}{2}(1 + e^{-2\lambda})$.

Question 4.2.18   If X is Poisson with parameter λ and the conditional pdf of Y given x is binomial with parameters x and p, show that the marginal pdf of Y is Poisson with parameter λp.

Question 4.2.19   Let Y denote the interval length between consecutive occurrences of an event described by a Poisson distribution. If the Poisson events are occurring at a rate of λ per unit time, show that

$$f_Y(y) = \lambda e^{-\lambda y} \qquad y > 0$$

Question 4.2.20 Among the most famous of all meteor showers are the Perseids, which occur each year in early August. In some areas the frequency of visible Perseids can be as high as 40 per hour. Assume that the sighting of these meteors is a Poisson event. Use Question 4.2.19 to find the probability that an observer will have to wait at least 5 min between successive sightings.

Question 4.2.21 Suppose that in a certain country commercial airplane crashes occur at the rate of 2.5 per year. Assuming the frequency of crashes per year to be a Poisson random variable, find the probability that four or more crashes will occur next year. Also, find the probability that the next two crashes will occur within three months of one another. (*Hint:* See Question 4.2.19.)

Question 4.2.22 Let X be a Bernoulli random variable where the probability of a success is equal to p. Suppose Y is such that $f_{Y|0}$ is Poisson with parameter λ and $f_{Y|1}$ is Poisson with parameter μ. Show that

$$f_Y(y) = pe^{-\lambda}\frac{\lambda^y}{y!} + (1-p)e^{-\mu}\frac{\mu^y}{y!}$$

Question 4.2.23 Suppose that X is Poisson with parameter λ and Y is Poisson with parameter μ. If X and Y are independent, show that $X + Y$ is Poisson with parameter $\lambda + \mu$.

Question 4.2.24 Let X and Y be two independent random variables, each described by a Poisson pdf with $\lambda = 2$. Let $Z = X + Y$. Show that the conditional pdf of X given z is binomial.

Question 4.2.25 Hasselblad (66) has tabulated the number of death notices of women over 80 appearing in the London *Times* each day for three years. These data, given below, cannot be modeled by a Poisson density. However, the death rate is higher during the winter months, so the density given in Question 4.2.22 might provide a suitable model. Good choices for the parameters can be obtained from the data using standard statistical techniques: $p = 0.29$, $\lambda = 1.1$, and $\mu = 2.6$. Using these parameters and f_Y from Question 4.2.22, find the expected frequencies.

Number of deaths	Observed frequency
0	162
1	267
2	271
3	185
4	111
5	61
6	27
7	8
8	3
9	1
	$\overline{1096}$

4.3 THE NORMAL DISTRIBUTION

The limit proposed by Poisson was not the only, or even the first, approximation to the binomial: DeMoivre had already derived a quite different one in his 1718 tract, *Doctrine of Chances*. Like Poisson's work, DeMoivre's theorem did not initially attract the attention it deserved; it did catch the eye of Laplace, though, who generalized it and included it in his influential *Théorie Analytique des Probabilités*, published in 1812.

> **Theorem 4.3.1** (DeMoivre–Laplace). Let X be a binomial random variable defined on n independent trials each having success probability p. For any numbers c and d,
>
> $$\lim_{n \to \infty} P\left(c < \frac{X - np}{\sqrt{npq}} < d\right) = \frac{1}{\sqrt{2\pi}} \int_c^d e^{-x^2/2}\, dx$$

Proof. We will prove a generalization of this result in Chapter 7. For a proof of Theorem 4.3.1 itself, see (41).

Comment

The integral in Theorem 4.3.1 cannot be expressed in closed form and must be approximated by numerical techniques. As early as 1783, Laplace recognized the pressing need for its tabulation and suggested an appropriate computational formula. The first substantial tables were due to the French physicist, C. Kramp. They were published in 1799. A table of

$$\frac{1}{\sqrt{2\pi}} \int_{-\infty}^d e^{-x^2/2}\, dx$$

for values of d ranging from -3.90 to $+3.90$ is given in Table A.1 in the Appendix.

Case Study 4.3.1 shows how the DeMoivre–Laplace theorem is typically put into practice. As with the leukemia clustering problem, we will assess the statistical significance of an observed event—this time the number of correct guesses in an ESP experiment—by computing the probability of getting, by chance, the number actually observed, *or more.*

CASE STUDY 4.3.1

Research in extrasensory perception has ranged from the slightly unconventional to the downright bizarre. Toward the latter part of the nineteenth century and even well into the twentieth century, much of what was done involved spiritualists and mediums. But beginning around 1910, experimenters moved out of the seance parlors and into the laboratory, where they began setting up controlled studies that could be analyzed statistically. In 1938, Pratt and Woodruff, working out of Duke University, did an experiment that became a prototype for an entire generation of ESP research (64).

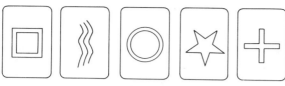

Figure 4.3.1 ESP symbols.

The investigator and a subject sat at opposite ends of a table. Between them was a screen with a large gap at the bottom. Five blank cards, visible to both participants, were placed side by side on the table beneath the screen. On the subject's side of the screen one of the standard ESP symbols (see Figure 4.3.1) was hung over each of the blank cards.

The experimenter shuffled a deck of ESP cards, picked up the top one, and concentrated on it. The subject tried to guess its identity: if he thought it was a *circle*, he would point to the blank card on the table that was beneath the circle card hanging on his side of the screen. The procedure was then repeated. Altogether, a total of 32 subjects, all students, took part in the experiment. They made a total of 60,000 guesses—and were correct 12,489 times.

With five denominations involved, the probability of a subject's making a correct identification just by chance was $\frac{1}{5}$. Assuming a binomial model, the expected number of correct guesses would be $60,000 \times \frac{1}{5}$, or 12,000. The question is, how "near" to 12,000 is 12,489? Should we write off the observed excess of 489 as nothing more than luck, or can we conclude that ESP has been demonstrated?

To effect a resolution, here, between the conflicting "luck" and "ESP" hypotheses, we need to compute the probability of the students' getting 12,489 or more correct answers *under the presumption that $p = \frac{1}{5}$*. Only if that probability is very small, can 12,489 be construed as evidence in support of ESP.

Let the random variable X denote the number of correct responses in 60,000 tries. Then

$$P(X \geq 12,489) = \sum_{x=12,489}^{60,000} \binom{60,000}{x}\left(\frac{1}{5}\right)^x \left(\frac{4}{5}\right)^{60,000-x} \tag{4.3.1}$$

Now, rather than compute the 47,512 binomial probabilities indicated in Equation 4.3.1 (or the 12,489 making up its complement), we will appeal to the DeMoivre–Laplace theorem:

$$P(X \geq 12,489) = P\left(\frac{X - np}{\sqrt{npq}} \geq \frac{12,489 - 60,000(1/5)}{\sqrt{60,000(1/5)(4/5)}}\right)$$

$$= P\left(\frac{X - np}{\sqrt{npq}} \geq 4.99\right)$$

$$\doteq \frac{1}{\sqrt{2\pi}} \int_{4.99}^{\infty} e^{-x^2/2} \, dx$$

$$= 0.0000003$$

this last value being obtained from a more extensive version of Table A.1 in the Appendix.

Here, the fact that $P(X \geq 12{,}489)$ is so extremely small makes the "luck" hypothesis ($p = \frac{1}{5}$) untenable. It would appear that something other than chance had to be responsible for the occurrence of so many correct guesses. Still, it does not follow that ESP has necessarily been demonstrated. Flaws in the experimental setup as well as errors in reporting the scores could have inadvertently produced what appears to be a statistically significant result. Suffice it to say that a great many scientists remain highly skeptical of ESP research in general and of the Pratt–Woodruff experiment in particular. [For a more thorough critique of the data we have just described, see (40).]

Question 4.3.1 Use Table A.1 in the Appendix to evaluate each of the following.

(a) $\dfrac{1}{\sqrt{2\pi}} \displaystyle\int_{1.25}^{2.50} e^{-x^2/2}\, dx$

(b) $\dfrac{1}{\sqrt{2\pi}} \displaystyle\int_{1.17}^{\infty} e^{-x^2/2}\, dx$

Question 4.3.2 Suppose that X is a binomial random variable with $n = 100$ and $p = 0.4$. Use Theorem 4.3.1 to approximate the probability that X is between 38 and 43.

Question 4.3.3 Suppose that X is a binomial random variable defined on n independent trials, each trial having a success probability of $p = \frac{1}{2}$. Find

$$P\left(-1 < \frac{X - n(1/2)}{\sqrt{n(1/2)(1/2)}} < +1 \right)$$

for $n = 2, 5, 8, 10, 12,$ and 15, and compare the results with the DeMoivre–Laplace limit.

Question 4.3.4 There is a theory embraced by certain parapsychologists that hypnosis can bring out a person's ESP. In an experiment designed to test that hypothesis (19), 15 students were hypnotized and then asked to guess the identity of an ESP card on which another person (also hypnotized) was concentrating. Each of the students made 100 guesses during the course of the experiment. The total number correct (out of 1500) was 326. Write the binomial expression giving the probability of the event $X \geq 326$ and then use the DeMoivre–Laplace theorem to approximate the probability numerically. What would your conclusion be regarding the efficacy of hypnosis as a technique for improving a person's ESP?

Question 4.3.5 Airlines A and B offer identical service on two flights leaving at the same time (meaning the probability of a passenger choosing either is $\frac{1}{2}$). Suppose that both airlines are competing for the same pool of 400 potential passengers. Airline A sells tickets to everyone who requests one, and the capacity of its plane is 230. Approximate the probability that airline A overbooks.

Question 4.3.6 A random sample of some 747 obituaries published in Salt Lake City newspapers revealed that 46% of the decedents died in the 3-month period following their birthdays (113). Assess the statistical significance of that finding by estimating the probability that 46% *or more* would die in that particular interval if deaths occurred randomly throughout the year. What would you conclude on the basis of your answer?

Question 4.3.7 A certain type of seed has a probability of 0.8 of germinating. In a package of 100 seeds, what is the probability that at least 75% will germinate?

Question 4.3.8 A political poll of some 200 registered voters shows the Democratic candidate in a gubernatorial race is favored by 110 voters and the Republican candidate by 90, a margin of 10%. But suppose sentiment for the two candidates in the general population is evenly split. Let Y denote the margin percentage in the polls. Use the DeMoivre–Laplace theorem to estimate the probability that Y is as large or larger than 0.10.

Question 4.3.9 A basketball team has a 70% foul-shooting percentage.
(a) Write a formula for the exact probability that out of their next 100 free throws, they will make between 75 and 80, inclusive. Assume the throws are independent.
(b) Using an appropriate approximation, estimate the probability asked for in part (a).

We saw in Theorem 4.2.2 that Poisson's limit to the binomial was, itself, a probability distribution. The same is true of the DeMoivre–Laplace limit. To verify this, it is necessary to show that

$$\frac{1}{\sqrt{2\pi}} \int_{-\infty}^{\infty} e^{-x^2/2} \, dx = 1$$

[Clearly, $(1/\sqrt{2\pi})e^{-x^2/2}$ is greater than or equal to 0 for all x.] The proof is not obvious and relies on the trick of changing to polar coordinates to show that the *square* of the integral is 1.

Theorem 4.3.2.

$$\frac{1}{\sqrt{2\pi}} \int_{-\infty}^{\infty} e^{-x^2/2} \, dx = 1$$

Proof. The theorem will be proved if we can show that

$$\int_{-\infty}^{\infty} e^{-x^2/2} \, dx \cdot \int_{-\infty}^{\infty} e^{-y^2/2} \, dy = 2\pi \qquad (4.3.2)$$

To begin, note that the product of the integrals in Equation 4.3.2 is equal to a double integral:

$$\int_{-\infty}^{\infty} e^{-x^2/2} \, dx \cdot \int_{-\infty}^{\infty} e^{-y^2/2} \, dy = \int_{-\infty}^{\infty} \int_{-\infty}^{\infty} e^{-(1/2)(x^2+y^2)} \, dx \, dy$$

Let $x = r \cos \theta$ and $y = r \sin \theta$, making $dx \, dy = r \, dr \, d\theta$. Then

$$\int_{-\infty}^{\infty} \int_{-\infty}^{\infty} e^{-(1/2)(x^2+y^2)} \, dx \, dy = \int_{0}^{2\pi} \int_{0}^{\infty} e^{-r^2/2} \, r \, dr \, d\theta$$

$$= \int_{0}^{\infty} re^{-r^2/2} \, dr \cdot \int_{0}^{2\pi} d\theta$$

$$= -e^{-r^2/2} \Big|_{0}^{\infty} \cdot \theta \Big|_{0}^{2\pi}$$

$$= 1(2\pi) = 2\pi$$

Definition 4.3.1 introduces a critical generalization of the density we have just looked at. Not having any parameters to lend it flexibility, $f_X(x) = (1/\sqrt{2\pi})e^{-x^2/2}$ would be of little practical value as a model for data. Transformed in the manner stated in Definition 4.3.1, though, it suddenly becomes the most useful of *all* probability models.

Definition 4.3.1. A random variable Z is said to have a *standard normal distribution* if its pdf is given by

$$f_Z(z) = \frac{1}{\sqrt{2\pi}} \, e^{-z^2/2} \qquad -\infty < z < \infty$$

A random variable X is said to have a *normal distribution with parameters μ and σ* if

$$f_X(x) = \frac{1}{\sqrt{2\pi}\,\sigma} \, e^{-(1/2)[(x-\mu)/\sigma]^2} \qquad -\infty < x < \infty$$

We will denote such a random variable by the symbol $N(\mu, \sigma^2)$. [It follows that a standard normal random variable would be denoted $N(0, 1)$.]

Comment

The two pdf's of Definition 4.3.1 are related through Theorem 3.5.1. If X is a $N(\mu, \sigma^2)$ random variable, and we define the transformation

$$Z = \frac{X - \mu}{\sigma}$$

then Z's pdf will be the $f_Z(z)$ for the standard normal. The symbol $\sim$ is sometimes used as an abbreviation for "is distributed as." Using that notation, we can summarize the effect of the Z *transformation* quite easily: If $X \sim N(\mu, \sigma^2)$, then $(X - \mu)/\sigma \sim N(0, 1)$.

Mathematicians and scientists were quick to recognize the usefulness of the two-parameter normal. One of its first applications was due to Gauss, who used it to model observational errors in astronomy. [It appeared in this context in Gauss' celebrated work of 1809, *Theoria Motus Corporum Coelestium in Sectionibus Conicis Solem Ambientium* (*Theory of the Motion of Heavenly Bodies Moving about the Sun in Conic Sections*).] By the 1830s, the normal distribution was in quite general use by physicists as well. It was Quetelet, though, who focused attention on the *scope* of the normal and, in the process, established its primacy among empirical distributions.

Lambert Adolphe Jacques Quetelet (1796–1874) was a Belgian scholar of broad training and interests—mathematics, statistics, astronomy, and writing poetry, just to name a few. In preparation for establishing the royal observatory at Brussels, he studied probability under Laplace in Paris during the winter of 1823–1824. It was his passion for collecting data, though, that made him such an effective champion for the normal distribution: he was able to demonstrate that

$$f_X(x) = \frac{1}{\sqrt{2\pi}\,\sigma}\, e^{-(1/2)[(x-\mu)/\sigma]^2}$$

described an incredibly wide range of sociological and anthropological phenomena. And being a member of more than one hundred scholarly societies allowed the peripatetic Quetelet to share his enthusiasm for the normal with the entire European scientific community.

Another scholar of the nineteenth century often associated with the normal distribution is the noted English biologist Sir Francis Galton. That Galton was influenced by the work of Quetelet is readily apparent from a statement he made in 1889 (52):

> I need hardly remind the reader that the Law of Error upon which these Normal Values are based, was excogitated for the use of astronomers and others who are concerned with extreme accuracy of measurement, and without the slightest idea until the time of Quetelet that they might be applicable to human measures. But Errors, Differences, Deviations, Divergencies, Dispersions, and individual Variations, all spring from the same kind of causes. Objects that bear the same name, or can be described by the same phrase, are thereby acknowledged to have common points of resemblance, and to rank as members of the same species . . . .

Before discussing one of Quetelet's applications of the normal distribution, it would be well to clarify the roles played by μ and σ. As the notation would suggest, one is related to the distribution's location; the other, to its dispersion.

Theorem 4.3.3. Let the random variable X have a normal distribution with parameters μ and σ. Then

$$E(X) = \mu \qquad \text{and} \qquad \text{Var}\,(X) = \sigma^2$$

Proof. By virtue of Theorems 3.9.1 and 3.10.2, it suffices to show that

$$E\left(\frac{X-\mu}{\sigma}\right) = 0 \qquad \text{and} \qquad \text{Var}\left(\frac{X-\mu}{\sigma}\right) = 1$$

The statement about μ follows immediately, since the integrand for the expected value is an odd function:

$$E\left(\frac{X - \mu}{\sigma}\right) = \int_{-\infty}^{\infty} y \cdot \frac{1}{\sqrt{2\pi}}\, e^{-y^2/2}\, dy = 0$$

To prove the second part of the theorem, we note that

$$\text{Var}\left(\frac{X - \mu}{\sigma}\right) = E\left(\left(\frac{X - \mu}{\sigma}\right)^2\right) - 0^2$$

$$= \int_{-\infty}^{\infty} y^2 \frac{1}{\sqrt{2\pi}}\, e^{-y^2/2}\, dy \qquad (4.3.3)$$

After an integration by parts, the right-hand side of Equation 4.3.3 reduces to

$$\int_{-\infty}^{\infty} y^2 \frac{1}{\sqrt{2\pi}}\, e^{-y^2/2}\, dy = \frac{1}{\sqrt{2\pi}}\, ye^{-y^2/2}\Big|_{-\infty}^{\infty} + \frac{1}{\sqrt{2\pi}}\int_{-\infty}^{\infty} e^{-y^2/2}\, dy$$

$$= 0 + 1 = 1$$

and the result follows.

CASE STUDY 4.3.2

Among Quetelet's many anthropometric applications of the normal was one where he fitted the distribution to the chest measurements of 5738 Scottish soldiers. Table 4.3.1 is a facsimile of Quetelet's data as they appeared in an

TABLE 4.3.1 QUETELET'S CHEST MEASUREMENTS OF SCOTTISH SOLDIERS

Measures de la poitrine	Nombre d'hommes	Nombre Proportionnel	Probabilité d'après L'observation	Rang dans La table	Rang d'après le calcul.	Probabilité d'après La table	Nombre d'observations calculé
Pouces							
33	3	5	0.5000			0.5000	7
34	18	31	0.4995	52	50	0.4993	29
35	81	141	0.4964	42.5	42.5	0.4964	110
36	185	322	0.4823	33.5	34.5	0.4854	323
37	420	732	0.4501	26.0	26.5	0.4531	732
38	749	1305	0.3769	18.0	18.5	0.3799	1333
39	1073	1867	0.2464	10.5	10.5	0.2466	1838
			0.0597	2.5	2.5	0.0628	
40	1079	1882	0.1285	5.5	5.5	0.1359	1987
41	934	1628	0.2913	13	13.5	0.3034	1675
42	658	1148	0.4061	21	21.5	0.4130	1096
43	370	645	0.4706	30	29.5	0.4690	560
44	92	160	0.4866	35	37.5	0.4911	221
45	50	87	0.4953	41	45.5	0.4980	69
46	21	38	0.4991	49.5	53.5	0.4996	16
47	4	7	0.4998	56	61.8	0.4999	3
48	1	2	0.5000			0.5000	1
	5738	1,0000					1,0000

1846 book of letters to the Duke of Saxe-Cobourg and Gotha (127). Column 1 lists, in inches, the range of possible chest measurements. Column 2 gives the corresponding observed frequencies, which are then divided by 5738 and reappear in column 3 as *relative* frequencies (the decimal points are omitted).

In the last column are the expected probabilities, assuming a normal distribution. Specifically, the parameter μ was estimated to be 39.8 inches and σ, 2.05 inches, so these final entries were based on the model,

$$f_X(x) = \frac{1}{\sqrt{2\pi}\,(2.05)}\, e^{-(1/2)[(x-39.8)/2.05]^2}$$

The expected probability, then, of a Scottish soldier's having a chest measurement, x, of, say, 42 inches would be gotten by integration:

$$P(X \text{ "equals" } 42) = P(41.5 < X < 42.5)$$

$$= P\left(\frac{41.5 - 39.8}{2.05} < \frac{X - \mu}{\sigma} < \frac{42.5 - 39.8}{2.05}\right)$$

$$= P\left(0.829 < \frac{X - \mu}{\sigma} < 1.317\right)$$

$$= \frac{1}{\sqrt{2\pi}} \int_{0.829}^{1.317} e^{-y^2/2}\, dy$$

$$= 0.1096$$

this latter figure being the tenth entry appearing in the last column.

Notice how closely the calculated probabilities in the last column agree with the empirical probabilities in column 3. The normal is clearly a very appropriate model for this particular measurement.

Question 4.3.10 The Stanford–Binet IQ test is scaled to give a mean score of 100 with a standard deviation of 16. Suppose that children having IQs of less than 80 or greater than 145 are deemed to need special attention. Given a population of 2000 children, what will be the expected demand for these additional services?

Question 4.3.11 A criminologist has developed a questionnaire for predicting whether a teenager will become a delinquent. Scores on the questionnaire can range from 0 to 100, with higher values reflecting a presumably greater criminal tendency. As a rule of thumb, the criminologist decides to classify a teenager as a potential delinquent if his or her score exceeds 75. The criminologist has already tested the questionnaire on a large sample of teenagers, both delinquent and nondelinquent. Among those considered nondelinquent, scores were normally distributed with a mean of 60 and a standard deviation of 10. Among those considered delinquent, scores were normally distributed with a mean of 80 and a standard deviation of 5.
(a) What proportion of the time will the criminologist misclassify a nondelinquent as a delinquent? Treat the scores as a continuous variable.

(b) What proportion of the time will the criminologist misclassify a delinquent as a non-delinquent? Treat the scores as a continuous variable.

Question 4.3.12 Assume that the number of miles a driver gets on a set of radial tires is normally distributed with a mean of 30,000 miles and a standard deviation of 5000 miles. Would the manufacturer of these tires be justified in claiming that 90% of all drivers will get at least 25,000 miles?

Question 4.3.13 The diameter of the connecting rod in the steering mechanism of a certain foreign sports car must be between 1.480 and 1.500 cm, inclusive, to be usable. The distribution of connecting-rod diameters produced by the manufacturing process is normal with a mean of 1.495 cm and a standard deviation of 0.005 cm. What percentage of rods will have to be scrapped?

Question 4.3.14 Show that the moment-generating function for the normal distribution with parameters μ and σ is

$$M_X(t) = e^{\mu t + \sigma^2 t^2 / 2}$$

Compute $M'_X(0)$ and $M''_X(0)$ and verify Theorem 4.3.3.

Question 4.3.15 If $e^{3t + 8t^2}$ is the moment-generating function for the random variable X, find $P(-1 \leq X \leq 9)$. (*Hint*: See Question 4.3.14.)

Question 4.3.16 Suppose that X is $N(\mu, \sigma^2)$. Use the properties of moment-generating functions to show that

$$Z = \frac{X - \mu}{\sigma}$$

is $N(0, 1)$.

Applications of statistical reasoning are not always confined to laboratory research or field surveys, nor do they always appear in technical journals. Case Study 4.3.3 is a case in point.

CASE STUDY 4.3.3
___

The following letter appeared in a well-known advice-to-the-lovelorn column (163):

Dear Abby: You wrote in your column that a woman is pregnant for 266 days. Who said so? I carried my baby for ten months and five days, and there is no doubt about it because I know the exact date my baby was conceived. My husband is in the Navy and it couldn't have possibly been conceived any other time because I saw him only once for an hour, and I didn't see him again until the day before the baby was born.

I don't drink or run around, and there is no way this baby isn't his, so please print a retraction about the 266-day carrying time because otherwise I am in a lot of trouble.

San Diego Reader

While the full implications of San Diego Reader's plight are beyond the scope of this text, it *is* possible to assess quantitatively the statistical likelihood of a pregnancy being 310 days long ($= 10$ months and 5 days). By the same reasoning used in Case Study 4.2.1, this would be done by computing the probability that, by chance alone, a pregnancy will be 310 days long *or longer*. That is, if Y denotes an arbitrary pregnancy duration, we want to compute $P(Y \geq 310)$—the smaller this probability is, the less credibility San Diego Reader has. Of course, despite our best-intentioned efforts to come up with a relevant and objective probability, the interpretation of that probability will remain somewhat subjective, since the particular value of $P(Y \geq 310)$ that delineates what we choose to accept as the truth from what we reject as nontruth is necessarily arbitrary.

According to well-documented norms, the mean and standard deviation for Y are 266 days and 16 days, respectively. If it can be assumed that the distribution of Y is normal,

$$P(Y \geq 310) = P\left(\frac{Y - 266}{16} \geq \frac{310 - 266}{16}\right) = P(Z \geq 2.75)$$

But, from Table A.1 in the Appendix,

$$P(Z \geq 2.75) = 1 - 0.9970$$

$$= 0.003$$

Figure 4.3.2 shows the two equivalent areas involved, the original one under the $N(266, (16)^2)$ distribution and the transformed one under the $N(0, 1)$.

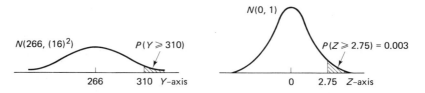

Figure 4.3.2 Equivalent areas under two normal distributions.

It is left to the reader to decide how the 0.003 should be interpreted in this particular setting. [Insurance companies sometimes use probability computations of this sort in deciding whether to reimburse married couples for maternity costs. Certain policies contain a clause to the effect that if Y is too much *less* than 266 (as measured from the couple's date of marriage) the company can refuse to pay.]

Example 4.3.1 is typical of an entire class of problems where tables of the standard normal distribution have to be used *backward*. Instead of finding the area associated with an interval, we need to find the interval associated with an area.

Example 4.3.1

Mensa is an international society whose membership is limited to persons having IQs above the general population's 98th percentile. It is well known that the average IQ for the general population is 100, the standard deviation is 16, and the distribution, itself, is normal (recall Question 4.3.10). What, then, is the *lowest* IQ that will qualify a person to belong to Mensa?

Let Y denote the IQ of a random person. By definition, the lowest qualifying value for Y, call it Y^*, must satisfy the equation

$$P(Y \geq Y^*) = 0.02$$

Standardizing Y as prescribed by the corollary gives

$$P(Y \geq Y^*) = P\left(\frac{Y - 100}{16} \geq \frac{Y^* - 100}{16}\right) = P\left(Z \geq \frac{Y^* - 100}{16}\right) = 0.02$$

From Table A.1 in the Appendix, though, the 98th percentile of the $N(0, 1)$ is 2.05:

$$P(Z \geq 2.05) = 0.02$$

Therefore, setting the "two" 98th percentiles equal gives

$$Y^* - \frac{100}{16} = 2.05$$

from which it follows that a person must have an IQ of at least 133 to be a card-carrying Mensan:

$$Y^* = 100 + 2.05(16) = 133$$

Question 4.3.17 Dental structure provides an effective criterion for classifying certain fossils. Not long ago, a baboon skull of unknown origin was discovered in a cave in Angola (106); the length of its third molar was 9.0 mm. Speculation arose that the baboon in question might be a "missing link" and belong to the genus *Papio*. Members of that genus have third molars measuring, on the average, 8.18 mm long with a standard deviation of 0.47 mm. Quantify the significance of the 9.0-mm molar. What would your inference be regarding the baboon's lineage?

Question 4.3.18 A college professor teaches Chemistry 101 each fall to a large class of first-year students. For tests, she uses standardized exams that she knows from past experience produce bell-shaped grade distributions with a mean (μ) of 70 and a standard deviation (σ) of 12. Her philosophy of grading is to impose standards that will yield, in the long run, 14% A's, 20% B's, 32% C's, 20% D's, and 14% F's. Where should the cutoff be between the A's and the B's? Between the B's and the C's?

Question 4.3.19 The systolic blood pressure of 18-year-old women is normally distributed with a mean of 120 mm Hg and a standard deviation of 12 mm Hg. What is the probability that the blood pressure of a randomly selected 18-year-old woman will be greater than 150? Less than 115? Between 110 and 130?

Question 4.3.20 At a certain Ivy League college, the average score of a first-year student on the verbal part of the SAT is 565, with a standard deviation of 75. If the distribution of scores is normal, what proportion of that school's first-year students have verbal SAT's over 650? Under 500?

Question 4.3.21 Find the 60th percentile for the verbal SAT scores described in Question 4.3.20. If there are 1000 first-year students at that particular school, what is the standard deviation of the number of students scoring between 570 and 590, inclusive?

Question 4.3.22 It is estimated that 80% of all 18-year-old women have weights ranging from 103.5 to 144.5 lb. Assuming the weight distribution can be adequately approximated by a normal curve and assuming that 103.5 and 144.5 are equal distances from the average weight μ, calculate σ.

Question 4.3.23 A recent year's traffic-death rates (fatalities per 100 million motor-vehicle miles) are given for each of the 50 states (102).

Ala.	6.4	La.	7.1	Ohio	4.5
Alaska	8.8	Maine	4.6	Okla.	5.0
Ariz.	6.2	Mass.	3.5	Oreg.	5.3
Ark.	5.6	Md.	3.9	Pa.	4.1
Calif.	4.4	Mich.	4.2	R.I.	3.0
Colo.	5.3	Minn.	4.6	S.C.	6.5
Conn.	2.8	Miss.	5.6	S.Dak.	5.4
Del.	5.2	Mo.	5.6	Tenn.	7.1
Fla.	5.5	Mont.	7.0	Tex.	5.2
Ga.	6.1	N.C.	6.2	Utah	5.5
Hawaii	4.7	N.Dak.	4.8	Va.	4.5
Idaho	7.1	Nebr.	4.4	Vt.	4.7
Ill.	4.3	Nev.	8.0	W.Va.	6.2
Ind.	5.1	N.H.	4.6	Wash.	4.3
Iowa	5.9	N.J.	3.2	Wis.	4.7
Kans.	5.0	N.Mex.	8.0	Wyo.	6.5
Ky.	5.6	N.Y.	4.7		

Make a histogram of these observations using as classes 2.0–2.9, 3.0–3.9, and so on. Assuming the distribution of rates can be approximated by a normal distribution, determine the expected frequencies for each of the classes. Let μ and σ have the values 5.3 and 1.3, respectively.

Question 4.3.24 The army is developing a new missile and is concerned about its precision. By observing points of impact, launchers can adjust the missile's initial trajectory, thereby controlling the mean of its impact distribution. If the standard deviation of that distribution is too large, though, the missile will be ineffective. Suppose the Pentagon requires that at least 95% of the missiles must fall within $\frac{1}{8}$ mile of the target when the missiles are aimed properly. What is the maximum allowable standard deviation for the impact distribution? (Assume that the latter is normal.)

4.4 THE GEOMETRIC DISTRIBUTION

Given a series of independent Bernoulli trials, we are accustomed to thinking of n and p as fixed, and x, the number of successes, as the (binomial) random variable. Suppose that the problem is turned around, though, and the question is asked, how many trials will be required in order to achieve the first success? Put this way, n is the random variable and x is what is fixed. Clearly, the first success will occur on the very first trial with probability p, on the second trial with probability $(1 - p)p$, and so on. In general, the probability function for N, the trial number of the first success, will be

$$f_N(n) = p(1 - p)^{n-1} = pq^{n-1} \qquad n = 1, 2, \ldots$$

This is called the *geometric distribution* (with parameter p).

Intuitively, the average number of trials required for the first success to appear—that is, $E(N)$—should be inversely proportional to p. The next theorem shows that, in fact, $E(N)$ is *exactly* $1/p$.

Theorem 4.4.1. Let N be a geometric random variable with parameter $p \, [= P(\text{success at any trial})]$. Then

$$E(N) = \frac{1}{p} \qquad \text{and} \qquad \text{Var}(N) = \frac{q}{p^2}$$

Proof. By definition,

$$E(N) = \sum_{n=1}^{\infty} npq^{n-1} = p \sum_{n=1}^{\infty} nq^{n-1} \qquad (4.4.1)$$

This last sum is not immediately recognizable but it can be evaluated by considering it as a function of q and taking the derivative of its integral. Set

$$h(q) = \sum_{n=1}^{\infty} nq^{n-1}$$

Then

$$\int h(q) \, dq = \sum_{n=1}^{\infty} q^n = \sum_{n=0}^{\infty} q^n - q^0$$

$$= \frac{1}{1 - q} - 1$$

$$= \frac{q}{1 - q}$$

By using the rule for the derivative of a quotient, we get that $h(q) = 1/p^2$:

$$h(q) = \frac{d}{dq} \int h(q) \, dq = \frac{(1 - q) - (-q)}{(1 - q)^2} = \frac{1}{(1 - q)^2} = \frac{1}{p^2}$$

Substituting $1/p^2$ into Equation 4.4.1 gives the first part of the theorem,

$$E(N) = p\left(\frac{1}{p^2}\right) = \frac{1}{p}$$

The proof that Var $(N) = q/p^2$ is left as an exercise.

Example 4.4.1

A grocery store is sponsoring a sales promotion where the cashiers give away one of the letters A, E, L, S, U, and V for each purchase. If a customer collects all six (spelling VALUES), he or she gets $10 worth of groceries free. What is the expected number of trips to the store a customer needs to make in order to get a complete set? Assume the different letters are given away randomly.

Let X_i denote the number of purchases necessary to get the ith different letter, $i = 1, 2, \ldots, 6$, and let X denote the number of purchases necessary to qualify for the $10. Then $X = X_1 + X_2 + \cdots + X_6$ (see Figure 4.4.1). Clearly,

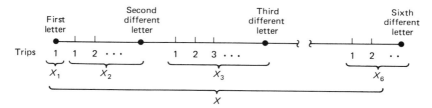

Figure 4.4.1

X_1 equals 1 with probability 1, so $E(X_1) = 1$. Having received the first letter, the chances of getting a different one are $\frac{5}{6}$ for each subsequent trip to the store. Therefore,

$$f_{X_2}(k) = P(X_2 = k) = \left(\frac{1}{6}\right)^{k-1} \frac{5}{6} \qquad k = 1, 2, \ldots$$

That is, X_2 is a geometric random variable with parameter $p = \frac{5}{6}$. By Theorem 4.4.1, $E(X_2) = \frac{6}{5}$. Similarly, the chances of getting a *third* different letter are $\frac{4}{6}$ (for each purchase), so

$$f_{X_3}(k) = P(X_3 = k) = \left(\frac{2}{6}\right)^{k-1} \frac{4}{6} \qquad k = 1, 2, \ldots$$

and $E(X_3) = \frac{6}{4}$. Continuing in this fashion, we can find the remaining $E(X_i)$'s. It follows that a customer will have to make 14.7 trips to the store, on the average, to collect a complete set of six letters:

$$E(X) = \sum_{i=1}^{6} E(X_i)$$

$$= 1 + \frac{6}{5} + \frac{6}{4} + \frac{6}{3} + \frac{6}{2} + \frac{6}{1}$$

$$= 14.7$$

A *wet spell of x days* is defined to be a "run" of x days on each of which measurable precipitation occurs. Under the assumption that the weather tomorrow depends only on the weather today, there is a probability p_0 that a wet day will be followed by a dry day. From this, the probability of an x-day-long wet spell is $p_0(1 - p_0)^{x-1} = p_0 q_0^{x-1}$—that is, a geometric distribution. We will let X denote the random variable representing the lengths of wet spells.

In a similar way, there is a random variable Y associated with the lengths of *dry* spells; furthermore, $f_Y(y) = p_1 q_1^{y-1}$, where p_1 is the probability of a dry day followed by a wet one. Since a given day's weather is affected only by the previous day's, it follows that X and Y are independent.

Now, a *weather cycle* will be defined as, say, a wet spell followed by a dry spell. If Z denotes the length of such a cycle, then $Z = X + Y$. We leave it to the reader to verify that

$$f_{X+Y}(z) = p_0 p_1 \frac{q_0^{z-1} - q_1^{z-1}}{q_0 - q_1} \qquad z = 2, 3, \dots$$

[*Hint*: Write $P(X + Y = z) = P(X = 1)P(Y = z - 1) + \cdots + P(X = z - 1)$ $P(Y = 1)$ and simplify.]

The lengths of 351 weather cycles in Tel Aviv were recorded during the months of December, January, and February from the winter of 1923–1924 through the winter of 1949–1950 (50). The proposed model's two parameters, p_0 and p_1, were estimated to be 0.338 and 0.250, respectively.

TABLE 4.4.1

Cycle length	Observed frequency	Expected frequency
2	33	30
3	38	42
4	33	44
5	50	42
6	39	37
7	26	32
8	30	26
9	24	21
10	16	17
11	14	14
12	15	11
13	9	8
14	3	6
15+	21	21

Table 4.4.1 lists both the observed and expected distributions for Z, the latter being computed from the model

$$f_{X+Y}(z) = (0.338)(0.250) \frac{(0.662)^{z-1} - (0.750)^{z-1}}{0.662 - 0.750}$$

Considering the simplicity of some of our assumptions, the agreement between the last two columns is quite good.

Question 4.4.1 A professional football team has three quarterbacks on its traveling squad. Suppose that each quarterback has an 80% chance of completing a game without getting injured. If the team plays a schedule of 12 games, what is the probability the third-string quarterback makes his first starting appearance in the eleventh game? What is the probability the team has to start a fourth quarterback before the season is over? (Assume that the probability of two or more injuries in any one game is zero.)

Question 4.4.2 A young couple plans to continue having children until they have their first girl. Suppose the probability that a child is a girl is $\frac{1}{2}$ and the outcome of each birth is an independent event. What is their expected family size?

Question 4.4.3 Show that a geometric random variable, X, is *memoryless*. That is, prove that

$$P(X = n - 1 + k \mid X > n - 1) = P(X = k)$$

Question 4.4.4 A somewhat uncoordinated man is attempting to get a driver's license. Having successfully completed the written exam, he needs only to pass the road test to achieve his objective. His abilities in that particular area, though, are less than exceptional: An unbiased observer would give him no more than a 10% chance of passing. What is the probability he will have to take the test at least seven times in order to get his license? Assume that his driving skills remain at the same level, regardless of how many times he fails the test. What is the expected number of times he will have to take the test?

Question 4.4.5 A fair die is tossed. What is the probability the first 5 occurs on the fourth roll?

Question 4.4.6 Find an expression for the cdf of a geometric random variable.

Question 4.4.7 Show that the moment-generating function for the geometric distribution is

$$M_N(t) = \frac{pe^t}{1 - qe^t}$$

and use $M_N(t)$ to find $E(N)$ and Var (N).

4.5 THE NEGATIVE BINOMIAL DISTRIBUTION

In this section we consider a natural generalization of the geometric distribution. Rather than focus on the number of Bernoulli trials required for the *first* success to occur, we now look at the number required for the occurrence of the rth success. Let X denote the number of trials necessary for this occurrence. It should be clear that the only possible scenario for X's being equal to, say, x is for a total of $r - 1$ successes [and $x - 1 - (r - 1)$ failures] to occur somewhere during the first $x - 1$ trials and for one additional success (the rth) to occur on the very next (i.e., the xth) trial. Assuming all trials to be independent, it follows that the distribution of X should be

$$f_X(x) = \binom{x - 1}{r - 1} p^{r - 1} q^{x - 1 - (r - 1)} p$$

$$= \binom{x - 1}{r - 1} p^r q^{x - r} \qquad x = r, r + 1, \ldots$$

In practice, the random variable X is often replaced by Y, where Y is the number of trials *in excess of* r required to produce the rth success. That is, $Y = X - r$. Letting n be the argument for Y, we can write

$$f_Y(n) = \binom{n + r - 1}{r - 1} p^r q^n = \binom{n + r - 1}{n} p^r q^n \qquad n = 0, 1, \ldots$$

Since we derived $f_Y(n)$ as a pdf, we fully expect its values to sum to one. It will be instructive, though, to prove that property independently. In so doing, we show that $f_Y(n)$ is a pdf even when r is not an integer. This last statement makes no sense unless $\binom{n + r - 1}{n}$ is defined for all real r. Up to this point, the symbol $\binom{k}{n}$ has been used as shorthand for the quantity $k(k - 1) \cdots (k - n + 1)/n!$ only when k was a non-negative integer. But $k(k - 1) \cdots (k - n + 1)/n!$ is well defined for *any* real k, so there is no problem in extending the definition of $\binom{k}{n}$.

Theorem 4.5.1.

$$\sum_{n=0}^{\infty} f_Y(n) = \sum_{n=0}^{\infty} \binom{n + r - 1}{n} p^r q^n = 1$$

Proof. We begin by establishing a combinatorial identity involving $\binom{n + r - 1}{n}$. Since

$$\binom{-r}{n} = \frac{-r(-r - 1) \cdots (-r - n + 1)}{n!}$$

$$= (-1)^n \frac{r(r + 1) \cdots (n + r - 1)}{n!}$$

$$= (-1)^n \binom{n + r - 1}{n}$$

we can write

$$\binom{n + r - 1}{n} = (-1)^n \binom{-r}{n} \tag{4.5.1}$$

With Equation 4.5.1 we have our key for evaluating $\sum_{n=0}^{\infty} f_Y(n)$. Note, first of all, that

$$\sum_{n=0}^{\infty} \binom{n + r - 1}{n} p^r q^n = p^r \sum_{n=0}^{\infty} \binom{n + r - 1}{n} q^n$$

$$= p^r \sum_{n=0}^{\infty} \binom{-r}{n} (-1)^n q^n$$

$$= p^r \sum_{n=0}^{\infty} \binom{-r}{n} (-q)^n \tag{4.5.2}$$

Recall Newton's binomial formula: For any real k,

$$(1 + t)^k = \sum_{n=0}^{\infty} \binom{k}{n} t^n \tag{4.5.3}$$

Applying Equation 4.5.3 to Equation 4.5.2 gives the result:

$$\sum_{n=0}^{\infty} \binom{n + r - 1}{n} p^r q^n = p^r (1 - q)^{-r}$$

$$= p^r p^{-r} = 1$$

What we have shown to be true about Y, the number of Bernoulli trials in excess of r required to produce the rth success, is stated formally in Definition 4.5.1. The mean and variance of Y are given in Theorem 4.5.2.

Definition 4.5.1. The random variable Y is said to have the negative binomial distribution with parameters r and p if

$$f_Y(n) = \binom{n + r - 1}{n} p^r q^n \qquad n = 0, 1, \ldots$$

Theorem 4.5.2. If the random variable Y has the negative binomial distribution with parameters r and p, then

$$E(Y) = \frac{rq}{p} \qquad \text{and} \qquad \text{Var}(Y) = \frac{rq}{p^2}$$

Proof. The theorem is true for arbitrary $r > 0$, but we will consider only the case where r is a positive integer. Let $X_1, X_2, \ldots, X_r$ be independent geometric random variables, each with the same parameter p. Since each X_i can be thought of as the number of trials required for the first success, it seems intuitively reasonable that their sum, $X = X_1 + X_2 + \cdots + X_r$, should have the same distribution as the number of trials required for the rth success—that is, $Y = X - r$ should be negative binomial with parameters r and p (see Question 4.5.4). It follows, then, from Theorem 4.4.1 and the properties of sums of independent random variables, that

$$E(X) = rE(X_1) = r\left(\frac{1}{p}\right) = \frac{r}{p}$$

and

$$\text{Var}(X) = r \text{ Var}(X_1) = r\left(\frac{q}{p^2}\right) = \frac{rq}{p^2}$$

Finally, the expected value of Y is $E(X - r) = E(X) - r = (r/p) - r = rq/p$, and the variance of Y is $\text{Var}(X - r) = \text{Var}(X)$.

Where r is a positive integer, its interpretation in the negative binomial model is clear. When r is positive but not an integer, we lose that physical interpretation, but the model itself may still be very useful. Case Study 4.5.1 is a case in point.

CASE STUDY 4.5.1

An early application of the negative binomial was due to the famous British statistician, Sir Ronald A. Fisher. Pursuing a question of somewhat less gravity than was his custom, Fisher was able to show that Y, the number of ticks found on a sheep, could be described remarkably well by a function of the form

$$f_Y(n) = \binom{n + r - 1}{n} p^r q^n$$

His data consisted of 60 sheep and their uninvited entourage of some 200 ticks. Table 4.5.1 shows the observed and expected tick distributions (45). The

TABLE 4.5.1 DISTRIBUTION OF TICKS ON SHEEP

Number of Ticks	Observed Frequency	Expected Frequency
0	7	6
1	9	10
2	8	11
3	13	10
4	8	8
5	5	5
6	4	4
7	3	2
8	0	1
9	1	1
10+	2	2
	$\overline{60}$	$\overline{60}$

parameters r and p were estimated to be 3.75 and 0.536, respectively, so the entries in column 3 were gotten by multiplying 60 times

$$\binom{n + 3.75 - 1}{n}(0.536)^{3.75}(0.464)^n$$

The row-by-row agreement, here, between columns 2 and 3 is obviously quite good—$f_Y(n)$ is a more-than-adequate model.

Question 4.5.1 Suppose that an underground military installation is fortified to the extent that it can withstand up to four direct hits from air-to-surface missiles and still function. Enemy aircraft can score direct hits with these particular missiles 7 times out of 10. What is the probability that a plane will require fewer than 8 shots to destroy the installation? Assume all firings are independent.

Question 4.5.2 Let Y be a negative binomial random variable. Prove directly that the expected value of $X = Y + r$ is r/p by evaluating the sum

$$\sum_{x=r}^{\infty} x\binom{x-1}{r-1}p^r q^{x-r}$$

(*Hint*: Reduce the sum to one involving negative binomial probabilities with parameters $r + 1$ and p.)

Question 4.5.3 Suppose that Y_1 and Y_2 are two independent negative binomial random variables, each with parameters r and p. Show that $Y_1 + Y_2$ is negative binomial with parameters $2r$ and p.

Question 4.5.4 Let $X_1, X_2, \ldots, X_r$ be r independent random variables, each having the geometric pdf with parameter p. Show that $X = X_1 + X_2 + \cdots + X_r$ has pdf

$$f_X(x) = \binom{x-1}{r-1}p^x q^{x-r} \qquad x = r, r+1, \ldots$$

and hence that $Y = X - r$ has the negative binomial pdf with parameters r and p.

Question 4.5.5 A door-to-door encyclopedia salesperson is required to document five in-home visits each day. Suppose that she has a 30% chance of being invited into any given home, with each home representing an independent trial. If she selects, ahead of time, the addresses of 10 households upon which to call, what is the probability her fifth success occurs on the tenth trial? What is the probability she requires fewer than 8 addresses to record her fifth success? Suppose she goes to only 7 homes: What is the probability she will meet or exceed her quota?

Question 4.5.6 The moment-generating function for the negative binomial pdf as given in Definition 4.5.1 is not the same as the moment-generating function for

$$f_X(x) = \binom{x-1}{r-1}p^r q^{x-r} \qquad x = r, r+1, \ldots$$

yet both pdf's relate to the occurrence of the rth success in a series of independent Bernoulli trials. Why are the two mgf's different?

Question 4.5.7 When a machine is improperly adjusted, it has probability 0.15 of producing a defective item. Each day the machine is run until three defective items are produced. Then it is stopped and checked for adjustment. What is the probability that an improperly adjusted machine will produce five or more items before being stopped? What is the average number of items the machine will produce before being stopped?

4.6 THE GAMMA DISTRIBUTION

In Section 4.2, the useful Poisson distribution was introduced as the limiting form of the binomial. The Poisson concerns the *number* of occurrences of an event during a fixed time period. Here we use the ideas of Section 4.2 to derive the distribution of the *time* required for a fixed number of events to occur.

Suppose that a certain event occurs homogeneously over a time interval of length x at an average rate of λ per unit time. We wish to examine the continuous variable X measuring the length of time required for r events to occur. Assume, as in Section 4.2, that the interval $[0, x]$ can be subdivided into small, independent subintervals so that the probability of more than one event in a given subinterval is negligible. Let W be the random variable counting the number of occurrences of the event in the interval $[0, x]$. Then W is a Poisson random variable with parameter λx. The desired cdf of X can be obtained using W as follows:

$$
\begin{aligned}
F_X(x) &= P(X \le x) \\
&= 1 - P(X > x) \\
&= 1 - P(\text{fewer than } r \text{ events occur in the interval}\,[0, x]) \\
&= 1 - F_W(r - 1) \\
&= 1 - \sum_{k=0}^{r-1} e^{-\lambda x} \frac{(\lambda x)^k}{k!}
\end{aligned}
$$

Therefore,

$$
\begin{aligned}
f_X(x) &= -\frac{d}{dx} \sum_{k=0}^{r-1} e^{-\lambda x} \frac{(\lambda x)^k}{k!} \\
&= -\left[-\lambda e^{-\lambda x} + \sum_{k=1}^{r-1} (-\lambda) e^{-\lambda x} \frac{(\lambda x)^k}{k!} + e^{-\lambda x}(\lambda) \frac{(\lambda x)^{k-1}}{(k-1)!} \right] \\
&= \sum_{k=0}^{r-1} \lambda e^{-\lambda x} \frac{(\lambda x)^k}{k!} - \sum_{k=1}^{r-1} \lambda e^{-\lambda x} \frac{(\lambda x)^{k-1}}{(k-1)!} \\
&= \sum_{k=0}^{r-1} \lambda e^{-\lambda x} \frac{(\lambda x)^k}{k!} - \sum_{k=0}^{r-2} \lambda e^{-\lambda x} \frac{(\lambda x)^k}{k!} \\
&= \lambda e^{-\lambda x} \frac{(\lambda x)^{r-1}}{(r-1)!} \\
&= \frac{\lambda^r}{(r-1)!} x^{r-1} e^{-\lambda x} \qquad\qquad (4.6.1)
\end{aligned}
$$

What we have just derived for $f_X(x)$ is a special case of a *gamma* density, a family of probability distributions that is formally introduced in Definition 4.6.2. Before examining the gamma in any detail, though, we generalize Equation 4.6.1 to include those cases where r is positive but not necessarily an integer. To do this, it is necessary to replace $(r-1)!$ with a continuous function of (nonnegative) r, $\Gamma(r)$, the latter reducing to $(r-1)!$ when r is a positive integer. Such a function was first defined in 1731 by Euler, who reexpressed it 50 years later in the form it takes in Definition 4.6.1. Legendre is credited with giving it the name *gamma function*.

Definition 4.6.1. For any real number $r > 0$, the gamma function (of r) is given by

$$\Gamma(r) = \int_0^\infty x^{r-1}e^{-x}\,dx$$

Six properties of the gamma function are stated in Theorem 4.6.1. The first five will be left as homework exercises; the sixth is proved in Appendix 4.1.

Theorem 4.6.1. Let $\Gamma(r) = \int_0^\infty x^{r-1}e^{-x}\,dx$. Then

(a) $\Gamma(1) = 1$.

(b) $\Gamma(\frac{1}{2}) = \sqrt{\pi}$.

(c) $\Gamma(r+1) = r\Gamma(r)$, for any positive real r.

(d) $\Gamma(r+1) = r!$, if r is a nonnegative integer.

(e) $\dbinom{n+r-1}{n} = \dfrac{\Gamma(n+r)}{\Gamma(n+1)\Gamma(r)}$

(f) $\dfrac{\Gamma(r)\Gamma(s)}{\Gamma(r+s)} = \int_0^1 u^{r-1}(1-u)^{s-1}\,du$.

Having found the appropriate generalization of $(r-1)!$ for arbitrary (nonnegative) r, we can now define the gamma family of probability functions.

Definition 4.6.2. Let X be a random variable such that

$$f_X(x) = \frac{\lambda^r}{\Gamma(r)}x^{r-1}e^{-\lambda x} \qquad x > 0$$

Then X is said to have a gamma distribution with parameters r and λ. Both r and λ must be greater than 0.

Records of daily rainfall in Sydney, Australia, for the period from October 17 to November 7 were assembled for the years 1859 through 1952 (30). It was postulated that these data might have a gamma distribution, the reason being that precipitation occurs only if water particles can coalesce around dust of sufficient mass, but the accumulation of such dust is similar to the "waiting-time" aspect implicit in the gamma model.

Table 4.6.1 lists the results. Estimates for r and λ were found to be 0.105 and 0.013, respectively.

TABLE 4.6.1 DISTRIBUTION OF RAINFALL

Rainfall (mm)	Observed Frequency	Expected Frequency
0–5	1631	1639
6–10	115	106
11–15	67	62
16–20	42	44
21–25	27	32
26–30	26	26
31–35	19	21
36–40	14	17
41–45	12	14
46–50	18	12
51–60	18	20
61–70	13	15
71–80	13	12
81–90	8	9
91–100	8	7
101–125	16	12
126–150	7	7
151–425	14	13

To compute the expected probability of rainfall measuring between, say, 0 and 5 mm, we compute

$$p = \frac{(0.013)^{0.105}}{\Gamma(0.105)} \int_0^{5.5} x^{0.105-1} e^{-0.013x} \, dx$$

or, making the substitution $u = 0.013x$,

$$p = \frac{1}{\Gamma(0.105)} \int_0^{0.0715} u^{0.105-1} e^{-u} \, du \qquad (4.6.2)$$

The integral in Equation 4.6.2 can be evaluated using Tables of the Incomplete Gamma Function [see (162)]. In this case, $p = 0.793$, so the expected number of days with rainfall between 0 and 5 mm is equal to 2068×0.793, or 1639.

Theorem 4.6.2. Let X be a random variable having a gamma distribution with parameters r and λ. Then

$$E(X) = \frac{r}{\lambda} \quad \text{and} \quad \text{Var } (X) = \frac{r}{\lambda^2}$$

Proof. By definition,

$$E(X) = \int_0^\infty x \, \frac{\lambda^r}{\Gamma(r)} \, x^{r-1} e^{-\lambda x} \, dx$$

$$= \frac{\lambda^r}{\Gamma(r)} \int_0^\infty x^r e^{-\lambda x} \, dx$$

Note that the integrand is the variable part of a gamma density with parameters $r + 1$ and λ. Therefore, the integral equals $\Gamma(r + 1)/\lambda^{r+1}$ and

$$E(X) = \frac{\lambda^r}{\Gamma(r)} \cdot \frac{\Gamma(r + 1)}{\lambda^{r+1}} = \frac{r}{\lambda}$$

The derivation of Var (X) is similar; we omit the details.

Question 4.6.1 Show that $\Gamma(\frac{7}{2}) = (15\sqrt{\pi})/8$.

Question 4.6.2 Show that $f_X(x)$ as given in Definition 4.6.2 is a true probability density; verify that

$$\int_0^\infty \frac{\lambda^r}{\Gamma(r)} \, x^{r-1} e^{-\lambda x} \, dx = 1$$

Question 4.6.3 Let X be the gamma random variable of Theorem 4.6.2. Show that $E(X^2) = (r + 1)r/\lambda^2$. Then Var $(X) = E(X^2) - [E(X)]^2 = r/\lambda^2$.

Question 4.6.4 Let X be a gamma random variable. Derive a formula for $E(X^m)$, where m is any positive integer. (*Hint*: Follow the technique used in the proof of Theorem 4.6.2.)

Question 4.6.5 For a gamma random variable X, show that $M_X(t) = [\lambda/(\lambda - t)]^r$. Use the moment-generating function to verify Theorem 4.6.2.

Question 4.6.6 Suppose that X is a gamma random variable with $E(X) = 2$ and Var $(X) = 7$. Find r and λ.

Question 4.6.7 Suppose that X_1 and X_2 are independent gamma random variables, X_1 with parameters r and λ and X_2 with parameters s and λ. Show that $X_1 + X_2$ is a gamma random variable with parameters $r + s$ and λ.

Question 4.6.8 Use Question 4.6.7 to show that the sum of r independent exponential random variables each with parameter λ is a gamma random variable with parameters r and λ.

Question 4.6.9 Suppose that an Antarctic weather station has three electronic wind gauges, an original and two backup gauges. The lifetime of each gauge is exponentially distributed with a mean of 1000 hours. What is the pdf for Y, the random variable measuring the time until the last gauge wears out?

Question 4.6.10 Let Z be a standard normal random variable. Show that Z^2 has the gamma pdf with parameters $r = \frac{1}{2}$ and $\lambda = \frac{1}{2}$.

APPENDIX 4.1 A PROPERTY OF THE GAMMA FUNCTION

Theorem A4.1.1. Let $r > 0$ and $s > 0$. Then

$$\frac{\Gamma(r)\Gamma(s)}{\Gamma(r + s)} = \int_0^1 u^{r-1}(1 - u)^{s-1} \, du$$

Proof. By definition,

$$\Gamma(r)\Gamma(s) = \int_0^\infty x^{r-1}e^{-x} \, dx \int_0^\infty y^{s-1}e^{-y} \, dy$$

$$= \int_0^\infty \int_0^\infty x^{r-1}y^{s-1}e^{-(x+y)} \, dx \, dy$$

Let $u = x/(x + y)$ so that $x = uy/(1 - u)$. Then $dx = y \, du/(1 - u)^2$ and

$$\Gamma(r)\Gamma(s) = \int_0^\infty \int_0^1 \left(\frac{uy}{1 - u}\right)^{r-1} y^{s-1}e^{-(y/(1-u))} \frac{y}{(1 - u)^2} \, du \, dy$$

Now, make the substitution $v = y/(1 - u)$, or, equivalently, $y = (1 - u)v$. This reduces the double integral to

$$\Gamma(r)\Gamma(s) = \int_0^\infty \int_0^1 (uv)^{r-1}(1 - u)^{s-1}v^{s-1}e^{-v}v \, du \, dv$$

$$= \int_0^\infty \int_0^1 u^{r-1}(1 - u)^{s-1}v^{r+s-1}e^{-v} \, du \, dv$$

$$= \left(\int_0^\infty v^{r+s-1}e^{-v} \, dv\right)\left[\int_0^1 u^{r-1}(1 - u)^{s-1} \, du\right]$$

$$= \Gamma(r + s) \int_0^1 u^{r-1}(1 - u)^{s-1} \, du$$

and the theorem follows.

5

Estimation

Ronald Aylmer Fisher (1890–1962)

A towering figure in the development of both applied and mathematical statistics, Fisher took formal training in mathematics and theoretical physics (he graduated from Cambridge in 1912). After a brief career as a teacher, he accepted a post in 1919 as statistician at the Rothamsted Experimental Station. There the day-to-day problems encountered in collecting and interpreting agricultural data led directly to much of his most important work in the theory of estimation and experimental design. Fisher was also a prominent geneticist and devoted considerable time to the development of a quantitative argument that would support Darwin's theory of natural selection. He returned to academia in 1933, succeeding Karl Pearson as the Galton Professor of Eugenics at the University of London. Fisher was knighted in 1952.

5.1 INTRODUCTION

The ability of probability functions to describe, or *model*, experimental data was demonstrated in numerous examples in Chapter 4. In Section 4.2, for example, the Poisson distribution was shown to predict very well the number of Prussian cavalry soldiers kicked to death by their horses. That same family of distributions was also seen to model the number of alpha emissions from a radioactive source as well as the annual number of outbreaks of war, worldwide, from 1500 to 1931. In Section 4.3 another probability function, the normal, was applied to phenomena as diverse as the chest measurements of Scottish soldiers and the performance of 106 randomly selected adults on a motivation test. Still other probability models fitted to data in Chapter 4 included the geometric, the negative binomial, and the gamma. In this chapter we want to scrutinize this relationship between probability models and data more mathematically. Beginning with the material presented here, our focus will shift away from the study of probability and toward the study of statistics.

The application of the methods of probability to the analysis and interpretation of empirical data is known as *statistical inference*. More specifically, statistical inference is the process by which we generalize from a particular sample—that is, from a set of n observations—to the theoretical population from which that sample came. Of course, the precise form of the generalization can vary considerably from situation to situation. It may be a single numerical estimate, a *range* of numerical estimates, or even a simple "yes" or "no."

A few examples may clarify some of these ideas. Imagine the chief programming executive at ABC trying to decide which shows to cancel and which to renew. As input, he may have the day-by-day logs of the programs that are watched by, say, two or three thousand specially selected families, representing various socioeconomic, age, and ethnic groups. His problem is to use that sample information to estimate the *total* number of viewers tuned to ABC's programs. (The "population" here is simply the aggregate of all TV viewers.) For a given program, his final inference might be that "21.6% of the potential market is tuned to program X" (i.e., a single numerical estimate), or it might take the form that "between 18.5% and 23.0% of the market is tuned to program X" (i.e., a range of estimates).

As a second example, suppose a zoologist (for reasons best left unsaid) would like to know whether *Desmodus rotundus*, a small South American vampire bat, prefers blood at room temperature (23°C) or at body temperature (38.5°C) [see (16)]. To get some data, he puts equal numbers of similar bats into two cages. The drinking tubes in cage A are supplied with blood kept at room temperature; those in cage B, with blood kept at body temperature. Suppose that after both groups have had enough time to drink as much as they wanted, he finds that the bats in cage A consumed 3% more blood than the bats in cage B. What should he conclude? Does the 3% excess "prove" that vampire bats prefer blood at room temperature, or is that figure too small to mean anything? Here, one way to phrase the inference would be a simple "yes" or "no": "Yes, the bats show a preference" or "No, they do not."

These two examples point up the two broad areas into which the subject of statistical inference is traditionally divided: *estimation* and *hypothesis testing*. In the former the inference is numerical; in the latter it becomes a yes or no decision between two conflicting theories. Both areas, as we will see, have wide applicability.

Chapter 5 presents the basic principles of estimation. This will be essentially a two-part endeavor—first it will be necessary to define the mathematical properties a "good" estimator should have; then we will look for procedures that yield estimates having these properties. Hypothesis testing will be taken up in Chapter 6.

5.2 DEFINITIONS

In this section we introduce some of the basic terminology that has come to be associated with statistical estimation. The place to begin is to recall that most probability models—particularly those general enough to be of any practical value—are indexed by (that is, are functions of) one or more constants (see Section 4.1). We call these constants *parameters*. Thus the Poisson

$$f_Y(y) = \frac{e^{-\lambda}\lambda^y}{y!}$$

is a function of the occurrence rate, λ. The normal,

$$f_Y(y) = \frac{1}{\sqrt{2\pi}\,\sigma}\, e^{-(1/2)[(y-\mu)/\sigma]^2}$$

is indexed by *two* parameters, the mean μ and the standard deviation σ. Similarly, the binomial depends on n, the number of trials, and p, the probability of success at any given trial. Each of these distributions, then, is actually a *family* of distributions, where each member of the family has a different parameter value (and a correspondingly different shape) (see Figure 5.2.1).

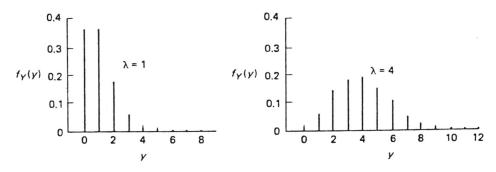

Figure 5.2.1 Two members of the Poisson family, $f_Y(y) = \dfrac{e^{-\lambda}\lambda^y}{y!}$.

The role that parameters play in the inference process is a crucial one. In many situations the family of probability models describing a phenomenon may be known (or at least assumed to be known) but the particular member of the family that *best* describes that phenomenon may be unknown. A criminologist, for example, may have good reason to believe that the distribution of a certain fingerprint abnormality in the general population follows a Poisson distribution, but only after he does a survey and gets some feeling for the rate at which the abnormality occurs (say, 19.5 times for every 10,000 people) does he know *which* Poisson is the most appropriate model. It is probably fair to say that estimating the unknown parameter (or parameters) of a presumed data model is an intermediate, if not a final, step in almost every inference problem.

By convention, various symbols have come to be associated with the parameters of different distributions: μ and σ with the normal, n and p with the binomial, and so on. However, to facilitate the discussion of parameters and estimates of parameters in general, it will be necessary to adopt a more flexible notation. Accordingly, we will let θ denote an arbitrary parameter. The symbol $f_Y(y; \theta_1, \theta_2, \ldots, \theta_k)$, then, will denote a probability function whose k parameters (known or unknown) are $\theta_1, \theta_2, \ldots, \theta_k$. Most of the problems we will deal with, though, involve only a single parameter. For these, the subscript on θ will be deleted and the pdf written $f_Y(y; \theta)$.

If in a given problem a particular θ_i is unknown, we must estimate it, using the sample data. This will be done through a function known as a *statistic*. By definition, a statistic, $W = h(Y_1, Y_2, \ldots, Y_n)$, is any function of the random variables, $Y_1, Y_2, \ldots, Y_n$, the set of Y_i's being a random sample of size n (see Section 3.4). Thus the mean, $\bar{Y} = (1/n) \sum_{i=1}^{n} Y_i$, the standard deviation, $S = \sqrt{(1/(n-1)) \sum_{i=1}^{n} (Y_i - \bar{Y})^2}$, and the range, $R = \max_i Y_i - \min_i Y_i$, are all examples of statistics.

Having defined the notions of "parameter" and "statistic," we can now rephrase the main objective of this chapter as outlined in Section 5.1: *we are seeking criteria—and, based on those criteria, procedures—for finding statistics that will serve as "good" estimators of unknown parameters.* Suppose, for example, that three light bulbs are put "on test" and kept lit continuously until they burn out. The resultant data would be the three observed lifetimes—say, 610, 1150, and 1570 hours. If it can be assumed (and there are physical reasons to justify this) that the density function describing the distribution of the lifetimes is of the form $f_Y(y; \lambda) = (1/\lambda)e^{-y/\lambda}$, how should we estimate λ? Should the estimator have the form

$$W = h(Y_1, Y_2, Y_3) = \sum_{i=1}^{3} Y_i \quad \text{or} \quad W = h(Y_1, Y_2, Y_3) = \sum_{i=1}^{3} \frac{3}{Y_i^2}$$

or something entirely different? We will see a little later that according to at least one principle of estimation, W should be $\frac{1}{3} \sum_{i=1}^{3} Y_i$, or, in this case, $3330/3 = 1110$.

Comment

In Chapter 3 the distinction was made between (1) random variables as generic designations, and (2) sample observations as particular realizations of random variables. The same duality exists in connection with estimation. We call the *form* of an expression for estimating an unknown parameter an *estimator*. In the example just described,

$$\frac{1}{3} \sum_{i=1}^{3} Y_i \quad \text{estimator for } \lambda$$

would be an estimator (for λ). However, once the random variables in the estimator are replaced by their sample values [such as $\frac{1}{3}(610 + 1150 + 1570)$], the resulting numerical quantity (1110) is referred to as an *estimate (for λ)*. In what follows, estimators will always be denoted by uppercase letters and estimates by the corresponding lowercase letters.

Basically, there are two formats into which an estimate can be cast: it can be expressed as a point or as an interval. As the name implies, a *point estimate* is a single number that represents, in some sense, our best guess as to the value of the unknown parameter. An *interval estimate*, on the other hand, is a range of numbers generated by a procedure having a high a priori probability (often 0.95 or 0.99) of including the unknown parameter. For the light bulb example, $w = 1110$ would be a point estimate for λ. Without going into the details here (see Question 7.5.17), it can be shown that the interval $\left(2 \sum_{i=1}^{3} Y_i/14.4, 2 \sum_{i=1}^{3} Y_i/1.24 \right)$ has a 95% chance of "containing" λ, in the sense that

$$P\left(\frac{2 \sum_{i=1}^{3} Y_i}{14.4} < \lambda < \frac{2 \sum_{i=1}^{3} Y_i}{1.24} \right) = 0.95$$

Substituting 3330 for $\sum_{i=1}^{3} Y_i$ gives (462, 5371) as the corresponding interval estimate for λ.

In one important way, interval estimates are superior to point estimates in terms of what they reveal about the data: specifically, the *length* of the interval provides a very useful measure of the estimator's precision. Merely stating, for example, that a point estimate for some θ is, say, $w = 50$, gives no indication of whether the *true* θ is likely to be within just a few units of 50 or whether it could easily be 17, or -35, or $+210$. However, knowing that an interval estimate for θ is (48, 52), we can safely infer that the possibility of the true θ's being something either much less than 48 or much greater than 52 is quite remote.

Because of their simplicity, point estimates will be discussed first. Sections 5.3 through 5.7 present a number of desirable properties to be looked for in an estimator of this type. Section 5.8 then describes two procedures for finding the actual *form* of $W = h(Y_1, Y_2, \ldots, Y_n)$ knowing only $f_Y(y; \theta)$. Interval estimation is introduced in Sections 5.9 and 5.10.

Comment

Applications of these concepts to the normal distribution will be deferred until Chapter 7, where they can be dealt with in a more organized way. Here we will rely primarily on the uniform, the exponential, the Poisson, and the binomial distributions to provide us with examples.

5.3 PROPERTIES OF POINT ESTIMATORS: UNBIASEDNESS AND EFFICIENCY

Before making any attempt to set down a list of qualities that a good W should possess, it is important to understand one very basic fact about the fundamental nature of estimators: every one, by virtue of its being a function of the sample data, is itself a random variable. This means that its behavior for different random samples will be described by a probability density function. Such a function, in keeping with the notation introduced in Chapter 3, will be denoted $f_W(w)$.

Example 5.3.1 examines the $f_W(w)$ associated with an estimator for the parameter of a uniform distribution.

Example 5.3.1

Suppose that a sample of size n is drawn from the uniform distribution

$$f_Y(y; \theta) = \begin{cases} \dfrac{1}{\theta} & 0 < y < \theta \\ 0 & \text{elsewhere} \end{cases}$$

and we decide to estimate θ with the maximum of the y_i's. That is,

$$W = h(Y_1, Y_2, \ldots, Y_n) = \max_i Y_i = Y_{\max} = Y'_n$$

From the corollary to Theorem 3.6.1, recall that the pdf for the largest of n order statistics is

$$f_{Y_{\max}}(y) = n f_Y(y)[F_Y(y)]^{n-1}$$

where $f_Y(y)$ and $F_Y(y)$ are the pdf and cdf, respectively, for Y. Here, since Y is uniform

$$F_Y(y; \theta) = \begin{cases} 0 & y \le 0 \\ \dfrac{y}{\theta} & 0 < y < \theta \\ 1 & y \ge \theta \end{cases}$$

The density function for the estimator $W = Y'_n = \hat{Y}_{\max}$ easily reduces, then, to

$$f_W(w; \theta) = \begin{cases} \dfrac{n w^{n-1}}{\theta^n} & 0 < w < \theta \\ 0 & \text{elsewhere} \end{cases}$$

Figure 5.3.1 shows $f_W(w)$ for the particular case where $\theta = 1$ and $n = 4$.

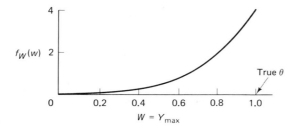

$f_W(w)$

$W = Y_{\max}$

True θ

Figure 5.3.1 Graph of $f_W(w)$.

Question 5.3.1 For the particular case of the estimator described in Figure 5.3.1—$\theta = 1$ and $n = 4$—find the probability that W will be within $\frac{1}{8}$ of θ.

Question 5.3.2 Let $Y_1, Y_2, \ldots, Y_n$ be a random sample of size n from an exponential pdf, $f_Y(y; \theta) = (1/\theta)e^{-y/\theta}$, $y > 0$, where θ is unknown. For reasons we will discuss later in this chapter, the usual estimator for θ is $W = (1/n) \sum_{i=1}^{n} Y_i$; another is $W = n \cdot Y_{\min}$. For the latter, find $f_W(w)$.

Realizing that each W has an associated $f_W(w)$ describing its behavior, it is not hard to think of some requirements for a good estimator. As a first condition, it seems reasonable to ask that $f_W(w)$ be somehow "centered" with respect to θ. If it is not, W will tend either to overestimate or underestimate θ, a condition hardly desirable. Figure 5.3.2(a) shows a W distribution that *does* meet this condition; Figure 5.3.2(b), one that does not. Of course, we have already seen an example of the latter—using $Y_{\max}$ to estimate the parameter of a uniform distribution. (The notion of W's being centered with respect to θ will be made more precise in the next section when we define *unbiasedness*.)

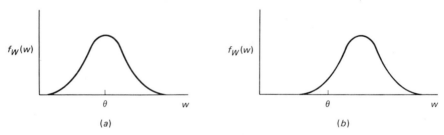

Figure 5.3.2 Centered and uncentered distributions.

A second property that a good estimator should possess is *precision*. We will say that an estimator is precise if the dispersion of its distribution is small. In Figure 5.3.3, both $f_{W_1}(w_1)$ and $f_{W_2}(w_2)$ appear to be centered with respect to the parameter, but clearly the better estimator is the one whose behavior is described by $f_{W_2}(w_2)$ because, of the two, it has a greater chance of being close to the true θ. (In Section 5.5 we will define formally what is meant by the *efficiency* of an estimator.)

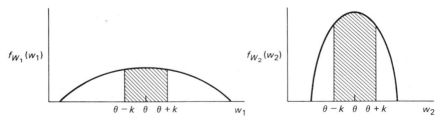

Figure 5.3.3 Two distributions with different dispersions.

Question 5.3.3 Let $Y_1, Y_2, \ldots, Y_n$ be a random sample of size n from the exponential pdf $f_Y(y; \theta) = (1/\theta)e^{-y/\theta}$, $y > 0$. Two estimators for θ were proposed in Question 5.3.2:

$$W_1 = (1/n) \sum_{i=1}^{n} Y_i \text{ and } W_2 = n \cdot Y_{\min}. \text{ Find Var } (W_1) \text{ and Var } (W_2). \text{ Which estimator}$$

is more precise?

$$\left[Hint: \int_0^\infty y^k (1/\theta)e^{-y/\theta} \, dy = k! \theta^k. \right]$$

5.4 UNBIASEDNESS

Although other criteria exist, by far the most common way to decide whether or not an estimator's distribution is suitably "centered" (relative to θ) is by looking at its expected value. If, *on the average*, W is equal to θ, we pronounce the estimator centered (and call it *unbiased*). Definition 5.4.1 gives the formal statement.

> **Definition 5.4.1.** Let $Y_1, Y_2, \ldots, Y_n$ be a random sample from $f_Y(y; \theta)$. An estimator $W = h(Y_1, Y_2, \ldots, Y_n)$ is said to be *unbiased* (for θ) if $E(W) = \theta$, for all θ.

Example 5.4.1

Consider again the problem of estimating the parameter of a uniform distribution with $W = Y_{\max}$. Since $Y_{\max}$ is always less than or equal to θ, it clearly cannot be unbiased. However, we can show that by multiplying it by an appropriate constant, we can transform it into a new statistic that *is* unbiased. From Example 5.3.1

$$E(W) = E(Y_{\max}) = \int_0^\theta w f_W(w) \, dw = \int_0^\theta w \frac{nw^{n-1}}{\theta^n} \, dw$$

$$= \frac{nw^{n+1}}{(n+1)\theta^n} \bigg|_0^\theta = \left(\frac{n}{n+1} \right) \theta$$

Suppose, now, that a new estimator, W_1, is defined, where

$$W_1 = \left(\frac{n+1}{n} \right) Y_{\max}$$

Since

$$E(W_1) = \left(\frac{n+1}{n} \right) E(Y_{\max}) = \theta$$

W_1 is unbiased.

The distribution of W_1 is easily derived. Recall from Chapter 3 that if Y is a continuous random variable with pdf $f_Y(y)$, the pdf of the transformed random variable X, where $X = aY$, is given by

$$f_X(x) = \left(\frac{1}{a}\right) f_Y\left(\frac{x}{a}\right)$$

Applying this result to $W_1 = [(n + 1)/n]W$ gives

$$f_{W_1}(w_1) = \frac{n}{n + 1} \cdot \frac{n\left[w_1 \Big/ \left(\dfrac{n + 1}{n}\right)\right]^{n-1}}{\theta^n}$$

$$= \frac{\dfrac{n^{n+1}}{(n + 1)^n}(w_1)^{n-1}}{\theta^n} \qquad \text{for} \quad 0 < w_1 < \left(\frac{n + 1}{n}\right)\theta$$

Figure 5.4.1 shows $f_{W_1}(w_1)$ for the special case described in Example 5.3.1: $\theta = 1$ and $n = 4$.

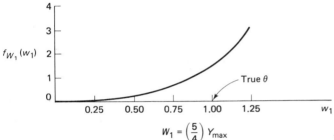

$$W_1 = \left(\frac{5}{4}\right)Y_{max}$$

Figure 5.4.1 Graph of $f_{W_1}(w_1)$.

Example 5.4.2

Let Y be the number of successes in n independent Bernoulli trials, where each trial has a success probability of p. Find an unbiased estimator for p based on Y.

Under the given assumptions, of course, Y is binomial and $E(Y) = np$. It follows from the latter that $W = Y/n$ is an unbiased estimator for p:

$$E(W) = E\left(\frac{Y}{n}\right) = \frac{1}{n} \cdot E(Y) = \frac{np}{n} = p$$

Thus, for the binomial model, the *sample* proportion of successes is an unbiased estimator for the *true* proportion of successes.

Is $W = Y/n$ the *only* unbiased estimator for p? No, not at all. Consider any of the Bernoulli random variables giving the number of successes at an individual trial:

$$Y_i = \begin{cases} 1 & \text{if } i\text{th trial ends in success} \\ 0 & \text{otherwise} \end{cases}$$

for $i = 1, 2, \ldots, n$. Since $E(Y_i) = 1 \cdot p + 0 \cdot (1 - p) = p$, any of the Y_i's is itself an unbiased estimator for p.

Example 5.4.3

If $\mu = E(Y)$ is the parameter being estimated, then the sample mean $\bar{Y} = (1/n) \cdot \sum_{i=1}^{n} Y_i$ is always an unbiased estimator. This is an immediate consequence of the distributive property of expected values:

$$E(\bar{Y}) = \frac{1}{n} \sum_{i=1}^{n} E(Y_i) = \frac{n\mu}{n} = \mu$$

Example 5.4.4

The statistic

$$\frac{1}{n} \sum_{i=1}^{n} (Y_i - \mu)^2$$

is an unbiased estimator for σ^2 (see Question 5.4.10). In many cases, though, we must estimate σ^2 without knowing μ. The statistic often used in that situation is the *sample variance* S^2, where

$$S^2 = \frac{1}{n-1} \sum_{i=1}^{n} (Y_i - \bar{Y})^2$$

Dividing the sum of the squared deviations by $n - 1$ (rather than n) is necessary to make S^2 unbiased. We can write

$$E\left[\sum_{i=1}^{n} (Y_i - \bar{Y})^2 \right] = E\left[\sum_{i=1}^{n} (Y_i^2 - 2Y_i\bar{Y} + \bar{Y}^2) \right]$$

$$= E\left[\left(\sum_{i=1}^{n} Y_i^2 - n\bar{Y}^2 \right) \right]$$

$$= \sum_{i=1}^{n} E(Y_i^2) - nE(\bar{Y}^2)$$

But

$$E(Y_i^2) = \sigma^2 + \mu^2$$

and

$$E(\bar{Y}^2) = \frac{\sigma^2}{n} + \mu^2 \qquad \text{(why?)}$$

Therefore,

$$E\left[\sum_{i=1}^{n} (Y_i - \bar{Y})^2 \right] = (n\sigma^2 + n\mu^2) - \sigma^2 - n\mu^2 = (n - 1)\sigma^2$$

so

$$E(S^2) = \frac{1}{n-1} (n - 1)\sigma^2 = \sigma^2$$

Comment

An equivalent expression for S^2 is

$$S^2 = \frac{n \sum\limits_{i=1}^{n} y_i^2 - \left(\sum\limits_{i=1}^{n} y_i \right)^2}{n(n-1)}$$

This provides a very convenient computing formula. Accumulating sums and sums of squares is much easier to do, particularly on a calculator, than computing sums of squares of differences. Rounding errors will also tend to be less of a problem with this second formulation.

Question 5.4.1 Epidemiologists at a county hospital located near a nuclear waste disposal plant are monitoring all admissions having a tentative diagnosis of chronic leukemia. The numbers of patients falling into that category for the past 12 years were 2, 1, 0, 0, 1, 1, 1, 4, 1, 0, 2, and 3. Assume that the occurrence of such patients (per year) can be described by a Poisson pdf. Estimate the relevant parameter. (*Hint*: See Example 5.4.3.)

Question 5.4.2 Let $Y_1, Y_2, \ldots, Y_n$ be a random sample from any pdf whose mean is μ. Under what conditions is the estimator

$$W = h(Y_1, Y_2, \ldots, Y_n) = \sum_{i=1}^{n} a_i Y_i$$

unbiased?

Question 5.4.3 Find an unbiased estimator for θ based on the smallest of n order statistics drawn from a uniform distribution defined over $(0, \theta)$.

Question 5.4.4 A drug reaction surveillance program was carried out in nine major hospitals (105). Out of a total of 11,526 patients monitored, it was found that 3240 experienced some form of adverse reaction to their medication. What mathematical model applies to these data? Estimate the relevant parameter. Is the estimator on which your estimate is based, unbiased?

Question 5.4.5 Suppose that 14, 10, 18, and 21 constitute a random sample of size four drawn from a uniform pdf defined over the interval $(0, \theta)$, where θ is unknown. Find an expression for an unbiased estimator for θ based on Y_3', the third order statistic. Evaluate the estimator using the data given. Is it possible that an estimate based on Y_3' would be clearly inappropriate in a given situation? Explain.

Question 5.4.6 A student takes a freshman chemistry exam that the professor claims has always resulted in approximately normal distributions when he has given the test in the past. If the student makes a grade of 66, estimate μ, the class average.

Question 5.4.7 Let W_1 and W_2 be any two unbiased estimators for the parameter θ. If $W = aW_1 + bW_2$ is also to be an unbiased estimator for θ, what must be true of a and b?

Question 5.4.8 A marketing research firm did two independent surveys of the proportion of viewers in a certain metropolitan area who watched "Hill Street Blues." The first survey revealed that 35 of the 100 persons questioned were tuned to the show. In the second survey, 100 of the 350 persons contacted watched the show. Give two unbiased estimators for the true proportion of people watching "Hill Street Blues." Make both of your estimators use the information contained in both surveys. Intuitively, what would be a reasonable way to combine the information in the two samples? Explain.

Question 5.4.9 Let Y have a uniform pdf over the interval $(0, \theta)$. Show that $2\bar{Y}$ is an unbiased estimator for θ.

Question 5.4.10 Suppose that $Y_1, Y_2, \ldots, Y_n$ is a random sample from a pdf with mean μ and variance σ^2. Show that $(1/n) \cdot \sum_{i=1}^{n} (Y_i - \mu)^2$ is an unbiased estimator for σ^2.

Question 5.4.11 Let X have a uniform pdf over the interval $(0, \theta)$. Find an unbiased estimator for θ^2 (*Hint*: Consider a function of X^2.)

Question 5.4.12 A secretary spends much of her morning trying to place long-distance calls for her supervisor. The people she is trying to reach are hard to find, and she has to make 16 calls before making connections (on the seventeenth call) with the fourth (and last) name on her list. Assuming each attempted call to be an independent trial with the same probability, p, of success, estimate p. See Question 5.4.13.

Question 5.4.13 Let Y have the pdf

$$f_Y(y) = \binom{y + r - 1}{y} p^r q^y \qquad y = 0, 1, \ldots$$

Show that $W = (r - 1)/(r + Y - 1)$ is an unbiased estimator for p.

Question 5.4.14 Let Y_i be the ith observation recorded in a sample of size n drawn from a uniform pdf defined over $(0, \theta)$. Is $W = Y_i$ unbiased? If not, construct a W_1 based on W that does have that property.

Question 5.4.15 An estimator $W_n = W_n(Y_1, Y_2, \ldots, Y_n)$ for the parameter θ is said to be *asymptotically unbiased* if

$$\lim_{n \to \infty} E(W_n) = \theta$$

Let $Y_1, Y_2, \ldots, Y_n$ be a random sample from a uniform distribution over $(0, \theta)$. Let $W_n = Y_{max}$. Is W_n asymptotically unbiased?

Question 5.4.16 Let X_1 and X_2 be a random sample of size 2 from an exponential pdf, $f_X(x) = (1/\lambda)e^{-x/\lambda}$, $x > 0$. Define Y to be the *geometric mean* of X_1 and X_2: $Y = \sqrt{X_1 X_2}$. Show that $4Y/\pi$ is an unbiased estimator for λ.

Question 5.4.17 As an alternative to imposing unbiasedness, an estimator's distribution can be "centered" by requiring that its *median* be equal to θ. If it is, W is said to be *median unbiased*. Let $Y_1, Y_2, \ldots, Y_n$ be a random sample from the uniform pdf over $(0, \theta)$. For arbitrary n, is $W_1 = [(n + 1)/n]Y'_n$ median unbiased? Is it *ever* median unbiased?

Question 5.4.18 We say that W is a *best linear unbiased estimator* for a parameter θ if (1) W is a linear function of the observations; (2) $E(W) = \theta$, for all θ; and (3) Var $(W) \leq$ Var (W^*), for all other estimators W^* satisfying conditions (1) and (2). Suppose that X is a random variable with mean μ and variance σ^2. Show that $\bar{X} = (1/n)(X_1 + X_2 + \cdots + X_n)$ is the best linear unbiased estimator for μ. (*Hint*: Write as a general linear estimator, $W = a_1 X_1 + \cdots + a_n X_n + c$; show that $c = 0$ and take partial derivatives with respect to the a_i's to show that $a_i = 1/n$.)

5.5 EFFICIENCY

Our discussion of unbiasedness has led to three important ideas in connection with point estimates: first, that it makes sense to talk about the expected value of an estimator and that imposing the requirement that $E(W)$ be equal to θ (for all θ) is one way of eliminating systematic bias; second, that even if a given W is *not* unbiased, it can often be easily transformed into a W_1 that is; and, finally, that for any given $f_Y(y; \theta)$, there may be many different W's that satisfy the criterion of unbiasedness. It is this latter observation that we want to examine more fully in this section.

To the practitioner facing the problem of estimating an unknown θ, the lack of uniqueness of unbiased estimators poses an obvious question: *which* unbiased estimator should be used? Are they all equivalent, or are some better than others? For an answer, we need to think back to the *second* property of good estimators that was mentioned at the beginning of Section 5.3: that the dispersion of any estimator should be small, so that the probability of its being close to the true θ will be large. As usual, the concept of dispersion will be translated mathematically into a statement about a variance.

Definition 5.5.1. Let W_1 and W_2 be two unbiased estimators for θ with variances Var (W_1) and Var (W_2), respectively. We will call W_1 *more efficient than* W_2 if

$$\text{Var } (W_1) < \text{Var } (W_2)$$

Also, the *relative efficiency* of W_1 with respect to W_2 will be defined as the ratio

$$\text{Var } (W_2)/\text{Var } (W_1)$$

Example 5.5.1

As a simple example of two unbiased estimators having very different precisions, consider the problem of estimating λ in an exponential distribution,

$$f_{Y_i}(y_i) = \left(\frac{1}{\lambda}\right) e^{-y_i/\lambda} \qquad y_i > 0; i = 1, 2, \ldots, n$$

Since $E(Y_i) = \lambda$, one unbiased estimator is $W_1 = \bar{Y}$ (recall Example 5.4.3); another is $W_2 = Y_1$, where Y_1 is the first observation recorded.

Finding the variance of W_2 is easy. Since W_2 has the same pdf as any of the Y_i's, it follows from the properties of the exponential that

$$\text{Var } (W_2) = \lambda^2$$

To find Var (W_1) we first need $f_{W_1}(w_1)$. Recall from Question 4.6.8 that the sum of n independent exponential random variables with parameter λ has the gamma distribution with parameters n and λ. Let

$$Y = \sum_{i=1}^{n} Y_i$$

and denote its pdf by $f_Y(y)$. Since $W_1 = \left(\dfrac{1}{n}\right)Y$, it follows that

$$f_{W_1}(w_1) = n \cdot f_Y(w_1 n)$$

$$= \frac{1}{\Gamma(n)(\lambda/n)^n} w_1^{n-1} e^{-w_1 n/\lambda} \qquad 0 < w_1 < \infty$$

That is, W_1 is gamma with parameters n and λ/n. Therefore, from Theorem 4.6.2,

$$\text{Var } (W_1) = \frac{\lambda^2}{n}$$

Quite clearly, W_1 is the better of the two estimators: Var (W_1) is smaller than Var (W_2) by a factor of n.

Example 5.5.2

For the problem of estimating the parameter of a uniform distribution, we have already seen that one unbiased estimator is

$$W_1 = \left(\frac{n+1}{n}\right) \cdot Y_{\max}$$

and that

$$f_{W_1}(w_1) = \frac{\dfrac{n^{n+1}}{(n+1)^n} \cdot (w_1)^{n-1}}{\theta^n} \qquad 0 < w_1 < \left(\frac{n+1}{n}\right)\theta$$

By following a similar sort of argument (see Question 5.4.3), we could base a second unbiased estimator for θ on $Y_1' = Y_{\min}$, the *smallest* of the n order statistics. In particular,

$$f_{Y_{\min}}(y; \theta) = \frac{n}{\theta}\left(1 - \frac{y}{\theta}\right)^{n-1} \qquad 0 < y < \theta$$

and

$$E(Y_{\min}) = \int_0^\theta \frac{ny}{\theta}\left(1 - \frac{y}{\theta}\right)^{n-1} dy = \left(\frac{1}{n+1}\right)\theta$$

From the latter result, it can be seen that

$$W_2 = (n + 1) \cdot Y_{\min}$$

would be a second unbiased estimator for θ. Also, it can easily be verified that the pdf for W_2 is given by

$$f_{W_2}(w_2; \theta) = \frac{n}{(n + 1)\theta} \left(1 - \frac{w_2}{(n + 1)\theta}\right)^{n - 1} \qquad 0 < w_2 < (n + 1)\theta$$

According to Definition 5.5.1, deciding which estimator is better—W_1 or W_2—requires a comparison of their respective variances. First, consider W_1:

$$E(W_1^2) = \int_0^{[(n + 1)/n]\theta} \frac{\dfrac{n^{n + 1}}{(n + 1)^n} (w_1)^{n + 1}}{\theta^n} \, dw_1$$

$$= \frac{(n + 1)^2}{n(n + 2)} \theta^2$$

in which case

$$\mathrm{Var}\,(W_1) = E(W_1^2) - [E(W_1)]^2$$

$$= \frac{(n + 1)^2}{n(n + 2)} \theta^2 - \theta^2$$

$$= \frac{\theta^2}{n(n + 2)}$$

Similarly (see Question 5.5.2),

$$\mathrm{Var}\,(W_2) = \frac{n\theta^2}{n + 2}$$

Since

$$\frac{\theta^2}{n(n + 2)} < \frac{n\theta^2}{n + 2} \qquad (\text{for } n > 1)$$

W_1 is more efficient than W_2. Actually, the superiority of W_1 is very pronounced—its relative efficiency is n^2:

$$\text{rel. eff. of } W_1 \text{ to } W_2 = \frac{\dfrac{n\theta^2}{n + 2}}{\dfrac{\theta^2}{n(n + 2)}} = n^2$$

CASE STUDY 5.5.1

On at least one occasion statistical estimation has been used for espionage purposes. During World War II, a very simple statistical procedure was developed for estimating German war production. It was based on *serial numbers* (in our terminology, order statistics) and proved to be highly effective.

Every piece of German equipment, whether it was a V-2 rocket, a tank, or just an automobile tire, was stamped with a serial number that indicated the order in which it was manufactured. If the total number of, say, Mark I tanks produced by a certain date was N, each would bear one of the integers from 1 to N. As the war progressed, some of these numbers became known to the Allies—either by the direct capture of a tank or from records seized when a command post was overrun. The problem was to estimate N using only the sample of "captured" serial numbers, $1 \leq Y'_1 < Y'_2 < \cdots < Y'_n \leq N$.

As a model, it was assumed that the Y'_i's were one of the $\binom{N}{n}$ possible ordered sets of n integers from 1 to N and that each set was equally probable. That is,

$$P(Y'_1 = y'_1 < Y'_2 = y'_2 < \cdots < Y'_n = y'_n) = \binom{N}{n}^{-1}$$

The procedure originally proposed for estimating N was to add to the largest order statistic the average "gap" in the Y'_i's. Specifically,

$$W_1 = Y'_n + \frac{1}{n-1} \sum_{i>j} (Y'_i - Y'_j - 1)$$

which reduces to

$$W_1 = Y'_n + \frac{Y'_n - Y'_1}{n-1} - 1$$

Thus, according to this statistic, if five tanks were captured and they bore the numbers 14, 28, 92, 146, and 298, the estimate for the total number of tanks produced would be 368:

$$w_1 = 298 + \frac{284}{4} - 1 = 368$$

The details will not be gone into here, but it can be shown that W_1 is unbiased and that

$$\text{Var } (W_1) = \frac{n(N-n)(N+1)}{(n-1)(n+1)(n+2)}$$

Another possible estimator for N, one more in keeping with what was used in Example 5.5.2 for the continuous case, would be an unbiased function of Y'_n. If

$$W_2 = \left(\frac{n+1}{n}\right) \cdot Y'_n - 1$$

it can be shown that $E(W_2) = N$. Also,

$$\text{Var } (W_2) = \frac{(N+1)(N-n)}{n(n+2)} \qquad \text{[see (58) for details]}$$

Both W_1 and W_2 are very good estimators, but the latter is better, the relative efficiency of W_1 to W_2 being slightly less than 1:

$$\frac{\text{Var } (W_2)}{\text{Var } (W_1)} = 1 - \frac{1}{n^2}$$

As n increases, though, the precision superiority of W_2 diminishes and, for all practical purposes, the two estimators become equivalent.

When the war was over and the official records of the Speer Ministry impounded, it was found that estimates derived from procedures similar to the one just described were far more accurate than those based on any other source of information. As an example, the serial-number estimate for German tank production in 1942 was 3400, which was very close to the actual figure. The "official" Allied estimate, based on information culled from intelligence reports and espionage activities, was 18,000. Errors of this magnitude were not uncommon. Often the reason for these inflated estimates was the Nazi propaganda machine. Efforts to create the impression that Germany was much stronger than it really was were highly successful. Only the completely objective serial-number procedure remained unaffected!

Question 5.5.1   Recall Example 5.5.1. Show that $W_3 = n \cdot Y_{\min}$ is also an unbiased estimator for λ, and find the relative efficiency of W_1 to W_3 and W_2 to W_3.

Question 5.5.2   In Example 5.5.2, carry out the details to verify the formulas for $E(Y_{\min})$, $f_{W_2}(w_2; \theta)$, and Var (W_2).

Question 5.5.3   Find the relative efficiency of $2Y$ with respect to $\left(\dfrac{n+1}{n}\right) Y_{\max}$ in Example 5.5.2.

Question 5.5.4   Two unbiased estimators for the binomial parameter p were discussed in Example 5.4.2: $W_1 = Y/n$ and $W_2 = Y_i$. Find the relative efficiency of W_2 with respect to W_1.

Question 5.5.5   Let Y_1, Y_2, Y_3 be a random sample from the exponential pdf $f_Y(y; \lambda) = (1/\lambda)e^{-y/\lambda}$, $y > 0$. Find the relative efficiency of $\frac{1}{4}(Y_1 + 2Y_2 + Y_3)$ with respect to $\bar{Y}$.

Question 5.5.6   Let Y_1, Y_2, ..., Y_n be a random sample from a Poisson pdf with parameter λ. Let $W_1 = Y_1$ and $W_2 = \bar{Y}$ be two unbiased estimators for λ. Compute their relative efficiency.

Question 5.5.7   It can be shown that if $2n + 1$ random observations are taken from a continuous symmetric pdf with mean μ, and if $f_X(\mu) \neq 0$, then the sample median, X'_{n+1}, is an unbiased estimator for μ, and, for large n, the variance of X'_{n+1} is approximately

$$\frac{1}{8[f_X(\mu)]^2 n}$$

[see (48)]. Let $X_1, X_2, \ldots, X_{2n+1}$ be a random sample from a uniform pdf over the interval $(0, \theta)$. Then $W_1 = X'_{n+1}$ is an unbiased estimator for $\theta/2$. Construct an unbiased estimator for $\theta/2$ using X'_{2n+1} and compute the relative efficiency of the latter to X'_{n+1}.

Question 5.5.8 The parameter for an exponential pdf can be estimated using a linear combination of selected order statistics:

$$W = \text{estimator for } \lambda = c_1 X'_{\alpha_1} + c_2 X'_{\alpha_2} + \cdots + c_k X'_{\alpha_k}$$

where $1 \le \alpha_1 < \alpha_2 < \cdots < \alpha_k \le n$, and n is the number of observations available. For various values of n and k, the optimal choices for the c_i's and the α_i's have been determined (89). For example, if $n = 30$, and we wish to estimate λ with a linear combination of *three* order statistics, the α_1, α_2, and α_3 that minimize the variance of the resulting unbiased estimator are 17, 26, and 30, respectively, and the estimator is given by

$$W = 0.450522 X'_{17} + 0.219090 X'_{26} + 0.053584 X'_{30}$$

Here the relative efficiency of W to $W^* = \bar{X}$ is 0.91. How many observations would be required to estimate λ using $\bar{X}$, but with the same precision afforded by W?

5.6 MINIMUM-VARIANCE ESTIMATORS: THE CRAMER–RAO LOWER BOUND

The concept of relative efficiency has solved one "problem" but, in so doing, raised another. What it provides is a working criterion for choosing between two competing estimators: given W_1 and W_2, both unbiased, we will elect as the "better" estimator the one whose variance is smaller. What it has *not* given us, though, is any reassurance that even the better of W_1 and W_2 is any good. How do we know, for example, that there is not a W_3 somewhere in the class of unbiased estimators whose variance is much smaller than either W_1's or W_2's? The next theorem offers at least a partial answer to that question in the form of a lower bound for the variance of an unbiased estimator.

Theorem 5.6.1. (Cramer–Rao Inequality) Let $Y_1, Y_2, \ldots, Y_n$ be a random sample from $f_Y(y; \theta)$, where $f_Y(y; \theta)$ has continuous first-order partial derivatives at all but a finite set of points. Suppose that the set where $f_Y(y; \theta) \ne 0$ does not depend on θ. Let $W = h(Y_1, Y_2, \ldots, Y_n)$ be any unbiased estimator for θ. Then the following hold:

(a)
$$\text{Var}(W) \ge \left\{ n \int_{-\infty}^{\infty} \left(\frac{\partial \ln f_Y(y; \theta)}{\partial \theta} \right)^2 f_Y(y; \theta) \, dy \right\}^{-1}$$

$$= \left\{ nE\left[\left(\frac{\partial \ln f_Y(Y; \theta)}{\partial \theta} \right)^2 \right] \right\}^{-1}$$

(b) Equality holds in part (a) if and only if

$$\sum_{i=1}^{n} \frac{\partial \ln f_Y(y_i; \theta)}{\partial \theta} = A(\theta)[h(y_1, \ldots, y_n) - \theta]$$

for all $y_1, y_2, \ldots, y_n$ (possibly excepting a set of probability 0), where $A(\theta)$ does not depend on the y_i's. In this case Var $(W) = [A(\theta)]^{-1}$.

(c) If $f_Y(y; \theta)$ also has continuous second partial derivatives, then the lower bound in part (a) can be written

$$\left[-n \int_{-\infty}^{\infty} \frac{\partial^2 \ln f_Y(y; \theta)}{\partial \theta^2} \cdot f_Y(y; \theta)\, dy \right]^{-1} = \left\{ -nE\left[\frac{\partial^2 f_Y(Y; \theta)}{\partial \theta^2} \right] \right\}^{-1}$$

Proof. See Appendix 5.1.

Example 5.6.1

Let $Y_1, Y_2, \ldots, Y_n$ be a random sample from the Bernoulli pdf,

$$f_{Y_i}(y_i; p) = p^{y_i}(1 - p)^{1 - y_i} \qquad y_i = 0, 1, \quad 0 < p < 1$$

(*Note*: We are writing the pdf as f_{Y_i} rather than f_Y to avoid confusing the pdf of the observations with the pdf of their *sum*, $Y = \sum_{i=1}^{n} Y_i$.) It has already been suggested (recall Example 5.4.2) that

$$W = \frac{Y}{n}$$

be used as an unbiased estimator for p. How does Var (W) compare with the Cramer–Rao lower bound?

Note, first, that

$$\text{Var } (W) = \frac{1}{n^2} \text{Var } (Y) = \frac{1}{n^2}\, np(1 - p) = \frac{p(1 - p)}{n} \tag{5.6.1}$$

To put Equation 5.6.1 in perspective, we can use part (c) of Theorem 5.6.1 to find the theoretical lower limit for the variance of an unbiased estimator for the Bernoulli parameter p.

To begin, we write

$$\ln f_{Y_i}(Y_i; p) = Y_i \cdot \ln p + (1 - Y_i) \cdot \ln (1 - p)$$

Therefore,

$$\frac{\partial \ln f_{Y_i}(Y_i; p)}{\partial p} = \frac{Y_i}{p} - \frac{1 - Y_i}{1 - p}$$

and

$$\frac{\partial^2 \ln f_{Y_i}(Y_i; p)}{\partial p^2} = -\frac{Y_i}{p^2} - \frac{1 - Y_i}{(1 - p)^2}$$

Taking the expected value of the second derivative gives

$$E\left[\frac{\partial^2 \ln f_{Y_i}(Y_i; p)}{\partial p^2}\right] = -\frac{p}{p^2} - \frac{(1-p)}{(1-p)^2} = -\frac{1}{p(1-p)}$$

It follows from the theorem, then, that the Cramer–Rao lower bound for this situation is

$$\frac{1}{-n\left[-\dfrac{1}{p(1-p)}\right]} = \frac{p(1-p)}{n} \tag{5.6.2}$$

Equations 5.6.1 and 5.6.2 show that the variance of $W = Y/n$ is *equal* to the Cramer–Rao lower bound. This means that W is "optimal" in the sense that among the class of unbiased estimators, none has a smaller variance.

We can also demonstrate that the variance of W achieves the Cramer–Rao bound by using part (b) of Theorem 5.6.1:

$$\sum_{i=1}^{n} \frac{\partial \ln f_{Y_i}(y_i; p)}{\partial p} = \sum_{i=1}^{n}\left(\frac{y_i}{p} - \frac{1-y_i}{1-p}\right)$$

$$= \frac{n}{p(1-p)}\left(\frac{\displaystyle\sum_{i=1}^{n} y_i}{n} - p\right)$$

Thus $A(p) = n/p(1-p)$, $W = (1/n)\displaystyle\sum_{i=1}^{n} Y_i$ has minimum variance, and Var $(W) = [A(p)]^{-1} = p(1-p)/n$.

Question 5.6.1 Let $Y_1, Y_2, \ldots, Y_n$ be a random sample from $f_Y(y; \lambda) = (1/\lambda)e^{-y/\lambda}$, $y > 0$. Find the Cramer–Rao lower bound for the variance of an unbiased estimator for λ. Also, compare the variance of $W = \bar{Y}$ to the Cramer–Rao lower bound.

The notion of optimality raised in Example 5.6.1 motivates the next two definitions and points up the usefulness of Theorem 5.6.1. Notice that what is defined here to be the *efficiency* of an unbiased estimator is simply its relative efficiency (in the terminology of Section 5.5) when compared to an estimator whose variance achieves the Cramer–Rao lower bound.

> **Definition 5.6.1.** Let U denote the class of all unbiased estimators, $W = h(Y_1, Y_2, \ldots, Y_n)$, for a given parameter. We say that W^* is a *best* (or *minimum-variance*) estimator if $W^* \in U$ and
>
> $$\text{Var}(W^*) \leq \text{Var}(W) \qquad \text{for all} \quad W \in U$$

Definition 5.6.2. An unbiased estimator, $W = h(Y_1, Y_2, \ldots, Y_n)$, is said to be *efficient* if and only if

$$\text{Var} (W) = \frac{1}{nE\left[\dfrac{\partial \ln f_Y(Y; \theta)}{\partial \theta}\right]^2}$$

Also, the *efficiency* of an unbiased W is defined to be the ratio

$$\text{efficiency of } W = \frac{\dfrac{1}{nE\left[\dfrac{\partial \ln f_Y(Y; \theta)}{\partial \theta}\right]^2}}{\text{Var} (W)}$$

Thus $W = Y/n$ is an *efficient* estimator for the binomial parameter, p. It is also a *best* estimator. Similarly, the sample mean, when used to estimate the parameter in an exponential distribution, is efficient and best (see Question 5.6.1).

Comment

The designations "efficient" and "best" are not synonymous. If the variance of an unbiased estimator is equal to the Cramer–Rao lower bound (as in the case of the binomial and the exponential), then that estimator by definition is a best estimator. The converse, though, is not always true. There are situations for which the variances of *no* unbiased estimators achieve the Cramer–Rao lower bound. Thus none of them is *efficient* but one (or more) of them would still be termed *best*. When the hypotheses of Theorem 5.6.1 are *not* satisfied, estimators may exist with variances less than the Cramer–Rao lower bound (see Question 5.6.5).

Question 5.6.2 Show that $W = \bar{Y}$ is an efficient estimator for the parameter λ in a Poisson distribution.

Question 5.6.3 Let $Y_1, Y_2, \ldots, Y_n$ be a random sample from a normal pdf with mean μ and variance σ^2. Assume that σ^2 is known. Show that $\bar{Y}$ is efficient.

Question 5.6.4 Let $Y_1, Y_2, \ldots, Y_n$ be a random sample from a gamma pdf with parameters r and $1/\theta$. Assume that r is known.
(a) Calculate both forms of the Cramer–Rao bound given in Theorem 5.6.1 for θ.
(b) Show that $(1/r)\bar{Y}$ is an efficient estimator for θ [do not forget to show that $(1/r)\bar{Y}$ is unbiased].

Question 5.6.5 Let $Y_1, Y_2, \ldots, Y_n$ be a random sample from the uniform pdf over $(0, \theta)$.
(a) Find the Cramer–Rao lower bound.
(b) Show that $[(n + 1)/n]Y_{\max}$ has variance less than the Cramer–Rao lower bound.

Question 5.6.6 Let $Y_1, Y_2, \ldots, Y_n$ be a random sample from the pdf $f_Y(y; \theta) = 2y/\theta^2$, $0 < y < \theta$.

(a) Show that $W = \frac{3}{2}\bar{Y}$ is an unbiased estimator for θ.

(b) Find the Cramer–Rao lower bound.

(c) Show that Var (W) is less than the Cramer–Rao lower bound.

5.7 FURTHER PROPERTIES OF ESTIMATORS: CONSISTENCY AND SUFFICIENCY

Although the notions of unbiasedness and efficiency lead to the most basic characterizations of point estimates, there are still deeper properties of W and $f_W(w)$ that merit examination. In this section we consider briefly two of these: the first, *consistency*, is an asymptotic property of an estimator, and the second, *sufficiency*, is a property related to the amount of "information" a given estimator contains.

To understand consistency—or, for that matter, any asymptotic property of an estimator—it is first necessary to think of W as really being W_n, the nth member of an infinite *sequence* of estimators, $W_1, W_2, \ldots, W_n, \ldots$. For example, if r observations were drawn from the exponential pdf, $f_Y(y; \lambda) = (1/\lambda)e^{-y/\lambda}$, we would probably estimate λ with

$$W = W_r = \frac{1}{r} \sum_{i=1}^{r} Y_i$$

where W_r itself is gamma distributed with parameters r and λ/r (see Example 5.5.1). But suppose that s observations were taken: then the estimator would become

$$W = W_s = \frac{1}{s} \sum_{i=1}^{s} Y_i$$

with W_s being gamma distributed with parameters s and λ/s. Thus we see that the *form* of the estimator remains the same (for different sample sizes) but the behavior of $f_{W_n}(w_n)$ may very well change. This suggests that it might prove instructive to examine the limiting behavior of $f_{W_n}(w_n)$ as n gets large. We may find—to W's credit—that $f_{W_n}(w_n)$ has some very desirable properties *in the limit* that it fails to possess for any finite n.

Consistency is one such desirable property of $f_{W_n}(w_n)$. Roughly speaking, an estimator is consistent if, as n gets large, the probability that W_n lies arbitrarily close to the parameter being estimated becomes, itself, arbitrarily close to 1. This has two immediate implications: (1) W_n is *asymptotically unbiased* (see Question 5.4.15), and (2) Var (W_n) converges to 0.

Definition 5.7.1. An estimator $W_n = h(Y_1, Y_2, \ldots, Y_n)$ is said to be *consistent* (for θ) if it converges in probability to θ—that is, if for all $\epsilon > 0$ and $\delta > 0$, there exists an $n(\epsilon, \delta)$ such that

$$P(|W_n - \theta| < \epsilon) > 1 - \delta \qquad \text{for} \quad n > n(\epsilon, \delta)$$

Example 5.7.1

Let $Y_1, Y_2, \ldots, Y_n$ be a random sample from

$$f_Y(y; \theta) = \frac{1}{\theta} \qquad 0 < y < \theta$$

and let $W_n = Y_{\max}$. We know that $Y_{\max}$ is biased (for θ), but it might still be consistent. The question is whether or not there is an $n(\epsilon, \delta)$ large enough so that

$$P(|W_n - \theta| < \epsilon) > 1 - \delta \qquad \text{for} \quad n > n(\epsilon, \delta)$$

Recall from Example 5.3.1 that

$$f_{W_n}(w_n) = \frac{n(w_n)^{n-1}}{\theta^n} \qquad 0 < w_n < \theta$$

Therefore,

$$P(|W_n - \theta| < \epsilon) = P(\theta - \epsilon < W_n < \theta)$$

$$= \int_{\theta - \epsilon}^{\theta} \frac{n(w_n)^{n-1}}{\theta^n} \, dw_n = \frac{w_n^n}{\theta^n} \bigg|_{\theta - \epsilon}^{\theta}$$

$$= 1 - \left(\frac{\theta - \epsilon}{\theta}\right)^n$$

Note that for any $\epsilon > 0$, it is possible to find an n making $[(\theta - \epsilon)/\theta]^n$ as small as desired (in particular, smaller than δ). By Definition 5.7.1, then, $W_n = Y_{\max}$ is consistent for θ.

Figure 5.7.1 shows the situation graphically. As n increases, the shape of $f_{W_n}(w_n)$ changes in such a way that it becomes increasingly concentrated within an ϵ-neighborhood of θ.

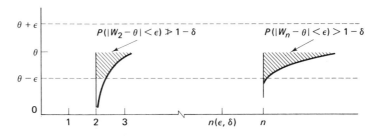

Figure 5.7.1 The consistency of $Y_{\max}$.

Example 5.7.2

The similarity between the definition of consistency and the statement of Chebyshev's inequality (see Chapter 3) suggests that it might be possible to use the latter to establish the former. As a case in point, consider our usual estimator for the binomial parameter, p:

$$W = Y/n = \text{sample proportion of successes}$$

Since W is a legitimate random variable and $E(W) = p$, it follows from Chebyshev's inequality that

$$P(|W - p| < \epsilon) > 1 - \frac{\text{Var } (W)}{\epsilon^2} \qquad (5.7.1)$$

Furthermore, since

$$\text{Var } (W) = \frac{p(1 - p)}{n} \leq \frac{1}{4n}$$

we can write Inequality 5.7.1 as

$$P(|W - p| < \epsilon) > 1 - \frac{1}{4n\epsilon^2}$$

From the latter expression, it can be seen that for $n > (1/4\epsilon^2\delta)$, the probability that W lies in an ϵ-neighborhood of p will be greater than $1 - \delta$. Thus Y/n is consistent for p.

Question 5.7.1 In Example 5.7.1, what is the smallest n for which $W_n = Y_{\max}$ has a probability of at least 0.95 of lying within 0.04 of θ? Assume that $\theta = 1$.

Question 5.7.2 Is the estimator $W = \bar{Y}$ consistent for the parameter of a Poisson distribution?

Question 5.7.3 Suppose that a sample of size n is drawn from an exponential distribution and we choose Y_1, the first observation recorded, to be an estimator for λ. Show that $W = Y_1$ is not consistent.

Question 5.7.4 Suppose that $Y_1, Y_2, \ldots, Y_n$ is a random sample from an exponential distribution. Show that $W = \sum_{i=1}^{n} Y_i$ is not consistent.

Question 5.7.5 Suppose that Y is a random variable with mean μ and variance σ^2. Let $\bar{Y}$ be the sample mean of a random sample from f_Y. Show that $\bar{Y}$ is consistent for μ.

Question 5.7.6 Let $W_1, W_2, \ldots, W_n, \ldots$ be any sequence of estimators for θ. Show that

$$E[(W_n - \theta)^2] = \text{Var } (W_n) + [\theta - E(W_n)]^2$$

Question 5.7.7 A sequence of estimators $W_1, W_2, \ldots, W_n, \ldots$ for θ is said to be *squared-error consistent* if

$$\lim_{n \to \infty} E[(W_n - \theta)^2] = 0 \qquad \text{for all } \theta$$

Show that if W_n is squared-error consistent, then W_n is also consistent in the sense of Definition 5.7.1.

Question 5.7.8 Let $Y_1, Y_2, \ldots, Y_n$ be a random sample from the uniform pdf over the interval $(0, \theta)$. Let $W_n = Y_{max}$. Is W_n squared-error consistent?

Question 5.7.9 Show that a squared-error consistent sequence of estimators is asymptotically unbiased (see Question 5.4.15). Does any estimator in the sequence have to be unbiased?

Question 5.7.10 Suppose that $X_1, X_2, \ldots, X_{2n+1}$ is a random sample taken from a symmetric continuous pdf with mean μ, where $f_X(\mu) \neq 0$. Show that the sample median X'_{n+1} is a consistent estimator for μ. (*Hint*: See Question 5.5.7.)

The next several results deal with the second property of estimators mentioned at the outset of this section, *sufficiency*. For reasons to be outlined shortly, the search for minimum-variance unbiased estimators can always be narrowed to a class of W's that share this particular feature. To develop the idea of sufficiency in any great detail, though, requires considerably more theoretical background than the previous chapters have provided. Consequently, all we can do here is offer a heuristic explanation and present a few simple examples.

If a sample of size n is drawn from some $f_Y(y; \theta)$, we know that the sample space is the totality of n-tuples, $(y_1, y_2, \ldots, y_n)$. It is also true that defining an estimator, W, has the effect of partitioning this sample space into a set of mutually exclusive and exhaustive subsets. For example, if two observations are drawn from a Poisson distribution one possible estimator for the occurrence rate, λ, is

$$W = \bar{Y} = \frac{1}{2} \sum_{i=1}^{2} Y_i$$

Note that W will equal, say, 3 for any of the following 2-tuples: (0, 6), (1, 5), (2, 4), (3, 3), (4, 2), (5, 1), or (6, 0); similarly, it will equal 2.5 if (0, 5), (1, 4), (2, 3), (3, 2), (4, 1), or (5, 0) is the sample outcome. Each possible outcome, then, can be put into a subset according to the value of W it produces.

Basically, whether or not W is sufficient for λ depends on whether the particular 2-tuple observed—say, (1, 5)—conveys any more information about λ than what is already contained in the statement "$W = 3$." If knowing which partition set an outcome belongs to reveals as much information about the unknown parameter as the particular outcome, itself, then the method of partitioning—that is, W—is said to be *sufficient* (for λ).

The binomial distribution can be used to illustrate this idea in a little more detail. Suppose that in four Bernoulli trials, two successes and two failures—in that order—are observed. That is, the sample outcome is (s, s, f, f) or, in terms of the usual Bernoulli random variable, (1, 1, 0, 0). From this, the conventional estimate for p would be 0.5:

$$w = \frac{\text{number of successes}}{n} = \frac{2}{4} = 0.5$$

Suppose, now, that we look at the conditional distribution of $(1, 1, 0, 0)$ given that $w = 0.5$. By definition,

$$P((1, 1, 0, 0) \mid w = 0.5) = \frac{P((1, 1, 0, 0) \; and \; W = 0.5)}{P(W = 0.5)}$$

$$= \frac{P(1, 1, 0, 0)}{P(\text{exactly 2 successes in 4 trials})}$$

$$= \frac{p^2(1 - p)^2}{\binom{4}{2} p^2 (1 - p)^2} = \binom{4}{2}^{-1}$$

$$= P((0, 1, 0, 1) \mid w = 0.5)$$

$$= P((0, 0, 1, 1) \mid w = 0.5) \qquad \text{and so on}$$

Notice that these conditional probabilities are independent of p. That is, knowing that the particular outcome is, say, $(1, 1, 0, 0)$ adds nothing to our information about p that we do not already know, having been told that $w = 0.5$. (Of course, it follows that any one-to-one function of W, say nW, $W + 3$, and so on, will enjoy this same property.)

Definition 5.7.2.   Let $Y_1, Y_2, \ldots, Y_n$ be a random sample from $f_Y(y; \theta)$. The statistic $W = h(Y_1, Y_2, \ldots, Y_n)$ is said to be *sufficient* for θ if, for all θ and all possible sample points, the conditional pdf of $Y_1, Y_2, \ldots, Y_n$ given w does not depend on θ, either in the function itself or in the function's domain. More precisely, W is sufficient if for each w with $f_W(w) \neq 0$ and for each n-tuple $(y_1, y_2, \ldots, y_n)$ with $h(y_1, \ldots, y_n) = w$,

$$\frac{f_{Y_1}(y_1; \theta) \cdots f_{Y_n}(y_n; \theta)}{f_W(w; \theta)}$$

does not involve θ.

In establishing that a particular statistic is sufficient, we do not usually use Definition 5.7.2 directly. Various factorization criteria are available that are much easier to work with. One of them, the Fisher–Neyman criterion, is outlined in Theorem 5.7.1. Its proof will not be given [see (178)].

Theorem 5.7.1 (Fisher–Neyman Criterion).   Let $Y_1, Y_2, \ldots, Y_n$ be a random sample from $f_Y(y; \theta)$. Then $W = h(Y_1, Y_2, \ldots, Y_n)$ is a sufficient statistic for θ if and only if the joint density of the Y_i's factors into the product of the pdf of W and a second function that does not depend on θ. That is, W is sufficient if and only if

$$f_Y(y_1; \theta) \cdots f_Y(y_n; \theta) = f_W(h(y_1, \ldots, y_n); \theta) \cdot s(y_1, \ldots, y_n)$$

Example 5.7.3

It is a simple matter to use the Fisher–Neyman criterion to show that

$$W = \sum_{i=1}^{n} Y_i = \text{total number of successes in } n \text{ trials}$$

is a sufficient statistic for the binomial parameter, p. The joint probability function of the Y_i's is the product of n Bernoullis,

$$f_Y(y_1; p) \cdots f_Y(y_n; p) = p^{\sum_{i=1}^{n} y_i}(1-p)^{n-\sum_{i=1}^{n} y_i}$$

and W is binomial with parameters n and p,

$$f_W(w; p) = \binom{n}{w} p^w (1-p)^{n-w} = \binom{n}{\sum_{i=1}^{n} y_i} p^{\sum_{i=1}^{n} y_i}(1-p)^{n-\sum_{i=1}^{n} y_i}$$

Therefore, by choosing

$$s(y_1, y_2, \ldots, y_n) = \binom{n}{\sum_{i=1}^{n} y_i}^{-1}$$

we have that

$$f_Y(y_1; p) \cdots f_Y(y_n; p) = f_W(w; p) \cdot s(y_1, y_2, \ldots, y_n)$$

proving that W is sufficient for p.

Comment

In any real problem, we would not use W itself to estimate p but, rather, W/n, the latter being unbiased. If W is sufficient for p, though, so is W/n.

Example 5.7.4

As a final example of sufficiency, we will prove that the largest order statistic, Y_{max}, is a sufficient statistic for the parameter of a uniform distribution, $f_Y(y; \theta) = 1/\theta, 0 < y < \theta$. First, note that the joint pdf of the y_i's is simply $(1/\theta)^n$ and recall that

$$f_W(w) = \begin{cases} \dfrac{nw^{n-1}}{\theta^n} & \text{for } 0 < w < \theta \\ 0 & \text{otherwise} \end{cases}$$

where $W = Y_{max}$. If we write

$$f_Y(y_1; \theta) \cdots f_Y(y_n; \theta) = \left(\frac{1}{\theta}\right)^n$$

$$= \frac{nw^{n-1}}{\theta^n} \cdot \frac{1}{nw^{n-1}}$$

$$= f_W(w) \cdot s(y_1, y_2, \ldots, y_n)$$

it follows that Y_{max} is a sufficient statistic.

Question 5.7.11 Show that $W = \bar{Y}$ is sufficient for λ in the exponential pdf, $f_Y(y; \lambda) = (1/\lambda)e^{-y/\lambda}$, $y > 0$.

Question 5.7.12 Show that $W = Y_{\min}$ is a sufficient statistic for the threshold parameter θ in the exponential pdf

$$f_Y(y; \theta) = e^{-(y-\theta)} \qquad \theta < y < \infty$$

Using the Fisher–Neyman criterion requires that $f_W(w; \theta)$ be explicitly identified as one of the two factors that multiply together to give the sample's joint pdf. An easier factorization for establishing sufficiency is described in Theorem 5.7.2.

> **Theorem 5.7.2. (Factorization Theorem)** Let $Y_1, Y_2, \ldots, Y_n$ be a random sample from $f_Y(y; \theta)$. Then $W = h(Y_1, Y_2, \ldots, Y_n)$ is a sufficient statistic for θ if and only if
>
> $$f_{Y_1}(y_1; \theta) \cdots f_{Y_n}(y_n; \theta) = g[h(y_1, y_2, \ldots, y_n); \theta]u(y_1, \ldots, y_n)$$
>
> The function g is dependent on the sample only through $h(y_1, \ldots, y_n)$, and u does not involve θ.

Proof. See (73).

In concluding this section, it is appropriate to comment briefly on the way sufficiency relates to two of the other properties discussed earlier—unbiasedness and efficiency. Basically, the importance of sufficiency derives from the fact that the minimum-variance unbiased estimator for any parameter will necessarily be a function of a sufficient statistic. It can be shown that contained in the set, U, of unbiased estimators for a given parameter is a subset, U_S, whose elements are functions of sufficient statistics. Furthermore, U_S has the property that each of its members has a smaller variance than every unbiased estimator in its complement. This means that in looking for a minimum-variance unbiased estimator we need only be concerned with W's that are functions of sufficient statistics. To further simplify matters, in many situations [depending on the nature of $f_Y(y; \theta)$] it can be proved that U_S has only *one* member. Thus, to construct a minimum-variance estimator (in these situations) we need simply find a sufficient statistic and perform whatever transformations are necessary to make it unbiased.

Comment

Explicit consideration of point estimation can be traced back to the time of Galileo, perhaps even earlier. Much of the structure that we now associate with the subject, however, was formulated in the early years of the present century by Sir Ronald A. Fisher. It was Fisher who introduced the notions of unbiasedness, efficiency, and sufficiency.

Question 5.7.13 Let $X_1, X_2, \ldots, X_n$ be a random sample from a geometric pdf with parameter p. Show that $W = X_1 + X_2 + \cdots + X_n$ is a sufficient statistic for p.

Question 5.7.14 Let $X_1, X_2, \ldots, X_n$ be a random sample from $f_X(x) = \theta x^{\theta - 1}$, $0 < x < 1$; $\theta > 0$. Use the factorization theorem to suggest a sufficient statistic for θ.

Question 5.7.15 Suppose $X_1, X_2, \ldots, X_n$ is a random sample from a pdf that can be written in *exponential form*:

$$f_X(x; \theta) = e^{K(x)p(\theta) + S(x) + q(\theta)}$$

where the range of x is independent of θ. Show that

$$W = \sum_{i=1}^{n} K(x_i)$$

is a sufficient statistic for θ.

Question 5.7.16 Write the exponential pdf $f_X(x) = (1/\lambda)e^{-x/\lambda}$, $x > 0$, in the form suggested in Question 5.7.15 and deduce a sufficient estimator for λ (given a random sample X_1, $X_2, \ldots, X_n$).

Question 5.7.17 Let $X_1, X_2, \ldots, X_n$ be a random sample from a *Pareto* distribution

$$f_X(x) = \frac{\theta}{(1 + x)^{\theta + 1}} \qquad 0 < x < \infty, \quad 0 < \theta < \infty$$

Write $f_X(x)$ in exponential form and deduce a sufficient statistic for θ. [*Note*: The Pareto distribution finds many economic applications. It models such phenomena as the distribution of incomes above a threshold value and the distribution of the amount of insurance claims above a threshold value (47).]

Question 5.7.18 Consider a series of three Bernoulli trials, each with the same unknown success probability p. Let X_i denote the number of successes on the ith trial ($X_i = 0$ or 1). It has already been established that $W = X_1 + X_2 + X_3$ is a sufficient estimator for p; show that $W' = X_1 + 2X_2 + 3X_3$ is not.

5.8 ESTIMATING PARAMETERS: THE METHOD OF MAXIMUM LIKELIHOOD AND THE METHOD OF MOMENTS

Thus far, our discussion has centered exclusively on the *properties* of estimators—how those estimators were arrived at has not been a concern. Given that an estimator for, say, the Poisson parameter is $W = \bar{Y}$, we have simply tried to establish, after the fact, whether or not it satisfies certain desirable criteria. Is it unbiased? efficient? consistent? sufficient? What is perhaps the more fundamental question, though, still remains: how was W chosen to be $\bar{Y}$ in the first place, as opposed to $\prod_{i=1}^{n} Y_i$, $Y_{\max}$, or any of the other infinitely many possible statistics?

In this section we answer that question. Two different methods of estimation will be described, the *method of maximum likelihood* and the *method of moments*. While these are not the only techniques for finding estimators, they are among the most common. From a theoretical standpoint, the method of maximum likelihood is particularly important, having as it does the property that the estimators it produces are always functions of sufficient statistics. From a practical standpoint, though, it is sometimes difficult to apply. For certain $f_Y(y; \theta)$ the equations that arise in carrying out the method of maximum likelihood are nonlinear and have solutions that can only be approximated. In these situations the method of moments can sometimes be of real value. Although "moment" estimators are not necessarily sufficient, efficient, or even unbiased, they are, nevertheless, "reasonable" and they can often be obtained with a minimum of mathematical difficulty.

The first procedure to be discussed will be the method of maximum likelihood. Let $Y_1, Y_2, \ldots, Y_n$ be a random sample from $f_Y(y; \theta)$ and let L be its joint pdf:

$$L = f_{Y_1, Y_2, \ldots, Y_n}(y_1, y_2, \ldots, y_n; \theta) = f_Y(y_1; \theta) \cdots f_Y(y_n; \theta)$$

$$= \prod_{i=1}^{n} f_Y(y_i; \theta)$$

Up to now, the joint pdf of a random sample has always been thought of as a function *of the sample*. That is, $L = L(y_1, y_2, \ldots, y_n) = L(y_1, y_2, \ldots, y_n; \theta)$, meaning that θ is fixed and the y_i's are variables. For the purposes of finding estimators, though, it will make sense to reverse the roles of the parameter and the data and think of L as a function of θ—with the y_i's considered fixed. Accordingly, we will write

$$L = L(\theta) = L(\theta; y_1, y_2, \ldots, y_n) = \prod_{i=1}^{n} f_Y(y_i; \theta)$$

Definition 5.8.1. Let $Y_1, Y_2, \ldots, Y_n$ be a random sample from $f_Y(y; \theta)$. If L, the joint pdf of the Y_i's, is thought of as being a function of θ and if the y_i's are thought of as being fixed, then

$$L = L(\theta) = \prod_{i=1}^{n} f_Y(y_i; \theta)$$

is called the *likelihood function*.

A simple numerical example will illustrate how we intend to make use of $L(\theta)$. Suppose three observations—$y_1 = 2.0$, $y_2 = 4.0$, $y_3 = 3.6$—are drawn from $f_Y(y; \theta) = (1/\theta)e^{-y/\theta}$, $y > 0$. By definition, the corresponding likelihood function is

$$L(\theta) = \prod_{i=1}^{3} \left(\frac{1}{\theta}\right)e^{-y_i/\theta}$$

$$= \left(\frac{1}{\theta^3}\right)e^{-(1/\theta)\sum_{i=1}^{3} y_i}$$

$$= \left(\frac{1}{\theta^3}\right)e^{-9.6/\theta}$$

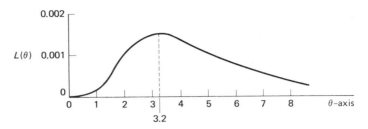

Figure 5.8.1 Graph of $L(\theta)$.

(see Figure 5.8.1). Notice that $L(\theta)$ is greatest when $\theta = 3.2$. That is, for the given three observations, the joint pdf is maximized when the parameter is assigned the value 3.2. This amounts to saying that, in terms of the joint pdf, the value $\theta = 3.2$ "best explains" or is "most compatible with" the data. It would not be unreasonable, then, to set W equal to 3.2 (which, in this case, equals $\bar{y}$). Definition 5.8.2 summarizes the "principle" of using $L(\theta)$ to locate a suitable estimate for θ.

> **Definition 5.8.2 (Maximum-Likelihood Estimation).** Let $Y_1, Y_2, \ldots,$ Y_n be a random sample from $f_Y(y; \theta)$ and let $L(\theta)$ be the corresponding likelihood function. As a general procedure for constructing estimators, we will look for the w that maximizes $L(\theta)$. Any value that does—that is, any w for which
>
> $$L(w) \geq L(\theta) \qquad \text{for all } \theta \neq w$$
>
> is said to be a *maximum-likelihood estimate* (or MLE) for θ. We will sometimes write $w = \hat{\theta}$.

In many cases of interest, $L(\theta)$ is differentiable and its maximization is a familiar calculus problem: the basic procedure is to differentiate $L(\theta)$ with respect to θ, set the derivative equal to 0, and solve for θ. The usual second derivative test can be used to verify that $\hat{\theta}$ is, in fact, *maximizing $L(\theta)$*. (An exception is noted in Example 5.8.3.)

Comment

Often it will be easier to maximize $\ln L(\theta)$ rather than $L(\theta)$. Of course, the monotonicity of the natural logarithm insures that the same w maximizes both functions.

Comment

If the presumed data model is a function of k unknown parameters, then the vector of maximum-likelihood estimates, $(w_1, w_2, \ldots, w_k)$, is gotten in most cases by solving simultaneously the system of equations,

$$\frac{\partial L(\theta_1, \theta_2, \ldots, \theta_k)}{\partial \theta_i} = 0 \qquad i = 1, 2, \ldots, k$$

Example 5.8.1

Let $Y_1, Y_2, \ldots, Y_n$ be a random sample from the Poisson distribution, $f_Y(y; \lambda) = e^{-\lambda}\lambda^y/y!$, $y = 0, 1, 2, \ldots$. The corresponding likelihood function, L, and its natural log are

$$L(\lambda; y_1, y_2, \ldots, y_n) = \prod_{i=1}^{n} \frac{e^{-\lambda}\lambda^{y_i}}{y_i!} = \frac{e^{-n\lambda}\lambda^{\sum_{i=1}^{n} y_i}}{\prod_{i=1}^{n} y_i!}$$

and

$$\ln L(\lambda; y_1, y_2, \ldots, y_n) = -n\lambda + \left(\sum_{i=1}^{n} y_i\right)\ln \lambda - \ln\left(\prod_{i=1}^{n} y_i!\right)$$

Setting the derivative of $\ln L$ (with respect to λ) equal to 0 gives

$$-n + \frac{1}{\lambda}\sum_{i=1}^{n} y_i = 0 \qquad (5.8.1)$$

Let w denote the value of λ that satisfies Equation 5.8.1. Then

$$w = \frac{1}{n}\sum_{i=1}^{n} y_i = \bar{y}$$

Thus the maximum-likelihood estimate for the parameter of the Poisson is the one we have been using all along. Note that $w = \bar{y}$ locates the *maximum* of $\ln L$ and not the minimum, because the second derivative of $\ln L$ is $-(1/\lambda^2)\sum_{i=1}^{n} y_i$, a number always less than or equal to 0.

Example 5.8.2

If a series of independent Bernoulli trials is terminated with the occurrence of the first success, the probability function for K, the length of the series, will be given by the *geometric distribution*,

$$f_K(k; p) = P(K = k) = (1 - p)^{k-1}p \qquad k = 1, 2, \ldots$$

where p is the (unknown) probability of success at any trial. If n such series are observed, the data will consist of n lengths, $k_1, k_2, \ldots, k_n$. To get the maximum-likelihood estimator for p, note, first of all, that

$$L(p; k_1, k_2, \ldots, k_n) = L(p) = (1 - p)^{\sum_{i=1}^{n} k_i - n} p^n \qquad (5.8.2)$$

Taking the ln of Equation 5.8.2 gives

$$\ln L(p) = \left(\sum_{i=1}^{n} k_i - n\right)\ln (1 - p) + n \ln p$$

from which it follows that

$$\frac{d}{dp}[\ln L(p)] = \left(\frac{1}{1-p}\right)\left(n - \sum_{i=1}^{n} k_i\right) + \frac{n}{p}$$

Let w denote the value of p for which $d/dp\, [\ln L(p)] = 0$. It can be easily seen that

$$w = \frac{n}{\displaystyle\sum_{i=1}^{n} k_i} = \frac{1}{\bar{k}}$$

Again, notice that

$$\frac{d^2}{dp^2}\, [\ln L(p)] = \frac{1}{(1-p)^2}\left(n - \sum_{i=1}^{n} k_i\right) - \frac{1}{p^2} < 0$$

guaranteeing that the differentiation has, indeed, produced a maximum.

Example 5.8.3

Maximum-likelihood estimators are not always gotten by setting a derivative equal to 0. Consider, for instance, a sample of size n drawn from the uniform distribution over $(0, \theta)$. The likelihood function is

$$L(\theta; y_1, y_2, \ldots, y_n) = \begin{cases} \left(\dfrac{1}{\theta}\right)^n & \text{for} \quad 0 \le y_i \le \theta \\ 0 & \text{otherwise} \end{cases}$$

and, since each y_i must be less than or equal to θ, the range of θ extends from y_{max} to infinity. To find the θ that maximizes $L(\theta)$ we need simply graph $(1/\theta)^n$. It is clear by inspection that, being monotonic, $L(\theta)$ is maximized at its left endpoint (see Figure 5.8.2). Thus

$$w = y_{max}$$

(recall Example 5.4.1).

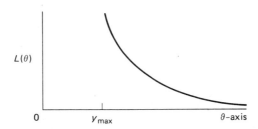

Figure 5.8.2 Graph of $(1/\theta)^n$.

Recall the data of Case Study 4.2.4, giving the annual number of outbreaks of war from 1500 to 1931 (Table 5.8.1).

If Y denotes the number of wars started in a given year—and if it is assumed that the distribution of Y can be described by a Poisson—then

$$P(Y = y) = \frac{e^{-\lambda}\lambda^y}{y!} \qquad y = 0, 1, 2, \ldots$$

TABLE 5.8.1 OUTBREAKS OF WAR

Number of Outbreaks in a Given Calendar Year	Number of Years
0	223
1	142
2	48
3	15
4	4
5+	0
	$\overline{432}$

The maximum-likelihood estimate for λ in this case is 0.69:

$$w = \bar{y} = \frac{1}{432} (223 \cdot 0 + 142 \cdot 1 + 48 \cdot 2 + 15 \cdot 3 + 4 \cdot 4)$$

$$= \frac{299}{432} = 0.69 \text{ "new" wars/year} \quad \textit{estimate}$$

The specific model being proposed, then, is

$$f_Y(y; \lambda) = f_Y(y; 0.69) = P(Y = y) = \frac{e^{-0.69}(0.69)^y}{y!} \quad y = 0, 1, \ldots$$

As Table 4.2.6 makes clear, the agreement between the observed war distribution and the Poisson predictions based on a w of 0.69 is remarkably good.

CASE STUDY 5.8.2

Interstate 66 joins Virginia's beautiful Shenandoah Valley with our nation's capital. The highway was begun in the rolling terrain near Strasburg, Virginia, in 1958, but the last 10 miles were not opened until December 22, 1982, some 24 years later. This last 10-mile stretch runs through suburban areas west of Washington, D.C., and its construction was bitterly opposed by many area residents concerned over pollution and disruption of their neighborhoods. Proponents of the expressway recognized the need to move a high volume of Washington commuter traffic. In 1977 then Secretary of Transportation William T. Coleman accepted a compromise reducing the roadway from six to eight lanes to four. To encourage more efficient use of these four lanes, restrictions were placed on rush-hour use. In the morning inbound cars had to have at least four occupants. The same rule applied to outbound afternoon traffic.

Limiting public access to a facility built with public funds is always a difficult proposition. Although the I66 dilemma was settled by legal and political means, one might wish for a more quantitative solution. How much is gained by restricting use to "high-occupancy vehicles"? To answer such a question, the distribution of the number of vehicle occupants would be helpful. Even though average occupancy figures are often cited, very little work has been done on finding the relevant pdf's.

Haight (63) has provided some insight into the distribution of vehicle occupancy. Table 5.8.2 gives his data for passenger cars observed at the corner of Wilshire and Bundy in Los Angeles during one hour of the day. The data suggest the geometric distribution as a model, although we shall see that the geometric pdf provides only a point of departure.

TABLE 5.8.2 PASSENGER CAR OCCUPANCY

Number of Occupants	Frequency
1	678
2	227
3	56
4	28
5	8
6+	14
	$\overline{1011}$

If Y denotes the number of occupants of a car, what is being proposed is that

$$f_Y(y; p) = P(Y = y) = (1 - p)^{y-1}p \qquad y = 1, 2, \ldots$$

It is difficult to establish a meaning for p beyond its being a parameter that gives flexibility to the pdf.

From Table 5.8.2 we compute $\bar{y}$ to be $1536/1011 = 1.519$ (if all the $6+$'s are set equal to 6):

$$\bar{y} = \frac{678 \cdot 1 + 227 \cdot 2 + \cdots + 14 \cdot 6}{1011} = \frac{1536}{1011}$$

$$= 1.519$$

By the result established in Example 5.8.2, the maximum likelihood estimator for p is

$$W = \frac{1}{\bar{Y}}$$

or in this case, $w = 1/1.519 = 0.658$. Table 5.8.3 shows the corresponding expected frequencies $[= 1011 \cdot f_Y(y; 0.658)]$.

Comment

Keep in mind that finding the maximum-likelihood estimator in no way validates the choice of the geometric model. In this particular case the suitability of the model is certainly open to question. Table 5.8.3 shows considerable disagreement between observed and expected frequencies in the $6+$ category.

TABLE 5.8.3 OBSERVED AND EXPECTED FREQUENCIES

Number of Occupants	Observed Frequency	Expected Frequency
1	678	665.4
2	227	227.4
3	56	77.7
4	28	26.6
5	8	9.1
6+	14	4.8
	1011	1011.0

One explanation for this is the inclusion in the data of limousines and passenger vans. The model can not account for this "jump" in the number of occupants, suggesting that these vehicles should be excluded in future studies.

Question 5.8.1   Find the maximum-likelihood estimate for the parameter, p, in a binomial distribution. Assume that a random sample $y_1, y_2, \ldots, y_n$ of Bernoulli random variables has been observed. Let $y = y_1 + y_2 + \cdots + y_n$ be the value of the corresponding binomial random variable.

Question 5.8.2   A statistically minded fraternity man keeps records on how many girls he has to ask before one of them says she will be his date for a Saturday football game. His school plays five home games and his five acceptances come on the third, sixth, fourth, second, and ninth girls he asks. Assume that the probability, p, that any girl he asks will accept his invitation is constant from girl to girl. Estimate p.

Question 5.8.3   Find the maximum-likelihood estimator for θ in the pdf $f_Y(y; \theta) = e^{-(y-\theta)}$, $y > \theta$. Assume a random sample of size n has been drawn.

Question 5.8.4   Let $X_1, X_2, \ldots, X_n$ be a random sample from

$$f_X(x) = \left(\frac{1}{\theta^2}\right)xe^{-x/\theta} \qquad 0 < x < \infty, \quad 0 < \theta < \infty$$

Find the MLE for θ.

Question 5.8.5   Experience has shown that in a certain state the distribution of voters supporting Democratic candidates for local offices is described quite well by the family of pdf's having the form:

$$f_P(p) = \theta p^{\theta - 1} \qquad 0 < p < 1, \quad 0 < \theta < \infty$$

where P is the percentage of registrants in a precinct who vote Democratic. Suppose that voters in five randomly selected precincts are polled immediately before an election and the proportions intending to vote for the Democratic candidate are 0.45, 0.68, 0.87, 0.36, and 0.54, respectively. Estimate θ.

Question 5.8.6 Consider a reliability inspection policy when n items are put "on test" and the first r failure times, $T_1 < T_2 < \cdots < T_r$, are recorded. Assume that the pdf describing the response variable is exponential with parameter λ:

$$f_T(t) = \left(\frac{1}{\lambda}\right) e^{-t/\lambda} \qquad t > 0$$

Write down the appropriate joint pdf, differentiate it, and show that the MLE for λ is given by

$$\hat{\lambda} = \frac{1}{r} \sum_{j=1}^{r} t_j + (n - r)t_r$$

Question 5.8.7 Maximum-likelihood estimators satisfy an *invariance* property: if $\hat{\theta}$ is the MLE for θ, then $g(\hat{\theta})$ is the MLE for $g(\theta)$. Use that property to find the MLE for the standard deviation of a Poisson distribution. Assume the data to be a random sample of size n.

Question 5.8.8 Show by a counterexample that unbiasedness is not preserved under invariance.

Question 5.8.9 A commuter's trip home consists of first riding a subway to a bus stop and then taking the bus home. The bus she would like to catch arrives uniformly over the interval (θ_1, θ_2). She would like estimates for both θ_1 and θ_2 so she has some idea of when she should be at the stop (θ_1) and when she is probably too late and will have to wait for the next bus (θ_2). Over an eight-day period, she makes certain to be at the stop early so as to not miss the bus and records the following arrival times: 5: 15, 5: 21, 5: 14, 5: 23, 5: 29, 5: 17, 5: 15, and 5: 18. Estimate θ_1 and θ_2. Also, give the maximum-likelihood estimate for the mean of the arrival distribution.

Question 5.8.10 Prove that the MLE for a parameter θ is a function of a sufficient statistic for that parameter.

A second procedure for estimating parameters, one that is often more tractable than the method of maximum likelihood in situations where $f_Y(y; \theta_1, \theta_2, \ldots, \theta_k)$ has a complicated form, is the *method of moments*. Using this method, the set of unknown parameters is estimated by equating the theoretical moments of Y to its corresponding sample moments.

Recall that the first k moments of Y, if they exist, are given by

$$\mu_{(j)} = E(Y^j) = \int_{-\infty}^{\infty} y^j f_Y(y; \theta_1, \theta_2, \ldots, \theta_k) \, dy \qquad j = 1, 2, \ldots, k$$

This implies that the $\mu_{(j)}$'s will be functions of the θ_j's. That is, we will be able to write

$$\mu_{(1)} = \mu_1(\theta_1, \theta_2, \ldots, \theta_k)$$
$$\mu_{(2)} = \mu_2(\theta_1, \theta_2, \ldots, \theta_k)$$
$$\vdots \qquad\qquad \vdots$$
$$\mu_{(k)} = \mu_k(\theta_1, \theta_2, \ldots, \theta_k)$$

Analogs to the $\mu_{(j)}$'s can be defined for the y_i's. Specifically, we will let the first k *sample* moments be denoted

$$m_{(j)} = \frac{1}{n} \sum_{i=1}^{n} y_i^j \qquad j = 1, 2, \ldots, k$$

These, of course, should be approximately equal to the corresponding $\mu_{(j)}$'s if the presumed $f_Y(y; \theta_1, \theta_2, \ldots, \theta_k)$ is a suitable model. It is precisely this observation that motivates the method of moments. Definition 5.8.3 gives the details.

> **Definition 5.8.3 (Moment Estimation).** Let $Y_1, Y_2, \ldots, Y_n$ be a random sample from $f_Y(y; \theta_1, \theta_2, \ldots, \theta_k)$. Let $\mu_{(j)}$ and $m_{(j)}, j = 1, 2, \ldots, k$, be the first k theoretical and sample moments, respectively. As a general procedure for estimating the θ_i's, we will solve simultaneously the system of equations,
>
> $$\mu_{(j)} = \mu_j(\theta_1, \theta_2, \ldots, \theta_k) = m_{(j)} \qquad j = 1, 2, \ldots, k$$
>
> The solutions, $w_1, w_2, \ldots, w_k$, will be called the method of moments estimates. By convention, w_i is often replaced by the symbol $\hat{\theta}_i$.

Comment

The method of moments was first proposed near the turn of the century by the great British statistician, Karl Pearson. The method of maximum likelihood goes back much further: both Gauss and Daniel Bernoulli made use of the technique—the latter as early as 1777 (121). Fisher, though, in the early years of the twentieth century, was the first to make a thorough study of the method's properties, and the procedure is often credited to him.

Example 5.8.4

Let $Y_1, Y_2, \ldots, Y_n$ be a random sample from

$$f_Y(y; \theta) = (\theta + 1)y^\theta \qquad 0 < y < 1$$

Suppose that we want to find the method of moments estimate for θ. First, a simple integration gives $\mu_{(1)}$:

$$\mu_{(1)} = E(Y) = \int_0^1 y \cdot (\theta + 1)y^\theta \, dy$$

$$= \frac{\theta + 1}{\theta + 2}$$

Then, setting $\mu_{(1)}$ equal to the first sample moment, we have

$$\frac{\theta + 1}{\theta + 2} = \frac{1}{n} \sum_{i=1}^{n} y_i = \bar{y} \tag{5.8.3}$$

Let w denote the value of θ that satisfies Equation 5.8.3. Then w is the method-of-moments estimate:

$$w = \frac{2\bar{y} - 1}{1 - \bar{y}}$$

Although hurricanes generally strike only the eastern and southern coastal regions of the United States, they do occasionally sweep inland before completely dissipating. The U.S. Weather Bureau confirms that in the period from 1900 to 1969 a total of 36 hurricanes moved as far as the Appalachians. In Table 5.8.4 are listed the maximum 24-hour precipitation levels recorded for those 36 storms during the time they were over the mountains (62).

TABLE 5.8.4 MAXIMUM 24-HOUR PRECIPITATION RECORDED FOR 36 INLAND HURRICANES (1900–1969)

Year	Name	Location	Maximum Precipitation (inches)
1969	Camille	Tye River, Va.	31.00
1968	Candy	Hickley, N.Y.	2.82
1965	Betsy	Haywood Gap, N.C.	3.98
1960	Brenda	Cairo, N.Y.	4.02
1959	Gracie	Big Meadows, Va.	9.50
1957	Audrey	Russels Point, Ohio	4.50
1955	Connie	Slide Mt., N.Y.	11.40
1954	Hazel	Big Meadows, Va.	10.71
1954	Carol	Eagles Mere, Pa.	6.31
1952	Able	Bloserville 1-N, Pa.	4.95
1949		North Ford #1, N.C.	5.64
1945		Crossnore, N.C.	5.51
1942		Big Meadows, Va.	13.40
1940		Rhodhiss Dam, N.C.	9.72
1939		Caesars Head, S.C.	6.47
1938		Hubbardston, Mass.	10.16
1934		Balcony Falls, Va.	4.21
1933		Peekamoose, N.Y.	11.60
1932		Caesars Head, S.C.	4.75
1932		Rockhouse, N.C.	6.85
1929		Rockhouse, N.C.	6.25
1928		Roanoke, Va.	3.42
1928		Caesars Head, S.C.	11.80
1923		Mohonk Lake, N.Y.	0.80
1923		Wappingers Falls, N.Y.	3.69
1920		Landrum, S.C.	3.10
1916		Altapass, N.C.	22.22
1916		Highlands, N.C.	7.43
1915		Lookout Mt., Tenn.	5.00
1915		Highlands, N.C.	4.58
1912		Norcross, Ga.	4.46
1906		Horse Cove, N.C.	8.00
1902		Sewanee, Tenn.	3.73
1901		Linville, N.C.	3.50
1900		Marrobone, Ky.	6.20
1900		St. Johnsbury, Vt.	0.67

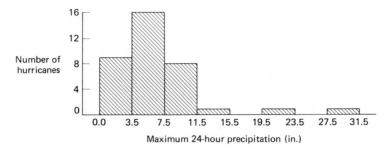

Number of hurricanes

Maximum 24-hour precipitation (in.)

Figure 5.8.3 Inland hurricane precipitation.

Figure 5.8.3 shows a histogram of the data: its skewed shape suggests that Y, the maximum 24-hour precipitation, might be well approximated by a member of the gamma family,

$$f_Y(y; \alpha, \beta) = \frac{1}{\alpha^\beta \Gamma(\beta)} y^{\beta-1} e^{-y/\alpha} \qquad y > 0$$

Here, α and β are the parameters to be estimated. The complexity of $f_Y(y; \alpha, \beta)$, though, makes the method of maximum likelihood unwieldy. As an alternative, we will find the method-of-moments estimates.

Recall from Theorem 4.6.2 that the first two moments of a gamma distribution are given by

$$\mu_{(1)} = E(Y) = \alpha\beta$$

and

$$\mu_{(2)} = E(Y^2) = \alpha^2 \beta(\beta + 1)$$

For the data of Table 5.8.4 the first two sample moments are

$$m_{(1)} = \frac{1}{36} \sum_{i=1}^{36} y_i = \frac{1}{36}(262.35)$$

$$= 7.29$$

and

$$m_{(2)} = \frac{1}{36} \sum_{i=1}^{36} y_i^2 = \frac{1}{36}(3081.2177)$$

$$= 85.59$$

The system of equations that derives from setting the $\mu_{(j)}$'s equal to their corresponding $m_{(j)}$'s reduces to

$$\alpha\beta = 7.29$$

$$\alpha^2 \beta(\beta + 1) = 85.59$$

Estimation   Chap. 5

Solving the first for β and substituting the result into the second gives

$$\alpha^2\left(\frac{7.29}{\alpha}\right)\left(\frac{7.29}{\alpha} + 1\right) = 7.29\alpha + 53.14 = 85.59$$

Therefore,

$$\hat{\alpha} = \frac{85.59 - 53.14}{7.29} = 4.45$$

and

$$\hat{\beta} = \frac{7.29}{\hat{\alpha}} = 1.64$$

Figure 5.8.4 shows the fitted model

$$f_Y(y; 4.45, 1.64) = \frac{1}{4.45^{1.64}\Gamma(1.64)}\, y^{0.64}e^{-y/4.45}$$

superimposed over the original data.

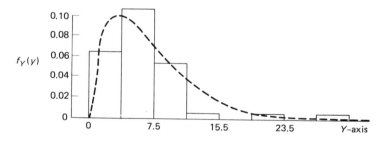

Figure 5.8.4   Original data with fitted model.

Considering the relatively small number of observations in the sample, the agreement is quite good. This would come as no surprise to a meteorologist: the gamma distribution is a very frequently used model for describing precipitation levels (recall Case Study 4.6.1).

Question 5.8.11   Find the maximum-likelihood estimator for the parameter θ in the pdf described in Example 5.8.3. Assume that the data consist of a random sample of size n. Also, suppose that $n = 3$ and the observations recorded are $y_1 = 0.5$, $y_2 = 0.4$, and $y_3 = 0.9$. Evaluate both the MLE and the method-of-moments estimate.

Question 5.8.12   Use the method of moments to estimate θ in the one-parameter beta distribution

$$f_X(x) = \theta x^{\theta - 1} \qquad 0 < x < 1$$

Question 5.8.13 A criminologist is searching through FBI files to document the frequency of occurrence of a rare double-whorl fingerprint. Among six consecutive sets of 100,000 prints scanned by a computer, the numbers of persons having this particular abnormality were 3, 0, 3, 4, 2, and 1. Assume that the occurrence of the double whorl is a Poisson phenomenon. Use the method of moments to estimate λ, its rate of occurrence.

Question 5.8.14 Let $X_1, X_2, \ldots, X_n$ be a random sample from a uniform pdf over the interval $(0, \theta)$. Find the method-of-moments estimator for θ. Compare the values of the method-of-moments estimate and the maximum-likelihood estimate for the following random sample of size five: 17, 92, 46, 39, and 56.

Question 5.8.15 Use the method of moments to estimate c and θ in the two parameter uniform density

$$f_Y(y; c, \theta) = \frac{1}{2\theta} \qquad c - \theta < y < c + \theta$$

Question 5.8.16 Find the method-of-moments estimates for μ and σ^2 based on a random sample of size n taken from an $N(\mu, \sigma^2)$ distribution.

5.9 INTERVAL ESTIMATION

It has already been mentioned that all point estimates, no matter how many desirable properties such as unbiasedness and sufficiency they may possess, share the same fundamental weakness: they provide no indication of their inherent precision. We know, for example, that $W = \bar{Y}$ is the "best" estimator for the Poisson parameter. But suppose a sample of size n is taken from a Poisson distribution and it is found that $w = \bar{y} = 3.8$. What exactly does this tell us about the true value of λ? Can we feel reasonably certain, for example, that λ lies somewhere *close* to w—say, in the interval from 3.7 to 3.9? Or, on the other hand, is W so variable that there is a good chance that $|W - \lambda|$ is fairly large?

These are questions any experimenter would be likely to raise; unfortunately, given only the fact that $w = 3.8$, there is no way to answer them. Clearly, what is needed is to combine a point estimate's numerical value with some sort of statement about its variability. This can be done in several ways. Perhaps the most obvious is simply to write the estimate as $w \pm \sqrt{\text{Var}(W)}$, thus letting the magnitude of the standard deviation reflect the precision of W. An even more informative approach, though, is a format known as a *confidence interval*.

We will introduce the notion of a confidence interval with an example. Consider again the problem of estimating the parameter of a uniform distribution with $W = [(n + 1)/n] \cdot Y_{\max}$. We know that W can range from 0 to $[(n + 1)/n]\theta$, and that it does so according to the pdf

$$f_W(w) = \frac{n^{n+1}}{(n + 1)^n} \frac{w^{n-1}}{\theta^n} \qquad \text{for} \quad 0 \le w \le [(n + 1)/n]\,\theta$$

Estimation   Chap. 5

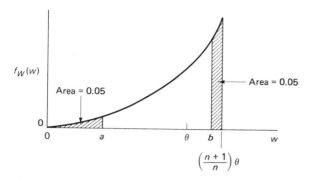

Figure 5.9.1 Tail areas of $f_W(w)$.

(see Example 5.4.1). Note that with $f_W(w)$ specified, it is possible to solve two different kinds of problems related to W: (1) we can find the probability that W lies between two specified numbers, and (2) we can find the numbers that "contain" W with a specified probability. The latter type will lead us to the definition of a confidence interval.

As an illustration, suppose that we want to find two limits, a and b, for which

$$P(a < W < b) = 0.90$$

and

$$P(W \le a) = P(W \ge b) = 0.05$$

That is, a and b are the numbers along the W-axis that "cut off" areas of 0.05 in either tail of $f_W(w)$ (see Figure 5.9.1). By inspection, we see that a is the solution of

$$\int_0^a \frac{n^{n+1}}{(n+1)^n} \frac{w^{n-1}}{\theta^n} \, dw = 0.05$$

After integrating,

$$\frac{n^{n+1}}{(n+1)^n} \frac{a^n}{n\theta^n} = 0.05$$

implying that

$$a = a(\theta) = \sqrt[n]{0.05} \left(\frac{n+1}{n} \right) \theta$$

Similarly, for the right-hand tail of $f_W(w)$,

$$\int_b^{[(n+1)/n]\theta} \frac{n^{n+1}}{(n+1)^n} \frac{w^{n-1}}{\theta^n} \, dw = 0.05$$

which gives

$$b = b(\theta) = \sqrt[n]{0.95} \left(\frac{n+1}{n} \right) \theta$$

Putting these results together gives the two limits we set out to find:

$$P\left[\sqrt[n]{0.05}\left(\frac{n+1}{n}\right)\theta < W < \sqrt[n]{0.95}\left(\frac{n+1}{n}\right)\theta\right] = 0.90$$

Our real interest, though, will not be in the interval $[a(\theta), b(\theta)]$ but in one of its related forms, where the endpoints do not involve θ. Mathematically, the statement

$$P\left[\sqrt[n]{0.05}\left(\frac{n+1}{n}\right)\theta < W < \sqrt[n]{0.95}\left(\frac{n+1}{n}\right)\theta\right] = 0.90$$

is the same as

$$P\left[\frac{W}{\sqrt[n]{0.95}\left(\frac{n+1}{n}\right)} < \theta < \frac{W}{\sqrt[n]{0.05}\left(\frac{n+1}{n}\right)}\right] = 0.90$$

This second formulation, though, has a much different interpretation. Imagine that a very large number of samples of size n have been drawn from the uniform distribution over $(0, \theta)$ and for each one, the interval

$$\left(\frac{w}{\sqrt[n]{0.95}\left(\frac{n+1}{n}\right)}, \frac{w}{\sqrt[n]{0.05}\left(\frac{n+1}{n}\right)}\right)$$

is computed. What the second statement says is that 90% of these intervals will include (or contain) the true θ as an interior point. For this reason, the set of estimates

$$\left(\frac{w}{\sqrt[n]{0.95}\left(\frac{n+1}{n}\right)}, \frac{w}{\sqrt[n]{0.05}\left(\frac{n+1}{n}\right)}\right)$$

is called a *90% confidence interval* (for θ).

CASE STUDY 5.9.1

Table 5.9.1 lists 30 random samples of size $n = 5$ drawn from the particular uniform distribution having θ equal to 1. Listed next to each sample is y_{max}, w, and the corresponding 90% confidence interval,

$$\left(\frac{w}{\sqrt[5]{0.95}\left(\frac{6}{5}\right)}, \frac{w}{\sqrt[5]{0.05}\left(\frac{6}{5}\right)}\right)$$

The last column indicates whether each of the intervals is "correct"—that is, whether it contains as an interior point the value $\theta = 1$. A tally of that column reveals that 27 of the intervals, or 90%, *do* contain the true θ.

TABLE 5.9.1 A SAMPLING EXPERIMENT

Sample	y_{max}	w	90% Interval	Contains θ?
(0.90, 0.90, 0.73, 0.75, 0.54)	0.90	1.08	(0.91, 1.64)	Yes
(0.08, 0.28, 0.53, 0.91, 0.89)	0.91	1.09	(0.92, 1.66)	Yes
(0.77, 0.19, 0.21, 0.51, 0.99)	0.99	1.19	(1.00+, 1.80)	No
(0.33, 0.85, 0.84, 0.56, 0.65)	0.85	1.02	(0.86, 1.55)	Yes
(0.38, 0.37, 0.97, 0.21, 0.73)	0.97	1.16	(0.98, 1.76)	Yes
(0.07, 0.60, 0.83, 0.10, 0.39)	0.83	1.00	(0.84, 1.51)	Yes
(0.59, 0.38, 0.30, 0.65, 0.27)	0.65	0.78	(0.66, 1.18)	Yes
(0.91, 0.68, 0.48, 0.06, 0.10)	0.91	1.09	(0.92, 1.66)	Yes
(0.12, 0.21, 0.19, 0.67, 0.60)	0.67	0.80	(0.67, 1.22)	Yes
(0.53, 0.24, 0.83, 0.16, 0.60)	0.83	1.00	(0.84, 1.51)	Yes
(0.32, 0.51, 0.47, 0.20, 0.66)	0.66	0.79	(0.67, 1.20)	Yes
(0.86, 0.69, 0.93, 0.68, 0.62)	0.93	1.12	(0.94, 1.69)	Yes
(0.93, 0.11, 0.44, 0.17, 0.87)	0.93	1.12	(0.94, 1.69)	Yes
(0.81, 0.01, 0.87, 0.47, 0.95)	0.95	1.14	(0.96, 1.73)	Yes
(0.03, 0.07, 0.06, 0.99, 0.43)	0.99	1.19	(1.00+, 1.80)	No
(0.75, 0.23, 0.94, 0.18, 0.13)	0.94	1.13	(0.95, 1.71)	Yes
(0.19, 0.84, 0.54, 0.42, 0.14)	0.84	1.01	(0.85, 1.53)	Yes
(0.65, 0.38, 0.65, 0.35, 0.07)	0.65	0.78	(0.66, 1.18)	Yes
(0.36, 0.74, 0.38, 0.36, 0.57)	0.74	0.89	(0.75, 1.35)	Yes
(0.13, 0.40, 0.51, 0.50, 0.12)	0.51	0.61	(0.52, 0.93)	No
(0.68, 0.55, 0.05, 0.94, 0.69)	0.94	1.13	(0.95, 1.71)	Yes
(0.51, 0.28, 0.73, 0.10, 0.34)	0.73	0.88	(0.74, 1.33)	Yes
(0.71, 0.56, 0.21, 0.64, 0.85)	0.85	1.02	(0.86, 1.55)	Yes
(0.96, 0.59, 0.05, 0.13, 0.64)	0.96	1.15	(0.97, 1.75)	Yes
(0.88, 0.90, 0.56, 0.49, 0.07)	0.90	1.08	(0.91, 1.64)	Yes
(0.36, 0.62, 0.35, 0.11, 0.91)	0.91	1.09	(0.92, 1.66)	Yes
(0.04, 0.31, 0.86, 0.79, 0.45)	0.86	1.03	(0.87, 1.56)	Yes
(0.58, 0.52, 0.07, 0.27, 0.11)	0.58	0.70	(0.58, 1.06)	Yes
(0.83, 0.91, 0.27, 0.95, 0.20)	0.95	1.14	(0.96, 1.73)	Yes
(0.04, 0.32, 0.28, 0.55, 0.48)	0.55	0.66	(0.56, 1.00+)	Yes

Question 5.9.1 Let $Y_1, Y_2, \ldots, Y_n$ be a random sample from the uniform distribution defined over $(0, \theta)$. Find the formula for a 90% confidence interval for θ based on the estimator $W_1 = (n + 1) \cdot Y_{min}$. How would you expect intervals based on W_1 to compare with those given in Table 5.9.1?

Question 5.9.2 Let X have an $N(\mu, \sigma^2)$ pdf. Find the value h having the property that

$$P(\mu - h < X < \mu + h) = 0.50$$

(In the physical sciences, the quantity h is referred to as the *probable error*.)

Question 5.9.3 Is there any information in a point estimate that may be "lost" in an interval estimate?

Most applications of confidence intervals involve the parameters of the normal distribution (an exception is noted in Section 5.10). For that reason, the methodology of interval estimation will be developed more fully in Chapter 7. Nevertheless, a few general comments are in order here. First, it should be noted that there is nothing fundamentally special about the figure *90%* as a measure of confidence. We could just as easily have defined a 50%, a 78%, or a 99.9% confidence interval. In practice, though, convention dictates that the confidence "coefficient" almost always be set equal to either 90%, 95%, or 99%. Finally, it must not be forgotten that the "confidence" in a confidence interval refers to the procedure and not to any particular realization of that procedure. *Before* any data are collected, it is true that the (random) interval

$$\left(\frac{W}{\sqrt[n]{0.95}\left(\frac{n+1}{n}\right)}, \frac{W}{\sqrt[n]{0.05}\left(\frac{n+1}{n}\right)} \right)$$

has a 90% chance of containing θ. *After* the sample is drawn, though, and w is computed, the resulting interval is no longer a random variable: all we can say is that that particular interval either contains θ or does not.

Comment

Robert Frost was certainly more familiar with iambic pentameters than he was with estimated parameters, but in 1942 he wrote a couplet that sounded very much like a poet's perception of a confidence interval:

> We dance round in a ring and suppose,
> But the Secret sits in the middle and knows.

[See (91).]

5.10 CONFIDENCE INTERVALS FOR THE BINOMIAL PARAMETER

The binomial distribution affords us the opportunity to apply the basic confidence-interval notions of Section 5.9 to a problem of considerable practical importance: given n independent Bernoulli trials, resulting in a total of y successes, construct an interval estimate for the unknown success probability, p. The fact that this is not an uncommon data structure is what makes the problem so important. We begin this section by deriving both approximate and exact interval estimates for p. The section concludes with a discussion of an important related problem—sample-size determination.

First, we recall an approximation—namely, the DeMoivre–Laplace statement that for large n the quantity

$$\frac{\frac{Y}{n} - p}{\sqrt{\frac{p(1-p)}{n}}}$$

behaves like a standard normal. Thus

$$P\left[-z_{\alpha/2} < \frac{\frac{Y}{n} - p}{\sqrt{\frac{p(1-p)}{n}}} < z_{\alpha/2}\right] \doteq 1 - \alpha \qquad (5.10.1)$$

where $P(Z \leq -z_{\alpha/2}) = P(Z \geq +z_{\alpha/2}) = \alpha/2$. As in Section 5.9, the problem is to isolate the unknown parameter in the center of the inequalities; the endpoints will then define an approximate $100(1 - \alpha)\%$ confidence interval for p.

Note that the statement inside the brackets in Equation 5.10.1 is the same as

$$\frac{\left|\frac{Y}{n} - p\right|}{\sqrt{\frac{p(1-p)}{n}}} < z_{\alpha/2}$$

which is equivalent to

$$\left(\frac{Y}{n} - p\right)^2 < z_{\alpha/2}^2\left[\frac{p(1-p)}{n}\right]$$

Expanded out and written in terms of powers of p, this becomes

$$p^2\left(1 + \frac{z_{\alpha/2}^2}{n}\right) - p\left(\frac{2Y}{n} + \frac{z_{\alpha/2}^2}{n}\right) + \left(\frac{Y}{n}\right)^2 < 0 \qquad (5.10.2)$$

The left-hand side of this inequality defines a parabola in p; the confidence interval we are looking for lies between its two roots (see Figure 5.10.1).

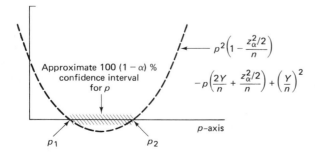

Approximate $100(1 - \alpha)\%$ confidence interval for p

$$\leftarrow p^2\left(1 - \frac{z_\alpha^2/2}{n}\right)$$

$$-p\left(\frac{2Y}{n} + \frac{z_\alpha^2/2}{n}\right) + \left(\frac{Y}{n}\right)^2$$

p-axis

p_1 p_2

Figure 5.10.1 Parabola defined by Inequality 5.10.2.

The quadratic formula applied to the left-hand side of Inequality 5.10.2 gives

$$p_1 = \frac{\dfrac{Y}{n} + \dfrac{z_{\alpha/2}^2}{2n} - \dfrac{z_{\alpha/2}}{\sqrt{n}} \sqrt{\left(\dfrac{Y}{n}\right)\left(1 - \dfrac{Y}{n}\right) + \dfrac{z_{\alpha/2}^2}{4n}}}{1 + \dfrac{z_{\alpha/2}^2}{n}}$$

and

$$p_2 = \frac{\dfrac{Y}{n} + \dfrac{z_{\alpha/2}^2}{2n} + \dfrac{z_{\alpha/2}}{\sqrt{n}} \sqrt{\left(\dfrac{Y}{n}\right)\left(1 - \dfrac{Y}{n}\right) + \dfrac{z_{\alpha/2}^2}{4n}}}{1 + \dfrac{z_{\alpha/2}^2}{n}}$$

The set of p values from p_1 to p_2 certainly qualify as a $100(1 - \alpha)\%$ confidence interval for p. In practice, though, all terms in $z_{\alpha/2}^2/n$ are usually dropped—on the grounds that $z_{\alpha/2}$ will be quite small relative to n. (If $z_{\alpha/2}$ is *not* much smaller than n, the DeMoivre–Laplace approximation of Equation 5.10.1 should not have been used in the first place.) Omitting the $z_{\alpha/2}^2/n$ terms leaves

$$\left(\frac{y}{n} - z_{\alpha/2}\sqrt{\frac{\dfrac{y}{n}\left(1 - \dfrac{y}{n}\right)}{n}}, \; \frac{y}{n} + z_{\alpha/2}\sqrt{\frac{\dfrac{y}{n}\left(1 - \dfrac{y}{n}\right)}{n}} \right)$$

as the approximate $100(1 - \alpha)\%$ confidence interval for p.

CASE STUDY 5.10.1

Surveys to determine consumer preferences are now a routine part of the development and marketing of many products. Most such surveys are conducted with a great deal less visibility than the live television taste tests sponsored by the Joseph Schlitz Brewing Company in the winter of 1980–1981. This expensive series of commercials was undertaken to counter a rapidly decreasing share of the premium beer market. In 1974 an ill-advised change in its brewing procedure coupled with a price increase caused Schlitz to fall from second in sales to a distant third by 1977. Even though the company then hired a master brewer as president, customers stayed away. The firm decided to take on its leading competitors via live television taste tests.

A typical commercial in the Schlitz campaign was broadcast on December 28, 1980, during halftime of an NFL playoff game. One hundred imbibers, who had signed affidavits that they drank at least two six-packs of Budweiser a week, were each presented with two identical unlabeled beer mugs. One mug contained Budweiser, the other Schlitz. Veteran NFL referee Tommy Bell, wearing his "zebra" shirt, presided over the test. Each participant sampled both beers off-camera—broadcasting regulations prohibit otherwise. Then the panel appeared live and, at the sound of referee Bell's whistle, pressed a button showing their preference. The vote was 46 of the 100 for Schlitz.

Even though the test was conducted by an independent market-survey firm, one might question the validity of data collected with such hoopla. Nevertheless, let us calculate a 95% confidence interval for p, the proportion of dedicated Bud drinkers who prefer Schlitz. Since even Schlitz's competitors admitted that Schlitz made a quality beer, and since many experts doubt the ability of beer drinkers to discriminate between brands, we should not be surprised if the confidence interval contains the value $p = 0.5$.

Substituting $z_{0.025} = 1.96$ and $y/n = 46/100$ into the expression for a 95% confidence interval gives

$$\left(0.46 - 1.96 \sqrt{\frac{(0.46)(0.54)}{100}}, \quad 0.46 + 1.96 \sqrt{\frac{(0.46)(0.54)}{100}} \right) \quad p.580$$

which reduces to the interval $(0.36, 0.56)$.

Comment

It *is* possible to find exact confidence intervals for a binomial parameter, but the computations are very tedious. Suppose that y successes are observed in n independent Bernoulli trials ($y \neq 0$ and $y \neq n$). If p_1 and p_2 are the true lower and upper $100(1 - \alpha)\%$ confidence limits for p, then

$$P(p_1 < p < p_2 \mid Y = y) = 1 - \alpha$$

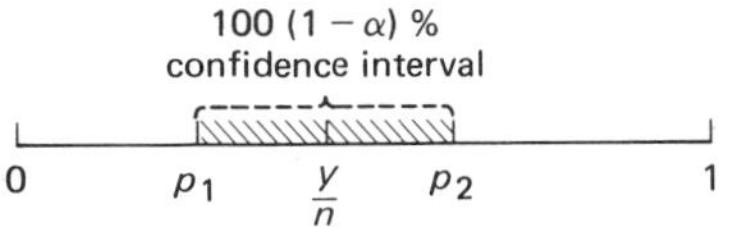

Figure 5.10.2 Exact confidence interval for binomial parameter.

(see Figure 5.10.2). Notice that if p_1 were moved further to the left it would be less compatible with the data (that is, with y). The largest p_1 that would *not* be included in the confidence interval is the solution of

$$\sum_{j=y}^{n} \binom{n}{j} p_1^j (1 - p_1)^{n-j} = \frac{\alpha}{2}$$

That is, any value less than or equal to p_1 would not be sufficiently compatible with the observed sample proportion of y/n to warrant its inclusion in the confidence interval. Similarly, the smallest value of p_2 lying outside the interval is the solution of

$$\sum_{j=0}^{y} \binom{n}{j} p_2^j (1 - p_2)^{n-j} = \frac{\alpha}{2}$$

Tables of (p_1, p_2) values for $1 - \alpha = 0.95$ and 0.99 and for values of n and y from 2 to 100 are readily available (149). If, for example, 33 successes are observed in 42 trials ($y/n = \frac{33}{42} = 0.786$), the exact 95% confidence interval for p is $(0.632, 0.897)$. [The approximate 95% confidence intervals are $(0.662, 0.910)$ *without* terms in $z_{\alpha/2}^2/n$ and $(0.641, 0.883)$ *with* terms in $z_{\alpha/2}^2/n$.]

Question 5.10.1 Food-poisoning outbreaks are quite often the result of contaminated salads. In 1967 the New York City Department of Health examined 220 tuna salads marketed by various retail and wholesale outlets. A total of 179 were found to be unsatisfactory according to their established criteria (153). Find a 95% confidence interval for p, the true proportion of contaminated tuna salads marketed in New York City.

Question 5.10.2 A food-processing company is considering marketing a new spice mix for Creole and Cajun cooking. They interview 200 consumers and find that 37 would purchase such a product. Find a 90% confidence interval for p, the true proportion of potential buyers.

Question 5.10.3 A baseball player has a .315 batting average based on 200 official at-bats. Let p denote his true probability of hitting safely on a typical trip to the plate. Construct a 95% confidence interval for p.

Question 5.10.4 Give a definition of a one-sided confidence interval for p. Find the two one-sided 95% confidence intervals for the consumer preference data described in Question 5.10.2.

Question 5.10.5 Suppose that $n = 3$ independent Bernoulli trials yield $y = 1$ success. Find the exact 95% confidence interval for p. (See the comment following Case Study 5.10.1.)

We conclude this section with an experimental "design" question that is directly related to the construction of a confidence interval for p. Suppose that a researcher wants to estimate the parameter p in a series of n independent Bernoulli trials, *where n is yet to be determined.* Of course, regardless of n, we know that the point estimator for p will be Y/n. We also know that the standard deviation of Y/n will decrease as n increases. Therefore, as the sample size increases, so does the precision of the estimation procedure. Unfortunately, the cost of any study will also increase with n. We are thus faced with a trade-off—the desire, on one hand, to have as precise an estimator as possible and the necessity, on the other, of keeping costs to a minimum. These two conflicting objectives will often force an experimenter to pose the following sort of question: "What is the smallest n that will 'guarantee' (with a probability of $1 - \alpha$) that Y/n will be within some specified distance, d, of p?"

Phrasing the question more formally, we are looking for the smallest n such that

$$P\left(\left|\frac{Y}{n} - p\right| < d\right) = 1 - \alpha \qquad (5.10.3)$$

But note that

$$\left|\frac{Y}{n} - p\right| < d$$

is the same as

$$-d < \frac{Y}{n} - p < d$$

or

$$\frac{-d}{\sqrt{\frac{p(1-p)}{n}}} < \frac{\frac{Y}{n} - p}{\sqrt{\frac{p(1-p)}{n}}} < \frac{d}{\sqrt{\frac{p(1-p)}{n}}}$$

We can approximate Equation 5.10.3, then, with the expression

$$P\left[\frac{-d}{\sqrt{\frac{p(1-p)}{n}}} < Z < \frac{d}{\sqrt{\frac{p(1-p)}{n}}}\right] = 1 - \alpha$$

However, since $P(-z_{\alpha/2} < Z < z_{\alpha/2}) = 1 - \alpha$, it must be true that

$$z_{\alpha/2} = \frac{d}{\sqrt{\frac{p(1-p)}{n}}}$$

or, solving for n,

$$n = \frac{z_{\alpha/2}^2 \, p(1-p)}{d^2} \tag{5.10.4}$$

Equation 5.10.4 is not an acceptable final answer, though, because it involves the unknown p. However, since $0 \le p \le 1$, the product $p(1-p)$ will always be less than or equal to $\frac{1}{4}$. Therefore, we can insure that Equation 5.10.3 is satisfied in even the most "difficult" of situations (when p is actually $\frac{1}{2}$) by choosing as the sample size the smallest n such that

$$n \ge \frac{z_{\alpha/2}^2}{4d^2} \tag{5.10.5}$$

As an application of Inequality 5.10.5 suppose that for the test described in Case Study 5.10.1 the sponsors want the sample proportion of those preferring Schlitz to be within 0.05 of the true proportion—with 95% confidence. That is, they wish to perform the smallest number of tests, n, such that

$$P\left(\left|\frac{Y}{n} - p\right| < 0.05\right) = 0.95$$

According to Inequality 5.10.5, n should be 385:

$$\frac{(1.96)^2}{4(0.05)^2} = 384.12$$

One final comment about this problem deserves mention. If any prior information about the value of p is available, it may be possible to reduce substantially the necessary sample size by *not* making the $p(1 - p) = 1/4$ assumption. For example, suppose a preliminary small-scale survey has indicated that no more than 25% of Bud drinkers are apt to switch. We would then determine n by using Equation 5.10.4 rather than Inequality 5.10.5. The unknown p would be replaced by 0.25, giving $n = 289$:

$$\frac{(1.96)^2(0.25)(0.75)}{(0.05)^2} = 288.12$$

In this case the minimum sample size was lowered from 385 to 289, a reduction of 25%.

Question 5.10.6 An attorney for a civil liberties group is interviewing female nurses in a large metropolitan hospital to see whether they feel they have been victims of sexist discrimination on the part of the male doctors they work with. Let p denote the true proportion of nurses who would consider themselves victims of such discrimination. How many nurses must the attorney interview in order to be 60% certain that the sample proportion of nurses who say, "Yes, I have been discriminated against," is within 0.05 of p?

Question 5.10.7 A public health survey is to be done in an inner-city area to estimate the proportion of children, ages 0 to 14, having adequate polio immunization. It is desired that the final estimate be within 0.05 of the true proportion with probability 0.98. What is the minimum sample size required?

Question 5.10.8 What would be the minimum sample size required to achieve the objectives stated in Question 5.10.7 if it could be assumed that the actual proportion of children adequately immunized against polio was at least 65%?

Question 5.10.9 Since graduating from Vanderbilt last semester, Vanessa has sought a full-time position with her alma mater. Unfortunately, her IQ of 130 overqualified her for the kinds of positions that were available. Desperate, and having no other recourse, she stole her personnel file, lowered her IQ by 40 points, and resubmitted her application. She was immediately hired as an Assistant Dean in the College of Arts and Sciences and given a joint appointment as a full professor in the Philosophy Department. Her first project now is to study the feasibility of a proposed Vanderbilt-in-Tibet summer program. She decides to take a random sample of Vanderbilt undergraduates and ask them whether they would be in favor of such a program. More or less arbitrarily, she decides to question enough students so that there will be an 80% chance that her estimate will be within 0.02 of the true proportion of undergraduates in favor of the program. What is the smallest number of students that she should interview?

APPENDIX 5.1 PROOF OF THE
CRAMER–RAO INEQUALITY

Theorem 5.6.1. Let Y_1, Y_2, ..., Y_n be a random sample from $f_Y(y; \theta)$. Suppose that $f_Y(y; \theta)$ has continuous first partial derivatives, except possibly at a finite set of points, and suppose that the set where $f_Y \neq 0$ does not depend on θ. Let $W = h(Y_1, ..., Y_n)$ be any unbiased estimator for θ. Then

(a) $\displaystyle \text{Var}(W) \geq \left[n \int_{-\infty}^{\infty} \left(\frac{\partial \ln f_Y(y; \theta)}{\partial \theta} \right)^2 f_Y(y; \theta) \, dy \right]^{-1}$

$$= \frac{1}{nE\left[\left(\dfrac{\partial \ln f_Y(Y; \theta)}{\partial \theta} \right)^2 \right]}$$

(b) Equality holds above if and only if

$$\sum_{i=1}^{n} \frac{\partial \ln f_Y(y_i; \theta)}{\partial \theta} = A(\theta)[h(y_1, ..., y_n) - \theta]$$

for all y_1, ..., y_n, except possibly for a set of probability 0. In this case $\text{Var}(W) = 1/A(\theta)$.

(c) If $f_Y(y; \theta)$ has continuous second partial derivatives, then the lower bound in part (a) can also be written

$$\left[-n \int_{-\infty}^{\infty} \frac{\partial^2 \ln f_Y(y; \theta)}{\partial \theta^2} \cdot f_Y(y; \theta) \, dy \right]^{-1}$$

$$= \frac{1}{-nE\left[\dfrac{\partial^2 \ln f_Y(Y; \theta)}{\partial \theta^2} \right]}$$

Proof. We will assume that the reader is familiar with the material in Section 10.2. The proof below is for continuous pdf's. Changing integrals into sums yields a proof for the discrete case.

First note that the hypotheses of the theorem allow the interchange of integral and derivative. Thus such transpositions will be made without further explanation.

The first result needed follows from an obvious fact:

$$1 = \int_{-\infty}^{\infty} f_Y(y; \theta) \, dy$$

Then

$$0 = \frac{\partial}{\partial \theta} \int_{-\infty}^{\infty} f_Y(y; \theta) \, dy$$

$$= \int_{-\infty}^{\infty} \frac{\partial}{\partial \theta} f_Y(y; \theta) \, dy$$

$$= \int_{-\infty}^{\infty} \left(\frac{\partial}{\partial \theta} f_Y(y; \theta) \right) \cdot \frac{1}{f_Y(y; \theta)} f_Y(y; \theta) \, dy$$

$$= \int_{-\infty}^{\infty} \left(\frac{\partial}{\partial \theta} \ln f_Y(y; \theta) \right) \cdot f_Y(y; \theta) \, dy$$

or

$$0 = E\left(\frac{\partial}{\partial \theta} \ln f_Y(Y; \theta) \right) \tag{A5.1.1}$$

Since W is unbiased, we have

$$\theta = E(W) = \int_{-\infty}^{\infty} \cdots \int_{-\infty}^{\infty} h(y_1, \ldots, y_n) f_Y(y_1; \theta)$$

$$\cdot \cdots \cdot f_Y(y_n; \theta) \, dy_1 \cdots dy_n$$

Differentiate both sides of the equation with respect to θ. Then

$$1 = \int_{-\infty}^{\infty} \cdots \int_{-\infty}^{\infty} h(y_1, \ldots, y_n) \frac{\partial}{\partial \theta} [f_Y(y_1; \theta) \cdots$$

$$f_Y(y_n; \theta)] \, dy_1 \cdots dy_n$$

$$= \int_{-\infty}^{\infty} \cdots \int_{-\infty}^{\infty} h(y_1, \ldots, y_n) \left(\sum_{i=1}^{n} \frac{\partial}{\partial \theta} \ln f_Y(y_i; \theta) \right)$$

$$f_Y(y_1; \theta) \cdots f_Y(y_n; \theta) \, dy_1 \cdots dy_n$$

$$= E\left(W \cdot \sum_{i=1}^{n} \frac{\partial \ln f_Y(Y_i; \theta)}{\partial \theta} \right) \tag{A5.1.2}$$

Note that in Equations A5.1.1 and A5.1.2 an unusual but acceptable kind of random variable emerges. Even though $f_Y(y; \theta)$ plays a special role as a density, it is still permissible to apply the real-valued function $\partial \ln f_Y(y; \theta)/\partial \theta$ to the random variable Y to obtain the random variable $\partial \ln f_Y(Y; \theta)/\partial \theta$.

For convenience, set

$$U = \sum_{i=1}^{n} \frac{\partial \ln f_Y(Y_i; \theta)}{\partial \theta}$$

Then Equation A5.1.1 gives immediately that $E(U) = 0$, and Equation A5.1.2 is the statement $E(U \cdot W) = 1$. The lower bound in part (a) is $1/\text{Var}(U)$ since

$$\text{Var}(U) = \sum_{i=1}^{n} \text{Var}\left(\frac{\partial \ln f_Y(Y_i; \theta)}{\partial \theta}\right)$$

$$= n \, \text{Var}\left(\frac{\partial \ln f_Y(Y; \theta)}{\partial \theta}\right)$$

$$= nE\left(\frac{\partial \ln f_Y(Y; \theta)}{\partial \theta}\right)^2 \qquad \text{(why?)}$$

Thus it suffices for part (a) to prove that

$$\text{Var}(W) \geq \frac{1}{\text{Var}(U)}$$

Recall that $\text{Cov}(U, W) = E(U \cdot W) - E(U)E(W)$, so $\text{Cov}(U, W) = 1$. Since $|\rho(U, W)| \leq 1$, we have

$$\frac{1}{\sqrt{\text{Var}(U) \cdot \text{Var}(W)}} \leq 1$$

or

$$\text{Var}(W) \geq \frac{1}{\text{Var}(U)}$$

Equality holds in part (a) if and only if $\text{Var}(W) = 1/\text{Var}(U)$ or, equivalently, $|\rho(U, W)| = 1$. By Theorem 10.2.4(b), the latter follows if and only if $U = aW + b$, for constants a and b. Taking expected values gives $0 = E(U) = a\theta + b$ or $b = -a\theta$. Thus $U = a(W - \theta)$. Since U and W depend on θ, so does a. Thus we write $a = A(\theta)$, and part (b) follows.

We leave part (c) as an exercise.

6

Hypothesis Testing

Pierre-Simon, Marquis de Laplace (1749–1827)

As a young man, Laplace went to Paris to seek his fortune as a mathematician, disregarding his father's wishes that he enter the clergy. He soon became a protégé of d'Alembert and at the age of 24 was elected to the Academy of Sciences. Laplace was recognized as one of the leading figures of that group for his work in physics, celestial mechanics, and pure mathematics. He also enjoyed some political prestige, and his friend, Napoleon Bonaparte, made him Minister of the Interior for a brief period. With the restoration of the Bourbon monarchy, Laplace renounced Napoleon for Louis XVIII, who later made him a marquis.

6.1 INTRODUCTION

Inferences, as we have seen, often take the form of numerical estimates, either as single points or as confidence intervals. But not always. In many experimental situations the conclusion to be drawn is *not* numerical and is more aptly phrased as a choice between two conflicting theories, or *hypotheses*. Thus a court psychiatrist is called upon to pronounce an accused murderer either "sane" or "insane"; the FDA "approves" or "rejects" an application by a pharmaceutical house to get a flu vaccine licensed; a geneticist has to decide whether the inheritance of eye color in *Drosophila melanogaster* "does" or "does not" follow classical Mendelian principles; a stockbroker "buys" or "sells" a certain stock. In this chapter we examine the statistical methodology involved in making decisions of this sort.

The process of dichotomizing the possible conclusions of an experiment and then using the theory of probability to choose between the two alternatives is known as *hypothesis testing*. The two competing propositions are called the *null hypothesis* (written H_0) and the *alternative hypothesis* (written H_1). How we go about choosing between H_0 and H_1 is conceptually similar to the way a jury deliberates in a court trial. The null hypothesis is analogous to the defendant: just as the latter is presumed innocent until proven guilty, so is the null hypothesis presumed true until the data argue overwhelmingly to the contrary.

In Sections 6.2 and 6.3 we develop some of the mathematical structure associated with H_0 and H_1. Particularly important will be the rationale that is followed in deciding which of the two hypotheses to accept. Section 6.4 introduces the notion of optimality (when is one hypothesis test "better" than another?) and then considers the rather formidable problem of actually finding a procedure for testing a given H_0 against a given H_1.

6.2 THE DECISION RULE

We will introduce the basic concepts of hypothesis testing with an example. Suppose that the army is considering installing a new heat-sensing guidance system (HSGS) in one of its surface-to-air missiles. Extensive combat records show that the old, radar-based system was on target 50% of the time. Changing over, though, would be extremely expensive, so before the army gives their go-ahead they will need to be convinced that the HSGS represents a genuine improvement. What they want to see is a field demonstration, so a mock attack is staged that calls for 18 of the missiles (equipped with the new guidance system) to be launched against a remote-controlled squadron of attacking fighters. When the smoke clears, only 6 of the fighters are still airborne—*12* (or, 67%) have been shot down. What should the army conclude?

At first glance, the new system has, indeed, performed better than what would have been expected of its predecessor: on the average, the 50% effective radar-based system would have shot down only *9* planes. But the fact that in this particular set of 18 trials, the HSGS has accounted for an additional 3 "successes" does not automatically imply that the army should acknowledge its superiority. Those 3 extra hits could be taken as evidence that the new system *is* better or they could be written off as normal variation for a system that isn't (the fact that a system is known to be on target 50% of the time does not mean that in every set of n trials exactly $n/2$ will end favorably). Our problem is deciding how to interpret what we observed.

Keep in mind the analogy between hypothesis testing and the courtroom. Here, the null hypothesis—which is typically a statement reflecting the status quo—is that the new guidance system is no better than the old one; the alternative says the new system is better. By agreement, we give H_0 (like the defendant) any benefit of the doubt. Thus, if the number of planes shot down is 9, *or "close" to 9*, we must conclude that the new system has *not* demonstrated its superiority. What we decide, then, hinges on what "close" means.

It will help at this point to formalize the problem a bit by introducing a probability model for Y, the number of missiles on target. From what has already been described, the obvious choice for $f_Y(y)$ is the binomial, with each missile being thought of as a Bernoulli trial. Let $p = P$ (missile hits its target). The two hypotheses in question, then, can be written

$$H_0: \quad p = \frac{1}{2} \qquad \text{(the new guidance system \textit{is not} better than the old guidance system)}$$

$$H_1: \quad p > \frac{1}{2} \qquad \text{(the new guidance system \textit{is} better than the old guidance system)}$$

Notice that the possible values of Y—the integers from 0 through 18—can be viewed as a credibility scale for H_0. Values of Y less than or equal to 9 are certainly grounds for *accepting* the null hypothesis; so are values a little larger than 9 (since we are committed to giving H_0 the benefit of the doubt). On the other hand, values of Y close to 18 should be considered strong evidence *against* the null hypothesis, leading to a decision of "reject H_0." It follows that somewhere between 9 and 18 there is a point—call it y^*—where, for all practical purposes, the credibility of H_0 ends. Phrasing our answer in courtroom terminology, we will say that a Y value greater than or equal to y^* implies that H_0 is false *beyond all reasonable doubt*.

> **Definition 6.2.1.** Any function of the observed data whose numerical value dictates whether we accept or reject H_0 is called a *test statistic*. The notation we will use follows the same conventions introduced for estimators in Chapter 5: the function itself (a random variable) is denoted W; after being evaluated with the sample data, it will be written w and called an *observed test statistic*. The pdf for a test statistic will be written $f_W(w)$. The set of values for the test statistic that result in the null hypothesis being rejected is called the *critical region*, and denoted C.

Comment

For the missile example, $W = Y$ and

$$C = \{y : y^* \le y \le 18, \text{ where } y \text{ is an integer}\}$$

In general, the point y^* separating the rejection region from the acceptance region (but *in* the rejection region) is called the *critical value*.

In practice, the critical region for a binomial hypothesis test is determined in a very direct fashion. The *rationale* that is followed, though, is decidedly indirect. What makes the problem nontrivial is having to translate "beyond all reasonable doubt" into something numerical. The next several paragraphs show the probabilistic consequences associated with several arbitrary choices for y^*. Interpreted properly, these calculations will suggest a specific answer to the missile question and at the same time motivate the general formulation of a binomial hypothesis test, as stated in Theorem 6.2.1.

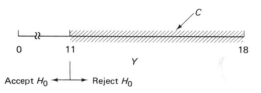

Accept H_0 ◄———► Reject H_0

Figure 6.2.1 Critical value set equal to 11.

Suppose, for example, that we set the critical value equal to 11 (see Figure 6.2.1). Under the assumption, then, that H_0 is true, it follows that the probability of Y's equaling or exceeding 11 is 0.24:

$$P(W \in C \,|\, H_0 \text{ is true}) = P(Y \geq y^* \,|\, H_0 \text{ is true}) = P(Y \geq 11 \,|\, p = \tfrac{1}{2})$$

$$= \sum_{y=11}^{18} \binom{18}{y}\left(\frac{1}{2}\right)^y\left(1 - \frac{1}{2}\right)^{18-y}$$

$$= 0.24$$

(see Figure 6.2.2). This implies that if 11 were chosen as the critical value, we would incorrectly reject H_0 almost 25% of the time! For most people, this would not be a satisfactory definition of reasonable doubt. No jury, for example, would convict a defendant knowing they had a 25% chance of being wrong.

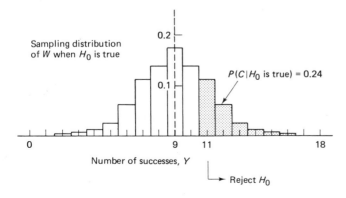

Sampling distribution
of W when H_0 is true

$P(C \,|\, H_0 \text{ is true}) = 0.24$

Number of successes, Y

Reject H_0

Figure 6.2.2 Sampling distribution of $W(y^* = 11)$.

Clearly, we must look for a critical value closer to 18. Consider what would happen if y^* were set equal to 15. The probability of Y's exceeding (or equaling) *this* number—assuming H_0 to be true—is 0.004:

$$P(Y \geq 15) = \sum_{y=15}^{18} \binom{18}{y}\left(\frac{1}{2}\right)^y\left(1 - \frac{1}{2}\right)^{18-y} = 0.004$$

(see Figure 6.2.3). Now we may have gone too far to the other extreme. Requiring the observation to be this far out in the tail of the Y distribution before we reject H_0 would be like a jury's not convicting a defendant unless the prosecution could produce a host of eyewitnesses, an obvious motive, and a signed confession!

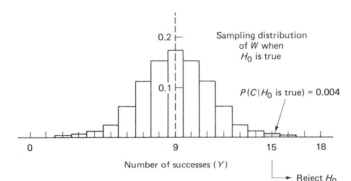

Figure 6.2.3 Sampling distribution of $W(y^* + 15)$.

What the consequences of these two possible choices for y^* seem to be implying is that a reasonable decision rule lies somewhere *between* "Reject H_0 if $y \geq 11$" and "Reject H_0 if $y \geq 15$." Or, what is equivalent, the probability of rejecting H_0 when H_0 is actually true should be something less than 0.24 but greater than 0.004. To be sure, there is no way to argue that this rejection probability *should be* 0.10 or 0.18 or 0.009—or anything else. *In many situations, researchers define the beginning of reasonable doubt as the value of the test statistic that is equaled or exceeded only 5% of the time (when H_0 is true).*

Invoking the "0.05 decision rule" here implies that y^* should be chosen so that

$$P(Y \geq y^* \mid H_0 \text{ is true}) = 0.05$$

More specifically, y^* should satisfy the equation

$$0.05 = \sum_{y=y^*}^{18} \binom{18}{y}\left(\frac{1}{2}\right)^y\left(1 - \frac{1}{2}\right)^{18-y} \tag{6.2.1}$$

(Keep in mind that the discreteness of Y may make it impossible to find a y^* with the property that $P(Y \geq y^* \mid H_0 \text{ is true})$ is *exactly* 0.05. What we will typically have to settle for is the y^* for which $P(Y \geq y^* \mid H_0 \text{ is true}) \leq 0.05$ and $P(Y \geq y^* - 1 \mid H_0 \text{ is true}) > 0.05$.) Here, a little trial and error shows that $y^* = 13$. This means that the army should reject $H_0: p = \frac{1}{2}$ if and only if 13 or more planes are shot down by missiles equipped with the new guidance system. Since only $y = 12$ missiles were on target, the proper decision is to *accept* H_0—the new guidance system has failed to demonstrate its superiority beyond all reasonable doubt (as measured by our 5% criterion).

Comment

Defining the critical region in terms of the W value exceeded only 5% of the time when H_0 is true is the most common way to quantify reasonable doubt, but there are others. A 1% rejection probability is sometimes used, and, to a lesser extent, so are 10% and 0.1%. More will be said about the implications of these different decision rules in Section 6.3.

Question 6.2.1. Find $y*$ for the problem just discussed if a 10% decision rule is used—that is, if $y*$ is chosen such that $P(Y \geq y* \mid H_0$ is true$) \leq 0.10$ and $P(Y \geq y* - 1 \mid H_0$ is true$) > 0.10$.

Question 6.2.2 (a) Suppose that $n = 12$. Find the 0.05 decision rule for testing

$$H_0: \quad p = \frac{1}{2}$$

versus

$$H_1: \quad p > \frac{1}{2}$$

(b) Suppose that $n = 14$. Find the 0.05 decision rule for testing

$$H_0: \quad p = \frac{3}{5}$$

versus

$$H_1: \quad p < \frac{3}{5}$$

(*Hint:* What sort of values for Y would be interpreted as strong evidence against H_0 and in favor of H_1?)

(c) Suppose that $n = 20$. Find the 0.01 decision rule for testing

$$H_0: \quad p = 0.3$$

versus

$$H_1: \quad p > 0.3$$

Question 6.2.3 A graphologist is given a set of 10 folders, each containing handwriting samples of two persons, one "normal" and the other schizophrenic. Her task is to identify which of the writings are the work of the schizophrenics. Set up an appropriate H_0 and H_1 for this situation and find the 0.05 decision rule. When this experiment was actually performed [see (119)] the graphologist made 6 correct identifications. Did she at that time demonstrate a statistically significant ability to distinguish the writing of a schizophrenic from the writing of a normal person?

When the sample size n is large, binomial hypothesis tests can be facilitated by using the DeMoivre–Laplace limit theorem—$P(Y \geq y^* \mid H_0$ is true), for example, is equivalent to $P[(Y - np)/\sqrt{npq} \geq (y^* - np)/\sqrt{npq} \mid H_0$ is true], which is approximately equal to $P(Z \geq (y^* - np)/\sqrt{npq} \mid H_0$ is true). Solving $0.05 = P(Z \geq (y^* - np)/\sqrt{npq} \mid H_0$ is true) for y^* is much easier than solving $0.05 = \sum\limits_{y=y^*}^{n} \binom{n}{y} \cdot p^y (1 - p)^{n-y}$. Case Study 6.2.1 shows the application of a *large-sample* binomial hypothesis test.

CASE STUDY 6.2.1

Anyone following sports knows that athletes and their coaches are a superstitious lot. Famed basketball coach Adolph Rupp always wore a brown suit to games; in his later years, legendary Alabama mentor Bear Bryant was never seen along the sidelines without his houndstooth hat. And players themselves often have lucky shirts or socks (that sometimes go unwashed for the duration of a winning streak).

Equally important in the lore of superstition is the notion of a jinx. Baseball players, for example, are notoriously afraid of the dreaded sophomore slump. Of more recent vintage is the apprehension that the appearance of a player or a team on the cover of *Sports Illustrated* leads to a subsequent decline in performance.

Gluckson and Leone (56) put the "*Sports Illustrated* jinx" to the test—specifically, to a binomial hypothesis test. Let p denote the probability that the performance level of a cover subject declines. If the definition of declining performance is such that in normal circumstances performance is as likely to decline as not, then the hypotheses that Gluckson and Leone set out to test can be written

$$H_0: \quad p = \frac{1}{2} \quad \text{(SI cover has no effect)}$$

versus

$$H_1: \quad p > \frac{1}{2} \quad \text{(SI jinx exists)}$$

Included in the study were some 271 subjects appearing on SI covers during the years 1954 through 1983. Let Y denote the number of subjects whose performance subsequently declined. If we define reasonable doubt by the 0.05 criterion, we should reject $H_0: p = \frac{1}{2}$ if $y \geq y^*$, where the critical value y^* satisfies the equation

$$P(Y \geq y^* \mid H_0 \text{ is true}) = 0.05 = \sum_{y=y^*}^{271} \binom{271}{y} \left(\frac{1}{2}\right)^y \left(1 - \frac{1}{2}\right)^{271-y} \quad (6.2.2)$$

As written, solving Equation 6.2.2 would be a difficult task, but the computations can be greatly simplified by appealing to the normal approximation to the binomial. Note that

$$P(Y \geq y^* \mid H_0 \text{ is true}) = P\left[\frac{Y - 271(1/2)}{\sqrt{271(1/2)(1/2)}} \geq \frac{y^* - 271(1/2)}{\sqrt{271(1/2)(1/2)}}\right]$$

and we know from the DeMoivre–Laplace limit theorem that the distribution of

$$\frac{Y - 271(1/2)}{\sqrt{271(1/2)(1/2)}}$$

is approximated by the standard normal (when H_0 is true). Therefore,

$$P(Y \geq y^* \mid H_0 \text{ is true}) \doteq P\left(Z \geq \frac{y^* - 135.5}{\sqrt{271(1/2)(1/2)}}\right)$$

Since $P(Z \geq 1.64) = 0.05$, it follows that y^* is the solution of the equation

$$1.64 = \frac{y^* - 135.5}{\sqrt{271(1/2)(1/2)}}$$

Specifically, $y^* = 149$.

By comparison, the *observed* number of declines—the test statistic—was found to be 114, a figure well to the left of the critical value. Our conclusion is clear: Accept the null hypothesis—on the basis of these data, there is no evidence of a *Sports Illustrated* jinx. (Not surprisingly, athletes—whether superstitious or not—are more than willing to accept the possibility of incurring a jinx in exchange for the national attention that comes from being featured in *Sports Illustrated*. As one of the magazine's officials put it, "I've never heard anyone say to me, 'Don't put me on the cover.'")

Comment

The two hypothesis tests we have seen thus far had *one-sided* alternatives (H_1: $p > \frac{1}{2}$). In both cases, the physical nature of the problem suggested that values of *p less than* the H_0 value were unrealistic and could be ignored. There are certainly situations, though, where parameter deviations on either side of the H_0 value are possible; in these situations the alternative should be *two-sided*. The distinction is important because the nature of H_1 has a direct effect on how we define the critical region. Specifically, if H_1 is two-sided, there are *two* critical regions, one in each tail of the *W* distribution. For example, if the hypotheses being tested were

$$H_0: \quad p = \frac{1}{2}$$

versus

$$H_1: \quad p \neq \frac{1}{2}$$

(and 0.05 were taken as the measure of reasonable doubt) the two critical values would be y_1^* and y_2^*, where

$$P(Y \le y_1^* \mid H_0 \text{ is true}) = 0.025$$

and

$$P(Y \ge y_2^* \mid H_0 \text{ is true}) = 0.025$$

That is, *half* the prespecified reasonable doubt probability is put into each tail (see Figure 6.2.4). The decision rule, then, calls for H_0 to be rejected if y is either less than or equal to y_1^* or greater than or equal to y_2^*.

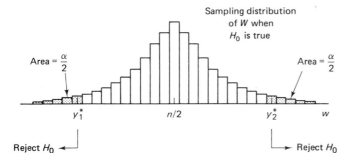

Figure 6.2.4 Critical region for a two-sided alternative.

As indicated in Case Study 6.2.1, the De Moivre–Laplace limit theorem is frequently used to expedite hypothesis tests for the binomial parameter, p. Theorem 6.2.1 gives a formal statement of the procedure. The proof is based on the principle introduced in Section 6.4, but the details will not be presented here.

Theorem 6.2.1. Suppose that a total of y successes are observed in a series of n independent Bernoulli trials, where y and n are sufficiently large.* Let $p = P$ (success occurs at any given trial) and let α be the probability associated with the definition of reasonable doubt. To test

$$H_0: \quad p = p_0$$

versus

$$H_1: \quad p \ne p_0$$

we should reject H_0 if

$$\frac{y - np_0}{\sqrt{np_0(1 - p_0)}} \text{ is either } \begin{cases} \le -z_{\alpha/2} \\ \text{or} \\ \ge +z_{\alpha/2} \end{cases}$$

where $P(Z \ge z_{\alpha/2}) = P(Z \le -z_{\alpha/2}) = \alpha/2$. To test

$$H_0: \quad p = p_0$$

versus

$$H_1: \quad p > p_0$$

we should reject H_0 if

$$\frac{y - np_0}{\sqrt{np_0(1 - p_0)}} \geq +z_\alpha$$

To test

$$H_0: \quad p = p_0$$

versus

$$H_1: \quad p < p_0$$

we should reject H_0 if

$$\frac{y - np_0}{\sqrt{np_0(1 - p_0)}} \leq -z_\alpha$$

* "Sufficiently large" is often taken to mean that

$$0 < y - 2\sqrt{n\left(\frac{y}{n}\right)\left(1 - \frac{y}{n}\right)} < y + 2\sqrt{n\left(\frac{y}{n}\right)\left(1 - \frac{y}{n}\right)} < n$$

If these inequalities are satisfied, the distribution of $(Y - np)/\sqrt{np(1 - p)}$ is adequately approximated by the standard normal.

Question 6.2.4 Efforts to find a genetic explanation for why certain people are right-handed and others, left-handed, have been largely unsuccessful. Reliable data are difficult to find because of environmental factors that also influence a child's "handedness." To avoid that complication, researchers often resort to studying the analogous problem of "pawedness" in animals, where genotypes can be partially controlled. In one such experiment, mice are put into a cage having a feeding tube that is equally accessible from the right or the left. Each mouse is then carefully watched over a number of feedings. If it uses its right paw more than half the time to activate the feeding tube, it is defined to be "right-pawed." Observations of this sort have shown that 67% of mice belonging to strain A/J are right-pawed. Collins (26) recorded similar information on 35 mice belonging to strain A/HeJ. Of those 35, a total of 18 proved to be right-pawed. Test whether the proportion of right-pawed mice found in the A/HeJ strain is significantly different from what is known about the A/J strain. Use a two-sided alternative and let 0.05 be the probability associated with the critical region.

Question 6.2.5 There is a theory [see (123)] that the anticipation of a birthday can prolong a person's life. In a study set up to examine that notion statistically, it was found that only 60 of 747 people whose obituaries were published in Salt Lake City in 1975 died in the three-month period preceding their birthday (113). Test the appropriate hypothesis. Let 0.01 be the probability associated with the critical region.

Question 6.2.6 Many species of insects are strongly attracted to light at a wavelength of 365 nm. This is in the near-ultraviolet range and is often the wavelength in outdoor light traps. The phototropic behavior elicited by longer wavelengths, though, is less understood. In a recent experiment (179) the reactions of a particular species of mosquitoes (*Anopheles stephensi*) to these longer wavelengths were studied in a laboratory environment. The experimental apparatus consisted of a metal tube, 56 cm long and 9 cm in diameter. At one end was an incandescent lamp; at the other end, a device that could produce monochromatic light of any desired wavelength. The two sources were adjusted so their intensities at the center of the tube were the same. Mosquitoes were introduced into the tube midway between the two ends. They were exposed simultaneously to both sources of light for $2\frac{1}{2}$ minutes, after which time the two ends of the tube were sealed off and the number of mosquitoes in each arm counted. Between 100 and 200 mosquitoes were placed in the tube for a given "trial"—and the entire procedure was replicated several times. For one series of trials, a total of 958 mosquitoes were exposed to *red* light (690 nm) coming from one end and "white" light from the other. After $2\frac{1}{2}$ minutes, 642 were found in the red end of the tube. Do the appropriate hypothesis test. Use the 0.01 decision rule. What are you assuming here that may not be true?

Question 6.2.7 Commercial fishermen working certain parts of the Atlantic Ocean sometimes have their efforts hindered by the presence of whales. The problem is to scare away the whales without frightening the fish. In the past, sonar operators have confirmed that 40% of all whales sighted left the area of their own accord, probably to get away from the noise of the boat. Recently, attempts have been made to increase that figure by transmitting underwater sounds of a killer whale. To date, the technique has been tried on 52 whales; of that number, a total of 24 immediately left the area.

(a) Set up a null and alternative hypothesis for testing whether broadcasting the cry of a killer whale is significantly more effective in "clearing" a fishing area than doing nothing.

(b) Test the hypothesis of part (a) using the 0.01 decision rule.

Example 6.2.1 shows the construction of a hypothesis test when $f_Y(y)$ is Poisson. In general, regardless of what distribution the Y_i's may have, hypothesis tests are set up following much the same approach that was used for the binomial. First, a sufficient statistic is found for the parameter being tested; then the critical region is defined as the set of W-values least favorable to $H_0 : \theta = \theta_0$ (but still admissible under H_1) and having the property that

$$
P(W \in C \,|\, H_0 \text{ is true}) = \alpha = \begin{cases} \displaystyle\int_C f_W(w; \theta_0)\, dw \\[6pt] \text{or} \\[6pt] \displaystyle\sum_C f_W(w; \theta_0) \end{cases} \tag{6.2.3}
$$

where α is usually set at either 0.05 or 0.01. When Equation 6.2.3 is solved for the limits of C, the test is completely specified.

Example 6.2.1

"But wait. There's more." Anyone who turns on a television set is likely to be assaulted by hard-sell commercials for records, knives, wrenches, and the like, promising ever more merchandise for an "amazingly low price." Businesses using this sort of promotional hype depend heavily on making sales via toll-free telephone calls. Having too few telephone lines or operators can mean lost sales. On the other hand, too much overhead diminishes profits. It is important, then, for these companies to know as much as possible about the distribution of incoming calls.

Recall from Section 4.2 that the arrival of calls at a switchboard is often Poisson: if Y is the number of calls received during a given minute, then $f_Y(y) = e^{-\mu}\mu^y/y!$, $y = 0, 1, 2, \ldots$, where μ is the average arrival rate. Suppose that a firm has sufficient capacity to handle an average of 3 calls per minute but is concerned that the arrival rate may actually be somewhat higher. What they need to do is test

$$H_0: \quad \mu = 3 \qquad (\mu \text{ mean } \# \text{ of calls/min})$$

versus

$$H_1: \quad \mu > 3 \qquad 1)\ \theta = \mu \text{ we choose}$$

and install additional equipment if H_0 is rejected. $\frac{1}{10}\sum_{i=1}^{10} Y_i$ as unbiased estimator

Suppose that the company monitors their switchboard for a total of ten 1-minute intervals, during which time they observe the following numbers of calls: 4, 3, 4, 1, 3, 4, 2, 5, 6, and 4. What should they conclude?

Even though we do not yet have any general techniques for choosing test statistics, we know that $\bar{Y}$ is an unbiased, sufficient, and efficient estimator for μ. That being the case, setting $W = \bar{Y}$ seems like a reasonable strategy. Values of $\bar{y}$ significantly larger than 3 would obviously be evidence against H_0 and in favor of H_1. In particular, we should reject H_0 (using the 0.05 criterion) if $\bar{y} \geq y^*$, where

$$P(\bar{Y} \geq y^* \mid H_0 \text{ is true}) = 0.05$$

or, equivalently, if

$$P\left(\sum_{i=1}^{10} Y_i \geq 10y^* \mid H_0 \text{ is true}\right) = 0.05$$

Note here that $\sum_{i=1}^{10} Y_i$ has a Poisson distribution with parameter 30 (recall Question 3.12.9). By trial and error,

$$\sum_{k=40}^{\infty} \frac{e^{-30}30^k}{k!} = 0.05$$

implying that $10y^* = 40$. Therefore, we should reject H_0 if $\bar{y} \geq 4.0$. Since the observed $\bar{y}$ is 3.6 $[=(4 + 3 + \cdots + 4)/10]$, the company should resist the temptation to expand their operation.

Question 6.2.8 Suppose that Y_1, Y_2, Y_3, and Y_4, a random sample of size $n = 4$, is taken from a Poisson pdf whose parameter is unknown. Furthermore, suppose that the hypotheses to be tested are

$$H_0: \quad \lambda = 1$$

versus

$$H_1: \quad \lambda > 1$$

and the decision rule is to be of the form

$$\text{Reject } H_0 \text{ if } \sum_{i=1}^{4} y_i \geq k$$

Find the value of k that gives an α approximately equal to 0.05.

Question 6.2.9 If Y is a Poisson random variable with parameter λ, it can be shown (see Section 7.4) that if λ is large

$$\frac{Y - \lambda}{\sqrt{\lambda}}$$

has approximately a standard normal pdf. This suggests that a test of

$$H_0: \quad \lambda = \lambda_0$$

versus

$$H_1: \quad \lambda > \lambda_0$$

can be set up by rejecting H_0 whenever $(y - \lambda_0)/\sqrt{\lambda_0}$ is too large. The following situation is a case in point. On the average, the number of babies born in Cleveland, Ohio, in September is 1472. On January 26, 1977, the city was immobilized by a blizzard. Nine months later, in September 1977, the number of recorded births was 1718, an increase of 246. If births are assumed to be Poisson events with a September average of 1472, use the normal approximation to test $H_0: \lambda = 1472$ versus $H_1: \lambda > 1472$. Use a 0.05 decision rule.

Question 6.2.10 A random sample of size 2 is drawn from a uniform pdf defined over the interval $[0, \theta]$. We wish to test

$$H_0: \quad \theta = 2$$

versus

$$H_1: \quad \theta < 2$$

by rejecting H_0 when $y_1 + y_2 \leq k$. Find the value for k that gives a level of significance of 0.05.

Question 6.2.11 Suppose that the hypotheses of Question 6.2.10 are to be tested with a decision rule of the form, "Reject H_0: $\theta = 2$ if $y_1 y_2 \leq k^*$." Find the value of k^* that gives a level of significance of 0.05 (see Example 3.5.5).

Question 6.2.12 Suppose that the critical region for a certain hypothesis test is of the form

$$\text{Reject } H_0 \text{ if } \sum_{i=1}^{n} x_i \geq k$$

However, because of the discrete nature of the random variable X,

$$P\left(\sum_{i=1}^{n} X_i \geq k_1 \mid H_0 \text{ is true}\right) = \alpha_1 > \alpha$$

and

$$P\left(\sum_{i=1}^{n} X_i \geq k_2 \mid H_0 \text{ is true}\right) = \alpha_2 < \alpha$$

where $k_1 < k_2$. How might a "weighted" decision rule be formulated giving an *exact* critical region probability of α?

6.3 TYPE I AND TYPE II ERRORS

Errors are an inevitable by-product of hypothesis testing. No matter what sort of mathematical facade is laid atop the decision making process, there is no way to avoid the possibility of drawing an incorrect inference. One kind of error—rejecting H_0 when H_0 is true—figured prominently in Section 6.2: it was argued that critical regions should be defined so as to keep the probability of making such errors small, say, on the order of 0.05. In this section we want to examine in more detail the structure of hypothesis testing from the standpoint of controlling and interpreting the probability of making errors.

In any hypothesis test there are two different kinds of errors that can be committed: we can reject H_0 when H_0 is true or we can accept H_0 when H_0 is false. These are called Type I and Type II errors, respectively. Similarly, there are two kinds of correct decisions: we can accept a true H_0 or reject a false one. Figure 6.3.1 shows the four possible "decision–state of nature" combinations.

	True State of Nature	
	H_0 is true	H_1 is true
Accept H_0	Correct decision	Type II error
Reject H_0	Type I error	Correct decision

Figure 6.3.1   The two types of error.

Once an inference is made, there is no way to know whether, in fact, an error was committed. It is possible, though, to compute the *probability* of having made an error. Recall the missile example developed in Section 6.2: given $n = 18$ observations on a Bernoulli random variable, we wanted to test $H_0: p = \frac{1}{2}$ versus $H_1: p > \frac{1}{2}$. The decision rule stated that H_0 should be rejected if y, the number of missiles on target, equaled or exceeded 13. It follows that the probability of committing a Type I error is 0.05: by definition,

$$P \text{ (Type I error)} = P \text{ (we reject } H_0 \mid H_0 \text{ is true)}$$

$$= P\left(Y \geq 13 \mid p = \frac{1}{2}\right)$$

$$= \sum_{y=13}^{18} \binom{18}{y}\left(\frac{1}{2}\right)^{y}\left(1 - \frac{1}{2}\right)^{18-y}$$

$$\doteq 0.05$$

Of course, that the probability of committing a Type I error equals 0.05 should come as no surprise. In our earlier discussion of how "beyond all reasonable doubt" should be interpreted numerically, we specifically set the critical value equal to 13 so that the probability associated with the critical region *would* be 0.05.

In general, the probability of committing a Type I error is referred to as a test's *level of significance*, and is denoted, α. The concept is a crucial one. When the results of any statistical test are summarized, two items of information absolutely must be included: (1) whether H_0 was accepted or rejected, and (2) the level of significance at which the test was carried out.[1] Knowing only the former tells us very little. What the level of significance adds is a single-number summary of the "rules" by which the decision process is being conducted: α is reflecting nothing less than the amount of evidence the experimenter is demanding to see before abandoning the null hypothesis. (For the missile demonstration, we would write, "H_0 is accepted at the 0.05 level of significance.")

Question 6.3.1 Suppose that $H_0: p = p_0$ is tested against $H_1: p > p_0$. If H_0 is rejected at the $\alpha = 0.05$ level of significance, will it necessarily be rejected at the $\alpha = 0.01$ level of significance?

Question 6.3.2 If $H_0: p = p_0$ is rejected in favor of $H_1: p > p_0$ at the $\alpha = 0.01$ level of significance, will it necessarily be rejected at the $\alpha = 0.05$ level of significance?

We just saw that computing the probability of committing a Type I error is a nonproblem: there are no calculations necessary, since the probability equals whatever value the experimenter sets *a priori* for α. A similar simplicity does not hold for Type II errors. First, Type II error probabilities are not specified explicitly by the experimenter; second, each hypothesis test has an entire range of Type II error probabilities, one for each value of the parameter admissible under H_1. It will be necessary, therefore, to condition any Type II error calculation on some specific H_1 parameter value.

---

[1] The *P value* of a result is sometimes quoted as a substitute for a test's level of significance (see Question 6.3.10).

As an example, suppose that we wanted to find the probability of committing a Type II error in the missile demonstration problem *if the true p* [$= P$ (missile with new guidance system is on target)] *were 0.7*. By definition,

$$P \text{ (Type II error} \mid p = 0.7) = P \text{ (we accept } H_0 \mid p = 0.7)$$

$$= P(Y \le 12 \mid p = 0.7)$$

$$= \sum_{y=0}^{12} \binom{18}{y}(0.7)^y(1 - 0.7)^{18-y}$$

$$= 0.47$$

This means that if p were actually 0.7, the superiority of the new system would go "undetected"—that is, we would erroneously accept H_0: $p = \frac{1}{2}$—47% of the time.

The symbol for the probability of a Type II error is β. Figure 6.3.2 shows the sampling distributions of $W = Y$ when $p = \frac{1}{2}$ and when $p = 0.7$; the areas corresponding to α and β are shaded in.

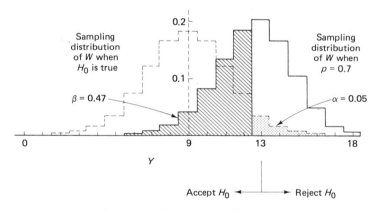

Figure 6.3.2   Sampling distributions of W, $p = 0.5$ and $p = 0.7$.

Clearly, the magnitude of β is a function of the presumed value for p. If, for example, the heat-sensing guidance system is so effective as to be on target *90%* of the time, the probability that our decision rule would lead us to make a Type II error is a much smaller 0.006:

$$\beta = P \text{ (Type II error} \mid p = 0.90)$$

$$= P \text{ (we accept } H_0 \mid p = 0.90) = P \text{ (}Y \le 12 \mid p = 0.90)$$

$$= \sum_{y=0}^{12} \binom{18}{y}(0.9)^y(1 - 0.9)^{18-y}$$

$$= 0.006$$

(see Figure 6.3.3).

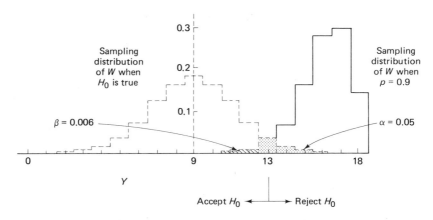

Figure 6.3.3 Sampling distributions of W, $p = 0.5$ and $p = 0.9$.

Comment

The basic notions of Type I and Type II errors first arose in a quality control context. The pioneering work was done at the Bell Telephone Laboratories: there the terms *producer's risk* and *consumer's risk* were introduced for what we now call α and β. Eventually, these ideas were generalized by Neyman and Pearson in the 1930s and evolved into the theory of hypothesis testing as we know it today.

If β is the probability that we *accept* H_0 when H_1 is true, then $1 - \beta$ is the probability of the complement, that we *reject* H_0 when H_1 is true. We call $1 - \beta$ the *power* of the test; it represents the ability of the decision rule to "recognize" (correctly) that H_0 is false. If W is the test statistic and C the critical region, then

$$1 - \beta = P(W \in C \mid H_1 \text{ is true})$$

The alternative hypothesis H_1 usually depends on a parameter, which makes $1 - \beta$ a function of that parameter. The relationship they share can be pictured by drawing a *power curve*, which is simply a graph of $1 - \beta$ versus the set of all possible parameter values. Figure 6.3.4 shows the power curve for testing

$$H_0: \quad p = \frac{1}{2}$$

versus

$$H_1: \quad p > \frac{1}{2}$$

where p is the probability of success in 18 Bernoulli trials and the decision rule is "Reject H_0 if $y \geq 13$." [The two marked points on the curve represent the $(p, 1 - \beta)$ pairs just determined, $(0.7, 0.53)$ and $(0.9, 0.994)$. Two other points can be gotten without doing any calculations: when $p = 1$, $1 - \beta = 1$, and when $p = p_0$ (the value specified by H_0), $1 - \beta = P(W \in C \mid H_0 \text{ is true}) = \alpha$.]

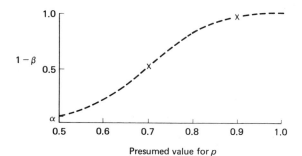

Presumed value for *p*

Figure 6.3.4 Power function.

To a mathematical statistician, power curves are very important. On the one hand, they completely characterize the "performance" that can be expected using a given test—in the binomial situation, for example, if the true value of p is presumed to be, say, p_1, where $p_1 \neq p_0$, the power curve shows precisely how much chance the test has of detecting that $H_0: p = p_0$ is not true. But power curves are also useful in comparing one test with another. There are many situations where more than one method can be devised for testing a given $H_0: \theta = \theta_0$. In such an event, how would we know which to choose? The answer is simple—take the one with the "steepest" power curve. Figure 6.3.5 shows two hypothetical power curves for testing $H_0: \theta = \theta_0$ versus $H_1: \theta \neq \theta_0$ at level of significance α. All other factors being equal (sample size required, difficulty of computation, and so on), method B is clearly the superior of the two, always having as it does a higher probability of correctly rejecting a false null hypothesis.

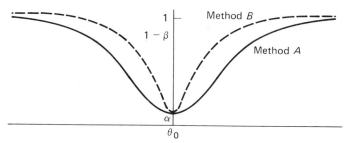

Figure 6.3.5 Two power curves

Question 6.3.3 Polygraphs used in criminal investigations typically measure five bodily functions: (1) thoracic respiration, (2) abdominal respiration, (3) blood pressure and pulse rate, (4) muscular movement and pressure, and (5) galvanic skin response. In principle, the magnitude of these responses when the subject is asked a relevant question ("Did you murder your wife?") indicate whether he is lying or telling the truth. The procedure, of course, is not infallible, as a recent study bore out (75). Seven experienced polygraph examiners were given a set of 40 records—20 were from innocent suspects and 20 from guilty suspects. The subjects had been asked 11 questions, on the basis of which each examiner was to make an overall judgment: "Innocent" or "Guilty." The results are shown below.

		Suspect's True Status	
		Innocent	Guilty
Examiner's Decision	"Innocent"	131	15
	"Guilty"	9	125

What would be the numerical values of α and β in this context? In a judicial setting, should Type I and Type II errors carry equal weight? Explain.

Question 6.3.4 Construct the power curve for the study described in Case Study 6.2.1. Use the DeMoivre–Laplace limit theorem to approximate $1 - \beta$ for p values of 0.30, 0.32, 0.35, and 0.40. Use your curve to estimate the probability of committing a Type II error if $p = 0.36$.

Question 6.3.5 An experiment consists of 12 independent Bernoulli trials. If $H_0: p = 1/3$ is tested against $H_1: p > 1/3$ by using the decision rule, "Reject H_0 if y, the number of successes, equals or exceeds 8," find α. Also, find β if $p = \frac{1}{2}$.

Question 6.3.6 A random sample of size $n = 1$ is drawn from a uniform pdf defined over the interval $[0, \theta]$. We decide to test

$$H_0: \quad \theta = 2$$
$$\text{versus}$$
$$H_1: \quad \theta \neq 2$$

by rejecting H_0 if either $x \leq 0.1$ or $x \geq 1.9$, where x is the value drawn. Find α. Also, find β if the true value of θ is 2.5.

Question 6.3.7 An urn contains 10 chips. An unknown number of the chips are white; the others are red. We wish to test

$$H_0: \quad \text{exactly half the chips are white}$$
$$\text{versus}$$
$$H_1: \quad \text{more than half the chips are white}$$

We will draw, without replacement, three chips and reject H_0 if two or more are white. Find α. Also, find β when the urn is (a) 60% white and (b) 70% white.

Question 6.3.8 Suppose that a random sample of size 5 is drawn from a uniform pdf

$$f_Y(y; \theta) = \begin{cases} \dfrac{1}{\theta} & 0 < y < \theta \\ 0 & \text{elsewhere} \end{cases}$$

We wish to test

$$H_0: \quad \theta = 2$$
$$\text{versus}$$
$$H_1: \quad \theta > 2$$

by rejecting the null hypothesis if $y_{max} \geq k$. Find the value of k that makes the probability of committing a Type I error equal to 0.05. Also, construct the power curve for the test.

Question 6.3.9 Sal is a pizza inspector for the city health department. Recently, he has received a number of complaints directed against a certain pizzeria for allegedly failing to comply with their advertisements. The pizzeria claims that on the average, each of their large pepperoni pizzas is topped with 2 ounces of pepperoni. The dissatisfied customers feel that the actual amount of pepperoni used is considerably less than that. To settle the matter, Sal decides to do a hypothesis test. First, he assumes that the distribution of pepperoni weights (per pizza) is normal with a mean of μ (ounces) and a standard deviation (σ) of 0.5 ounce. The hypotheses to be tested are

$$H_0: \quad \mu = 2.00$$

versus

$$H_1: \quad \mu < 2.00$$

Sal decides to take one large pizza at random and weigh the pepperoni it contains. If y, the pepperoni weight, is less than or equal to 1.3 ounces, he will reject H_0.
(a) Compute the α that corresponds to Sal's decision rule.
(b) Compute β if the true average pepperoni weight per pizza is 1.8 ounces.
(c) What should Sal's decision rule be if he wants α to be 0.01?
(d) Find β for the decision rule in part (c) and a true average weight of 1.8 ounces?

Question 6.3.10 When the results of statistical hypothesis tests are presented in journals or technical reports, any mention of a level of significance is often deleted—instead, a test's *P value* is quoted. By definition, the *P* value is the smallest α for which H_0 could be rejected. Find the *P* value for the whale data described in Question 6.2.7.

6.4 A NOTION OF OPTIMALITY: THE GENERALIZED LIKELIHOOD RATIO

In the next several chapters we will be looking closely at some of the particular hypothesis tests that statisticians most often need to use in dealing with real-world problems. All of these have the same conceptual heritage—a very fundamental notion known as the *generalized likelihood ratio*, or GLR. More than just a principle, the generalized likelihood ratio is a working criterion for actually *suggesting* test procedures. In a sense, the GLR does for hypothesis testing what the principle of maximum likelihood does for estimation.

We will illustrate the generalized likelihood ratio by deriving a test for the parameter of a uniform pdf. More substantive applications, ones involving the normal distribution, are detailed in Chapters 7 and 8.

Let $Y_1, Y_2, \ldots, Y_n$ be a random sample from a uniform distribution defined over $(0, \theta)$, where θ is unknown. Our objective is to test

$$H_0: \quad \theta = \theta_0$$

versus

$$H_1: \quad \theta < \theta_0$$

at a specified level of significance α.

As a starting point, it will be necessary to define two parameter spaces, ω and Ω, where ω is a particular subset of Ω. The former is the set of unknown parameter values admissible under H_0. Here, the only unknown parameter is θ, and the null hypothesis restricts it to a single point. In set notation,

$$\omega = \{\theta: \theta = \theta_0\}$$

The second parameter space, Ω, is the set of *all* possible values of all unknown parameters. In this case,

$$\Omega = \{\theta: 0 < \theta \le \theta_0\}$$

Now, recall the definition of the likelihood function, L, from Definition 5.8.1. Given a sample from a uniform distribution,

$$L = L(\theta) = \prod_{i=1}^{n} f_Y(y_i; \theta) = \begin{cases} \left(\dfrac{1}{\theta}\right)^n & \text{for } 0 \le y_i \le \theta \\ 0 & \text{otherwise} \end{cases}$$

For reasons the following will make clear, we will want to maximize $L(\theta)$ twice, once under ω and again under Ω. Since θ can take on only one value—θ_0—under ω,

$$\max_{\omega} L(\theta) = L(\theta_0) = \begin{cases} \left(\dfrac{1}{\theta_0}\right)^n & \text{for } 0 \le y_i \le \theta_0 \\ 0 & \text{otherwise} \end{cases}$$

Of course, maximizing $L(\theta)$ under Ω—that is, with *no* restrictions—is accomplished by substituting the maximum-likelihood estimator for θ into $L(\theta)$. For the uniform distribution, $W = Y_{\max}$ is the maximum-likelihood estimator (see Example 5.8.3). Therefore,

$$\max_{\Omega} L(\theta) = \left(\frac{1}{Y_{\max}}\right)^n$$

For notational simplicity, we denote $\max_{\omega} L(\theta)$ and $\max_{\Omega} L(\theta)$ by $L(\hat{\omega})$ and $L(\hat{\Omega})$, respectively.

Definition 6.4.1. Let $Y_1, Y_2, \ldots, Y_n$ be a random sample from $f_Y(y; \theta_1, \ldots, \theta_k)$. The generalized likelihood ratio, Λ, is defined to be

$$\Lambda = \frac{\max\limits_{\omega} L(\theta_1, \ldots, \theta_k)}{\max\limits_{\Omega} L(\theta_1, \ldots, \theta_k)} = \frac{L(\hat{\omega})}{L(\hat{\Omega})}$$

Note that Λ is a random variable, having been defined as a function of the random sample. In keeping with our random variable notation, values of Λ for a particular sample are denoted by lowercase λ.

For the uniform distribution,

$$\lambda = \frac{(1/\theta_0)^n}{(1/y_{max})^n} = \left(\frac{y_{max}}{\theta_0}\right)^n$$

Note that, in general, Λ will always be positive but never greater than 1 (why?). Furthermore, values of the likelihood ratio close to 1 suggest that the data are very compatible with H_0. That is, the observations are "explained" almost as well by the H_0 parameters as by *any* parameters [as measured by $L(\hat{\omega})$ and $L(\hat{\Omega})$]. For these values of Λ we should *accept* H_0. Conversely, if $L(\hat{\omega})/L(\hat{\Omega})$ were close to 0, the data would not be very compatible with the parameter values in ω and it would make sense to *reject* H_0. This is the rationale behind the generalized-likelihood-ratio *principle* as stated in the next definition.

Definition 6.4.2. A generalized-likelihood-ratio test (GLRT) is one that rejects H_0 whenever

$$0 < \lambda \leq \lambda*$$

where $\lambda*$ is chosen so that

$$P(0 < \Lambda \leq \lambda* \mid H_0 \text{ is true}) = \alpha$$

If we knew the distribution under H_0 of the random variable $\Lambda, f_\Lambda(\lambda \mid H_0)$, the critical value $\lambda*$ (and, hence, the critical region, C) could be determined by solving the equation

$$\alpha = \int_0^{\lambda*} f_\Lambda(\lambda \mid H_0) \, d\lambda$$

(see Figure 6.4.1). In most situations, though, $f_\Lambda(\lambda \mid H_0)$ is *not* known, and it becomes necessary to show that Λ is a monotonic function of some quantity W, where the distribution of W *is* known. Once we have found such a statistic, any test based on W will be equivalent to one based on Λ.

Figure 6.4.1 Rejection region for GLRT.

Here, a suitable W is easy to find. Note that

$$P(\Lambda \leq \lambda* \mid H_0) = \alpha = P\left[\left(\frac{Y_{max}}{\theta_0}\right)^n \leq \lambda* \mid H_0\right]$$

$$= P\left(\frac{Y_{max}}{\theta_0} \leq \sqrt[n]{\lambda*} \mid H_0\right)$$

Let $W = Y_{max}/\theta_0$ and $w^* = \sqrt[n]{\lambda^*}$. Then

$$P(\Lambda \leq \lambda^* \mid H_0) = P(W \leq w^* \mid H_0) \qquad (6.4.1)$$

Here the right-hand side of Equation 6.4.1 can be evaluated from what we already know about the density function for the largest order statistic from a uniform distribution. Let $f_{Y_{max}}(y; \theta_0)$ be the density function for Y_{max}. Then

$$f_W(w; \theta_0) = \theta_0 \, f_{Y_{max}}(\theta_0 w)$$

which, from Example 5.3.1, reduces to

$$\frac{\theta_0 \, n(\theta_0 w)^{n-1}}{\theta_0^n} = nw^{n-1} \qquad 0 \leq w \leq 1$$

Therefore,

$$P(W \leq w^* \mid H_0) = \int_0^{w^*} nw^{n-1} \, dw = (w^*)^n = \alpha$$

implying that the critical value for W is

$$w^* = \sqrt[n]{\alpha}$$

That is, the GLRT calls for H_0 to be rejected if

$$w = \frac{y_{max}}{\theta_0} \leq \sqrt[n]{\alpha}$$

Comment

The GLR is applied to other hypothesis-testing situations in a manner very similar to what was described here: first we find $L(\hat{\omega})$ and $L(\hat{\Omega})$, then Λ, and finally W. The algebra involved, though, usually becomes considerably more formidable. For example, in the "normal" model taken up in Chapters 7 and 8, both parameter spaces are two-dimensional and the likelihood function is a product of densities of the form

$$f_Y(y; \mu, \sigma^2) = \frac{1}{\sqrt{2\pi}\,\sigma} e^{-(1/2)[(y-\mu)/\sigma]^2}$$

Question 6.4.1 Let $X_1, X_2, \ldots, X_n$ be a random sample from the geometric probability function

$$f_X(x, p) = pq^{x-1} \qquad x = 1, 2, \ldots$$

where $q = 1 - p$. Find Λ, the generalized likelihood ratio for testing $H_0: p = p_0$ versus $H_1: p \neq p_0$.

Question 6.4.2 Let Y_1, Y_2, ..., Y_{10} be a random sample from an exponential pdf with unknown parameter λ. Find the form of the GLRT for H_0: $\lambda = \lambda_0$ versus H_1: $\lambda \neq \lambda_0$. What integral would have to be evaluated to determine the critical value if α were equal to 0.05?

Question 6.4.3 Let Y_1, Y_2, ..., Y_n be a random sample from a normal pdf with unknown mean μ and variance 1. Find the form of the GLRT for H_0: $\mu = \mu_0$ versus H_1: $\mu \neq \mu_0$.

7

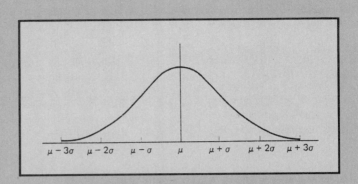

The Normal Distribution

Francis Galton

I know of scarcely anything so apt to impress the imagination as the wonderful form of cosmic order expressed by the "Law of Frequency of Error" (the normal distribution). The law would have been personified by the Greeks and deified, if they had known of it. It reigns with serenity and in complete self effacement amidst the wildest confusion. The huger the mob, and the greater the anarchy, the more perfect is its sway. It is the supreme law of Unreason.

7.1 INTRODUCTION

Finding probability distributions to describe—and, ultimately, to predict—empirical data is one of the most important contributions a statistician can make to the research scientist. Already we have seen a number of "models" pressed into this sort of service; the binomial was an obvious choice as a model for the number of correct responses in an ESP experiment; there were physical reasons why the gamma would provide a good fit for rainfall data; and the Poisson was seen to apply to situations as diverse as the number of alpha particles emitted from a radioactive source to the number of major labor strikes occurring weekly in the United Kingdom. But by far the most widely used probability model in statistics is the *normal* distribution,

$$f_Y(y) = \frac{1}{\sqrt{2\pi}\,\sigma}\, e^{-(1/2)[(y-\mu)/\sigma]^2} \qquad -\infty < y < \infty \qquad (7.1.1)$$

In Chapter 4 we recounted some of the history surrounding this more than 250-year-old function: how it first appeared as a limiting form of the binomial, only to be quickly popularized as a model in its own right by the likes of Gauss, Quetelet, and Galton. Today we recognize that the unique prominence enjoyed by this distribution actually derives from two sources. On the one hand, many real-world phenomena *do* behave according to the bell-shaped pattern that Equation 7.1.1 prescribes. A second, and more compelling, reason is the *central limit theorem*: the vast majority of inference procedures in common use are based on averages of independent, identically distributed random variables—precisely the sort of quantity, according to the theorem, that tends to be approximated by the normal. The result is that when the sample size, n, is sufficiently large, many inference procedures can be reduced to a probability statement about an appropriate normal variable.

In this chapter we examine some of the more basic hypothesis-testing and estimation problems associated with the normal distribution. What is important in this material is the theory, not the applications. Later chapters will investigate problems of far more practical significance. The solutions of those problems, though, will depend on relatively simple extensions of the results proved in the next several sections.

7.2 POINT ESTIMATES FOR μ AND σ^2

As we observed in Chapter 4, the normal distribution is a two-parameter family—μ being a measure of location and σ^2 a measure of dispersion. Finding point estimates for these parameters will be our first objective. Because of the normal's relative simplicity, both $\hat{\mu}$ and $\hat{\sigma}^2$ can be easily obtained using the method of maximum likelihood.

Comment

Recall that if Y is a random variable and its density function is Equation 7.1.1, we will write $Y \sim N(\mu, \sigma^2)$. If $\mu = 0$ and $\sigma^2 = 1$, the variable is said to be a *standard normal* and will be denoted by the letter Z.

Theorem 7.2.1. Let $Y_1, Y_2, \ldots, Y_n$ be a random sample from an $N(\mu, \sigma^2)$ distribution. The maximum-likelihood estimates for μ and σ^2 are

$$\hat{\mu} = \frac{1}{n} \sum_{i=1}^{n} y_i = \bar{y}$$

and

$$\hat{\sigma}^2 = \frac{1}{n} \sum_{i=1}^{n} (y_i - \bar{y})^2$$

Proof. The likelihood function of the y_i's can be written

$$L(\mu, \sigma^2) = \prod_{i=1}^{n} f_Y(y_i; \mu, \sigma^2)$$

$$= \frac{1}{\sqrt{2\pi}\,\sigma} e^{-(1/2)[(y_1 - \mu)/\sigma]^2} \cdots \frac{1}{\sqrt{2\pi}\,\sigma} e^{-(1/2)[(y_n - \mu)/\sigma]^2}$$

$$= \left(\frac{1}{2\pi\sigma^2}\right)^{n/2} e^{-(1/2\sigma^2)\sum_{i=1}^{n}(y_i - \mu)^2}$$

Also,

$$\ln L(\mu, \sigma^2) = -\frac{n}{2} \ln 2\pi\sigma^2 - \frac{1}{2\sigma^2} \sum_{i=1}^{n} (y_i - \mu)^2$$

Setting the two partial derivatives of $\ln L(\mu, \sigma^2)$ equal to 0 gives

$$\frac{\partial \ln L(\mu, \sigma^2)}{\partial \mu} = \frac{1}{\sigma^2} \sum_{i=1}^{n} (y_i - \mu) = 0 \tag{7.2.1}$$

and

$$\frac{\partial \ln L(\mu, \sigma^2)}{\partial \sigma^2} = -\frac{n}{2\sigma^2} + \frac{1}{2\sigma^4} \sum_{i=1}^{n} (y_i - \mu)^2 = 0 \tag{7.2.2}$$

The simultaneous solution of Equations 7.2.1 and 7.2.2 gives the statement of the theorem:

$$\hat{\mu} = \frac{1}{n} \sum_{i=1}^{n} y_i = \bar{y}$$

and

$$\hat{\sigma}^2 = \frac{1}{n} \sum_{i=1}^{n} (y_i - \bar{y})^2$$

Theorems 7.2.2 and 7.2.3 list some of the properties of $\hat{\mu}$ and $\hat{\sigma}^2$. All are straightforward applications of the definitions and theorems of Chapter 5.

Proof. The unbiasedness of $\bar{Y}$ is a special case of Example 5.4.3. To show that $\bar{Y}$ is efficient, note that

$$\frac{\partial \ln f_Y(y; \mu, \sigma^2)}{\partial \mu} = \frac{y - \mu}{\sigma^2}$$

and

$$\frac{\partial^2 \ln f_Y(y; \mu, \sigma^2)}{\partial \mu^2} = -\frac{1}{\sigma^2}$$

From Theorem 5.6.1, then, the Cramer–Rao lower bound for the variance of an unbiased estimator for μ is

$$\frac{1}{-n\left(-\dfrac{1}{\sigma^2}\right)} = \frac{\sigma^2}{n}$$

But the variance of the maximum-likelihood estimator *does* achieve that bound: $\text{Var}(\bar{Y}) = \sigma^2/n$, so the efficiency of $\bar{Y}$ is established.

Chebyshev's inequality can be used to prove that $\bar{Y}$ is consistent. By the statement of the inequality,

$$P(|\bar{Y} - \mu| < \epsilon) > 1 - \frac{\text{Var}(\bar{Y})}{\epsilon^2}$$

or, after substituting for $\text{Var}(\bar{Y})$,

$$P(|\bar{Y} - \mu| < \epsilon) > 1 - \frac{\sigma^2}{n\epsilon^2}$$

from which it is clear that $\bar{Y}$ converges in probability to μ (i.e., is consistent): for any $\delta > 0$, $\bar{Y}$ will be in an ϵ-neighborhood of μ a minimum of $100(1 - \delta)\%$ of the time, provided $n = n(\epsilon, \delta) \geq \sigma^2/\epsilon^2\delta$.

The sufficiency of $\bar{Y}$ follows from the Fisher–Neyman criterion (Theorem 5.7.1) and the corollary to Theorem 7.3.1. The details will be left as an exercise.

The second parameter in a normal pdf, σ^2, is usually estimated not by the $\hat{\sigma}^2$ of Theorem 7.2.1 but by the sample variance, s^2, where

$$s^2 = \left(\frac{n}{n-1}\right)\hat{\sigma}^2 = \frac{1}{n-1}\sum_{i=1}^{n}(y_i - \bar{y})^2$$

(recall Example 5.4.4).

Theorem 7.2.3. Given that $Y_1, Y_2, \ldots, Y_n \sim N(\mu, \sigma^2)$,

$$S^2 = \frac{1}{n-1} \sum_{i=1}^{n} (Y_i - \bar{Y})^2$$

is unbiased and consistent.

Proof. To establish the consistency of S^2 requires a result proved later in this chapter (Theorem 7.5.2) and will be left as an exercise. That the estimator is unbiased is a special case of Example 5.4.4.

CASE STUDY 7.2.1

In the eighth century B.C., the Etruscan civilization was the most advanced in all of Italy. Its art forms and political innovations were destined to leave indelible marks on the entire Western world. Originally located along the western coast between the Arno and Tiber rivers (the region now known as Tuscany), it spread quickly across the Apennines and eventually overran much of Italy. But as quickly as it came, it faded. Militarily it was to prove no match for the burgeoning Roman legions, and by the dawn of Christianity it was all but gone.

No chronicles of the Etruscan empire have ever been found, and to this day its origins remain shrouded in mystery. Were the Etruscans native Italians, or were they immigrants? And if they were immigrants, where did they come from? Much of what *is* known has come from archeological investigations and anthropometric studies—the latter involving the use of body measurements to determine racial characteristics and ethnic origins. The data presented here are an example of one such study.

Research has shown that the maximum head breadth of modern Italian males averages 132.4 mm. Listed in Table 7.2.1 are the maximum head breadths recorded for 84 male Etruscan skulls uncovered in various archeological digs throughout Italy (4).

TABLE 7.2.1 MAXIMUM HEAD BREADTHS (MM) OF 84 ETRUSCAN MALES

141	148	132	138	154	142	150
146	155	158	150	140	147	148
144	150	149	145	149	158	143
141	144	144	126	140	144	142
141	140	145	135	147	146	141
136	140	146	142	137	148	154
137	139	143	140	131	143	141
149	148	135	148	152	143	144
141	143	147	146	150	132	142
142	143	153	149	146	149	138
142	149	142	137	134	144	146
147	140	142	140	137	152	145

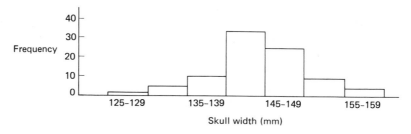

Figure 7.2.1 Graph of Etruscan head-breadth data.

A graph of these data (Figure 7.2.1) indicates that the population from which they are a sample may very well be normal. Lacking any other information, we would probably choose $N(\bar{y}, s^2)$ as the particular normal having the best chance of fitting the data. For these 84 observations,

$$\sum_{i=1}^{84} y_i = 12,077$$

and

$$\sum_{i=1}^{84} y_i^2 = 1,739,315$$

Therefore,

$$\bar{y} = \frac{12,077}{84} = 143.8$$

and, using the formula given in Chapter 5,

$$s = \sqrt{\frac{84(1,739,315) - (12,077)^2}{84(83)}} = 6.0$$

Figure 7.2.2 shows the $N(143.8, (6.0)^2)$ density superimposed over the histogram of Figure 7.2.1. The fit seems to justify our assumption of normality.

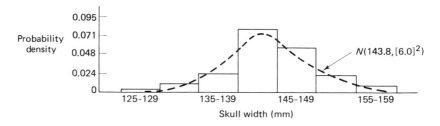

Figure 7.2.2 Normal curve fitted to histogram.

Comment

From a hypothesis-testing standpoint, it would make sense to test

$$H_0: \quad \mu = 132.4$$

versus

$$H_1: \quad \mu \neq 132.4$$

where μ is the true average maximum head breadth for Etruscans. Rejecting H_0 could be interpreted as evidence (although evolutionary shifts would have to be taken into account) that the Etruscans and the Italians are of different ethnic origins.

Question 7.2.1 Assuming the model

$$f_Y(y) = \frac{1}{\sqrt{2\pi}\,(6.0)}\, e^{-(1/2)[(y-143.8)/6.0]^2}$$

compute the expected number of skulls having widths in the range 135–139 mm (see Case Study 4.3.2).

Question 7.2.2 In the home, the amount of radiation emitted by a color television set does not pose a health problem of any consequence. The same may not be true, though, in department stores, where as many as 15 or 20 sets may be turned on at the same time and in a relatively confined area. The following readings were taken at ten different department stores, each having at least five sets in their display areas. The figure shown for each store is an average radiation level (in milliroentgens per hour) based on readings taken at several different locations within each area (83).

Location	Net mr/hour
1	0.40
2	0.48
3	0.60
4	0.15
5	0.50
6	0.80
7	0.50
8	0.36
9	0.16
10	0.89

(a) Compute the sample mean and the sample standard deviation for these data.
(b) The recommended safety limit for exposure of this sort has been set by the National Council on Radiation Protection at 0.5 mr/hour. Assume that the $\bar{y}$ and s calculated in part (a) are the true μ and σ, respectively, characterizing radiation exposure in department-store display areas. If the distribution of exposures from store to store follows a normal pdf, estimate the proportion of stores giving exposures in excess of the recommended safety limit.

Question 7.2.3 For data coming from a normal pdf, the proportion of observations in the interval $(\bar{y} - s, \bar{y} + s)$ is approximately 0.68, and the proportion in the interval $(\bar{y} - 2s, \bar{y} + 2s)$ is approximately 0.95. [These are estimates based on the fact that $P(\mu - \sigma < X < \mu + \sigma) = 0.68$ and $P(\mu - 1.96\sigma < X < \mu + 1.96\sigma) = 0.95$ if $X \sim N(\mu, \sigma^2)$.] Determine the one- and two-standard-deviation intervals for the traffic-death rates given in Question 4.3.23 and compute the proportion of observations they each contain.

Question 7.2.4 What percentage of normally distributed observations would be expected to lie in the interval $(\bar{y} - s, \bar{y} + 2s)$? "Verify" your answer using the Etruscan skull data given in Case Study 7.2.1.

Question 7.2.5 The effects of age on various characteristics of a person's blood are not entirely understood. Whether or not age affects the platelet count was the reason for collecting the data listed (161). The subjects were 24 female rest-home patients. In a more typical, nongeriatric population, platelet counts are expected to range from 140 to 440 (thousands per cubic millimeter of blood).

Subject	Count	Subject	Count
1	125	13	180
2	170	14	180
3	250	15	280
4	270	16	240
5	144	17	270
6	184	18	220
7	176	19	110
8	100	20	176
9	220	21	280
10	200	22	176
11	170	23	188
12	160	24	176

Assuming platelet counts to be normally distributed, estimate the percentage of elderly women whose counts would be considered abnormal. Draw a histogram of these data. Does the normality assumption seem tenable?

Question 7.2.6 At a certain southern university, the 1975 freshman class numbered 2670, of whom 72.5% were males. The men averaged 631 on the verbal part of the SAT and 691 on the quantitative part. The women's averages were 623 and 640, respectively. Find the average freshman SAT score at this university, where the average score is defined to be one-half times the sum of the verbal and quantitative scores.

Question 7.2.7 For a random sample of size n, find the method-of-moments estimators for μ and σ^2.

Question 7.2.8 In Question 5.4.15, the notion of an estimator's being *asymptotically unbiased* was introduced. If $Y_1, Y_2, \ldots, Y_n \sim N(\mu, \sigma^2)$, is the sample variance asymptotically unbiased for σ^2?

Question 7.2.9 Show that the $N(\mu, \sigma^2)$ pdf, where μ is known, can be written in exponential form (see Question 5.7.15). Use that fact to suggest a sufficient statistic for σ^2.

7.3 LINEAR COMBINATIONS OF NORMAL RANDOM VARIABLES

As pointed out in Section 7.1, it is the rule rather than the exception for estimation and hypothesis-testing procedures to be based on *averages*, either of the original data or of some function of the original data. The methodology that has evolved for the normal distribution is no different. We will see in a later section that inferences focusing on μ will be based on $\bar{y}$, while those concerning σ^2 will derive from s^2, which is just an average of $(y_i - \bar{y})^2$ terms. What this implies is that the place to begin our discussion is with the sampling behavior of linear combinations of independent normals.

The main result of this section is Theorem 7.3.1, which gives the distribution of the sum of two independent normals—and, hence, by induction, of n independent normals. The proof is accomplished by writing the sum as a convolution and then doing the indicated integration. To simplify the algebra, we will consider a special case first; from this and Definition 4.3.1 the general statement will follow immediately.

Comment

A much easier way to prove Theorem 7.3.1, using moment-generating functions, is indicated in Question 7.3.1. The lengthier proof given here is still noteworthy, though, because of the manipulative techniques it uses—techniques that remain applicable in situations where moment-generating functions are of no help.

Lemma If $Y_1 \sim N(0, \sigma^2)$ and $Y_2 \sim N(0, 1)$, and if Y_1 and Y_2 are independent,

$$Y_1 + Y_2 \sim N(0, \sigma^2 + 1)$$

Proof. Written as a convolution, the probability function for $Y_1 + Y_2$ is the integral of the product of two normal densities:

$$f_{Y_1 + Y_2}(t) = \int_{-\infty}^{\infty} f_{Y_1}(y) f_{Y_2}(t - y) \, dy$$

$$= \int_{-\infty}^{\infty} \frac{1}{\sqrt{2\pi}\,\sigma} e^{-y^2/2\sigma^2} \frac{1}{\sqrt{2\pi}} e^{-(1/2)(t-y)^2} \, dy$$

$$= \frac{1}{2\pi\sigma} \int_{-\infty}^{\infty} e^{-(1/2)[y^2/\sigma^2 + (t-y)^2]} \, dy$$

The quantity in the brackets of the exponent can be simplified to give

$$\frac{y^2}{\sigma^2} + y^2 - 2ty + t^2 = \left(\frac{1 + \sigma^2}{\sigma^2}\right) y^2 - 2ty + t^2 \qquad (7.3.1)$$

If we let

$$c = \frac{\sigma}{\sqrt{1 + \sigma^2}}$$

the right-hand side of Equation 7.3.1 becomes

$$\frac{1}{c^2} y^2 - 2ty + t^2 = \frac{1}{c^2} y^2 - 2ty + c^2t^2 + t^2 - c^2t^2$$

$$= \left(\frac{y}{c} - ct\right)^2 + (1 - c^2)t^2$$

$$= \left(\frac{y - c^2t}{c}\right)^2 + (1 - c^2)t^2$$

Therefore,

$$f_{Y_1 + Y_2}(t) = \frac{1}{2\pi\sigma} \int_{-\infty}^{\infty} e^{-(1/2)[(y - c^2t)/c]^2} e^{-(1/2)(1 - c^2)t^2} \, dy$$

$$= \frac{1}{\sqrt{2\pi} \, (\sigma/c)} e^{-(1/2)(1 - c^2)t^2} \frac{1}{\sqrt{2\pi} \, c} \int_{-\infty}^{\infty} e^{-(1/2)[(y - c^2t)/c]^2} \, dy$$

Note that the last factor is 1, being the integral over the entire range of a normal variable with mean c^2t and variance c^2. Furthermore, since

$$1 - c^2 = 1 - \frac{\sigma^2}{1 + \sigma^2} = \frac{1}{1 + \sigma^2}$$

and

$$\frac{\sigma}{c} = \sqrt{1 + \sigma^2}$$

the first factor is a normal density with mean 0 and variance $1 + \sigma^2$.

Theorem 7.3.1. Let $Y_1 \sim N(\mu_1, \sigma_1^2)$ and $Y_2 \sim N(\mu_2, \sigma_2^2)$. If Y_1 and Y_2 are independent, then $Y_1 + Y_2 \sim N(\mu_1 + \mu_2, \sigma_1^2 + \sigma_2^2)$.

Proof. We can write

$$Y_1 + Y_2 = \sigma_2\left(\frac{Y_1 - \mu_1}{\sigma_2} + \frac{Y_2 - \mu_2}{\sigma_2}\right) + \mu_1 + \mu_2$$

An application of Theorem 3.5.1 gives

$$\frac{Y_1 - \mu_1}{\sigma_2} \sim N\left(0, \frac{\sigma_1^2}{\sigma_2^2}\right) \qquad \text{and} \qquad \frac{Y_2 - \mu_2}{\sigma_2} \sim N(0, 1)$$

By the lemma,

$$\frac{Y_1 - \mu_1}{\sigma_2} + \frac{Y_1 - \mu_2}{\sigma_2} \sim N\left(0, \frac{\sigma_1^2}{\sigma_2^2} + 1\right)$$

Again using Theorem 3.5.1, we obtain

$$Y_1 + Y_2 \sim N(\mu_1 + \mu_2, \sigma_1^2 + \sigma_2^2)$$

The most important special case of Theorem 7.3.1 concerns the distribution of the sample mean, which, of course, can be viewed as a linear combination of n identically distributed, independent normals, each having the same mean μ, variance σ^2, and coefficient $1/n$.

> **Corollary**   If $\bar{Y} = (1/n)\sum_{i=1}^{n} Y_i$ is the sample mean of n independent $N(\mu, \sigma^2)$ random variables, then $\bar{Y} \sim N(\mu, \sigma^2/n)$.

Comment

Because of Theorem 7.2.2, we already know that $\bar{Y}$ is a "best" estimator—that is, among the class of unbiased estimators for μ, $\bar{Y}$ has the smallest variance. But now with the corollary to Theorem 7.3.1 we have the *distribution* of $\bar{Y}$, which makes it possible to quantify, in a probabilistic sense, $\bar{Y}$'s precision. For example, suppose a single observation is drawn from an $N(100, (5)^2)$ distribution. The probability that Y lies within ± 5 (one standard deviation of Y) of its mean is 0.682:

$$P(95 \leq Y \leq 105) = P\left(\frac{95 - 100}{5} \leq \frac{Y - \mu}{\sigma} \leq \frac{105 - 100}{5}\right)$$

$$= P(-1.00 \leq Z \leq 1.00) = 0.682$$

If, on the other hand, a random sample of, say, size 4 had been drawn from the same distribution, the probability of the resulting *mean's* falling between 95 and 105 would have increased to 0.955:

$$P(95 \leq \bar{Y} \leq 105) = P\left(\frac{95 - 100}{5/\sqrt{4}} \leq \frac{\bar{Y} - \mu}{\sigma/\sqrt{4}} \leq \frac{105 - 100}{5/\sqrt{4}}\right)$$

$$= P(-2.00 \leq Z \leq 2.00) = 0.955$$

Table 7.3.1 lists the probability of $\bar{Y}$'s lying in the interval $\mu - \sigma$ to $\mu + \sigma$ for sample sizes ranging from 1 to 9.

TABLE 7.3.1 PROBABILITIES FOR NINE SAMPLE SIZES

n	$P(\mu - \sigma \leq \bar{Y} \leq \mu + \sigma) = P(-\sqrt{n} \leq Z \leq \sqrt{n})$
1	0.682
2	0.841
3	0.916
4	0.955
5	0.975
6	0.986
7	0.992
8	0.995
9	0.997

Comment

Notice above how the denominator of the Z transformation is modified, depending on whether the probability statement is about Y or $\bar{Y}$. The expression $\sigma/\sqrt{n}$ appearing in the latter is often referred to as the *standard error of the mean*.

Question 7.3.1 Prove Theorem 7.3.1 by examining the product of the moment-generating functions for Y_1 and Y_2.

Question 7.3.2 Use the corollary to Theorem 7.3.1 and the Fisher–Neyman criterion of Chapter 5 to prove that $\bar{Y}$ is a sufficient statistic for μ.

Question 7.3.3 The IQs of nine randomly selected persons are recorded. Let $\bar{Y}$ denote their average. Assuming the distribution from which the Y_i's were drawn is normal with a mean of 100 and a standard deviation of 16, what is the probability that $\bar{Y}$ will exceed 103? What is the probability that any arbitrary Y_i will exceed 103? What is the probability that exactly three of the Y_i's will exceed 103?

Question 7.3.4 The cylinders and pistons for a certain internal combustion engine are manufactured by a process that gives a normal distribution of cylinder diameters with a mean of 41.5 cm and a standard deviation of 0.4 cm. Similarly, the distribution of piston diameters is normal with a mean of 40.5 cm and a standard deviation of 0.3 cm. If the piston diameter is greater than the cylinder diameter, the former can be reworked until the two "fit." What proportion of cylinder–piston pairs will need to be reworked?

Question 7.3.5 A circuit contains three resistors wired in series. Each is rated at 6 ohms. Suppose, however, that the true resistance of each one is a normally distributed random variable with a mean of 6 ohms and a standard deviation of 0.3 ohm. What is the probability that the combined resistance will exceed 19 ohms? How "precise" would the manufacturing process have to be to make the probability less than 0.005 that the combined resistance of the circuit would exceed 19 ohms?

Question 7.3.6 The personnel department of a large corporation gives two aptitude tests to job applicants. One measures verbal ability; the other, quantitative ability. From many years' experience, the company has found that the verbal scores tend to be normally distributed with a mean of 50 and a standard deviation of 10. The quantitative scores are normally distributed with a mean of 100 and a standard deviation of 20, and they appear to be independent of the verbal scores. A composite score, C, is assigned to each applicant, where

$$C = 3(\text{verbal score}) + 2(\text{quantitative score})$$

If company policy prohibits hiring anyone whose C score is below 375, what percentage of applicants will be summarily rejected?

Question 7.3.7 Let $Y_1, Y_2, \ldots, Y_n$ be a random sample from an $N(2, 4)$ pdf. How large must n be in order that

$$P(1.9 \leq \bar{Y} \leq 2.1) \geq 0.99 \quad ?$$

Question 7.3.8 How large a random sample needs to be taken from a normal pdf in order for the probability to be at least 0.90 that the sample mean will be within one standard deviation of μ? Within two standard deviations of μ?

Question 7.3.9 An elevator in the athletic dorm at State University has a stated load capacity of 2400 pounds. Suppose that 10 football players get on at the twentieth floor. If the distribution of weights of football players at State is normally distributed with a mean of 220 pounds and a standard deviation of 20 pounds, what is the probability that there will be 10 fewer players at the next practice?

Question 7.3.10 Let $Y_1, Y_2, \ldots, Y_9$ be a random sample from an $N(0, 9)$ distribution. On the same axes, draw (a) the original pdf and (b) the pdf for $\bar{Y}$. Find $P(Y_i \geq 1)$ and $P(\bar{Y} \geq 1)$.

Question 7.3.11 Let $X_1, X_2, \ldots, X_9$ be an independent random sample from an $N(2, 4)$ pdf and Y_1, Y_2, Y_3, Y_4 an independent random sample from an $N(1, 1)$ pdf. Find $P(\bar{X} \geq \bar{Y})$.

7.4 THE CENTRAL LIMIT THEOREM

We have now established the normality of a sum of normal random variables. It would be too much to ask that the sum of nonnormal variables also have a normal distribution. Even so, under very general conditions such a sum can be *approximately* normal. The first theorem stating this sort of result was the now familiar Theorem 4.3.1 of DeMoivre and Laplace. There was a great deal of work on this problem during the nineteenth century, particularly by the Russian school of probabilists [see (97)], and it culminated in a rather general result by A. M. Lyapunov in 1901. Theorem 7.4.1 gives a version of this result strong enough for most statistical applications. The name *central limit theorem* is due to G. Polya ("Über den zentralen Grenzwertsatz der Wahrscheinlichkeitsrechnung und das Momentenproblem," *Mathematische Zeitschrift*, vol. 8, 1920, pp. 171–181).

> **Theorem 7.4.1 (Central Limit Theorem).** Let $Y_1, Y_2, \ldots$ be an infinite sequence of independent random variables, each with the same distribution. Suppose that the mean μ and the variance σ^2 of $f_Y(y)$ are both finite. For any numbers c and d,
>
> $$\lim_{n \to \infty} P\left(c < \frac{Y_1 + \cdots + Y_n - n\mu}{\sqrt{n}\,\sigma} < d\right) = \frac{1}{\sqrt{2\pi}} \int_c^d e^{-(1/2)y^2}\, dy$$

If the Y_i's are Bernoulli random variables, Theorem 7.4.1 reduces to the DeMoivre–Laplace theorem of Chapter 4. A proof for that special case requiring only modest analytical techniques can be found in (41). More general cases require background more elaborate than we can develop here. See (90).

It *is* possible to sketch a proof by strengthening the hypotheses to assume that the moment-generating functions of the Y_i's exist. Then, not only are the mean and variance finite, but the common moment-generating function, $M(t)$, of the Y_i's has $M^{(r)}(0)$ defined for all $r \geq 0$ and all moments exist.

The use of the hypothesized moment-generating functions is via the following lemma, whose proof is beyond the scope of this text.

Lemma Suppose that $Y, Y_1, Y_2, \ldots$ are random variables with $\lim_{n \to \infty} M_{Y_n}(t) = M_Y(t)$ for all t in some interval about 0. Then $\lim_{n \to \infty} F_{Y_n}(y) = F_Y(y)$ for all real values y.

Thus it is necessary to establish that

$$\lim_{n \to \infty} M_{(Y_1 + \cdots + Y_n - n\mu)/(\sqrt{n}\,\sigma)}(t) = M_Z(t) = e^{(1/2)t^2}$$

where $Z \sim N(0, 1)$.

For notational convenience write

$$\frac{Y_1 + \cdots + Y_n - n\mu}{\sqrt{n}\,\sigma} = \frac{W_1 + \cdots + W_n}{\sqrt{n}}$$

where $W_i = (Y_i - \mu)/\sigma$. Of course, $E(W_i) = 0$ and $\text{Var}(W_i) = 1$. From part (a) of Theorem 3.12.3 it follows that the moment-generating functions for the W_i's exist, since they exist for the Y_i's. Applying both parts of Theorem 3.12.3, we find that

$$M_{(W_1 + \cdots + W_n)/\sqrt{n}}(t) = \left[M\left(\frac{t}{\sqrt{n}}\right) \right]^n$$

where $M(t)$ represents the common moment-generating function of W_i.

From the normalization of the W_i's we obtain $M(0) = 1$, $M^{(1)}(0) = E(W_i) = 0$, and $M^{(2)}(0) = \text{Var}(W_i) = 1$. Applying Taylor's theorem with remainder to $M(t)$ gives

$$M(t) = 1 + M^{(1)}(0)t + \tfrac{1}{2}M^{(2)}(r)t^2 = 1 + \tfrac{1}{2}t^2 M^{(2)}(r)$$

for some number r, $|r| < |t|$. Thus

$$\lim_{n \to \infty} \left[M\left(\frac{t}{\sqrt{n}}\right) \right]^n = \lim_{n \to \infty} \left[1 + \frac{t^2}{2n} M^{(2)}(s) \right]^n \qquad |s| < \frac{|t|}{\sqrt{n}}$$

$$= \exp \lim_{n \to \infty} n \ln \left[1 + \frac{t^2}{2n} M^{(2)}(s) \right]$$

$$= \exp \lim_{n \to \infty} \frac{t^2}{2} \cdot M^{(2)}(s) \cdot \frac{\ln \left[1 + \dfrac{t^2}{2n} M^{(2)}(s) \right] - \ln (1)}{\dfrac{t^2}{2n} M^{(2)}(s)}$$

As remarked previously, the existence of $M(t)$ implies the existence of all its derivatives. In particular $M^{(3)}(t)$ exists, so $M^{(2)}(t)$ is continuous. Hence $\lim_{t \to 0} M^{(2)}(t) = M^{(2)}(0) = 1$. Since $|s| < |t|/\sqrt{n}$, $s \to 0$ as $n \to \infty$, so

$$\lim_{n \to \infty} M^{(2)}(s) = M^{(2)}(0) = 1$$

Also, as $n \to \infty$, the quantity $(t^2/2n)M^{(2)}(s) \to 0 \cdot 1 = 0$, so it plays the role of "Δx" in the definition of the derivative. Hence we obtain

$$\lim_{n \to \infty} \left[M\left(\frac{t}{\sqrt{n}}\right) \right]^n = \exp \frac{t^2}{2} \cdot 1 \cdot \ln^{(1)}(1) = e^{(1/2)t^2}$$

Since this last expression is the moment-generating function for a standard normal variable, the theorem is proved.

CASE STUDY 7.4.1

Theorem 7.4.1 describes an asymptotic result—as n goes to infinity, the limiting value of

$$P\left(c < \frac{Y_1 + \cdots + Y_n - n\mu}{\sqrt{n}\,\sigma} < d \right)$$

is an integral of the standard normal pdf. Of perhaps greater importance to statistical practitioners, though, is the *rate* at which the random variable $(Y_1 + \cdots + Y_n - n\mu)/\sqrt{n}\,\sigma$ converges to Z. Specifically, how large does n have to be before $\int_c^d e^{-(1/2)y^2}\, dy$ is a reasonably good approximation to $P(c < (Y_1 + \cdots + Y_n - n\mu)/\sqrt{n}\,\sigma < d)$?

As might be expected, meaningful statements about the "small sample" behavior of $(Y_1 + \cdots + Y_n - n\mu)/\sqrt{n}\,\sigma$ are difficult to formulate. Any such pronouncements need to take into account not only the value of n but also the nature of the Y_i's.

In general, pdf's that are symmetric with one local maximum admit the most rapid convergences to the normal. We should not find that surprising since these pdf's look rather like the normal to begin with. As the distribution being sampled departs from that shape, the rate of convergence usually slows down.

But even when a pdf is markedly nonnormal, convergence will sometimes proceed quite rapidly. A case in point are sums of uniform random variables. Suppose that we define

$$S_{15} = Y_1 + Y_2 + \cdots + Y_{15}$$

where $Y_1, Y_2, \ldots, Y_{15}$ are independent random variables, each uniformly distributed over $(0, 1)$. The central limit theorem suggests that S_{15} is approximately distributed as an $N(7.5, 15/12)$ variable [recall that $E(Y_i) = \frac{1}{2}$ and $\mathrm{Var}\,(Y_i) = \frac{1}{12}$]. Two hundred observations of S_{15} were generated by an IBM

TABLE 7.4.1

Value of S_{15}	Observed frequency	Expected frequency
<5	4	2.5
5–5.9	21	15.4
6–6.9	43	47.5
7–7.9	70	69.1
8–8.9	45	47.5
9–9.9	13	15.4
>10	4	2.5
	$\overline{200}$	

3081 computer: they appear in Table 7.4.1, grouped into intervals. The last column in the table gives the expected number of observations for each interval, based on the $N(7.5, 15/12)$ approximation. Agreement between the second and third columns is obviously quite good. Similar studies have shown that the distribution of sums of as few as *12* independent uniform random variables can be approximated quite well by a normal pdf.

Question 7.4.1 A student makes 100 check transactions between receiving two consecutive bank statements. Rather than subtract each amount exactly, he rounds each entry off to the nearest dollar. Let Y_i denote the round-off error associated with the *i*th transaction. [It can be assumed that Y_i has a uniform pdf over the interval $(-\frac{1}{2}, \frac{1}{2})$.] Use the central limit theorem to approximate the probability that the student's total accumulated error (either positive or negative) after 100 transactions exceeds \$5. Compare your answer here with the Chebyshev inequality upper bound worked out in Example 3.13.1.

Question 7.4.2 Suppose that a fair coin is tossed 200 times. Let $Y_i = 1$ if the *i*th toss comes up heads and $Y_i = 0$, otherwise, $i = 1, 2, \ldots, 200$.
(a) Use Chebyshev's inequality to get a lower bound for the probability that Y, the sum of the Y_i's, is within 5 of $E(Y)$.
(b) Use the DeMoivre–Laplace form of the central limit theorem to approximate $P[E(Y) - 5 \le Y \le E(Y) + 5]$.

Question 7.4.3 Suppose that 100 dice are tossed. Let Y_i be the number showing on the *i*th die. Estimate the probability that the sum of the Y_i's exceeds 370.

Question 7.4.4 Let $Y_1, Y_2, \ldots, Y_{15}$ be a random sample of size 15 from the pdf

$$f_Y(y) = 3(1 - y)^2 \qquad 0 < y < 1$$

Use the central limit theorem to approximate $P(\frac{1}{8} < \bar{Y} < \frac{3}{8})$.

Question 7.4.5 An agency of the Commerce Department in a certain state wishes to check the accuracy of weights in supermarkets. They decide to weigh 9 packages of ground meat labeled as 1 pound packages. They will investigate any supermarket where the average weight of the packages is less than 15.5 oz. What is the probability they will investigate an honest market? Assume that the standard deviation of package weights is 0.6 oz.

Question 7.4.6 In Chapter 4 it is shown that the sum of independent Poisson random variables is Poisson and that the parameter of the sum is the sum of the parameters. Thus any Poisson random variable Y with parameter λ can be written $Y = \sum_{i=1}^{n} X_i$, where the X_i's are independent Poisson variables, each with parameter λ/n. If λ is "large," the central limit theorem applies, and Y is approximately normal. Standardizing Y gives that $(Y - \lambda)/\sqrt{\lambda}$ is approximately $N(0, 1)$. Suppose that an electronics firm receives, on the average, 50 orders per week for a particular silicon chip. Assume that the distribution of weekly orders for this item is Poisson. If the firm has 60 chips on hand, what is the probability it will be unable to fill its orders for the week?

Question 7.4.7 Suppose that the annual number of earthquakes registering 2.5 or more on the Richter scale and having an epicenter within 40 miles of downtown Memphis follows a Poisson distribution with an average rate of $\lambda = 6.5$ per year. Use the central limit theorem to approximate the probability that there will be nine or more such earthquakes in the coming year. Use the Poisson model to find the "exact" answer.

Question 7.4.8 Recently there has been considerable controversy over the possible after-effects of a nuclear weapons test conducted in Nevada in 1957. Protocol for the test included some 3000 military and civilian "observers." Now, more than 25 years later, eight leukemia cases have been diagnosed among those 3000. The expected number of cases, adjusted for the demographic characteristics of the observers, was three.

(a) Use the central limit theorem to assess the significance of the eight cases. Assume that the occurrence of leukemia cases follows a Poisson distribution. What would you conclude?

(b) Find the "exact" probably asked for in part (a) by using the Poisson pdf directly.

Question 7.4.9 The *law of proportionate effect* states that changes in certain kinds of measurements, such as weight, are proportional to the current measurement. It can be shown that measurements of this sort can be modeled by a random variable Y where Y is a *product:* $Y = X_1 X_2 \cdots X_n$, with the X_i's being independent and identically distributed. This implies that log Y is a sum, and the central limit

Salary (pounds per week)	Observed frequency (thousands)	Expected frequency (thousands)
≤ 9	160.39	263.91
9–12	1079.20	
12–15	1337.35	
15–18	959.24	
18–21	519.83	
21–24	234.15	
24–27	94.19	
27–30	33.77	
30–35	17.33	
≥ 35	7.55	

theorem might provide an approximation. The table above [adopted from (165)] gives the weekly salary in pounds for 4,443,000 English male manual workers. If Y represents a worker's weekly salary, one proposed model assumes that log $Y \sim N(2.65, 0.0845)$. Use this model to complete the expected frequency column.

7.5 THE χ^2 DISTRIBUTION; INFERENCES ABOUT σ^2

Having derived the sampling distribution for the maximum-likelihood estimator for μ, we will now turn our attention to the second parameter of the normal distribution, σ^2. Before testing $H_0: \sigma^2 = \sigma_0^2$, though, or constructing confidence intervals for σ^2, we will need to investigate the properties of sums of the form $\sum_{i=1}^{n} Z_i^2$, where the Z_i's are independent standard normals. Central to these properties is a very important distribution we will be seeing for the first time, the chi square. As later chapters will make abundantly clear, the utility of the chi square extends far beyond the problems being dealt with here.

Definition 7.5.1. A random variable Y is said to have the *chi square distribution with n degrees of freedom* if

$$f_Y(y) = \frac{1}{2^{n/2}\Gamma(n/2)} y^{(n/2)-1} e^{-y/2} \qquad y > 0$$

For notational simplicity, we will write $Y \sim \chi_n^2$.

Note that χ_n^2 is a special case of the gamma density presented in Chapter 4. Specifically, a χ_n^2 variable is identical to a gamma variable (with parameters r and λ) when the parameters of the latter are set equal to $n/2$ and $\frac{1}{2}$, respectively. Figure 7.5.1 shows how the shape of a χ_n^2 distribution varies with n.

Comment

Table A.3 in the Appendix lists selected percentiles of the χ^2 distribution for degrees of freedom ranging from 1 to 50. For $n > 50$, the percentiles can be easily approximated:

$$\chi_{p,n}^2 \doteq n\left(1 - \frac{2}{9n} + z_{1-p}\sqrt{\frac{2}{9n}}\right)^3$$

where $\chi_{p,n}^2$ is the unknown $(100p)$th percentile of the χ_n^2 distribution and z_{1-p} is the $(100p)$th percentile of the $N(0, 1)$ [see (33)].

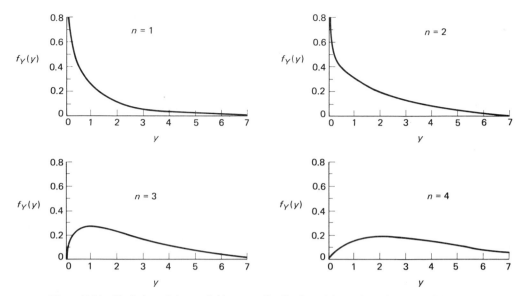

Figure 7.5.1 Variation of shape of chi square distribution with number of degrees of freedom.

Question 7.5.1 Show directly—without appealing to the fact that a χ^2 variable is a gamma variable—that $f_Y(y)$ as stated in Definition 7.5.1 is a true probability density.

Question 7.5.2 Show that $E(\chi_n^2) = n$ and $\text{Var}(\chi_n^2) = 2n$.

Like certain other distributions we have encountered—notably, the binomial, Poisson, and normal—the chi square reproduces itself. That is, the sum of two independent chi squares is itself a chi square.

> **Theorem 7.5.1.** Let $Y_1 \sim \chi_n^2$ and $Y_2 \sim \chi_m^2$ and let Y_1 and Y_2 be independent. Then
> $$Y_1 + Y_2 \sim \chi_{n+m}^2$$

Proof. The proof is a special case of Question 4.6.7.

The next theorem shows how the normal and χ^2 distributions are related. This is an important result, one that gives rise to a wide variety of inference procedures.

> **Theorem 7.5.2.** Let $Z_1, Z_2, \ldots, Z_n$ be n independent standard normal random variables. Then
> $$\sum_{i=1}^{n} Z_i^2 \sim \chi_n^2$$

Proof. The proof follows by induction. First, suppose that $n = 1$. Then

$$F_{Z_1^2}(t) = P(Z_1^2 \leq t) = P(-\sqrt{t} \leq Z_1 \leq \sqrt{t})$$
$$= 2P(0 \leq Z_1 \leq \sqrt{t})$$
$$= \frac{2}{\sqrt{2\pi}} \int_0^{\sqrt{t}} e^{-z^2/2} \, dz$$

Differentiation gives the density function for Z_1^2:

$$f_{Z_1^2}(t) = F'_{Z_1^2}(t) = \frac{2}{\sqrt{2\pi}} \frac{1}{2\sqrt{t}} e^{-t/2} = \frac{1}{2^{1/2}\Gamma(1/2)} t^{(1/2)-1} e^{-t/2}$$

Therefore, by inspection, $Z_1^2 \sim \chi_1^2$.

Now, suppose that the theorem is true for the first $n - 1$ random variables. We can write the sum of the first n Z_i^2's as

$$Z_1^2 + \cdots + Z_n^2 = Z_1^2 + (Z_2^2 + \cdots + Z_n^2)$$

By what we have just proved, $Z_1^2 \sim \chi_1^2$; also, by the induction hypothesis, $\sum_{i=2}^n Z_i^2 \sim \chi_{n-1}^2$. The final result then follows from Theorem 7.5.1.

Example 7.5.1

Tables have been compiled that list random samples from an $N(0, 1)$ distribution [see, for example, (33)]. By taking independent sets of n of these numbers and adding together their squares, we can simulate the conclusion of Theorem 7.5.2. Shown in Table 7.5.1 are the $\sum_{i=1}^4 Z_i^2$ values computed from 100 independent sets of four $N(0, 1)$ variables.

TABLE 7.5.1 $\sum_{i=1}^4 Z_i^2$ VALUES: A SAMPLING EXPERIMENT

3.472	6.472	8.347	13.025	1.483	7.832
0.772	5.449	1.037	4.744	4.940	2.186
2.920	1.083	3.047	5.627	4.091	1.031
4.532	1.033	3.146	4.004	2.685	4.379
1.510	0.964	1.519	4.668	12.723	2.018
6.018	3.820	4.900	3.300	3.147	5.741
6.613	9.386	4.874	9.775	5.290	6.854
8.992	3.330	2.574	0.611	0.870	1.152
0.738	8.630	5.233	0.579	1.653	1.237
8.484	3.643	2.118	5.813	5.168	4.255
1.079	3.145	4.541	2.052	6.846	0.570
0.476	2.151	0.391	0.758	3.700	2.476
2.680	0.756	3.549	2.694	10.884	1.630
2.392	3.084	2.577	4.354	3.785	2.232
1.348	1.840	6.208	10.938	2.217	1.264
1.330	1.808	1.642	3.434	3.596	4.687
2.650	6.203	1.830	4.865		

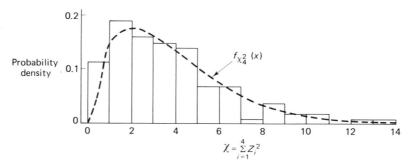

Figure 7.5.2   Fitting a χ_4^2 distribution.

As the theorem predicts, these numbers do seem to be described by a χ_4^2 distribution: Figure 7.5.2 shows a good fit between the function

$$f_{\chi_4^2}(x) = \frac{1}{2^{4/2}\Gamma(4/2)}\, x^{(4/2)-1}e^{-x/2}$$

and the histogram of $\sum_{i=1}^{4} Z_i^2$ values.

Question 7.5.3   Approximate the number of $\sum_{i=1}^{4} Z_i^2$ values expected to fall between 2.0 and 3.0. Assume that 100 sums (of four) are to be recorded.

One final result is necessary before we can derive any inference procedures for σ^2. Theorem 7.5.3 gives the sampling distribution of the ratio $(n-1)S^2/\sigma^2$, the significance of which will become evident in Theorems 7.5.4 and 7.5.5.

Theorem 7.5.3.   Let $Y_1, Y_2, \ldots, Y_n$ be n independent $N(\mu, \sigma^2)$ random variables. As defined earlier, let

$$S^2 = \frac{1}{n-1}\sum_{i=1}^{n}(Y_i - \bar{Y})^2$$

be their sample variance. Then

$$\frac{(n-1)S^2}{\sigma^2} \sim \chi_{n-1}^2$$

Proof. The proof is given in Appendix 7.2.

With the aid of Theorem 7.5.3, it becomes a simple matter to invert the variance ratio and construct confidence intervals for σ^2. Theorem 7.5.4 states the formal result. The proof will be left as an exercise. [Recall that the symbol $\chi_{p,n}^2$ denotes the $(100p)$th percentile of the χ_n^2 distribution.]

Comment

Inference procedures for σ^2, whether they be confidence intervals or hypothesis tests, are far less important in real-world applications than those for μ. They are presented here first only because the background results they require have already been developed. The usual inference procedures involving μ require a distribution (the Student t) whose derivation is best delayed until Section 7.6.

> **Theorem 7.5.4.** Let $Y_1, Y_2, \ldots, Y_n \sim N(\mu, \sigma^2)$. A $100(1 - \alpha)\%$ confidence interval for σ^2 is given by
>
> $$\left(\frac{(n-1)s^2}{\chi^2_{1-\alpha/2,\, n-1}}, \frac{(n-1)s^2}{\chi^2_{\alpha/2,\, n-1}} \right)$$

CASE STUDY 7.5.1

The chain of events that we call the geological evolution of the earth began hundreds of millions of years ago. Fossils have played a key role in documenting the *relative* times these events occurred, but to establish an *absolute* chronology, scientists rely primarily on radioactive decay. In this example, we look at the variability associated with a dating technique based on a mineral's potassium-argon ratio.

Almost all minerals contain potassium (K) as well as certain of its isotopes, including ^{40}K. But ^{40}K is unstable and decays into isotopes of argon and calcium, ^{40}Ar and ^{40}Ca. By knowing the rates at which these daughter products are formed, and by measuring the amounts of ^{40}Ar and ^{40}K present, we can estimate the age of the mineral.

Table 7.5.2 gives the estimated ages of 19 mineral samples collected from the Black Forest in southeastern Germany (101). Each of the estimates was based on the sample's potassium–argon ratio. A primary concern in this problem would be the *precision* of the potassium–argon procedure—that is, how close together would repeated age determinations be on the exact same sample? The most direct way of getting at this question would be to construct a confidence interval for σ^2, the procedure's true variance.

For the data given,

$$\sum_{i=1}^{19} y_i = 5261$$

$$\sum_{i=1}^{19} y_i^2 = 1{,}469{,}945$$

so the sample variance is 733.4:

$$s^2 = \frac{19(1{,}469{,}945) - (5261)^2}{19(18)} = 733.4$$

TABLE 7.5.2 POTASSIUM–ARGON DATES
OF SAMPLES FROM THE BLACK FOREST

Specimen	Estimated Age (millions of years)
1	$y_1 = 249$
2	$y_2 = 254$
3	$y_3 = 243$
4	$y_4 = 268$
5	$y_5 = 253$
6	$y_6 = 269$
7	$y_7 = 287$
8	$y_8 = 241$
9	$y_9 = 273$
10	$y_{10} = 306$
11	$y_{11} = 303$
12	$y_{12} = 280$
13	$y_{13} = 260$
14	$y_{14} = 256$
15	$y_{15} = 278$
16	$y_{16} = 344$
17	$y_{17} = 304$
18	$y_{18} = 283$
19	$y_{19} = 310$

From Table A.3 in the Appendix we find that

$$P(8.23 < \chi_{18}^2 < 31.53) = 0.95$$

Therefore, a 95% confidence interval for σ^2 is

$$\left(\frac{(19 - 1)(733.4)}{31.53}, \frac{(19 - 1)(733.4)}{8.23} \right)$$

$$= (418.7 (\text{million years})^2, 1604.0 (\text{million years})^2)$$

Comment

For the applied statistician, σ is a more meaningful measure of variation than σ^2 (since σ is in the same units as the data). To get a confidence interval for σ, though, we simply take the square root of the confidence limits for σ^2. Here, a 95% confidence interval for σ would be

$$(\sqrt{418.7}, \sqrt{1604.0}) = (20.5 \text{ million years}, 40.0 \text{ million years})$$

We conclude this section by applying the generalized-likelihood-ratio criterion to the problem of testing hypotheses about σ^2. Theorem 7.5.5 gives an approximate GLRT.

Theorem 7.5.5. Let $Y_1, Y_2, \ldots, Y_n \sim N(\mu, \sigma^2)$. To test $H_0: \sigma^2 = \sigma_0^2$ versus $H_1: \sigma^2 \neq \sigma_0^2$ at the α level of significance, reject the null hypothesis if $(n - 1)s^2/\sigma_0^2$ is either (a) $\leq \chi_{\alpha/2, n-1}^2$ or (b) $\geq \chi_{1-\alpha/2, n-1}^2$.

The Normal Distribution   Chap. 7

Proof. The two parameter spaces relevant to this problem are

$$\omega = \{(\mu, \sigma^2): -\infty < \mu < \infty, \sigma^2 = \sigma_0^2\}$$

and

$$\Omega = \{(\mu, \sigma^2): -\infty < \mu < \infty, 0 \le \sigma^2\}$$

In both, the MLE for μ is $\bar{y}$. In ω, the MLE for σ^2 is simply σ_0^2; in Ω, $\hat{\sigma}^2 = (1/n)\sum_{i=1}^{n}(y_i - \bar{y})^2$ (see Section 7.2). Therefore, the two likelihood functions, maximized over ω and over Ω, are

$$L(\hat{\omega}) = \left(\frac{1}{2\pi\sigma_0^2}\right)^{n/2} \exp\left[-\frac{1}{2}\sum_{i=1}^{n}\left(\frac{y_i - \bar{y}}{\sigma_0}\right)^2\right]$$

and

$$L(\hat{\Omega}) = \left[\frac{n}{2\pi\sum_{i=1}^{n}(y_i - \bar{y})^2}\right]^{n/2} \exp\left\{-\frac{n}{2}\sum_{i=1}^{n}\left[\frac{y_i - \bar{y}}{\sqrt{\sum_{i=1}^{n}(y_i - \bar{y})^2}}\right]^2\right\}$$

$$= \left[\frac{n}{2\pi\sum_{i=1}^{n}(y_i - \bar{y})^2}\right]^{n/2} e^{-n/2}$$

It follows that the generalized-likelihood-ratio criterion is given by

$$\lambda = \frac{L(\hat{\omega})}{L(\hat{\Omega})}$$

$$= \left[\frac{\sum_{i=1}^{n}(y_i - \bar{y})^2}{n\sigma_0^2}\right]^{n/2} \exp\left[-\frac{1}{2}\sum_{i=1}^{n}\left(\frac{y_i - \bar{y}}{\sigma_0}\right)^2 + \frac{n}{2}\right]$$

$$= \left(\frac{\hat{\sigma}^2}{\sigma_0^2}\right)^{n/2} e^{-(n/2)(\hat{\sigma}^2/\sigma_0{}^2) + (n/2)}$$

We need to know the behavior of λ, considered as a function of $(\hat{\sigma}^2/\sigma_0^2)$. For simplicity, let $x = (\hat{\sigma}^2/\sigma_0^2)$. Then $\lambda = x^{n/2}e^{-(n/2)x + n/2}$ and the inequality $\lambda \le \lambda^*$ is equivalent to $xe^{-x} \le e^{-1}(\lambda^*)^{2/n}$. The right-hand side is again an arbitrary constant, say κ^*. Graphing $y = xe^{-x}$ is a standard exercise in calculus, and the curve is given in Figure 7.5.3. What the graph shows is that values

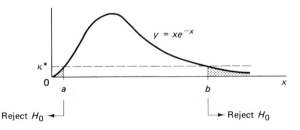

Figure 7.5.3 Critical region for testing $H_0: \sigma^2 = \sigma_0^2$ versus $H_1: \sigma^2 \ne \sigma_0^2$ (in terms of λ).

of $x = (\hat{\sigma}^2/\sigma_0^2)$ for which $xe^{-x} \leq \kappa^*$, and equivalently $\lambda \leq \lambda^*$, fall into two regions, one for values of $\hat{\sigma}^2/\sigma_0^2$ close to 0 and the other for values of $\hat{\sigma}^2/\sigma_0^2$ much larger than 1 (why 1?) According to the likelihood-ratio principle, we should reject H_0 for any $\lambda \leq \lambda^*$, where $P(\Lambda \leq \lambda^* \mid H_0) = \alpha$. But λ^* determines (via κ^*) numbers a and b so that the critical region is $C = \{(\hat{\sigma}^2/\sigma_0^2): (\hat{\sigma}^2/\sigma_0^2) \leq a$ or $(\hat{\sigma}^2/\sigma_0^2) \geq b\}$.

Comment

At this point it is necessary to make a slight approximation. Just because $P(\Lambda \leq \lambda^* \mid H_0) = \alpha$, it does not follow that

$$P\left[\frac{(1/n) \sum\limits_{i=1}^{n} (Y_i - \bar{Y})^2}{\sigma_0^2} \leq \alpha\right] = \frac{\alpha}{2} = P\left[\frac{(1/n) \sum\limits_{i=1}^{n} (Y_i - \bar{Y})^2}{\sigma_0^2} \geq b\right]$$

and, in fact, the two tails of the critical regions will *not* have exactly the same probability. Nevertheless, the two are numerically close enough so that we will not substantially compromise the likelihood-ratio criterion by setting each one equal to $\alpha/2$.

Note that

$$P\left[\frac{(1/n) \sum\limits_{i=1}^{n} (Y_i - \bar{Y})^2}{\sigma_0^2} \leq a\right] = P\left[\frac{\sum\limits_{i=1}^{n} (Y_i - \bar{Y})^2}{\sigma_0^2} \leq na\right]$$

$$= P\left[\frac{(n-1)S^2}{\sigma_0^2} \leq na\right]$$

$$= P(\chi_{n-1}^2 \leq na)$$

and, similarly,

$$P\left[\frac{(1/n) \sum\limits_{i=1}^{n} (Y_i - \bar{Y})^2}{\sigma_0^2} \geq b\right] = P(\chi_{n-1}^2 \geq nb)$$

Thus we will choose as critical values $\chi_{\alpha/2, \, n-1}^2$ and $\chi_{1-\alpha/2, \, n-1}^2$ and reject H_0 if either

$$\frac{(n-1)s^2}{\sigma_0^2} \leq \chi_{\alpha/2, \, n-1}^2$$

or

$$\frac{(n-1)s^2}{\sigma_0^2} \geq \chi_{1-\alpha/2, \, n-1}^2$$

(see Figure 7.5.4).

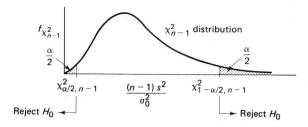

Reject H_0

$\chi^2_{\alpha/2,\,n-1}$ $\dfrac{(n-1)\,s^2}{\sigma_0^2}$ $\chi^2_{1-\alpha/2,\,n-1}$

Reject H_0

Figure 7.5.4 Critical region for testing $H_0\colon \sigma^2 = \sigma_0^2$ versus $H_1\colon \sigma^2 \neq \sigma_0^2$ (in terms of χ^2_{n-1}).

Comment

One-sided tests for dispersion are set up in a similar fashion. In the case of

$$H_0\colon \quad \sigma^2 = \sigma_0^2$$

versus

$$H_1\colon \quad \sigma^2 < \sigma_0^2$$

H_0 is rejected if

$$\frac{(n-1)s^2}{\sigma_0^2} \le \chi^2_{\alpha,\,n-1}$$

For

$$H_0\colon \quad \sigma^2 = \sigma_0^2$$

versus

$$H_1\colon \quad \sigma^2 > \sigma_0^2$$

H_0 is rejected if

$$\frac{(n-1)s^2}{\sigma_0^2} \ge \chi^2_{1-\alpha,\,n-1}$$

Example 7.5.2

Another procedure was used for dating rocks radioactively before the advent of the potassium–argon method described in Case Study 7.5.1. Based on a mineral's lead content, this earlier method was known to be capable of yielding estimates for this time period that had a standard deviation of 30.4 million years. Recall that the sample standard deviation for the 19 Black Forest rocks (i.e., for the potassium–argon method) was *less*—specifically, it was $\sqrt{733.4}$, or 27.1 million years. Can this be taken as "proof" that the potassium–argon procedure is significantly more precise than its competitor? That is, if σ^2 denotes the (true) variance for the potassium–argon method, we want to test

$$H_0\colon \quad \sigma^2 = (30.4)^2$$

versus

$$H_1\colon \quad \sigma^2 < (30.4)^2$$

Let $\alpha = 0.05$. For a χ^2 variable with 18 degrees of freedom,

$$P(\chi^2_{18} \le 9.39) = 0.05$$

Thus H_0 should be rejected if $(n-1)s^2/\sigma_0^2 \le 9.39$. But

$$\frac{(19-1)(733.4)}{(30.4)^2} = 14.3$$

meaning that the potassium–argon method has *not* been shown to be significantly more precise.

Question 7.5.4 Assume that the w/l ratios of Shoshoni rectangles are $N(\mu, \sigma^2)$ random variables (see Case Study 1.2.2). Construct a 95% confidence interval for σ^2.

Question 7.5.5 The A above middle C is the note given to an orchestra, usually by the oboe, for tuning purposes. Its pitch is defined to be the sound of a tuning fork vibrating at 440 cycles per second (cps). No tuning fork, of course, will always vibrate at *exactly* 440 cps; rather, the pitch, Y, is a random variable. Suppose that Y is normally distributed with $\mu = 440$ and variance σ^2. (Here the parameter σ^2 is a measure of the quality of the tuning fork.) With the standard manufacturing process, $\sigma^2 = 1.1$. A new production technique has just been suggested, however, and its proponents claim it will yield values of σ^2 significantly less than 1.1. To test the claim, six tuning forks are made according to the new procedure. The resulting vibration frequencies are 440.8, 440.3, 439.2, 439.8, 440.6, and 441.1 cps. Do the appropriate hypothesis test. Let $\alpha = 0.05$.

Question 7.5.6 One of the occupational hazards of being an airplane pilot is the hearing loss that results from being exposed to high noise levels. To document the magnitude of the problem, a team of researchers (88) measured the cockpit noise levels of 18 commercial aircraft. The results (in decibels) are listed below.

Plane	Noise Level (dB)	Plane	Noise Level (dB)
1	74	10	72
2	77	11	90
3	80	12	87
4	82	13	73
5	82	14	83
6	85	15	86
7	80	16	83
8	75	17	83
9	75	18	80

Assume that the distribution of cockpit noise levels from plane to plane is normally distributed. Find a 95% confidence interval for σ^2. Also find a 95% confidence interval for σ.

Question 7.5.7 Confidence intervals are typically two-sided, but the inversion technique of Section 5.9 works just as well to set up *one-sided* confidence intervals. Find the two 95% one-sided confidence intervals for the cockpit-noise-level data of Question 7.5.6.

Question 7.5.8 Potential differences (PD) measured across certain parts of the intestinal tract have proven to be useful in diagnosing gastrointestinal disorders. With that as a precedent, an experiment was set up to see whether PDs could, in a similar way, differentiate diseases involving the esophagus (170). There was some concern, however, about the variability of the procedure. By examining a large number of patients with no diseases of the esophagus, it was found that the variance of their PDs was 9.0. For 10 patients with clinically established esophagitis, the proposed procedure gave the following esophageal PDs (in millivolts): 26, 15, 18, 32, 12, 26, 39, 16, 10, and 19. Test $H_0: \sigma^2 = 9.0$ versus $H_1: \sigma^2 \neq 9.0$. Let $\alpha = 0.05$.

Question 7.5.9 A sheet-metal firm intends to run quality control tests on the shear strength of spot welds. As a guideline for choosing a sample size, they decide to take the smallest number of observations for which there is at least a 95% chance that the ratio of the sample variance to the true variance will be less than 2. Find the minimum n.

Question 7.5.10 The chi square distribution is asymptotically normal as n, the number of degrees of freedom, goes to infinity. (This should come as no surprise, since a χ_n^2 random variable can be written as a sum of n independent squared normals.) Find an expression approximating the pth percentile of a χ_n^2 distribution in terms of the pth percentile for an $N(0, 1)$. Use your formula to approximate the 95th percentile of the χ_{150}^2 distribution.

Question 7.5.11 Suppose that $X \sim \chi_{60}^2$. Use the asymptotic normality of the chi square pdf to approximate $P(45 < X < 65)$.

Question 7.5.12 Use the asymptotic normality of the chi square pdf to suggest a large-sample confidence interval for σ^2. Apply your procedure to the data given in Case Study 7.2.1.

Question 7.5.13 Let Y be a random variable whose pdf involves k parameters, $\theta_1, \theta_2, \ldots, \theta_k$. Based on a random sample of size n, we wish to test

$$H_0: \quad \theta_1 = \theta_{1_0}, \ldots, \theta_r = \theta_{r_0}, r \leq k$$

versus

$$H_1: \quad H_0 \text{ is false}$$

Let λ denote the generalized likelihood ratio as given in Definition 6.4.1. Under quite general conditions, it can be shown that $-2 \ln \Lambda$ has approximately a χ^2 distribution with r degrees of freedom and that H_0 should be rejected if $-2 \ln \lambda \geq \chi_{1-\alpha, r}^2$. (This is a very useful result in situations where the pdf of the generalized likelihood ratio is difficult to find.) Suppose that 100 independent Bernoulli trials result in 60 successes. Test $H_0: p = 0.5$ versus $H_1: p \neq 0.5$ at the $\alpha = 0.05$ level of significance with a critical region based on $-2 \ln \Lambda$.

Question 7.5.14 Let $Y_1, Y_2, \ldots, Y_n$ be a random sample from an $N(\mu, \sigma^2)$ pdf. Show that the variance of the sample variance is given by

$$\text{Var}(S^2) = \frac{2\sigma^4}{n-1}$$

Question 7.5.15 Given that $Y_1, Y_2, \ldots, Y_n$ is an independent $N(\mu, \sigma^2)$ random sample, prove that the sample variance, S^2, is consistent for σ^2 (see Question 7.5.14).

Question 7.5.16 Prove Theorem 7.5.4.

Question 7.5.17 Let $X_1, X_2, \ldots, X_n$ be a random sample of size n from an exponential pdf

$$f_X(x) = \left(\frac{1}{\theta}\right) e^{-x/\theta} \qquad x > 0, \theta > 0$$

(a) Use moment-generating functions to show that $Y = 2n\bar{X}/\theta$ has a χ^2_{2n} pdf.

(b) Use the result of part (a) to derive a formula for a $100(1 - \alpha)\%$ confidence interval for θ. Verify the statement mode in Section 5.2 that (462, 5371) is a 95% confidence interval for the exponential parameter if the sample observations are $x_1 = 610$, $x_2 = 1150$, and $x_3 = 1570$.

7.6 THE F DISTRIBUTION AND t DISTRIBUTION

There are several important probability distributions intimately related to the normal. The χ^2 density introduced in the previous section is one; in this section we look at two others, the *F distribution* (named after the great British statistician, Sir Ronald A. Fisher) and the *Student t* distribution (named after W. S. Gosset, who published under the pseudonym "Student").[1] Applications of the t and the F will be deferred to later sections. Here we will simply derive their distributions.

Theorem 7.6.1. Suppose $U \sim X^2_m$ and $V \sim X^2_n$. Let U and V be independent. Define

$$F = \frac{U/m}{V/n}$$

Then

$$f_F(z) = \frac{\Gamma\left(\dfrac{m+n}{2}\right)}{\Gamma\left(\dfrac{m}{2}\right)\Gamma\left(\dfrac{n}{2}\right)} \frac{m^{m/2}n^{n/2}z^{(m/2)-1}}{(n+mz)^{(m+n)/2}}$$

The random variable F is said to have an F distribution with m and n degrees of freedom.

Proof. We will begin by finding the density function for U/V. For notational simplicity, let $c = m/2$ and $d = n/2$. From Theorem 7.5.2,

$$f_U(u) = \frac{1}{\Gamma(c)2^c} u^{c-1} e^{-u/2} \qquad u > 0$$

[1] Gosset's first gainful employment after leaving Oxford in 1899 was with Messrs. Guinness, a Dublin brewery. Because of company policy, which forbade publication by employees, it was necessary for Gosset to publish his scientific papers under a pen name. The pseudonym he chose was "Student."

and

$$f_V(v) = \frac{1}{\Gamma(d)2^d} \, v^{d-1} e^{-v/2} \qquad v > 0$$

For the quotient U/V,

$$F_{U/V}(z) = P(U/V \le z) = P(U/z \le V)$$

(see Figure 7.6.1).

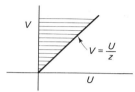

Figure 7.6.1   Region of integration for $U/V \le z$.

Therefore,

$$F_{U/V}(z) = \frac{1}{\Gamma(c)\Gamma(d)2^{c+d}} \int_0^\infty \left(\int_0^{zv} u^{c-1} e^{-u/2} \, du \right) v^{d-1} e^{-v/2} \, dv$$

Thus

$$f_{U/V}(z) = \frac{1}{\Gamma(c)\Gamma(d)2^{c+d}} \int_0^\infty [(zv)^{c-1} e^{-zv/2} v] v^{d-1} e^{-v/2} \, dv$$

Note that to obtain the above, we have interchanged the outer integral and d/dz, then used the Fundamental Theorem of Calculus (where does the extra v come from?). Simplification of the above integral gives

$$f_{U/V}(z) = \frac{1}{\Gamma(c)\Gamma(d)2^{c+d}} \, z^{c-1} \int_0^\infty v^{c+d-1} e^{-[(1+z)/2]v} \, dv$$

We recognize the integrand as the variable part of a gamma density with $r = c + d$ and $\lambda = (1 + z)/2$, so the integral has value $2^{c+d}\Gamma(c + d)/(1 + z)^{c+d}$.
Therefore,

$$f_{U/V}(z) = \frac{\Gamma(c + d)}{\Gamma(c)\Gamma(d)} \cdot \frac{z^{c-1}}{(1 + z)^{c+d}}$$

The statement of the theorem follows immediately from the fact that

$$f_{(n/m)(U/V)}(z) = f_F(z) = \frac{m}{n} \, f_{U/V}\left(\frac{m}{n} z\right)$$

Comment

The notation $F \sim F_{m,n}$ will denote that the random variable F has an F distribution with m and n degrees of freedom, with m being the degrees of freedom associated with the numerator. Table A.4 in the Appendix gives selected percentiles of $F_{m,n}$ for various m and n.

Question 7.6.1   If $F \sim F_{m, n}$ show that $1/F \sim F_{n, m}$.

Question 7.6.2   Divide the 100 entries in Table 7.5.1 into 50 groups of 2 and compute the ratio for each group. Compare the upper percentiles of the distribution of 50 quotients with the appropriate upper 0.10, 0.05, and 0.01 percentiles listed in Table A.4.

Question 7.6.3   A random variable Y is said to have a *beta distribution* with parameters r and s if

$$f_Y(y) = \frac{\Gamma(r + s)}{\Gamma(r)\Gamma(s)} \, y^{r - 1}(1 - y)^{s - 1} \qquad 0 < y < 1, \, r > 0, \, s > 0$$

Use the fact that $\int_0^1 f_Y(y) \, dy = 1$ to find $E(U/V)$, where U and V are independent χ^2 random variables with m and n degrees of freedom, respectively (recall the proof of Theorem 7.6.1). [*Hint:* In the integral defining $E(U/V)$, make the substitution $z = w/(1 - w)$.]

Question 7.6.4   Use the expression for $E(U/V)$ derived in Question 7.6.3 to find the expected value of an F distribution.

Question 7.6.5   Let Y be a beta random variable with parameters r and s. Find $E(Y)$ and Var (Y).

Question 7.6.6   Evaluate:

(a) $\displaystyle\int_0^{\pi/2} \cos^4 \theta \sin^6 \theta \, d\theta$

(b) $\displaystyle\int_0^4 x(4 - x)^{10} \, dx$

(see Question 7.6.3).

The next series of definitions and theorems introduce Gosset's best-known contribution to statistical theory, the Student t distribution. This is a critically important result, being the foundation on which confidence intervals for μ and hypothesis tests for μ are based.

Definition 7.6.1.   Let $Z \sim N(0, 1)$, $V \sim \chi_n^2$, and let Z and V be independent. The Student t ratio with n degrees of freedom will be defined as the quotient

$$T = \frac{Z}{\sqrt{V/n}}$$

Lemma $f_T(t)$ is symmetric: $f_T(t) = f_T(-t)$, for all t.

Proof. Note that

$$-T = \frac{-Z}{\sqrt{V/n}}$$

being the ratio of an $N(0, 1)$ to a random variable that is independent of Z, has the same distribution as T. But $-T$ has density $f_T(-t)$ at t. Thus $f_T(-t) = f_T(t)$.

Theorem 7.6.2. The density function for a Student t random variable with n degrees of freedom is

$$f_T(t) = \frac{\Gamma\left(\dfrac{n+1}{2}\right)}{\sqrt{n\pi}\,\Gamma\left(\dfrac{n}{2}\right)\left(1 + \dfrac{t^2}{n}\right)^{(n+1)/2}} \qquad -\infty < t < \infty$$

Proof. Observe that $T^2 = Z^2/(V/n) \sim F_{1,n}$. Therefore,

$$f_{T^2}(z) = \frac{n^{n/2}\,\Gamma\left(\dfrac{n+1}{2}\right)}{\Gamma\left(\dfrac{1}{2}\right)\Gamma\left(\dfrac{n}{2}\right)}\, z^{-1/2}\, \frac{1}{(n+z)^{(n+1)/2}}$$

Suppose that $t > 0$. By the symmetry of $f_T(t)$,

$$F_T(t) = P(T \le t) = \frac{1}{2} + P(0 \le T \le t)$$

$$= \frac{1}{2} + \frac{1}{2}\, P(-t \le T \le t)$$

$$= \frac{1}{2} + \frac{1}{2}\, P(0 \le T^2 \le t^2)$$

$$= \frac{1}{2} + \frac{1}{2}\, F_{T^2}(t^2)$$

Differentiating $F_T(t)$ gives the stated result:

$$f_T(t) = F'_T(t) = tf_{T^2}(t^2)$$

$$= t\, \frac{n^{n/2}\,\Gamma\left(\dfrac{n+1}{2}\right)}{\Gamma\left(\dfrac{1}{2}\right)\Gamma\left(\dfrac{n}{2}\right)}\, (t^2)^{-(1/2)}\, \frac{1}{(n+t^2)^{(n+1)/2}}$$

$$= \frac{\Gamma\left(\dfrac{n+1}{2}\right)}{\sqrt{n\pi}\,\Gamma\left(\dfrac{n}{2}\right)} \cdot \frac{1}{\left[1 + \left(\dfrac{t^2}{n}\right)\right]^{(n+1)/2}}$$

Comment

Over the years, the lower-case t has come to be the accepted symbol for the random variable of Definition 7.6.1. Here, we will use the lowercase t when the context makes its meaning clear, but in mathematical statements about distributions we will be consistent with random-variable notation and denote the ratio as T. In some instances it is desirable to know how many degrees of freedom are associated with the distribution. When this is the case, we will write t_n (or T_n).

Comment

The t distribution looks very much like the standard normal, although the t is somewhat flatter and has more area in its tails. As n increases, though, the family of t_n distributions converges to the $N(0, 1)$. To illustrate this, Table 7.6.1 shows the abscissa values cutting off areas of 0.01 and 0.05 in the right-hand tails of Z and t_n for various n between 1 and 100.

TABLE 7.6.1 COMPARISON OF 95TH AND 99TH PERCENTILES: t_n AND Z

n	0.05		0.01	
	t_n	Z	t_n	Z
1	6.314	1.64	31.821	2.33
2	2.920	1.64	6.965	2.33
3	2.353	1.64	4.541	2.33
4	2.132	1.64	3.747	2.33
5	2.015	1.64	3.365	2.33
10	1.812	1.64	2.764	2.33
15	1.753	1.64	2.602	2.33
20	1.725	1.64	2.528	2.33
25	1.708	1.64	2.485	2.33
30	1.697	1.64	2.457	2.33
40	1.684	1.64	2.423	2.33
50	1.676	1.64	2.403	2.33
75	1.666	1.64	2.377	2.33
100	1.660	1.64	2.364	2.33

Comment

Values of $t_{p, n}$, the $100(1 - p)$th percentile of the Student t distribution with n degrees of freedom, are tabulated in Table A.2 in the Appendix. The p values included are 0.20, 0.15, 0.10, 0.05, 0.025, 0.01, and 0.005; n ranges from 1 to 100. For $n > 100$, the Student t and the standard normal are sufficiently similar that percentiles of the normal may be used to approximate the same percentiles of the Student t.

Theorem 7.6.3 gives a particularly important special case of Definition 7.6.1.

Theorem 7.6.3. Let $Y_1, Y_2, \ldots, Y_n \sim N(\mu, \sigma^2)$. Then

$$\frac{\bar{Y} - \mu}{S/\sqrt{n}} \sim T_{n-1}$$

Proof. We can rewrite $(\bar{Y} - \mu)/(S/\sqrt{n})$ in the form

$$\frac{\bar{Y} - \mu}{S/\sqrt{n}} = \frac{\dfrac{\bar{Y} - \mu}{\sigma/\sqrt{n}}}{\sqrt{\dfrac{(n-1)S^2}{\sigma^2(n-1)}}}$$

But

$$\frac{\bar{Y} - \mu}{\sigma/\sqrt{n}} \sim N(0, 1) \qquad \text{and} \qquad \frac{(n-1)S^2}{\sigma^2} \sim X_{n-1}^2$$

Also, from Appendix 7.2,

$$\frac{\bar{Y} - \mu}{\sigma/\sqrt{n}} \quad \text{and} \quad \frac{(n-1)S^2}{\sigma^2}$$

are independent, so the result follows immediately from Definition 7.6.1.

By inverting the t ratio of Theorem 7.6.3, we can construct confidence intervals for μ. Theorem 7.6.4 gives the result. The proof is left as an exercise (see Question 7.6.16).

Theorem 7.6.4. Let $Y_1, Y_2, \ldots, Y_n \sim N(\mu, \sigma^2)$. A $100(1 - \alpha)\%$ confidence interval for μ is given by

$$\left(\bar{y} - t_{\alpha/2,\, n-1} \frac{s}{\sqrt{n}}, \quad \bar{y} + t_{\alpha/2,\, n-1} \frac{s}{\sqrt{n}} \right)$$

CASE STUDY 7.6.1

To hunt flying insects, bats emit high-frequency sounds and then listen for their echoes. Until an insect is located, these pulses are emitted at intervals of from 50 to 100 milliseconds. When an insect *is* detected, the pulse-to-pulse interval suddenly decreases—sometimes to as low as 10 milliseconds—thus enabling the bat to pinpoint its prey's position. This raises an interesting question: how far apart are the bat and the insect when the bat first senses that the insect is there? Or, put another way, what is the effective range of a bat's echolocation system?

The technical problems that had to be overcome in measuring the bat-to-insect detection distance were far more complex than the statistical problems involved in analyzing the actual data. The procedure that finally evolved was to put a bat into an 11-by-16-foot room, along with an ample supply of fruit flies, and record the action with two synchronized 16 mm sound-on-film cameras. By examining the two sets of pictures frame by frame, scientists could follow the bat's flight pattern and, at the same time, monitor its pulse frequency. For each insect that was caught, it was therefore possible to estimate the distance between the bat and the insect at the precise moment the bat's pulse-to-pulse interval decreased. Table 7.6.2 shows the results (60).

TABLE 7.6.2 BAT-TO-INSECT DETECTION DISTANCES FOR 11 "CATCHES"

Catch Number	Detection Distance (cm)
1	62
2	52
3	68
4	23
5	34
6	45
7	27
8	42
9	83
10	56
11	40

Letting $y_1 = 62$, $y_2 = 52$, ..., $y_{11} = 40$, we have that

$$\sum_{i=1}^{11} y_i = 532 \quad \text{and} \quad \sum_{i=1}^{11} y_i^2 = 29{,}000$$

Therefore,

$$\bar{y} = \frac{532}{11} = 48.4 \text{ cm}$$

and

$$s = \sqrt{\frac{11(29{,}000) - (532)^2}{11(10)}} = 18.1 \text{ cm}$$

If the population from which the y_i's are being drawn is normal, the behavior of

$$\frac{\bar{Y} - \mu}{S/\sqrt{n}}$$

will be described by a Student t curve with 10 degrees of freedom. From Table A.2 in the Appendix,

$$P(-2.23 < t_{10} < 2.23) = 0.95$$

Accordingly, the 95% confidence interval for μ is

$$\left(\bar{y} - 2.23\left(\frac{s}{\sqrt{11}}\right), \bar{y} + 2.23\left(\frac{s}{\sqrt{11}}\right)\right)$$

$$= \left(48.4 - 2.23\left(\frac{18.1}{\sqrt{11}}\right), 48.4 + 2.23\left(\frac{18.1}{\sqrt{11}}\right)\right)$$

$$= (36.2 \text{ cm}, 60.6 \text{ cm}).$$

Comment

This example is typical of the sort of situation where inferences take the form of confidence intervals rather than hypothesis tests. One of the primary objectives here would be to estimate μ, the true average bat-to-insect detection distance. However, there is no way to phrase such an objective in terms of a hypothesis test, because there is no "standard" value for μ that we can single out to associate with H_0. Presumably, this is the first time measurements such as these have ever been made.

Question 7.6.7 Use Table A.2 to find $t_{0.15, 24}$.

Question 7.6.8 What is $P(T < 1.7823)$, where T has a t distribution with 12 degrees of freedom?

Question 7.6.9 Find $P(-0.883 < T < 1.383)$, where T has a t distribution with 9 degrees of freedom.

Question 7.6.10 Show that as $n \to \infty$, the pdf of a Student t random variable with n degrees of freedom converges to the pdf for an $N(0, 1)$. (*Hint:* To show that the constant term in the pdf for T converges to $1/\sqrt{2\pi}$, use Stirling's formula:

$$n! \doteq \sqrt{2\pi n}\, n^n e^{-n}$$

Question 7.6.11 Use Definition 7.6.1 to get an expression for the even moments of a Student t random variable with n degrees of freedom. Are all the even moments defined?

Question 7.6.12 Evaluate the integral

$$\int_0^\infty \frac{1}{1 + x^2}\, dx$$

using the Student t distribution.

Question 7.6.13 Staffs from 15 different hospitals participated in a surveillance program to monitor the number of patients experiencing adverse reactions to prescribed medication. The percentages for the 15 hospitals were 5.8, 5.3, 4.5, 3.9, 4.6, 5.4, 7.9, 8.2, 6.9, 5.7, 4.6, 6.3, 8.4, 4.6, and 7.3. Let μ denote the true average percentage represented by these figures. Construct a 95% confidence interval for μ.

Question 7.6.14   Great discoveries in science tend to be made by persons who are quite young. Listed below are 12 major scientific breakthroughs from the middle of the sixteenth century to the early part of the twentieth century (183).

Discovery	Discoverer	Date	Age
Earth goes around sun	Copernicus	1543	40
Telescope, basic laws of astronomy	Galileo	1600	34
Principles of motion, gravitation, calculus	Newton	1665	23
Nature of electricity	Franklin	1746	40
Burning is uniting with oxygen	Lavoisier	1774	31
Earth evolved by gradual processes	Lyell	1830	33
Evidence for natural selection controlling evolution	Darwin	1858	49
Field equations for light	Maxwell	1864	33
Radioactivity	Curie	1896	34
Quantum theory	Planck	1901	43
Special theory of relativity, $E = mc^2$	Einstein	1905	26
Mathematical foundations for quantum theory	Schrödinger	1926	39

Let μ denote the true average age at which great scientific discoveries are made.
(a) Construct a 95% confidence interval for μ.
(b) Before constructing a confidence interval for a set of observations extending over a long period, we should be convinced that the x_i's exhibit no biases or trends. If, for example, the age at which scientists made major discoveries decreased from century to century, then the parameter μ would no longer be a constant and the confidence interval would be meaningless. Plot "date" versus "age" for the 12 discoveries given above. Put "Date of discovery" on the abscissa. Does the variability in the ages appear to be random with respect to time?

Question 7.6.15   Construct a 99% confidence interval for the true average department-store radiation exposure level using the data of Question 7.2.2.

Question 7.6.16   Prove Theorem 7.6.4. [*Hint:* Start with the expression

$$P(-t_{\alpha/2,\,n-1} < T_{n-1} < t_{\alpha/2,\,n-1}) = 1 - \alpha$$

and make use of the fact that

$$\frac{\bar{Y} - \mu}{S/\sqrt{n}} \sim T_{n-1}.]$$

Question 7.6.17   Let $Y_1, Y_2, \ldots, Y_n$ be a random sample from an $N(\mu, \sigma^2)$ pdf where σ^2 is known. Find an expression for a 95% confidence interval for μ. What is the smallest sample size that will yield an interval having a length less than L?

Question 7.6.18 Six readings, 4.3, 5.8, 8.4, 3.7, 5.2, and 5.1, are taken on a response variable known to be normally distributed with an unknown mean, μ, but a known variance of 1.50. Construct a 90% confidence interval for μ. Construct a 50% confidence interval for μ.

Question 7.6.19 Theorem 7.6.4 gives an expression for a $100(1 - \alpha)\%$ confidence interval for μ based on a random sample of size n from an $N(\mu, \sigma^2)$ pdf. Question 7.6.17 gives another. Do you think that both expressions are equally likely to come up in real-world applications? Explain.

Question 7.6.20 For random samples from an $N(\mu, \sigma^2)$ pdf, $\bar{Y}$ and S^2 are independent (see Theorem A7.2.1). It follows that if

$$P\left(-z_{\alpha/2} < \frac{\bar{Y} - \mu}{\sigma/\sqrt{n}} < +z_{\alpha/2}\right) = 1 - \alpha$$

and

$$P\left(\chi^2_{\alpha/2,\, n-1} < \frac{(n-1)S^2}{\sigma^2} < \chi^2_{1-\alpha/2,\, n-1}\right) = 1 - \alpha,$$

then

$$P\left[-z_{\alpha/2} < \frac{\bar{Y} - \mu}{\sigma/\sqrt{n}} < +z_{\alpha/2} \quad and \quad \chi^2_{\alpha/2,\, n-1} < \frac{(n-1)S^2}{\sigma^2} < \chi^2_{1-\alpha/2,\, n-1}\right] = (1 - \alpha)^2$$

Use the latter expression to set up boundaries in the $\mu\sigma^2$ plane for a *joint* $100(1 - \alpha)^2\%$ confidence "region" for μ and σ^2.

Question 7.6.21 Use the result of Question 7.6.20 to construct a joint 90% confidence interval for the μ and the σ^2 associated with the cockpit-noise-level data of Question 7.5.6. Graph your answer.

7.7 THE ONE-SAMPLE t TEST

Suppose that a random sample of size n has been drawn from a distribution presumed to be normal but whose mean and variance are unknown. Often the nature of the data will be such that our immediate concern is the unknown location parameter—that is, we wish to choose between $H_0: \mu = \mu_0$ and $H_1: \mu \neq \mu_0$. An example of this sort of objective was described at some length in Case Study 1.2.2. The question is whether or not the Shoshoni Indians had the same aesthetic standards regarding rectangles as did the ancient Greeks. If μ denotes the width-to-length ratio preferred by the Shoshonis, the problem reduces to a choice between $H_0: \mu = 0.618$ and $H_1: \mu \neq 0.618$ where 0.618 is the approximate w/l ratio for golden rectangles.

In deriving a test procedure for $H_0: \mu = \mu_0$ versus $H_1: \mu \neq \mu_0$, once again we appeal to the generalized-likelihood-ratio criterion. Theorem 7.7.1 gives the decision rule.

Theorem 7.7.1. Let $Y_1, Y_2, \ldots, Y_n \sim N(\mu, \sigma^2)$. The GLRT for

$$H_0: \quad \mu = \mu_0$$

versus

$$H_1: \quad \mu \neq \mu_0$$

requires that H_0 be rejected at the α level of significance if

$$\frac{\bar{y} - \mu_0}{s/\sqrt{n}}$$

is either (a) $\leq -t_{\alpha/2, n-1}$ or (b) $\geq +t_{\alpha/2, n-1}$.

Proof. Since σ^2 is assumed to be unknown, the parameter spaces restricted to $H_0(\omega)$ and $H_0 \cup H_1(\Omega)$ are

$$\omega = \{(\mu, \sigma^2) : \mu = \mu_0; 0 \leq \sigma^2 < \infty\}$$

and

$$\Omega = \{(\mu, \sigma^2) : -\infty < \mu < \infty; 0 \leq \sigma^2 < \infty\}$$

Without elaborating the details (see Section 7.2 for a very similar problem), it can be readily shown that, under ω,

$$\hat{\mu} = \mu_0 \quad \text{and} \quad \hat{\sigma}^2 = \frac{1}{n} \sum_{i=1}^{n} (y_i - \mu_0)^2$$

under Ω,

$$\hat{\mu} = \bar{y} \quad \text{and} \quad \hat{\sigma}^2 = \frac{1}{n} \sum_{i=1}^{n} (y_i - \bar{y})^2$$

Therefore, since

$$L(\mu, \sigma^2) = \left(\frac{1}{\sqrt{2\pi}\,\sigma}\right)^n \exp\left[-\frac{1}{2} \sum_{i=1}^{n} \left(\frac{y_i - \mu}{\sigma}\right)^2\right]$$

direct substitution gives

$$L(\hat{\omega}) = \left[\frac{\sqrt{n}}{\sqrt{2\pi}\,\sqrt{\sum_{i=1}^{n} (y_i - \mu_0)^2}}\right]^n e^{-n/2}$$

$$= \left[\frac{ne^{-1}}{2\pi \sum_{i=1}^{n} (y_i - \mu_0)^2}\right]^{n/2}$$

and

$$L(\hat{\Omega}) = \left[\frac{ne^{-1}}{2\pi \sum_{i=1}^{n} (y_i - \bar{y})^2}\right]^{n/2}$$

From $L(\hat{\omega})$ and $L(\hat{\Omega})$ we get the likelihood ratio:

$$\lambda = \frac{L(\hat{\omega})}{L(\hat{\Omega})} = \left[\frac{\sum\limits_{i=1}^{n} (y_i - \bar{y})^2}{\sum\limits_{i=1}^{n} (y_i - \mu_0)^2} \right]^{n/2} \qquad 0 < \lambda \le 1$$

As is often the case, it will prove to be more convenient to base a test on a monotonic function of λ, rather than on λ itself. We begin by rewriting the ratio's denominator:

$$\sum_{i=1}^{n} (y_i - \mu_0)^2 = \sum_{i=1}^{n} [(y_i - \bar{y}) + (\bar{y} - \mu_0)]^2$$

$$= \sum_{i=1}^{n} (y_i - \bar{y})^2 + n(\bar{y} - \mu_0)^2$$

Therefore,

$$\lambda = \left[1 + \frac{n(\bar{y} - \mu_0)^2}{\sum\limits_{i=1}^{n} (y_i - \bar{y})^2} \right]^{-n/2}$$

$$= \left(1 + \frac{t^2}{n-1} \right)^{-n/2}$$

where

$$t = \frac{\bar{y} - \mu_0}{s/\sqrt{n}}$$

Observe that as t^2 increases, λ decreases. This implies that the original GLRT—which, by definition, would have rejected H_0 for any λ that was too small, say, less than λ^*—is equivalent to a test that rejects H_0 whenever t^2 is too large. But t is an observation of the random variable

$$T = \frac{\bar{Y} - \mu_0}{S/\sqrt{n}} \sim T_{n-1} \qquad \text{(by Theorem 7.6.3)}$$

Thus "too large" translates numerically into $t_{\alpha/2,\, n-1}$:

$$0 < \lambda \le \lambda^* \Leftrightarrow t^2 \ge (t_{\alpha/2,\, n-1})^2$$

But

$$t^2 \ge (t_{\alpha/2,\, n-1})^2 \Leftrightarrow t \le -t_{\alpha/2,\, n-1} \quad \text{or} \quad t \ge t_{\alpha/2,\, n-1}$$

and the theorem is proved.

Example 7.7.1

Recall Case Study 1.2.2. Whether or not the Shoshoni Indians embrace the golden rectangle as an aesthetic standard reduces to a hypothesis test of

$$H_0: \quad \mu = 0.618$$

versus

$$H_1: \quad \mu \neq 0.618$$

If the level of significance is set at $\alpha = 0.05$, the decision rule should be:

$$\text{Reject } H_0 \text{ if } \frac{\bar{y} - 0.618}{s/\sqrt{20}} \text{ is either } \begin{cases} \text{(a)} & \leq -2.093 \; (= -t_{0.025,\,19}) \\ \text{(b)} & \geq +2.093 \; (= t_{0.025,\,19}) \end{cases}$$

From Table 1.2.3,

$$\sum_{i=1}^{20} y_i = 13.210 \qquad \text{and} \qquad \sum_{i=1}^{20} y_i^2 = 8.8878$$

making

$$\bar{y} = \frac{13.210}{20} = 0.661 \qquad \text{and} \qquad s = \sqrt{\frac{20(8.8878) - (13.210)^2}{20(19)}} = 0.093$$

The "observed" t ratio is thus 2.05, and we should accept H_0:

$$\frac{0.661 - 0.618}{0.093/\sqrt{20}} = 2.05$$

(see Figure 7.7.1).

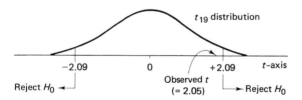

Figure 7.7.1 Distribution of $(\bar{Y} - 0.618)/S/\sqrt{20}$ when H_0 is true.

Comment

Although the conclusion at the $\alpha = 0.05$ level is to accept H_0, the proximity of the observed t to the upper cutoff value should serve as a warning that the evidence in favor of the null hypothesis was far from overwhelming. In fact, had α been chosen to be 0.10, the critical values would have been ± 1.73, and H_0 would have been rejected.

Example 7.7.2

An environmentalist group wants to monitor the temperature rise in the water 50 yards downstream from a nuclear reactor. They are concerned that the rise not exceed $3°F$. If it does, the ecological balance of the stream will be irreparably damaged. They decide to test

$$H_0: \quad \mu = 3°F$$

versus

$$H_1: \quad \mu > 3°F$$

where μ is the true average temperature increase in the water. Their plan is to collect 16 water samples and choose between H_0 and H_1 on the basis of the average temperature rise, $\bar{y}$. Similar studies suggest that the standard deviation (σ) of the Y_i's will be $1°F$. What decision rule should they use to give $\alpha = 0.05$?

It can be shown that in a case such as this where σ is known the GLRT requires that H_0 be rejected if $(\bar{y} - \mu_0)/\sigma/\sqrt{n} > z_\alpha$ (see Question 7.7.6). Here we reject H_0 if $(\bar{y} - 3)/1/\sqrt{16} > 1.64$, or if $\bar{y} > 3 + 1.64/4 = 3.41$.

When σ is known, power calculations are quite straightforward. For example, let us find the probability of committing a Type II error when the true $\mu = 3.8°F$. By definition,

$$\beta = P(\bar{Y} < 3.41 \mid \mu = 3.8) = P\left(\frac{\bar{Y} - 3.8}{\frac{1}{4}} < \frac{3.41 - 3.8}{\frac{1}{4}} \middle| \mu = 3.8\right)$$

$$= P(Z < -1.56) = 0.0594$$

(what does β equal when $\mu = 3°F$?)

Question 7.7.1 The production of a nationally marketed detergent results in certain workers receiving prolonged exposures to *Bacillus subtilis* enzyme. Nineteen workers were tested to determine the effects of these exposures on various respiratory functions (156). One such function, airflow rate, is measured by computing the ratio of a person's forced expiratory volume (FEV_1) to his vital capacity (VC). (Vital capacity is the maximum volume of air a person can exhale after taking as deep a breath as possible. FEV_1 is the maximum volume of air a person can exhale in 1 second.) In persons with no lung dysfunction, the "norm" for the FEV_1/VC ratio is 0.80. It can be assumed that if the enzyme does have an effect, it will be to *reduce* the FEV_1/VC ratio. The data are listed below.

Subject	FEV_1/VC	Subject	FEV_1/VC
RH	0.61	WS	0.78
RB	0.70	RV	0.84
MB	0.63	EN	0.83
DM	0.76	WD	0.82
WB	0.67	FR	0.74
RB	0.72	PD	0.85
BF	0.64	EB	0.73
JT	0.82	PC	0.85
PS	0.88	RW	0.87
RB	0.82		

Do the appropriate hypothesis test. Let $\alpha = 0.025$.

Question 7.7.2 A psychologist would like to know whether persons with high IQs (over 130) are significantly faster at working a certain puzzle than persons with average IQs. The puzzle requires that five irregularly shaped pieces be fitted together to form a cube. Based on considerable past experience, it is known that persons with average IQs complete the cube in a mean time of 4.6 minutes. When the psychologist presents the puzzle to four persons with IQs over 130, she records the following completion times (in minutes): 4.0, 4.2, 4.6, and 4.2. State the appropriate H_0 and H_1 for this problem and carry out the test at the $\alpha = 0.05$ level of significance.

Question 7.7.3 A contractor makes large purchases of cement from a local manufacturer. The bags of cement are supposed to weigh 94 pounds. The contractor decides to test a sample of bags to see if he is getting his money's worth. He weighs 10 bags and gets the following weights: 94.1, 93.4, 92.8, 93.4, 95.4, 93.5, 94.0, 93.8, 92.9, and 94.2. Make the appropriate hypothesis test at the $\alpha = 0.05$ level of significance.

Question 7.7.4 State the GLRTs for testing one-side alternatives: $H_0: \mu = \mu_0$ versus $H_1: \mu < \mu_0$ and $H_0: \mu = \mu_0$ versus $H_1: \mu > \mu_0$.

Question 7.7.5 Construct a power curve for the hypothesis test of Example 7.7.2.

Question 7.7.6 Let $Y_1, Y_2, \ldots, Y_n$ be a random sample from an $N(\mu, \sigma^2)$ pdf where σ^2 is known. Find the GLRT for

$$H_0: \quad \mu = \mu_0$$
$$\text{versus}$$
$$H_1: \quad \mu \neq \mu_0$$

Let $\alpha = 0.05$. Compare your result with the decision rule for a one-sample t test.

Question 7.7.7 A manufacturer of pipe for laying underground electrical cables is concerned about the pipe's rate of corrosion and whether a special coating may retard that rate. As a way of measuring corrosion, they examine a short length of pipe and record the depth of the maximum pit. Their tests have shown that in a year's time in the particular kind of soil they have to deal with the average depth of the maximum pit in a foot of pipe is 0.0042 inch. The standard deviation is 0.0003 inch. Ten pipes are coated with a new plastic and buried in the same soil. After one year, the following maximum pit depths are recorded: 0.0039, 0.0041, 0.0038, 0.0044, 0.0040, 0.0036, 0.0034, 0.0046, 0.0035, and 0.0036. Can it be concluded at the $\alpha = 0.05$ level that the plastic coating is beneficial?

Question 7.7.8 The melting point of a certain alloy is required to be 1000°C to meet engineering specifications. Should the true melting point (μ) deviate by more than 30°C, the production process should be stopped and reset. If the standard deviation of the melting point temperatures is 20°C, and we elect to test $H_0: \mu = 1000°C$ versus $H_1: \mu \neq 1000°C$, what will be the probability of detecting a shift of 30°C if 10 alloy samples are examined? Assume that $\alpha = 0.05$.

Question 7.7.9 A quality control engineer needs to monitor a machine that puts cereal into boxes. In particular, he is concerned that the machine may not be putting enough cereal in the boxes. According to the label, each box contains 16 ounces, so he wants to test (at the $\alpha = 0.05$ level of significance)

$$H_0: \quad \mu = 16$$
$$\text{versus}$$
$$H_1: \quad \mu < 16$$

where μ is the true average weight. Also, he wants the test to detect a true average weight as small as 15 ounces 95% of the time. From past experience with this machine, he knows that the weight, Y, per box is normally distributed with a standard deviation, σ, of 0.9 ounce. What is the smallest sample size that will meet his requirements?

Question 7.7.10 The pH of the catalyst in a chemical reaction is to be measured by a standard titration procedure. In order for the reaction to proceed at the desired rate, the catalyst pH must be very close to 6.4. It is known that the titration procedure has an associated standard deviation of 0.03. Suppose that the chemist wishes to test

$$H_0: \quad \mu = 6.4$$

versus

$$H_1: \quad \mu \neq 6.4$$

at the $\alpha = 0.01$ level of significance and wants to have at least a 95% chance of rejecting H_0 if the true pH is outside the interval (6.36, 6.44). How many readings should she take?

APPENDIX 7.1 POWER CALCULATIONS FOR A ONE-SAMPLE t TEST

By definition, the probability of committing a Type I error with the GLRT of Theorem 7.7.1 is α. But what about a Type II error? If $\mu = \mu_1$, where $\mu_1 \neq \mu_0$, what are the chances that the observed $\bar{y}$ and s will take on values that will "deceive" the decision maker into *accepting* H_0? To answer that, we need the following result, which gives the sampling distribution of the t ratio when H_1, rather than H_0, is true.

Theorem A7.1.1. Let $Y_1, Y_2, \ldots, Y_n \sim N(\mu_1, \sigma^2)$. Let

$$T_\Delta = \frac{\bar{Y} - \mu_1 + \Delta}{S/\sqrt{n}} = \frac{\bar{Y} - \mu_0}{S/\sqrt{n}}$$

where $\Delta = \mu_1 - \mu_0$. Then T_Δ is said to be a noncentral T variable with $n - 1$ degrees of freedom and noncentrality parameter γ, where

$$\gamma = \frac{\sqrt{n}\,\Delta}{\sigma}$$

Also,

$$f_{T_\Delta}(t) = \frac{e^{-(1/2)\gamma^2}\, \Gamma\!\left(\dfrac{n}{2}\right)}{\sqrt{\pi(n-1)}\, \Gamma\!\left(\dfrac{n}{2} - \dfrac{1}{2}\right)} \left(1 + \frac{t^2}{n-1}\right)^{-n/2} g(t)$$

where

$$g(t) = \sum_{r=0}^{\infty} \left(\frac{\sqrt{2}\,\gamma t}{\sqrt{n-1}}\right)^r \left[1 + \left(\frac{t^2}{n-1}\right)\right]^{-r/2} \frac{\Gamma[(1/2)(n+r)]}{r!\, \Gamma(n/2)}$$

The probability of committing a Type II error, given that $\mu = \mu_1$, is the integral of $f_{T_\Delta}(t)$ between the two critical values defined by the original decision rule:

$P(\text{accept } H_0 \,|\, H_1 \text{ is true and } \mu = \mu_1)$

$$= P\left[-t_{\alpha/2,\, n-1} < \frac{\bar{Y} - \mu_0}{S/\sqrt{n}} < t_{\alpha/2,\, n-1} \,|\, \mu = \mu_1 \right]$$

$$= P\left[-t_{\alpha/2,\, n-1} < \frac{\bar{Y} - \mu_1 + \Delta}{S/\sqrt{n}} < t_{\alpha/2,\, n-1} \,|\, \mu = \mu_1 \right]$$

$$= \int_{-t_{\alpha/2,\, n-1}}^{t_{\alpha/2,\, n-1}} f_{T_\Delta}(t)\, dt = \beta$$

Therefore, the *power* of the test (at $\mu = \mu_1$ and for some specified noncentrality parameter γ) is the sum of two integrals:

$$1 - \beta = \int_{-\infty}^{-t_{\alpha/2,\, n-1}} f_{T_\Delta}(t)\, dt + \int_{t_{\alpha/2,\, n-1}}^{\infty} f_{T_\Delta}(t)\, dt \qquad (A7.1.1)$$

Figure A7.1.1 (10) gives solutions to Equation A7.1.1 for a broad range of sample sizes and noncentrality parameters. The ordinate on the graph is simply $1 - \beta$; the abscissa is expressed in terms of ϕ, where $\phi = \gamma/\sqrt{2} = \sqrt{n/2}\,(\Delta/\sigma)$. Making up the body of the table are two sets of curves corresponding to two-sided tests at the $\alpha = 0.01$ and $\alpha = 0.05$ levels of significance. (If H_1 is one-sided, these same two sets of curves give $1 - \beta$ for tests at levels 0.005 and 0.025, respectively.)

As an illustration of how Figure A7.1.1 is used, suppose we wanted to construct the power function for the two-sided, $\alpha = 0.05$ test described in Case Study 1.2.1—H_0: $\mu = 0.618$ versus H_1: $\mu \neq 0.618$. (Recall from Example 7.7.1 that the decision rule required that H_0 be rejected if t was either less than or equal to -2.093 or greater than or equal to $+2.093$). We might begin, for example, by computing the probability that the decision rule will reject H_0 if μ has shifted 0.5 standard deviations (of Y) to the *right* of μ_0. This is equivalent to setting Δ/σ equal to 0.5, in which case $\phi = \sqrt{n/2}\,(\Delta/\sigma) = \sqrt{20/2}\,(0.5) = 1.58$. Note that if an $\alpha = 0.05$ curve corresponding to 19 degrees of freedom had been included in Figure A7.1.1, its ordinate for an abscissa of 1.58 would be approximately 0.57 (see Figure A7.1.2). Thus

$$P\left(\text{reject } H_0 \,\Big|\, \frac{\Delta}{\sigma} = 0.5 \right) = 1 - \beta$$

$$= \int_{-\infty}^{-2.093} f_{T_\Delta}\left(t; \frac{\Delta}{\sigma} = 0.5\right) dt + \int_{2.093}^{\infty} f_{T_\Delta}\left(t; \frac{\Delta}{\sigma} = 0.5\right) dt$$

$$\doteq 0.57$$

Proceeding in this same fashion, suppose that μ has shifted 0.75 standard deviations to the right of μ_0. Then $\phi = \sqrt{20/2}\,(0.75) = 2.37$ and, from Figure A7.1.1, $1 - \beta = 0.89$ (see Figure A7.1.3).

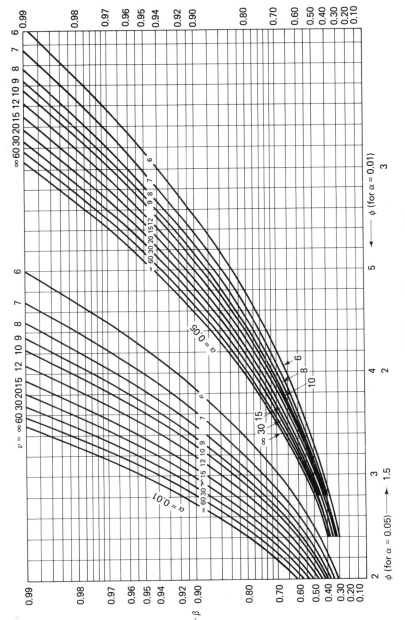

Figure A7.1.1 Tail areas of the noncentral T.

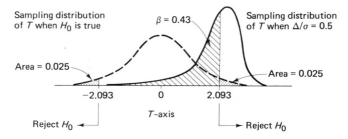

Figure A7.1.2   Noncentral T distributions ($\Delta/\sigma = 0$ and $\Delta/\sigma = 0.5$).

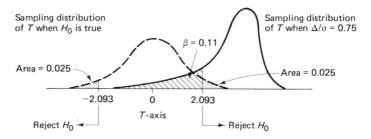

Figure A7.1.3   Noncentral T distributions ($\Delta/\sigma = 0$ and $\Delta/\sigma = 0.75$).

Figure A7.1.4 shows the complete power function for this particular test. The two X's on the curve correspond to Δ/σ values of 0.50 and 0.75. [That the power function is symmetric around $\Delta/\sigma = 0$ follows immediately from the functional representation of $f_{T_\Delta}(t)$ as given in Theorem A7.1.1].

In practice, an experimenter is likely to use Figure A7.1.1 in a way slightly different from the one we have just described. Rather than determine the power function for a test whose decision rule (and sample size) has already been decided upon, he will usually turn the problem around and look for the smallest sample size that guarantees that β will not exceed a certain value for a given Δ/σ.

For example, suppose that an experimenter has decided to test H_0: $\mu = \mu_0$ versus H_1: $\mu \neq \mu_0$ at the $\alpha = 0.05$ level of significance. How large a sample should

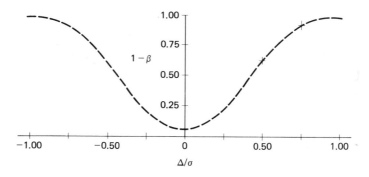

Figure A7.1.4   Power function for Shoshoni hypothesis test.

he take? Should n be 10 or 20 or 250—or something even larger? Clearly, this is a fundamental question in the design of every experiment. Unfortunately, it has no simple answer, but with the help of Figure A7.1.1 we can at least make some ballpark estimates.

Before n can even be estimated, though, the experimenter needs to specify a minimum $1 - \beta$ he is willing to accept for a given amount of shift away from μ_0. For instance, if the true μ is 0.5 standard deviations or more away from μ_0, he may want the test procedure to be capable of rejecting H_0 at least 80% of the time. In symbols, if Δ/σ is greater than or equal to 0.5, $1 - \beta$ should be greater than or equal to 0.80.

Comment

We know, of course, that for any fixed $\Delta/\sigma \neq 0$, $1 - \beta$ approaches 1 as n approaches infinity. The problem is to find the *smallest* (i.e., the most economical) n that will make $1 - \beta$ at least as large as 0.8 when μ shifts 0.5 standard deviations away from μ_0.

Note that if n were 20, ϕ would be $\sqrt{20/2}\,(0.5) = 1.58$. But, from Figure A7.1.1, $1 - \beta$ for $\phi = 1.58$ and $n = 20$ is approximately 0.56, implying that $n = 20$ is too small. Similarly, if n were 30, ϕ would be $\sqrt{30/2}\,(0.5) = 1.94$, and the corresponding $1 - \beta$ would be 0.75—still too small. But for $n = 40$, $\phi = \sqrt{40/2}\,(0.5) = 2.24$ and $1 - \beta = 0.87$. Therefore, the n we are looking for lies somewhere between 30 and 40.

APPENDIX 7.2 SOME DISTRIBUTION RESULTS FOR $\bar{Y}$ AND S^2

Theorem A7.2.1. Let $Y_i \sim N(\mu, \sigma^2)$, $i = 1, 2, \ldots, n$, and let the Y_i's be independent. Define

$$\bar{Y} = \frac{1}{n} \sum_{i=1}^{n} Y_i \quad \text{and} \quad S^2 = \frac{1}{n-1} \sum_{i=1}^{n} (Y_i - \bar{Y})^2$$

Then

(a) $\bar{Y}$ and S^2 are independent
(b) $(n-1)S^2/\sigma^2 \sim \chi_{n-1}^2$

Proof. The proof of this theorem relies on certain linear algebra techniques as well as a change-of-variables formula for multiple integrals. Definition A7.2.1 and the lemma that follows review the necessary background results. For further details, see (42) or (192).

Set $X_i = (Y_i - \mu)/\sigma$ for $i = 1, 2, \ldots, n$. Then all the X_i's are $N(0, 1)$. Let A be an $n \times n$ orthogonal matrix whose last row is $(1/\sqrt{n}, 1/\sqrt{n}, \ldots, 1/\sqrt{n})$. Let $\vec{X} = (X_1, \ldots, X_n)^T$ and define $\vec{Z} = (Z_1, Z_2, \ldots, Z_n)^T$ by the transformation $\vec{Z} = A\vec{X}$. [Note that $Z_n = (1/\sqrt{n})X_1 + \cdots + (1/\sqrt{n})X_n = \sqrt{n}\,\bar{X}$.]

For any set D,

$$P(\vec{Z} \in D) = P(A\vec{X} \in D) = P(\vec{X} \in A^{-1}D)$$

$$= \int_{A^{-1}D} f_{X_1, \ldots, X_n}(x_1, \ldots, x_n)\, dx_1 \cdots dx_n$$

$$= \int_D f_{X_1, \ldots, X_n}(g(\vec{z}))\, \det J(g)\, dz_1 \cdots dz_n$$

$$= \int_D f_{X_1, \ldots, X_n}(A^{-1}\vec{z}) \cdot 1 \cdot dz_1 \cdots dz_n$$

where $g(\vec{z}) = A^{-1}\vec{z}$. But A^{-1} is orthogonal, so, setting $(x_1, \ldots, x_n)^T = A^{-1}z$, we have that

$$x_1^2 + \cdots + x_n^2 = z_1^2 + \cdots + z_n^2$$

Thus

$$f_{X_1, \ldots, X_n}(\vec{x}) = (2\pi)^{-n/2} e^{-(1/2)(x_1^2 + \cdots + x_n^2)}$$

$$= (2\pi)^{-n/2} e^{-(1/2)(z_1^2 + \cdots + z_n^2)}$$

From this we conclude that

$$P(\vec{Z} \in D) = \int_D (2\pi)^{-n/2} e^{-(n/2)(z_1^2 + \cdots + z_n^2)} \, dz_1 \cdots dz_n$$

implying that the Z_j's are independent standard normals.

Finally,

$$\sum_{j=1}^{n} Z_j^2 = \sum_{j=1}^{n-1} Z_j^2 + n\bar{X}^2 = \sum_{j=1}^{n} X_j^2 = \sum_{j=1}^{n} (X_j - \bar{X})^2 + n\bar{X}^2$$

Therefore,

$$\sum_{j=1}^{n-1} Z_j^2 = \sum_{j=1}^{n} (X_j - \bar{X})^2$$

and the conclusion follows for standard normal variables. Also, since $\bar{Y} = \sigma\bar{X} + \mu$ and $\sum_{i=1}^{n} (Y_i - \bar{Y})^2 = \sigma^2 \sum_{i=1}^{n} (X_i - \bar{X})^2$, the conclusion follows for $N(\mu, \sigma^2)$ variables.

Comment

As part of the proof presented above, we established a version of *Fisher's lemma:*

Let $X_1, X_2, \ldots, X_n$ be independent standard normal random variables and let A be an orthogonal matrix. Define $(Z_1, \ldots, Z_n)^T = A(X_1, \ldots, X_n)^T$. Then the Z_i's are independent standard normal random variables.

8

Two-Sample Problems

William Sealy Gosset ("Student") (1876–1937)

After earning an Oxford degree in mathematics and chemistry, Gosset began working in 1899 for Messrs. Guinness, a Dublin brewery. Fluctuations in materials and temperature and the necessarily small-scale experiments inherent in brewing convinced him of the necessity for a new small-sample theory of statistics. Writing under the pseudonym, "Student," he published work with the t ratio that was destined to become a cornerstone of modern statistical methodology.

8.1 INTRODUCTION

The simplicity of the one-sample model makes it the logical starting point for any discussion of statistical inference, but it also limits its applicability to the real world. Very few experiments involve just a single treatment or a single set of conditions. On the contrary, researchers almost invariably design experiments to compare responses to *several* treatments—or, at the very least, to compare a single treatment with a control.

In this chapter we examine the simplest of these multitreatment designs, the *two-sample problem*. Structurally, the two-sample problem always falls into one of two different formats: either two (presumably) different treatments are applied to two independent sets of similar subjects or the same treatment is applied to two (presumably) different kinds of subjects. Comparing the effectiveness of germicide A relative to that of germicide B by measuring the zones of inhibition each one produces in two sets of similarly cultured Petri dishes would be an example of the first type. Another would be testing whether monkeys raised by themselves (treatment X) react differently in a stress situation from monkeys raised with siblings (treatment Y). On the other hand, examining the bones of 60-year-old men and 60-year-old women, all lifelong residents of the same city, to see whether both sexes absorb environmental strontium-90 at the same rate would be an example of the second type.

Inference in two-sample problems usually reduces to a comparison of *location* parameters. We might assume, for example, that the population of responses associated with, say, treatment X is normally distributed with mean μ_X and standard deviation σ_X while the Y distribution is normal with mean μ_Y and standard deviation σ_Y. Comparing location parameters, then, reduces to testing $H_0: \mu_X = \mu_Y$. As always, the alternative may be either one-sided, $H_1: \mu_X < \mu_Y$ or $H_1: \mu_X > \mu_Y$, or two-sided, $H_1: \mu_X \neq \mu_Y$. (If the data are binomial, the location parameters are p_X and p_Y, the true "success" probabilities for treatments X and Y, and the null hypothesis takes the form $H_0: p_X = p_Y$.)

Sometimes, although much less frequently, it becomes more relevant to compare the *variabilities* of two treatments, rather than their locations. A food company, for example, trying to decide which of two types of machines to buy for filling cereal boxes would naturally be concerned about the *average* weights of the boxes filled by each type, but they would also want to know something about the *variabilities* of the weights. Obviously, a machine that produced high proportions of "underfills" and "overfills" would be a distinct liability. In a situation of this sort, the appropriate null hypothesis is $H_0: \sigma_X^2 = \sigma_Y^2$.

For comparing the means of two normal populations, the standard procedure is the *two-sample t test*. As described in Section 8.2, this is a relatively straightforward extension of Chapter 7's one-sample t test. For comparing variances, though, it will be necessary to introduce a completely new test—this one based on the F distribution of Section 7.6. The binomial version of the two-sample problem, testing $H_0: p_X = p_Y$, is taken up in Section 8.4.

It was mentioned in connection with one-sample problems that certain inferences, for various reasons, are more aptly phrased in terms of confidence intervals rather than hypothesis tests. The same is true of two-sample problems. In Section 8.5, confidence intervals are constructed for the location *difference* of two populations, $\mu_X - \mu_Y$ (or $p_X - p_Y$), and the variability *quotient*, σ_X^2/σ_Y^2.

8.2 TESTING $H_0: \mu_X = \mu_Y$— THE TWO-SAMPLE t TEST

We will suppose the data for a given experiment consist of two independent random samples, $X_1, X_2, \ldots, X_n$ and $Y_1, Y_2, \ldots, Y_m$, representing either of the models referred to in Section 8.1. Furthermore, the two populations from which the X's and Y's are drawn will be presumed normal. Let μ_X and μ_Y denote their means. Our problem will be to derive a procedure for testing $H_0: \mu_X = \mu_Y$. Of course, to accept H_0 is to accept the equivalence of the two treatments (or the two sets of subjects), at least in terms of the average effects they elicit.

As it turns out, the precise form of the test we are looking for depends on the variances of the X and Y populations. If it can be assumed that σ_X^2 and σ_Y^2 are equal, it is a relatively straightforward task to produce the GLRT for $H_0: \mu_X = \mu_Y$. (This is, in fact, what we will do in Theorem 8.2.2.) But if the variances of the two populations are *not* equal, the problem becomes much more complex. This second case, known as the Behrens-Fisher problem, is more than 50 years old and remains one of the more famous "unsolved" problems in statistics. What headway investigators *have* made has been confined to approximate solutions [see, for example, Sukhatme (159) or Cochran (24)]. These however, will not be discussed here; we will restrict our attention to testing $H_0: \mu_X = \mu_Y$ when it can be assumed that $\sigma_X^2 = \sigma_Y^2$.

For the one-sample test that $\mu = \mu_0$, the GLRT was shown to be a function of a special case of the t ratio introduced in Definition 7.6.1 (recall Theorem 7.6.3). We begin this section with a theorem that gives still another special case of Definition 7.6.1. This one will do for the two-sample GLRT what $\sqrt{n}(\bar{Y} - \mu_0)/S$ did for the one-sample GLRT.

Theorem 8.2.1. Let $X_1, X_2, \ldots, X_n \sim N(\mu_X, \sigma^2)$ and $Y_1, Y_2, \ldots, Y_m \sim N(\mu_Y, \sigma^2)$ and let the X's and Y's be independent. Let S_X^2 and S_Y^2 be the two sample variances, and S_p^2, the *pooled variance*, where

$$S_p^2 = \frac{(n-1)S_X^2 + (m-1)S_Y^2}{n+m-2} = \frac{\displaystyle\sum_{i=1}^{n}(X_i - \bar{X})^2 + \sum_{i=1}^{m}(Y_i - \bar{Y})^2}{n+m-2}$$

Then

$$T = \frac{\bar{X} - \bar{Y} - (\mu_X - \mu_Y)}{S_p\sqrt{\dfrac{1}{n} + \dfrac{1}{m}}} \sim T_{n+m-2}$$

Proof. The method of proof here is very similar to what was used for Theorem 7.6.3. Note that an equivalent formulation of T would be

$$T = \frac{\dfrac{\bar{X} - \bar{Y} - (\mu_X - \mu_Y)}{\sigma\sqrt{\dfrac{1}{n} + \dfrac{1}{m}}}}{\sqrt{S_p^2/\sigma^2}}$$

$$= \frac{\dfrac{\bar{X} - \bar{Y} - (\mu_X - \mu_Y)}{\sigma\sqrt{\dfrac{1}{n} + \dfrac{1}{m}}}}{\sqrt{\dfrac{1}{n+m-2}\left[\sum_{i=1}^{n}\left(\dfrac{X_i - \bar{X}}{\sigma}\right)^2 + \sum_{i=1}^{m}\left(\dfrac{Y_i - \bar{Y}}{\sigma}\right)^2\right]}}$$

But $E(\bar{X} - \bar{Y}) = \mu_X - \mu_Y$ and Var $(\bar{X} - \bar{Y}) = \sigma^2/n + \sigma^2/m$, so the numerator of the ratio is clearly an $N(0, 1)$. In the denominator,

$$\sum_{i=1}^{n}\left(\frac{X_i - \bar{X}}{\sigma}\right)^2 = \frac{(n-1)S_X^2}{\sigma^2} \sim \chi^2_{n-1}$$

and

$$\sum_{i=1}^{m}\left(\frac{Y_i - \bar{Y}}{\sigma}\right)^2 = \frac{(m-1)S_Y^2}{\sigma^2} \sim \chi^2_{m-1}$$

so that by Theorem 7.5.1,

$$\sum_{i=1}^{n}\left(\frac{X_i - \bar{X}}{\sigma}\right)^2 + \sum_{i=1}^{m}\left(\frac{Y_i - \bar{Y}}{\sigma}\right)^2 \sim \chi^2_{n+m-2}$$

Also, from Appendix 7.2, it follows that the numerator and denominator of the ratio are independent. By Definition 7.6.1, then,

$$\frac{\bar{X} - \bar{Y} - (\mu_X - \mu_Y)}{S_p\sqrt{\dfrac{1}{n} + \dfrac{1}{m}}} \sim \frac{N(0, 1)}{\sqrt{\dfrac{\chi^2_{n+m-2}}{n+m-2}}} \sim T_{n+m-2}$$

Question 8.2.1 It was easy to verify from first principles that the numerator of the t ratio has a mean of 0 and a variance of 1. From what result, though, does the *normality* of the numerator derive?

Theorem 8.2.2 gives the GLRT for testing the equality of two normal means against a two-sided alternative. The modifications for one-sided alternatives should be readily apparent. Case Studies 8.2.1 and 8.2.2 show how Theorem 8.2.2 is applied.

> **Theorem 8.2.2.** Let $X_1, X_2, \ldots, X_n \sim N(\mu_X, \sigma^2)$ and $Y_1, Y_2, \ldots, Y_m \sim N(\mu_Y, \sigma^2)$ and let the X's and Y's be independent. At the α level of significance, the GLRT for $H_0: \mu_X = \mu_Y$ versus $H_1: \mu_X \neq \mu_Y$ calls for H_0 to be rejected if
>
> $$t = \frac{\bar{x} - \bar{y}}{s_p \sqrt{\dfrac{1}{n} + \dfrac{1}{m}}} \quad \text{is either} \quad \begin{cases} \leq -t_{\alpha/2, \, n+m-2} \\ \quad\quad \text{or} \\ \geq +t_{\alpha/2, \, n+m-2} \end{cases}$$

Proof. (See Appendix 8.1.)

CASE STUDY 8.2.1

Cases of disputed authorship are not very common but when they do occur, they can be very difficult to resolve. Speculation has persisted for several hundred years that some of Shakespeare's works were written by Sir Francis Bacon. And whether it was Alexander Hamilton or James Madison who wrote certain of the Federalist Papers is still an open question. A similar, though more recent, dispute centers around Mark Twain (15).

In 1861, a series of 10 essays appeared in the *New Orleans Daily Crescent*. Signed "Quintus Curtius Snodgrass," the essays purported to chronicle the author's adventures as a member of the Louisiana militia. While historians generally agree that the accounts referred to actually did happen, there seems to be no record of anyone named Quintus Curtius Snodgrass. Adding to the mystery is the fact that the style of the pieces bears unmistakable traces—at least to some critics—of the humor and irony that made Mark Twain so famous.

Most typically, efforts to unravel these sorts of "yes, he did—no, he didn't" controversies rely heavily on literary and historical clues. But not always. There is also a statistical approach to the problem. Studies have shown that authors are remarkably consistent in the extent to which they use words of a certain length. That is, a given author will use roughly the same proportion of, say, three-letter words in something he writes this year as he did in whatever he wrote last year. The same holds true for words of any length. *But,* the proportion of three-letter words that author A consistently uses will very likely be different from the proportion of three-letter words that author B uses. It follows that by comparing the proportions of words of a certain length in essays known to be the work of Mark Twain to the proportions found in the 10 Snodgrass essays, we should be able to assess the likelihood of the two authors' being one and the same.

Table 8.2.1 shows the proportions of three-letter words found in eight Twain essays and in the 10 Snodgrass essays. (Each of the Twain works was written at approximately the same time the Snodgrass essays appeared.)

TABLE 8.2.1 PROPORTION OF THREE-LETTER WORDS

Twain	Proportion	QCS	Proportion
Sergeant Fathom letter	0.225	Letter I	0.209
Madame Caprell letter	0.262	Letter II	0.205
Mark Twain letters in		Letter III	0.196
Territorial Enterprise		Letter IV	0.210
First letter	0.217	Letter V	0.202
Second letter	0.240	Letter VI	0.207
Third letter	0.230	Letter VII	0.224
Fourth letter	0.229	Letter VIII	0.223
First *Innocents Abroad* letter		Letter IX	0.220
First half	0.235	Letter X	0.201
Second half	0.217		

If $x_1 = 0.225$, $x_2 = 0.262$, ..., $x_8 = 0.217$, and $y_1 = 0.209$, $y_2 = 0.205$, ..., $y_{10} = 0.201$, then

$$\bar{x} = \frac{1.855}{8} = 0.232 \quad \text{and} \quad \bar{y} = \frac{2.097}{10} = 0.210$$

To analyze these data, we need to decide what the magnitude of the difference between the sample means, $\bar{x} - \bar{y} = 0.232 - 0.210 = 0.022$, actually tells us. Let μ_X and μ_Y denote the proportions of three-letter words in *all* essays written by Twain and by Snodgrass, respectively. Of course, not having examined the complete works of the two authors, we have no way of evaluating either μ_X or μ_Y, so they become the unknown parameters of the problem. What needs to be decided, then, is whether an observed *sample* difference (in the proportions of three-letter words) as large as 0.022 implies that the two *true* proportions, μ_X and μ_Y, are themselves not the same. Or is 0.022 small enough to still be compatible with the hypothesis that they are? Put more formally, we must choose between

$$H_0: \quad \mu_X = \mu_Y$$

and

$$H_1: \quad \mu_X \neq \mu_Y$$

Since

$$\sum_{i=1}^{8} x_i^2 = 0.4316 \quad \text{and} \quad \sum_{i=1}^{10} y_i^2 = 0.4406$$

the two sample variances are

$$s_X^2 = \frac{8(0.4316) - (1.855)^2}{8(7)}$$

$$= 0.0002103$$

and

$$s_Y^2 = \frac{10(0.4406) - (2.097)^2}{10(9)}$$

$$= 0.0000955$$

Combined, they give a pooled standard deviation of 0.012:

$$S_p = \sqrt{\frac{\sum\limits_{i=1}^{8} (x_i - 0.232)^2 + \sum\limits_{i=1}^{10} (y_i - 0.210)^2}{n + m - 2}}$$

$$= \sqrt{\frac{(n-1)s_X^2 + (m-1)s_Y^2}{n + m - 2}}$$

$$= \sqrt{\frac{7(0.0002103) + 9(0.0000955)}{8 + 10 - 2}}$$

$$= \sqrt{0.0001457}$$

$$= 0.012$$

According to Theorem 8.2.1, if $H_0: \mu_X = \mu_Y$ is true, the sampling distribution of

$$T = \frac{\bar{X} - \bar{Y}}{S_p \sqrt{\frac{1}{8} + \frac{1}{10}}}$$

is described by a Student t curve with 16 ($=8 + 10 - 2$) degrees of freedom. If we elect to test H_0 versus H_1 at the $\alpha = 0.01$ level of significance, the null hypothesis should be rejected if either

(a) $t \geq t_{\alpha/2, n+m-2} = t_{0.005, 16} = 2.92,$

or

(b) $t \leq -t_{0.005, 16} = -2.92$ (see Figure 8.2.1).

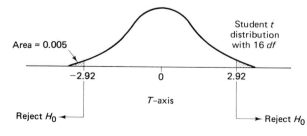

Area = 0.005

Student t distribution with 16 df

-2.92 0 2.92

T-axis

Reject H_0 Reject H_0

Figure 8.2.1   Distribution of $\dfrac{\bar{X} - \bar{Y}}{S_p \sqrt{\frac{1}{8} + \frac{1}{10}}}$ when H_0 is true.

But

$$t = \frac{0.232 - 0.210}{0.012\sqrt{1/8 + 1/10}}$$

$$= 3.86$$

a value falling considerably to the right of $t_{0.005, 16}$. Therefore, we *reject* H_0—it would appear that Twain and Snodgrass were not the same person.

Comment

The X_i's and Y_i's in this problem, being proportions, are not normal random variables, so the basic assumption of Theorem 8.2.2 is not met. Nevertheless, the probabilistic behavior of the t ratio is often only minimally affected by the nonnormality of the two populations being sampled. (More will be said about this property, known as *robustness*, in Chapter 13.) Suffice it to say that for this set of data, the t test provides a very adequate analysis.

Question 8.2.2 Should the alternative hypothesis here be one-sided or two-sided? Explain.

CASE STUDY 8.2.2

Poverty Point is the name given to a number of widely scattered archaeological sites throughout Louisiana, Mississippi, and Arkansas. These are the remains of a society thought to have flourished during the period from 1700 to 500 B.C. Among their characteristic artifacts are ornaments that were fashioned out of clay and then baked. Described in this example is a method for "dating" the various Poverty Point sites by using a property of baked clay known as thermoluminescence (78).

When certain substances are heated, they emit light in proportion to the amount of radiation to which they have been exposed. This is thermoluminescence. Furthermore, when these same substances are heated to a high enough temperature (i.e., *annealed*) they "lose" whatever exposure they had previously accumulated. This is what happened to the clay ornaments when they were first baked—over 2000 years ago. Each of them now provides a record of the total cosmic and background radiation it received since the time of its baking. By calibrating samples of clay to determine how much thermoluminescence is produced for a given amount of incident radiation, scientists can estimate the age of the artifacts.

Table 8.2.2 gives the estimated ages of eight clay ornaments, four each found at two geographically separated Poverty Point sites, Terral Lewis and Jaketown. The question to be answered is whether the technologies at these

TABLE 8.2.2 THERMOLUMINESCENT
DATES (YEARS B.C.)

Terral Lewis Estimates x_i	Jaketown Estimates, y_i
1492	1346
1169	942
883	908
988	858

two sites evolved at similar rates, as measured by when they were capable of making these ornaments.

Suppose that μ_X and μ_Y denote the true average thermoluminescent dates for all Terral Lewis and Jaketown artifacts, respectively. The hypotheses to be tested are

$$H_0: \quad \mu_X = \mu_Y$$
versus
$$H_1: \quad \mu_X \neq \mu_Y$$

(The alternative is two-sided because there is no a priori reason for anticipating which of the two sets of dates, if either, will be older than the other.) We will choose $\alpha = 0.05$ as the level of significance.

Since the total sample size is 8, the decision rule calls for H_0 to be rejected if

$$|t| = \frac{|\bar{x} - \bar{y}|}{s_p\sqrt{\frac{1}{4} + \frac{1}{4}}} \geq t_{\alpha/2,\, n+m-2} = t_{0.025,\, 6} = 2.45$$

But

$$\sum_{i=1}^{4} x_i = 4532 \qquad \sum_{i=1}^{4} x_i^2 = 5{,}348{,}458$$

and

$$\sum_{i=1}^{4} y_i = 4054 \qquad \sum_{i=1}^{4} y_i^2 = 4{,}259{,}708$$

so that

$$\bar{x} = \frac{4532}{4} = 1133.0$$

$$\bar{y} = \frac{4054}{4} = 1013.5$$

$$s_X^2 = \frac{4(5{,}348{,}458) - (4532)^2}{4(3)} = 71{,}234.0$$

$$s_Y^2 = \frac{4(4{,}259{,}708) - (4054)^2}{4(3)} = 50{,}326.3$$

and

$$s_p = \sqrt{\frac{3(71,234.0) + 3(50,326.3)}{4 + 4 - 2}} = 246.5$$

Therefore,

$$|t| = \frac{|1133.0 - 1013.5|}{246.5\sqrt{\frac{1}{4} + \frac{1}{4}}} = 0.68$$

and our conclusion is to *accept* H_0: if there was a difference in the rate of technological advancement between Terral Lewis and Jaketown, these data do not show it.

Comment

It occasionally happens that an experimenter wants to test H_0: $\mu_X = \mu_Y$ and *knows* the values of σ_X^2 and σ_Y^2. For those situations the t test of Theorem 8.2.2 is inappropriate. If the n X_i's and m Y_i's are normally distributed, it follows from Theorem 7.3.1 that

$$Z = \frac{\bar{X} - \bar{Y} - (\mu_X - \mu_Y)}{\sqrt{\frac{\sigma_X^2}{n} + \frac{\sigma_Y^2}{m}}} \qquad (8.2.1)$$

has a standard normal distribution. Any test, then, of H_0: $\mu_X = \mu_Y$ should be based on an observed Z ratio rather than an observed t ratio. In practice, applications of Equation 8.2.1 are not particularly common; any experimenter *not* knowing μ_X and μ_Y—and, hence, wanting to test H_0: $\mu_X = \mu_Y$—is not likely to know σ_X^2 and σ_Y^2.

Question 8.2.3 The use of carpeting in hospitals, while having obvious esthetic merits, raises an obvious question: are carpeted floors sanitary? One way to get at an answer is to compare bacterial levels in carpeted and uncarpeted rooms. Airborne bacteria can be counted by passing room air at a known rate over a growth medium, incubating that medium, and then counting the number of bacterial colonies that form. In one such study done in a Montana hospital (173), room air was pumped over a Petri dish at the rate of 1 cubic foot per minute. This procedure was repeated in 16 patient rooms, 8 carpeted and 8 uncarpeted. The results, expressed in terms of "bacteria per cubic foot of air," are listed in the table below.

Carpeted Rooms	Bacteria/ft^3	Uncarpeted Rooms	Bacteria/ft^3
212	11.8	210	12.1
216	8.2	214	8.3
220	7.1	215	3.8
223	13.0	217	7.2
225	10.8	221	12.0
226	10.1	222	11.1
227	14.6	224	10.1
228	14.0	229	13.7

For the carpeted rooms,

$$\sum_{i=1}^{8} x_i = 89.6 \quad \text{and} \quad \sum_{i=1}^{8} x_i^2 = 1053.70$$

For the uncarpeted rooms,

$$\sum_{i=1}^{8} y_i = 78.3 \quad \text{and} \quad \sum_{i=1}^{8} y_i^2 = 838.49$$

Test whether carpeting has any effect on the level of airborne bacteria in patient rooms. Let $\alpha = 0.05$.

Question 8.2.4 The pre-med advisor at State University is trying to decide whether he should encourage juniors to sign up for a private review course as a means of preparing for the MCAT's. Among the 15 students who had taken the exam most recently, five had enrolled in the review course and ten had not. The average MCAT scores for the first group were 10, 8, 9, 8, and 11; the average scores for the other 10 were 8, 7, 7, 9, 10, 8, 7, 11, 8, and 8. What should the pre-med advisor conclude?

Question 8.2.5 A time-study engineer is assigned the problem of comparing two different work sequences in a garment factory for measuring the shear strength of polyester fibers. To collect some data he randomly divides 12 workers into two groups. The first group measures the shear strength using work sequence A, the second group, work sequence B. The data recorded were the 12 completion times.

COMPLETION TIMES (SECONDS)

Work Sequence A	Work Sequence B
220	247
235	223
214	215
197	219
206	207
214	236

Test whether the difference in average completion times is significantly different. Let $\alpha = 0.05$.

Question 8.2.6 Serotonin is a substance found in the blood that may or may not be related to psychiatric disorders. Also, its concentration may or may not be affected by chronic LSD usage. A two-sample experiment was set up to see if any relationship could be established between LSD usage and serotonin formation in an animal population (32). Twenty-six rats were given a daily oral dose of 20 μg of LSD-25 per kilogram of body weight. The LSD was dissolved in 1 ml of water. (This particular amount was thought to be comparable in effect to the dosage a person might take.) A similar procedure was followed with a control group of 25 rats, but their "treatment" consisted of just the water. After 30 days, the animals were sacrificed and the concentrations of serotonin in their brains were measured. The results are summarized below.

SEROTONIN CONCENTRATION (NMOLE/G)

Control Group	LSD Group
$\bar{x} = 2.84$	$\bar{y} = 3.20$
$s_X/\sqrt{n} = 0.06$	$s_Y/\sqrt{m} = 0.15$

Test whether the two average serotonin concentrations are significantly different. Let $\alpha = 0.05$. Assume that the *true* variances are equal to the sample variances.

Question 8.2.7 Thrombocytopenia is a condition characterized by a chronically lowered blood platelet count. Among its most effective treatments is a splenectomy—the surgical removal of the patient's spleen. The success of such an operation, however, may be influenced by the patient's spleen weight. Listed below are the spleen weights (in grams) of 14 persons for whom a splenectomy was ultimately successful and of 5 persons for whom the operation was unsuccessful (117).

SPLEEN WEIGHTS (GRAMS)

Splenectomy Was Successful		Splenectomy Was Unsuccessful
150	136	70
142	122	110
160	200	85
110	160	90
120	102	210
240	152	
152	280	

Let μ_X and μ_Y denote the true average spleen weights of persons for whom the operation would be successful and unsuccessful, respectively. At the $\alpha = 0.05$ level of significance, test $H_0: \mu_X = \mu_Y$ versus $H_1: \mu_X \neq \mu_Y$.

Question 8.2.8 A drug is tested to determine whether it can lower the blood glucose level of diabetic rats. Six rats are given the drug, while five others are used as controls. The blood glucose levels (in mg/ml) of the treated group were 2.02, 1.71, 2.04, 1.50, 1.83, and 1.64; for the controls, 2.15, 1.92, 1.78, 2.04, and 2.22. Is the drug effective? Let $\alpha = 0.05$.

Question 8.2.9 Prove that the Z ratio given in Equation 8.2.1 has a standard normal distribution.

Question 8.2.10 A businessman has two basic routes he can take to and from work each day. The first involves going by the interstate; the second requires that he drive through town. On the average, it takes him 33 minutes to get to work via the interstate and 35 minutes by going through town. The standard deviations for the two routes are 6 minutes and 5 minutes, respectively. Assume the distributions of times for the two routes are both normally distributed.

(a) What is the probability that on a given day driving through town would be the quicker of his two alternatives?

(b) What is the probability that driving through town each way for an entire work week (ten trips) would yield a lower average time than taking the interstate for the entire week?

Hint: Use Equation 8.2.1.

Question 8.2.11 If $X_1, X_2, \ldots, X_n \sim N(\mu_X, \sigma^2)$ and $Y_1, Y_2, \ldots, Y_m \sim N(\mu_Y, \sigma^2)$, show that the pooled variance, S_p^2, as defined in Theorem 8.2.1, is an unbiased estimator for σ^2.

Question 8.2.12 Let $X_i \sim N(\mu_X, \sigma^2)$, $i = 1, 2, \ldots, n$ and $Y_j \sim N(\mu_Y, \sigma^2)$, $j = 1, 2, \ldots, m$, where σ^2 is known. Use the generalized-likelihood-ratio criterion to derive a test procedure for $H_0: \mu_X = \mu_Y$ versus $H_1: \mu_X \neq \mu_Y$. Compare your procedure to the two-sample t test of Theorem 8.2.2.

8.3 TESTING $H_0: \sigma_X^2 = \sigma_Y^2$—THE F TEST

Although by far the majority of two-sample problems are set up to detect possible shifts in location parameters, situations sometimes arise where it is equally important—perhaps even more important—to compare variability parameters. Two machines on an assembly line, for example, may be producing items whose *average* dimensions (μ_X and μ_Y) of some sort—say, thickness—are not significantly different but whose variabilities (as measured by σ_X^2 and σ_Y^2) are. This becomes a critical piece of information if the increased variability results in an unacceptable proportion of items from one of the machines falling outside the engineering specifications (see Figure 8.3.1).

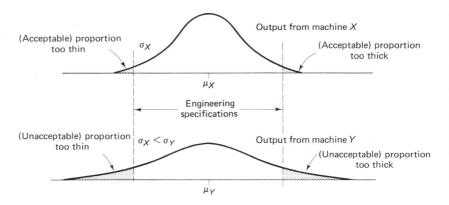

Figure 8.3.1 Variability of machine outputs.

In this section we will examine the generalized-likelihood-ratio test of $H_0: \sigma_X^2 = \sigma_Y^2$ versus $H_1: \sigma_X^2 \neq \sigma_Y^2$. The data will consist of two independent random samples of sizes n and m: the first—$X_1, X_2, \ldots, X_n$—is assumed to have come from a normal distribution having mean μ_X and variance σ_X^2; the second—$Y_1, Y_2, \ldots,$

Y_m—from a normal distribution having mean μ_Y and variance σ_Y^2. (All four parameters are assumed to be unknown.) Theorem 8.3.1 gives the test procedure that will be used. The proof will not be given, but it follows the same basic pattern we have seen in other GLRTs; the important step is showing that the likelihood ratio is a monotonic function of the F distribution defined in Theorem 7.6.1.

Comment

Tests of H_0: $\sigma_X^2 = \sigma_Y^2$ arise in another, more routine, context. Recall that the procedure for testing the equality of μ_X and μ_Y depended on whether or not the two population variances were equal. This implies that a test of H_0: $\sigma_X^2 = \sigma_Y^2$ should precede every test of H_0: $\mu_X = \mu_Y$. If the former is accepted, the t test on μ_X and μ_Y is done according to Theorem 8.2.2; but if H_0: $\sigma_X^2 = \sigma_Y^2$ is rejected, Theorem 8.2.2 is inappropriate and either the Sukhatme, Cochran, or some other approximation to the two-sample t test must be used.

Theorem 8.3.1. Let $X_1, X_2, \ldots, X_n \sim N(\mu_X, \sigma_X^2)$ and $Y_1, Y_2, \ldots, Y_m \sim N(\mu_Y, \sigma_Y^2)$ and let the X's and Y's be independent. An approximate GLRT for

$$H_0: \quad \sigma_X^2 = \sigma_Y^2$$

versus

$$H_1: \quad \sigma_X^2 \neq \sigma_Y^2$$

at the α level of significance calls for H_0 to be rejected if

$$\frac{S_Y^2}{S_X^2} \quad \text{is either} \quad \begin{cases} \leq F_{\alpha/2,\, m-1,\, n-1} \\ \text{or} \\ \geq F_{1-\alpha/2,\, m-1,\, n-1} \end{cases}$$

Comment

The GLRT described in Theorem 8.3.1 is *approximate* for the same sort of reason the GLRT for H_0: $\sigma^2 = \sigma_0^2$ was approximate (see Theorem 7.5.5). The distribution of the test statistic, S_Y^2/S_X^2, is not symmetric, and the two ranges of variance ratios yielding λ's less than or equal to λ^* (i.e., the left tail and right tail of the critical region) have slightly different areas. For the sake of convenience, though, it is customary to choose the two critical values so that each cuts off the same area, $\alpha/2$.

CASE STUDY 8.3.1

Electroencephalograms are records showing fluctuations of electrical activity in the brain. Among the several different kinds of brain "waves" produced, the dominant ones are usually *alpha* waves. These have a characteristic frequency of anywhere from 8 to 13 cycles per second.

The objective of the experiment described in this example was to see whether sensory deprivation over an extended period of time has any effect on the alpha-wave pattern. The subjects were 20 inmates in a Canadian prison. They were randomly split into two equal-sized groups. Members of one group were placed in solitary confinement; those in the other group were allowed to remain in their own cells. Seven days later, alpha-wave frequencies were measured for all 20 subjects (53), as shown in Table 8.3.1.

TABLE 8.3.1 ALPHA-WAVE FREQUENCIES (CPS)

Nonconfined, x_i	Solitary Confinement, y_i
10.7	9.6
10.7	10.4
10.4	9.7
10.9	10.3
10.5	9.2
10.3	9.3
9.6	9.9
11.1	9.5
11.2	9.0
10.4	10.9

Judging from the graph (Figure 8.3.2), there was an apparent *decrease* in alpha-wave frequency for persons in solitary confinement. There also appears to have been an *increase* in the variability for that group. We will use the F test to determine whether the observed difference in variability ($s_X^2 = 0.21$ versus $s_Y^2 = 0.36$) is statistically significant.

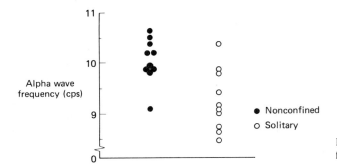

Figure 8.3.2 Alpha-wave frequencies (cps).

Let σ_X^2 and σ_Y^2 denote the true variances of alpha-wave frequencies for nonconfined and solitary-confined prisoners, respectively. The hypotheses to be tested are

$$H_0: \quad \sigma_X^2 = \sigma_Y^2$$

versus

$$H_1: \quad \sigma_X^2 \neq \sigma_Y^2$$

Let $\alpha = 0.05$ be the level of significance. Given that

$$\sum_{i=1}^{10} x_i = 105.8 \qquad \sum_{i=1}^{10} x_i^2 = 1121.26$$

$$\sum_{i=1}^{10} y_i = 97.8 \qquad \sum_{i=1}^{10} y_i^2 = 959.70$$

the sample variances become

$$s_X^2 = \frac{10(1121.26) - (105.8)^2}{10(9)} = 0.21$$

and

$$s_Y^2 = \frac{10(959.70) - (97.8)^2}{10(9)} = 0.36$$

Dividing the sample variances gives an observed F ratio of 1.71:

$$F = \frac{s_Y^2}{s_X^2} = \frac{0.36}{0.21} = 1.71$$

Both n and m are 10, so we would expect S_Y^2/S_X^2 to behave like an F random variable with 9 and 9 degrees of freedom (assuming H_0: $\sigma_X^2 = \sigma_Y^2$ is true). From Table A.4 in the Appendix, we see that the values cutting off areas of 0.025 in either tail of that distribution are 0.248 and 4.03 (see Figure 8.3.3).

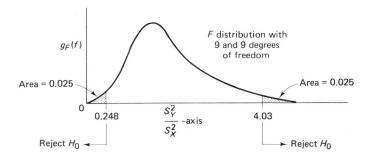

Figure 8.3.3 Distribution of S_Y^2/S_X^2 when H_0 is true.

Our conclusion, then, is to *accept* H_0. (In light of the Comment preceding Theorem 8.3.1, it would now be appropriate to test H_0: $\mu_X = \mu_Y$ using the two-sample t test of Section 8.2.)

Question 8.3.1 Mercury pollution is recognized as a serious ecological problem. Much of the mercury released into the environment originates as a by-product of coal burning and other industrial processes. It does not become really dangerous until it falls into large bodies of water where microorganisms change it into methylmercury, an organic form that is particularly toxic. Fish are the intermediaries: they ingest and absorb the methylmercury and are, in turn, eaten by humans. In a study designed to investigate the metabolism of methylmercury in humans (103), a group of six females and nine males were given an oral administration of protein-bound methylmercury. Listed in the table are the resulting half-lives (in days) of the methylmercury in their systems.

Females	CH_3^{203} half-life	Males	CH_3^{203} half-life
AE	52	RE	72
EH	69	AH	88
LJ	73	PK	87
AN	88	JK	74
KR	87	MM	78
LU	56	JM	70
		VM	78
		RP	93
		HT	74

Test $H_0: \sigma_X^2 = \sigma_Y^2$ versus $H_1: \sigma_X^2 \neq \sigma_Y^2$. Let $\alpha = 0.05$. Would it be appropriate to use Theorem 8.2.2 to test $H_0: \mu_X = \mu_Y$?

Question 8.3.2 Among the standard personality inventories used by psychologists is the thematic apperception test (TAT). A subject is shown a series of pictures and is asked to make up a story about each one. Interpreted properly, the content of the stories can provide valuable insights into the subject's mental well-being. The data below show the TAT results for 40 women, 20 of whom were the mothers of normal children and 20 the mothers of schizophrenic children. In each case the subject was shown the same set of ten pictures. The figures recorded were the numbers of stories (out of 10) that revealed a *positive* parent–child relationship, one where the mother was clearly capable of interacting with her child in a flexible, open-minded way (176).

TAT SCORES

Mothers of Normal Children					Mothers of Schizophrenic Children				
8	4	6	3	1	2	1	1	3	2
4	4	6	4	2	7	2	1	3	1
2	1	1	4	3	0	2	4	2	3
3	2	6	3	4	3	0	1	2	2

(a) Test $H_0: \sigma_X^2 = \sigma_Y^2$ versus $H_1: \sigma_X^2 \neq \sigma_Y^2$, where σ_X^2 and σ_Y^2 are the variances of the scores of mothers of normal children and scores of mothers of schizophrenic children, respectively. Let $\alpha = 0.05$.

(b) If $H_0: \sigma_X^2 = \sigma_Y^2$ is accepted in part (a), test $H_0: \mu_X = \mu_Y$ versus $H_1: \mu_X \neq \mu_Y$. Set α equal to 0.05.

Question 8.3.3 In a study designed to investigate the effects of a strong magnetic field on the early development of mice (5), 10 cages, each containing three 30-day-old albino female mice, were subjected for a period of 12 days to a magnetic field having an average strength of 80 Oe/cm. Thirty other mice, housed in 10 similar cages, were not put in the magnetic field and served as controls. Listed in the table are the weight gains, in grams, for each of the 20 sets of mice.

In magnetic field		Not in magnetic field	
Cage	Weight gain (g)	Cage	Weight gain (g)
1	22.8	11	23.5
2	10.2	12	31.0
3	20.8	13	19.5
4	27.0	14	26.2
5	19.2	15	26.5
6	9.0	16	25.2
7	14.2	17	24.5
8	19.8	18	23.8
9	14.5	19	27.8
10	14.8	20	22.0

Test whether the variances of the two sets of weight gains are significantly different. Let $\alpha = 0.05$.

Question 8.3.4 Raynaud's syndrome is characterized by the sudden impairment of blood circulation in the fingers, a condition which results in discoloration and heat loss. The magnitude of the problem is evidenced in the following data where 20 subjects (10 "normals" and 10 with Raynaud's syndrome) immersed their right forefingers in water kept at 19°C. The heat output (in cal/cm^2/minute) of the forefinger was then measured with a calorimeter (95).

Normal Subjects		Subjects with Raynaud's Syndrome	
Patient	Heat Output (cal/cm^2/min)	Patient	Heat Output (cal/cm^2/min)
W.K.	2.43	R.A.	0.81
M.N.	1.83	R.M.	0.70
S.A.	2.43	F.M.	0.74
Z.K.	2.70	K.A.	0.36
J.H.	1.88	H.M.	0.75
J.G.	1.96	S.M.	0.56
G.K.	1.53	R.M.	0.65
A.S.	2.08	G.E.	0.87
T.E.	1.85	B.W.	0.40
L.F.	2.44	N.E.	0.31

Test that the heat-output variances for normal subjects and those with Raynaud's syndrome are the same. Use a two-sided alternative and the 0.05 level of significance.

Question 8.3.5 Show that the generalized likelihood ratio for testing $H_0: \sigma_X^2 = \sigma_Y^2$ versus $H_1: \sigma_X^2 \neq \sigma_Y^2$ as described in Theorem 8.3.1 is given by

$$\lambda = \frac{L(\hat{\omega})}{L(\hat{\Omega})} = \frac{(m+n)^{(n+m)/2}}{n^{n/2}m^{m/2}} \frac{\left[\sum_{i=1}^{n} (x_i - \bar{x})^2 \right]^{n/2} \left[\sum_{j=1}^{m} (y_j - \bar{y})^2 \right]^{m/2}}{\left[\sum_{i=1}^{n} (x_i - \bar{x})^2 + \sum_{j=1}^{m} (y_j - \bar{y})^2 \right]^{(m+n)/2}}$$

Question 8.3.6 Let $X_1, X_2, \ldots, X_n \sim N(\mu_X, \sigma_X^2)$ and $Y_1, Y_2, \ldots, Y_m \sim N(\mu_Y, \sigma_Y^2)$, where μ_X and μ_Y are known. Derive the GLRT for

$$H_0: \quad \sigma_X^2 = \sigma_Y^2$$

versus

$$H_1: \quad \sigma_X^2 > \sigma_Y^2$$

Compare your answer with the procedure given in Theorem 8.3.1.

8.4 BINOMIAL DATA: TESTING $H_0: p_X = p_Y$

Up to this point, the data considered in Chapter 8 have been independent random samples of sizes n and m drawn from two *continuous* distributions—in fact, from two *normal* distributions. Obviously, many other sorts of data might have to be dealt with. The X's and Y's might represent continuous random variables, for example, but have density functions other than the normal. Or they might be *discrete*. In this section we consider the most common example of this latter type: situations where the two sets of data are *binomial*.

Suppose that n Bernoulli trials related to treatment X have resulted in x successes, and m (independent) Bernoulli trials related to treatment Y in y successes. We wish to test whether p_X and p_Y, the *true* probabilities of success for treatment X and treatment Y, are equal:

$$H_0: \quad p_X = p_Y \quad (=p)$$

versus

$$H_1: \quad p_X \neq p_Y$$

The level of significance will be α.

Here the two parameter spaces are given by

$$\omega = \{(p_X, p_Y): 0 \leq p_X = p_Y \leq 1\}$$

and

$$\Omega = \{(p_X, p_Y): 0 \leq p_X \leq 1, 0 \leq p_Y \leq 1\}$$

Furthermore, the likelihood function can be written

$$L = p_X^x (1 - p_X)^{n-x} \cdot p_Y^y (1 - p_Y)^{m-y}$$

Setting the derivative of $\ln L$ with respect to $p \, (= p_X = p_Y)$ equal to 0 and solving for p gives a not too surprising result—namely,

$$\hat{p} = \frac{x + y}{n + m}$$

That is, the MLE for p under H_0 is the pooled success proportion. Similarly, solving $\partial \ln L / \partial p_X = 0$ and $\partial \ln L / \partial p_Y = 0$ gives the two original sample proportions as the unrestricted MLEs for p_X and p_Y:

$$\hat{p}_X = \frac{x}{n} \qquad \hat{p}_Y = \frac{y}{m}$$

Putting $\hat{p}$, $\hat{p}_X$, and $\hat{p}_Y$ back into L gives the generalized likelihood ratio:

$$\lambda = \frac{L(\hat{\omega})}{L(\hat{\Omega})} = \frac{[(x + y)/(n + m)]^{x+y} [1 - (x + y)/(n + m)]^{n+m-x-y}}{(x/n)^x [1 - (x/n)]^{n-x} (y/m)^y [1 - (y/m)]^{m-y}} \qquad (8.4.1)$$

Equation 8.4.1 is such a difficult function to work with that it is necessary to find an approximation to the usual generalized-likelihood-ratio test. There are several available. It can be shown, for example, that $-2 \ln \Lambda$ for this problem has an asymptotic χ^2 distribution with 1 degree of freedom (178). Thus, an approximate two-sided, $\alpha = 0.05$ test is to reject H_0 if $-2 \ln \lambda \geq 3.84$.

Another approach, and the one most often used, is to appeal to the central limit theorem and make the observation that

$$\frac{\dfrac{X}{n} - \dfrac{Y}{m} - E\left(\dfrac{X}{n} - \dfrac{Y}{m}\right)}{\sqrt{\text{Var}\left(\dfrac{X}{n} - \dfrac{Y}{m}\right)}}$$

has an approximate $N(0, 1)$ distribution. Of course, under H_0,

$$E\left(\frac{X}{n} - \frac{Y}{m}\right) = 0$$

and

$$\text{Var}\left(\frac{X}{n} - \frac{Y}{m}\right) = \frac{p(1 - p)}{n} + \frac{p(1 - p)}{m}$$

$$= \frac{(n + m)p(1 - p)}{nm}$$

If p is now replaced by $(x + y)/(n + m)$, its MLE under ω, we get the statement of Theorem 8.4.1. Details of the proof will be omitted.

Sec. 8.4   Binomial Data: Testing $H_0 \colon p_X = p_Y$ **379**

Theorem 8.4.1. Let x and y denote the numbers of successes observed in two independent sets of n and m Bernoulli trials, respectively. Let p_X and p_Y denote the true success probabilities associated with each set of trials. An approximate GLRT at the α level of significance for

$$H_0: \quad p_X = p_Y$$

$$\text{versus}$$

$$H_1: \quad p_X \neq p_Y$$

is gotten by rejecting H_0 whenever

$$\frac{\dfrac{x}{n} - \dfrac{y}{m}}{\sqrt{\dfrac{\left(\dfrac{x+y}{n+m}\right)\left(1 - \dfrac{x+y}{n+m}\right)(n+m)}{nm}}} \text{ is either } \begin{cases} \leq -z_{\alpha/2} \\ \text{or} \\ \geq +z_{\alpha/2} \end{cases}$$

The utility of Theorem 8.4.1 actually extends far beyond the scope we have just described. Any continuous variable can always be dichotomized and "transformed" into a Bernoulli variable. For example, blood pressure can be recorded in terms of "mm Hg," a continuous variable, or as simply "normal" or "abnormal," a Bernoulli variable. The next two case studies illustrate these two sources of binomial data. In the first, the variables begin and end as Bernoullis, while in the second, the initial measurement of "number of nightmares per month" is immediately dichotomized into "often" and "seldom."

CASE STUDY 8.4.1

Law and order was a key issue for all the presidential contenders in 1968 but particularly so for George Wallace. He and his supporters were the sharpest critics of the courts and the strongest advocates of a renewed commitment to vigorous law enforcement. Whenever one segment of society moralizes to another, though, there is a natural tendency for the accused to question whether the accusers are, themselves, above reproach. "Wallaceites" talked a good game of law and order, but did they practice what they preached? In Nashville, Tennessee, a team of sociologists carried out a rather unorthodox survey that seemed to indicate that maybe they didn't (185).

Before the general election of 1968, the government of Nashville-Davidson County had passed a law requiring all locally operated vehicles to display on their windshield a "Metro sticker" (costing $15). It was far from clear, though, how strictly the law would be enforced. Many motorists thought they could get by without one.

For several days following the sticker deadline (November 1) the investigators made spot checks of various parking lots in and around Nashville to see whether supporters of the various candidates differed significantly in their compliance with the law. Table 8.4.1 shows some of the results.

TABLE 8.4.1 PARKING-LOT SURVEYS OF BUMPER STICKERS

In Support of[a]	Number of Cars	Number with Stickers
Humphrey	$n_H = 178$	$x_H = 154$
Wallace	$n_W = 361$	$w_W = 270$

[a] A car was assumed to be owned by a Humphrey supporter, for example, if it displayed a Humphrey bumper sticker.

The sample proportions of Humphrey and Wallace supporters obeying the law were $154/178 = 0.865$ and $270/361 = 0.748$, respectively. Presumably, these two figures are unbiased estimates of p_H and p_W, the *true* proportions of Humphrey and Wallace cars bearing a Metro sticker. Can we conclude that p_H and p_W are not equal, as the two sample proportions would indicate, or is the observed difference of 0.117 ($=0.865 - 0.748$) well within the normal bounds of sampling variability? Put more formally, the hypotheses to be tested are

$$H_0: \quad p_H = p_W \ (=p)$$

versus

$$H_1: \quad p_H \neq p_W$$

Let $\alpha = 0.01$.

If H_0 is true, a pooled estimate of p would be the overall sample proportion of cars with Metro stickers. That is,

$$\hat{p} = \frac{154 + 270}{178 + 361} = \frac{424}{539} = 0.787$$

According to Theorem 8.4.1, then, the test statistic is equal to 3.12:

$$\frac{0.865 - 0.748}{\sqrt{\dfrac{(0.787)(0.213)(539)}{(178)(361)}}} = 3.12$$

But $\pm z_{0.005} = \pm 2.58$. Thus, we should reject the null hypothesis: it would appear that the proportion of Wallace supporters breaking the law by not having a Metro sticker was significantly higher than the corresponding proportion for Humphrey supporters.

CASE STUDY 8.4.2

Over the years, numerous studies have sought to characterize the nightmare sufferer. Out of these has emerged the stereotype of someone with high anxiety, low ego strength, feelings of inadequacy, and poorer-than-average physical health. What is not so well known, though, is whether men fall into this pattern with the same frequency as women. To this end, a recent investigation (72) looked at nightmare frequencies for a sample of 160 men and 192 women. Each subject was asked whether he (or she) experienced nightmares "often" (at least once a month) or "seldom" (less than once a month). The findings are summarized in Table 8.4.2.

TABLE 8.4.2 FREQUENCY OF NIGHTMARES

	Men	Women	Total
Nightmares often	55	60	115
Nightmares seldom	105	132	237
Totals	160	192	

If p_M and p_W denote the true proportions of men having nightmares often and women having nightmares often, respectively, what we want to test is

$$H_0: \quad p_M = p_W$$

versus

$$H_1: \quad p_M \neq p_W$$

Suppose that α is set equal to 0.05. This makes the critical values equal to $\pm z_{0.025}$, or ± 1.96. Substituting the pooled estimate for the "nightmares often" probability, $\frac{115}{352}$, into the expression for the test statistic gives

$$\frac{55/160 - 60/192}{\sqrt{\dfrac{(115/352)(1 - 115/352)(352)}{(160)(192)}}} = 0.64$$

The conclusion, then, is clear: we should accept the null hypothesis—these data provide no convincing evidence that the frequency of nightmares is different for men than for women.

Question 8.4.1 The phenomenon of handedness has been extensively studied in human populations. The percentages of adults who are right-handed, left-handed, and ambidextrous are well documented. What is not so well known is that a similar phenomenon is present in lower animals. Dogs, for example, can be either right-pawed or left-pawed. Suppose that in a random sample of 200 beagles it is found that 55 are left-pawed and that in a random sample of 200 collies 40 are left-pawed. Can we conclude that the true proportion of collies that are left-pawed is significantly different from the true proportion of beagles that are left-pawed? Let $\alpha = 0.05$.

Question 8.4.2 In a study designed to see whether a controlled diet could retard the process of arteriosclerosis, a total of 846 randomly chosen persons were followed over an eight-year period. Half were instructed to eat only certain foods; the other half could eat whatever they wanted. At the end of eight years, 66 persons in the diet group were found to have died of either myocardial infarction or cerebral infarction, as compared to 93 deaths of a similar nature in the control group (182). Do the appropriate analysis. Let $\alpha = 0.05$.

Question 8.4.3 The kittiwake is a seagull whose mating behavior is basically monogamous. Normally, the birds separate for several months after the completion of one breeding season and reunite at the beginning of the next. Whether or not the birds actually do reunite, though, may be affected by the success of their "relationship" the season before. A total of 769 kittiwake pair-bonds were studied (27) over the course of two breeding seasons; of those 769, some 609 successfully bred during the first season; the remaining 160 were unsuccessful. The following season, 175 of the previously successful pair-bonds "divorced," as did 100 of the 160 whose prior relationship left something to be desired.

	Breeding in previous year	
	Successful	Unsuccessful
Number divorced	175	100
Number not divorced	434	60
Total	609	160
Percent divorced	29	63

Can we conclude that the two divorce rates (29% and 63%) are significantly different at the 0.05 level?

Question 8.4.4 A utility infielder for a National League club batted 0.260 last season in 300 trips to the plate. This year he hit 0.250 in 200 at-bats. The owners are trying to cut his pay for next year on the grounds that his output has deteriorated. The player argues, though, that his performances the last two seasons have not been significantly different, so his salary should not be reduced. Who is right?

Question 8.4.5 Compute $-2 \ln \lambda$ (see Equation 8.4.1) for the bumper-sticker data of Case Study 8.4.1 and use it to test the equality of p_H and p_W. Let $\alpha = 0.01$.

Question 8.4.6 Recall Case Study 8.4.1. What other explanation(s), besides the obvious one that Wallace supporters are less law-abiding than Humphrey supporters with respect to local tax ordinances, might account for the observed difference between $\hat{p}_H$ and $\hat{p}_W$?

Question 8.4.7 Compute the P-value associated with the data analyzed in Case Study 8.4.2 (see Question 6.3.10).

8.5 CONFIDENCE INTERVALS
FOR THE TWO-SAMPLE PROBLEM

Implicit in Theorems 8.2.2, 8.3.1, and 8.4.1 has been the sampling distribution *when H_0 is true* of $\bar{X} - \bar{Y}$, S_Y^2/S_X^2, and $X/n - Y/m$. More generally, when the null hypothesis is not necessarily true, these test statistics take the form given below. [*Note:* Statement (a) is included here only for the sake of completeness—it has already appeared as Theorem 8.2.1. Statement (c) represents a slight modification of the expression that appeared in Theorem 8.4.1, the difference being that the denominator here is the square root of the unpooled, rather than the pooled, estimator for the variance of $X/n - Y/m$. See (73) for a proof that the asymptotic normality still holds.]

(a)
$$\frac{\bar{X} - \bar{Y} - (\mu_X - \mu_Y)}{S_p\sqrt{\dfrac{1}{n} + \dfrac{1}{m}}} \sim T_{n+m-2}$$

(b)
$$\frac{S_Y^2/\sigma_Y^2}{S_X^2/\sigma_X^2} \sim F_{m-1,\,n-1}$$

(c)
$$\frac{\dfrac{X}{n} - \dfrac{Y}{m} - (p_X - p_Y)}{\sqrt{\dfrac{\left(\dfrac{X}{n}\right)\left(1 - \dfrac{X}{n}\right)}{n} + \dfrac{\left(\dfrac{Y}{m}\right)\left(1 - \dfrac{Y}{m}\right)}{m}}} \sim N(0,\,1),\ \text{approximately}$$

By inverting these expressions, it is possible to derive $100(1 - \alpha)\%$ confidence intervals for $\mu_X - \mu_Y$, σ_X^2/σ_Y^2, and $p_X - p_Y$. Theorems 8.5.1, 8.5.2, and 8.5.3 state the results. The proofs are very straightforward and will be left as exercises.

> **Theorem 8.5.1.** Let $X_1, X_2, \ldots, X_n \sim N(\mu_X, \sigma^2)$ and $Y_1, Y_2, \ldots, Y_m \sim N(\mu_Y, \sigma^2)$ and let the X's and Y's be independent. Let s_p be the pooled standard deviation. Then a $100(1 - \alpha)\%$ confidence interval for $\mu_X - \mu_Y$ is given by
>
> $$\left(\bar{x} - \bar{y} - t_{\alpha/2,\,n+m-2}\, s_p\sqrt{\frac{1}{n} + \frac{1}{m}},\ \bar{x} - \bar{y} + t_{\alpha/2,\,n+m-2}\, s_p\sqrt{\frac{1}{n} + \frac{1}{m}}\right)$$

CASE STUDY 8.5.1

Occasionally in forensic medicine, or in the aftermath of a bad accident, identifying the sex of a victim can be a very difficult task. In some of these cases, dental structure provides a useful criterion, since individual teeth will remain in good condition long after other tissues have deteriorated. Furthermore, studies have shown that female teeth and male teeth have different physical and chemical characteristics.

The extent to which X-rays can penetrate tooth enamel, for instance, is different for men than it is for women. Listed in Table 8.5.1 are "spectropenetration gradients" for eight female teeth and eight male teeth (49). These numbers are measures of the rate of change in the amount of X-ray penetration through a 500-micron section of tooth enamel at a wavelength of 600 nm as opposed to 400 nm.

TABLE 8.5.1 ENAMEL SPECTROPENETRATION GRADIENTS

Male, x_i	Female, y_i
4.9	4.8
5.4	5.3
5.0	3.7
5.5	4.1
5.4	5.6
6.6	4.0
6.3	3.6
4.3	5.0

Let μ_X and μ_Y be the population means of the spectropenetration gradients associated with male teeth and with female teeth, respectively. Note that

$$\sum_{i=1}^{8} x_i = 43.4 \qquad \sum_{i=1}^{8} x_i^2 = 239.32$$

from which

$$\bar{x} = \frac{43.4}{8} = 5.4$$

and

$$s_X^2 = \frac{8(239.32) - (43.4)^2}{8(7)} = 0.55$$

Similarly,

$$\sum_{i=1}^{8} y_i = 36.1 \qquad \sum_{i=1}^{8} y_i^2 = 166.95$$

so that

$$\bar{y} = \frac{36.1}{8} = 4.5$$

and

$$s_Y^2 = \frac{8(166.95) - (36.1)^2}{8(7)} = 0.58$$

Therefore, the pooled standard deviation is equal to 0.75:

$$s_p = \sqrt{\frac{7(0.55) + 7(0.58)}{8 + 8 - 2}} = \sqrt{0.565} = 0.75$$

We know that the ratio

$$\frac{\bar{X} - \bar{Y} - (\mu_X - \mu_Y)}{S_p\sqrt{\frac{1}{8} + \frac{1}{8}}}$$

will be approximated by a Student t curve with 14 degrees of freedom. Since $t_{0.025, 14} = 2.14$, the 95% confidence interval for $\mu_X - \mu_Y$ is given by

$$(\bar{x} - \bar{y} - 2.14s_p\sqrt{\tfrac{1}{8} + \tfrac{1}{8}}, \; \bar{x} - \bar{y} + 2.14s_p\sqrt{\tfrac{1}{8} + \tfrac{1}{8}})$$

$$= (5.4 - 4.5 - 2.14(0.75)\sqrt{0.25}, \; 5.4 - 4.5 + 2.14(0.75)\sqrt{0.25})$$

$$= (0.1, \; 1.7)$$

Comment

Here the 95% confidence interval does not include the value 0. This means that had we tested

$$H_0: \quad \mu_X = \mu_Y$$

versus

$$H_1: \quad \mu_X \neq \mu_Y$$

at the $\alpha = 0.05$ level of significance, H_0 would have been rejected.

Theorem 8.5.2. Let $X_1, X_2, \ldots, X_n \sim N(\mu_X, \sigma_X^2)$ and $Y_1, Y_2, \ldots, Y_m \sim N(\mu_Y, \sigma_Y^2)$ and let the X's and Y's be independent. A $100(1 - \alpha)\%$ confidence interval for the variance ratio, σ_X^2/σ_Y^2, is given by

$$\left(\frac{s_X^2}{s_Y^2} F_{\alpha/2, \, m-1, \, n-1}, \; \frac{s_X^2}{s_Y^2} F_{1-\alpha/2, \, m-1, \, n-1} \right)$$

CASE STUDY 8.5.2

The easiest way to measure the movement, or flow, of a glacier is with a camera. First a set of reference points is marked off at various sites near the glacier's edge. Then these points, along with the glacier, are photographed from an airplane. The problem is this: How long should the time interval be between photographs? If too *short* a period has elapsed, the glacier will not have moved very far and the errors associated with the photographic technique will be relatively large. If too *long* a period has elapsed, parts of the glacier might be deformed by the surrounding terrain, an eventuality that could introduce substantial variability into the point-to-point velocity estimates.

In this example, two sets of flow rates for the Antarctic's Hoseason Glacier are examined (107), one based on photographs taken *three* years apart, the other, *five* years apart. Both sets of data were taken under identical conditions. Also, on the basis of other considerations, it can be assumed that the "true" flow rate for the glacier was constant for the eight years in question. The data are listed in Table 8.5.2.

TABLE 8.5.2 FLOW RATES ESTIMATED FOR THE HOSEASON GLACIER (METERS PER DAY)

Three-Year Span, x_i	Five-Year Span, y_i
0.73	0.72
0.76	0.74
0.75	0.74
0.77	0.72
0.73	0.72
0.75	
0.74	

The objective here is to assess the relative variabilities associated with the three- and five-year time periods. One way to do this—assuming the data to be normal—is to construct, say, a 95% confidence interval for the variance ratio. If that interval does not contain the value "1", we infer that the two time periods lead to flow rate estimates of significantly different precision.

From Table 8.5.2,

$$\sum_{i=1}^{7} x_i = 5.23 \quad \text{and} \quad \sum_{i=1}^{7} x_i^2 = 3.9089$$

·so that

$$s_X^2 = \frac{7(3.9089) - (5.23)^2}{7(6)} = 0.000224$$

Similarly,

$$\sum_{i=1}^{5} y_i = 3.64 \quad \text{and} \quad \sum_{i=1}^{5} y_i^2 = 2.6504$$

making

$$s_Y^2 = \frac{5(2.6504) - (3.64)^2}{5(4)} = 0.000120$$

The two critical values come from Table A.4 in the Appendix:

$$F_{0.025, 4, 6} = 0.109 \qquad \text{and} \qquad F_{0.975, 4, 6} = 6.23$$

When all of these quantities are substituted into the statement of Theorem 8.5.2, we get a 95% confidence interval for σ_X^2/σ_Y^2:

$$\left(\frac{0.000224}{0.000120} \, 0.109, \ \frac{0.000224}{0.000120} \, 6.23 \right) = (0.203, \ 11.629)$$

Thus, although the three-year data had a larger *sample* variance than the five-year data, no conclusions can be drawn about the *true* variances being different, because the ratio $\sigma_X^2/\sigma_Y^2 = 1$ is contained in the confidence interval.

Theorem 8.5.3. Let x and y denote the numbers of successes observed in two independent sets of n and m Bernoulli trials, respectively. If p_X and p_Y denote the true success probabilities, an approximate $100(1 - \alpha)\%$ confidence interval for $p_X - p_Y$ is given by

$$\left(\frac{x}{n} - \frac{y}{m} - z_{\alpha/2} \sqrt{ \frac{\left(\frac{x}{n}\right)\left(1 - \frac{x}{n}\right)}{n} + \frac{\left(\frac{y}{m}\right)\left(1 - \frac{y}{m}\right)}{m} }, \right.$$

$$\left. \frac{x}{n} - \frac{y}{m} + z_{\alpha/2} \sqrt{ \frac{\left(\frac{x}{n}\right)\left(1 - \frac{x}{n}\right)}{n} + \frac{\left(\frac{y}{m}\right)\left(1 - \frac{y}{m}\right)}{m} } \right)$$

CASE STUDY 8.5.3

Until almost the end of the nineteenth century the mortality associated with surgical operations—even minor ones—was extremely high. The major problem was infection. The germ theory as a model for disease transmission was still unknown, so there was no concept of sterilization. As a result, many patients died from postoperative complications.

The major breakthrough that was so desperately needed finally came when Joseph Lister, a British physician, began reading about some of the work done by Louis Pasteur. In a series of classic experiments, Pasteur had succeeded in demonstrating the part that yeasts and bacteria play in fermentation. What Lister conjectured was that human infections might have a similar organic origin. To test his theory he began using carbolic acid as an operating-room disinfectant. The data in Table 8.5.3 show the outcomes of 75 amputations performed by Lister, 35 without the aid of carbolic acid and 40 with it (180).

TABLE 8.5.3 MORTALITY RATES—LISTER'S AMPUTATIONS

		Carbolic acid used?		
		No	Yes	Total
Patient · lived?	Yes	19	34	53
	No	16	6	22
	Total	35	40	

Let p_W (estimated by $\frac{34}{40}$) and $p_{W/O}$ (estimated by $\frac{19}{35}$) denote the true survival probabilities for patients amputated "with" and "without" the use of carbolic acid, respectively. To construct a 95% confidence interval for $p_W - p_{W/O}$ we note that $z_{\alpha/2} = 1.96$; then Theorem 8.5.3 reduces to

$$\left(\frac{34}{40} - \frac{19}{35} - 1.96\sqrt{\frac{(\frac{34}{40})(1 - \frac{34}{40})}{40} + \frac{(\frac{19}{35})(1 - \frac{19}{35})}{35}},\right.$$

$$\left.\frac{34}{40} - \frac{19}{35} + 1.96\sqrt{\frac{(\frac{34}{40})(1 - \frac{34}{40})}{40} + \frac{(\frac{19}{35})(1 - \frac{19}{35})}{35}}\right)$$

$$= (0.31 - 1.96\sqrt{0.0103}, \, 0.31 + 1.96\sqrt{0.0103})$$

$$= (0.11, 0.51)$$

Since $p_W - p_{W/O} = 0$ is not included in the interval, we would conclude that the presence or absence of carbolic acid *does* constitute a significant effect. Specifically, patients on whom carbolic acid is used have a better chance of recovery.

Question 8.5.1 Because of its association with a variety of serious physical disorders, caffeine is a closely-monitored food additive. Listed below are measurements made using high performance liquid chromatography of the caffeine content (g per 100 g of dry matter) found in 12 brands of instant coffee, 8 spray-dried and 4 freeze-dried (168).

Spray-dried, x_i	Freeze-dried, y_i
4.8	3.7
4.0	3.4
3.8	2.8
4.3	3.7
3.9	
4.6	
3.1	
3.7	

Construct a 90% confidence interval for $\mu_X - \mu_Y$. What do the endpoints of your interval imply about the outcome of testing $H_0: \mu_X = \mu_Y$ versus $H_1: \mu_X \neq \mu_Y$ at the $\alpha = 0.10$ level of significance?

Question 8.5.2 Construct a 99% confidence interval for the true average difference in alpha-wave frequencies, $\mu_X - \mu_Y$, as described in Case Study 8.3.1.

Question 8.5.3 Prove Theorem 8.5.1.

Question 8.5.4 Let $X_1, X_2, \ldots, X_n$ and $Y_1, Y_2, \ldots, Y_m$ be independent random samples from pdf's $N(\mu_X, \sigma_X^2)$ and $N(\mu_Y, \sigma_Y^2)$, respectively. Assume that σ_X^2 and σ_Y^2 are known. Derive an expression for a $100(1 - \alpha)\%$ confidence interval for $\mu_X - \mu_Y$.

Question 8.5.5 The effectiveness of charcoal filters was investigated with an experiment involving protozoa (175). *Paramecium aurelia* were suspended in a hanging drop inside a smoke chamber. Every 60 seconds, a 6-second puff of smoke was drawn through the chamber. The movements of the paramecia were watched through a stereomicroscope. The variable recorded was the length of time from the start of the experiment to when the last paramecium died. Altogether, the experiment was replicated 12 times. Six of those times the smoke came from a nonfilter cigarette, the other six times from a cigarette with a charcoal filter. Use the data shown in the table to construct a 95% confidence interval for σ_X^2/σ_Y^2.

SURVIVAL TIME (min)

Nonfilter, x_i	Charcoal filter, y_i
8	21
7	37
11	24
8	27
9	19
8	14

Question 8.5.6 One of the parameters used in evaluating myocardial function is the end diastolic volume (EDV). Shown in the table are EDVs recorded for eight persons considered to have normal cardiac function and for six with constrictive pericarditis (172).

END DIASTOLIC VOLUME (ml/m²)

Normal, x_i	Constrictive pericarditis, y_i
62	24
60	56
78	42
62	74
49	44
67	28
80	
48	

Construct a 99% confidence interval for σ_X^2/σ_Y^2.

Question 8.5.7 Construct a 90% confidence interval for σ_X^2/σ_Y^2 using the Mark Twain–Quintus Curtius Snodgrass data of Case Study 8.2.1.

Question 8.5.8 Prove Theorem 8.5.2.

Question 8.5.9 Construct a 70% confidence interval for $p_M - p_W$ in the nightmare-frequency data summarized in Case Study 8.4.2.

Question 8.5.10 The possible effects of a mouse's early environment on its aggressiveness later in life were studied in a recent experiment (76). Two groups of mice were tested: one group had been raised by their natural mothers, the other group by "foster" mice. (A foster mouse was a female whose own litter had been removed shortly after birth.) When a mouse was three months old it was placed in a box divided into two compartments by a partition. On the other side of the partition was another mouse, one that had had no previous contact with the "test" mouse. The partition was then removed and the behavior of the mice was observed for the next 6 minutes. Altogether the experiment was done 307 times. Of the 167 mice raised by their natural mothers, a total of 27 began fighting with the mouse on the other side of the partition. Among the remaining 140 mice, each of which had been raised by a foster mouse, a total of 47 began fighting. Let p_X and p_Y denote the *true* proportions of "natural" mice and "foster" mice showing hostility. Construct a 95% confidence interval for $p_X - p_Y$.

Question 8.5.11 Prove Theorem 8.5.3.

APPENDIX 8.1 A DERIVATION OF THE TWO-SAMPLE t TEST (A PROOF OF THEOREM 8.2.2)

To begin, we note that both the restricted and unrestricted parameter spaces, ω and Ω, are three-dimensional:

$$\omega = \{(\mu_X, \mu_Y, \sigma): -\infty < \mu_X = \mu_Y < \infty, 0 < \sigma < \infty\}$$

and

$$\Omega = \{(\mu_X, \mu_Y, \sigma): -\infty < \mu_X < \infty, -\infty < \mu_Y < \infty, 0 < \sigma < \infty\}$$

Since the X's and Y's are independent (and normal),

$$L(\omega) = \prod_{i=1}^{n} f_X(x_i) \prod_{j=1}^{m} f_Y(y_j)$$

$$= \left(\frac{1}{\sqrt{2\pi}\,\sigma}\right)^{n+m} \exp\left\{-\frac{1}{2\sigma^2}\left[\sum_{i=1}^{n}(x_i - \mu)^2 + \sum_{j=1}^{m}(y_j - \mu)^2\right]\right\} \quad \text{(A8.1.1)}$$

where $\mu = \mu_X = \mu_Y$. If we take $\ln L(\omega)$ and solve $\partial \ln L(\omega)/\partial\mu = 0$ and $\partial \ln L(\omega)/\partial\sigma^2 = 0$ simultaneously, the solutions will be the restricted maximum-likelihood estimates:

$$\hat{\mu} = \frac{\sum\limits_{i=1}^{n} x_i + \sum\limits_{j=1}^{m} y_j}{n + m} \tag{A8.1.2}$$

and

$$\hat{\sigma}^2 = \frac{\sum\limits_{i=1}^{n} (x_i - \hat{\mu})^2 + \sum\limits_{j=1}^{m} (y_j - \hat{\mu})^2}{n + m} \tag{A8.1.3}$$

Substituting Equations A8.1.2 and A8.1.3 into Equation A8.1.1 gives the numerator of the generalized likelihood ratio:

$$L(\hat{\omega}) = \left(\frac{e^{-1}}{2\pi\hat{\sigma}^2}\right)^{(n+m)/2}$$

Similarly, the likelihood function unrestricted by the null hypothesis is

$$L(\Omega) = \left(\frac{1}{\sqrt{2\pi}\,\sigma}\right)^{n+m} \exp\left\{-\frac{1}{2\sigma^2}\left[\sum_{i=1}^{n}(x_i - \mu_X)^2 + \sum_{j=1}^{m}(y_j - \mu_Y)^2\right]\right\} \tag{A8.1.4}$$

Here, solving

$$\frac{\partial \ln L(\Omega)}{\partial\mu_X} = 0 \qquad \frac{\partial \ln L(\Omega)}{\partial\mu_Y} = 0 \qquad \frac{\partial \ln L(\Omega)}{\partial\sigma^2} = 0$$

gives

$$\hat{\mu}_X = \bar{x} \qquad \hat{\mu}_Y = \bar{y}$$

$$\hat{\sigma}_\Omega^2 = \frac{\sum\limits_{i=1}^{n}(x_i - \bar{x})^2 + \sum\limits_{j=1}^{m}(y_j - \bar{y})^2}{n + m}$$

If these estimates are substituted into Equation A8.1.4, the maximum value for $L(\Omega)$ simplifies to

$$L(\hat{\Omega}) = (e^{-1}/2\pi\hat{\sigma}_\Omega^2)^{(n+m)/2}$$

It follows, then, that the generalized likelihood ratio, λ, is equal to

$$\lambda = \frac{L(\hat{\omega})}{L(\hat{\Omega})} = \left(\frac{\hat{\sigma}_\Omega^2}{\hat{\sigma}^2}\right)^{(n+m)/2}$$

or, equivalently,

$$\lambda^{2/(n+m)} = \frac{\sum\limits_{i=1}^{n}(x_i - \bar{x})^2 + \sum\limits_{j=1}^{m}(y_j - \bar{y})^2}{\sum\limits_{i=1}^{n}\left[x_i - \left(\dfrac{n\bar{x} + m\bar{y}}{n + m}\right)\right]^2 + \sum\limits_{j=1}^{m}\left[y_j - \left(\dfrac{n\bar{x} + m\bar{y}}{n + m}\right)\right]^2}$$

Using the identity

$$\sum_{i=1}^{n} \left(x_i - \frac{n\bar{x} + m\bar{y}}{n+m} \right)^2 = \sum_{i=1}^{n} (x_i - \bar{x})^2 + \frac{m^2 n}{(n+m)^2} (\bar{x} - \bar{y})^2$$

we can write $\lambda^{2/(n+m)}$ as

$$\lambda^{2/(n+m)} = \frac{\displaystyle\sum_{i=1}^{n} (x_i - \bar{x})^2 + \sum_{j=1}^{m} (y_j - \bar{y})^2}{\displaystyle\sum_{i=1}^{n} (x_i - \bar{x})^2 + \sum_{j=1}^{m} (y_j - \bar{y})^2 + \frac{nm}{n+m} (\bar{x} - \bar{y})^2}$$

$$= \frac{1}{1 + \dfrac{(\bar{x} - \bar{y})^2}{\left[\displaystyle\sum_{i=1}^{n} (x_i - \bar{x})^2 + \sum_{j=1}^{m} (y_j - \bar{y})^2 \right] \left(\dfrac{1}{n} + \dfrac{1}{m} \right)}}$$

$$= \frac{n + m - 2}{n + m - 2 + \dfrac{(\bar{x} - \bar{y})^2}{s_p^2[(1/n) + (1/m)]}}$$

where s_p^2 is the pooled variance:

$$s_p^2 = \frac{1}{n+m-2} \left[\sum_{i=1}^{n} (x_i - \bar{x})^2 + \sum_{j=1}^{m} (y_j - \bar{y})^2 \right]$$

Therefore, in terms of the observed t ratio, $\lambda^{2/(n+m)}$ simplifies to

$$\lambda^{2/(n+m)} = \frac{n+m-2}{n+m-2+t^2} \tag{A8.1.5}$$

At this point the proof is almost complete. The generalized-likelihood-ratio criterion, rejecting $H_0: \mu_X = \mu_Y$ when $0 < \lambda \le \lambda^*$, is clearly equivalent to rejecting the null hypothesis when $0 < \lambda^{2/(n+m)} \le \lambda^{**}$. But both of these, from Equation A8.1.5, are the same as rejecting H_0 when t^2 is too large. Thus the decision rule in terms of t^2 is

Reject $H_0: \mu_X = \mu_Y$ in favor of $H_1: \mu_X \ne \mu_Y$ if $t^2 \ge t^{*2}$

Or, phrasing this in still another way, we should reject H_0 if either $t \ge t^*$ or $t \le -t^*$, where

$$P(-t^* < T < t^* \,|\, H_0: \mu_X = \mu_Y \text{ is true}) = 1 - \alpha$$

By Theorem 8.2.1, though, $T \sim T_{n+m-2}$, which makes $\pm t^* = \pm t_{\alpha/2,\, n+m-2}$, and the theorem is proved.

APPENDIX 8.2 POWER CALCULATIONS
FOR A TWO-SAMPLE t TEST

Power calculations for a two-sample t test proceed along the same lines established for the one-sample t test and described in Appendix 7.1. Basically, there are two different questions that can be answered: (1) Given n, m, and α, what is the probability that $H_0: \mu_X = \mu_Y$ will be rejected if, in fact, the X and Y distributions have shifted apart a distance Δ? (2) Given α, what are the smallest values of n and m for which the probability of making a Type II error is no larger than β—for some fixed $\mu_X - \mu_Y$? [In both these questions the location shift $(\Delta = \mu_X - \mu_Y)$ is usually expressed in terms of standard deviations (Δ/σ).] Figure A7.1.1 can be used to approximate both answers.

As an example of the first situation, imagine testing

$$H_0: \quad \mu_X = \mu_Y$$

versus

$$H_1: \quad \mu_X \neq \mu_Y$$

with $n = 13$, $m = 9$, and $\alpha = 0.01$. We might have reason to ask the following: if μ_X has shifted 1.5 standard deviations to the right of μ_Y $[\Delta/\sigma = (\mu_X - \mu_Y)/\sigma = 1.5]$, what is the probability that H_0 will be rejected? Recall that the abscissa in Figure A7.1.1 is scaled in terms of ϕ, where

$$\phi = \frac{\Delta}{\sigma_{\hat{\Delta}}} \left(\frac{1}{\sqrt{2}} \right)$$

and $\sigma_{\hat{\Delta}}$ is the standard deviation of the sample estimator for Δ—namely, $\bar{X} - \bar{Y}$. Therefore,

$$\sigma_{\hat{\Delta}} = \sigma \sqrt{\frac{1}{n} + \frac{1}{m}} = \sigma \sqrt{\frac{1}{13} + \frac{1}{9}}$$

and

$$\phi = \frac{\Delta}{\sigma \sqrt{1/13 + 1/9}} \frac{1}{\sqrt{2}}$$

$$= \frac{\Delta}{\sigma} \frac{1}{\sqrt{1/13 + 1/9}} \frac{1}{\sqrt{2}}$$

$$= (1.5) \left(\frac{1}{0.434} \right) \left(\frac{1}{1.414} \right)$$

$$= 2.4$$

With the combined sample size totaling 22, the estimator for $\sigma - S_p$—will have 20 degrees of freedom. The probability, then, of rejecting H_0 is gotten by entering Figure A7.1.1 with a ϕ of 2.4 and reading off $1 - \beta$ from the $\alpha = 0.01$ curve having $\nu = 20$ degrees of freedom:

$$1 - \beta = P\left(\text{reject } H_0 \left| \frac{\Delta}{\sigma} = 1.5 \right. \right) = 0.71$$

The second problem, choosing n and m to satisfy requirements imposed on α and β, is likely to be much more relevant to an experimenter than the first procedure. As a numerical illustration of this second problem, suppose it has been decided that the hypotheses to be tested are

$$H_0: \quad \mu_X = \mu_Y$$

versus

$$H_1: \quad \mu_X \neq \mu_Y$$

and that the level of significance should be 0.05. The question is, how large should n and m be?

To simplify matters, we will assume that n and m are to be equal. Finally, as a precision requirement, we will insist that the sample size be large enough to enable the test to reject H_0 at least 80% of the time if $|\mu_X - \mu_Y|/\sigma \geq 1.75$. Accordingly, ϕ reduces to

$$\phi = \frac{\Delta}{\sigma_{\hat{\Delta}}} \times \left(\frac{1}{\sqrt{2}}\right) = \frac{\Delta}{\sigma \sqrt{(2/n)}} \times \frac{1}{\sqrt{2}} = \frac{\sqrt{n}}{2} \frac{\Delta}{\sigma} = \frac{\sqrt{n}}{2} \times 1.75$$

$$= 0.875 \sqrt{n}.$$

Now, suppose that n were 9. Then $\phi = 0.875 \sqrt{9} = 2.625$ and $v = 9 + 9 - 2 = 16$. From Figure A7.1.1, the probability of rejecting H_0 under these circumstances (two samples of size 9 and $\Delta/\sigma = 1.75$) is approximately 0.94. This figure, however, considerably exceeds our power requirement of 0.80, implying that a smaller sample size would be adequate. So, suppose that n were 4. Then $\phi = 0.875 \sqrt{4} = 1.75$, $v = 6$, and $1 - \beta = 0.55$. But now the test would not be precise enough. Table A8.2.1 lists $1 - \beta$ for sample sizes ranging from $n = 4$ to $n = 9$.

TABLE A8.2.1 VALUES OF $1 - \beta$ AS A FUNCTION OF n

n	ϕ	v	$1 - \beta$
4	1.75	6	0.55
5	1.96	8	0.67
6	2.14	10	0.78
7	2.31	12	0.85
8	2.47	14	0.90
9	2.62	16	0.94

Notice that $1 - \beta$ exceeds 0.80 for the first time when $n = 7$. This means that the experimenter should take two samples of size 7.

Question A8.2.1 Construct a power curve for the experiment described in Question 8.2.3.

Question A8.2.2 An experiment is to be conducted to determine whether vampire bats prefer blood at room temperature or at body temperature. Equal numbers of bats are to be given access to drinking tubes attached to a supply of blood kept at one of the two temperatures. The response variable will be the amount of blood (in milliliters) that each bat drinks [see (16)]. The experimenter wants to test whether the average amounts of room-temperature and body-temperature blood consumed are equal (against a two-sided alternative that they are not). If α is going to be set at 0.05 and if the experimenter wants to have at least an 85% chance of rejecting H_0 when, in fact, $|\mu_X - \mu_Y|/\sigma \geq 1.50$, what is the minimum number of bats that should be put in each group?

9

Goodness-of-Fit Tests

Karl Pearson (1857–1936)

Called by some the founder of twentieth-century statistics, Pearson received his university education at Cambridge, concentrating on physics, philosophy, and law. He was called to the bar in 1881 but never practiced. In 1911 Pearson resigned his chair of applied mathematics and mechanics at University College, London, and became the first Galton Professor of Eugenics, as was Galton's wish. Together with Weldon, Pearson founded the prestigious journal *Biometrika* and served as its principal editor from 1901 until his death.

9.1 INTRODUCTION

The give and take between the mathematics of probability and the empiricism of statistics should be, by now, a theme comfortably familiar. Time and time again we have seen repeated measurements, no matter what their source, exhibiting a regularity of pattern that can be well approximated by one or more of the handful of probability functions introduced in Chapter 4. Until now, all the inferences resulting from this interfacing have been parameter specific, a fact to which the many hypothesis tests about means, variances, and binomial proportions paraded forth in Chapters 6, 7, and 8 bear ample testimony. Still, there are other situations where the basic *form* of $f_Y(y)$, rather than the value of its parameters, is the most important question at issue. These situations are the focus of Chapter 9.

A geneticist, for example, might want to know whether the inheritance of a certain set of traits follows the same set of ratios as those prescribed by Mendelian theory. The objective of a psychologist, on the other hand, might be to confirm or refute a newly proposed model for cognitive serial learning. Probably the most habitual users of inference procedures directed at the entire $f_Y(y)$, though, are statisticians themselves: as a prelude to doing any sort of hypothesis test or confidence interval, an attempt should be made, sample size permitting, to verify that the data are, indeed, representative of whatever distribution that procedure presumes. Usually, this will mean testing to see whether or not the Y_i's are normal.

In general, any procedure that seeks to determine whether a set of data could reasonably have originated from some given probability distribution, or *class* of probability distributions, is called a *goodness-of-fit* test. The principle behind the particular goodness-of-fit test we will look at is very straightforward: first the observed Y_i's are grouped, more or less arbitrarily, into k classes; then each class's "expected" occupancy is calculated on the basis of the presumed model. If it should happen that the set of observed and expected frequencies show considerable disagreement (as measured by the appropriate statistic), our conclusion will be that the supposed $f_Y(y)$ was incorrect.

In practice, the method has two variants, depending on the specificity of the null hypothesis. Section 9.3 describes the version to use when both the form of the presumed $f_Y(y)$ and the values of all its parameters are given. The more typical situation occurs when $f_Y(y)$ is designated but its parameters need to be estimated; this is taken up in Section 9.4.

A somewhat different application of this same idea is the subject of Section 9.5. There the null hypothesis is one of *independence*: that $f_{X,Y}(x, y) = f_X(x) \cdot f_Y(y)$. Such tests are extremely practical—to the extent that they may be the most often used inference procedure in the applied statistician's repertoire.

9.2 THE MULTINOMIAL DISTRIBUTION

Their diversity notwithstanding, many goodness-of-fit tests are based on essentially the same statistic, one whose asymptotic distribution is a chi square. The underlying structure of that statistic, though, derives from the *multinomial distribution*, a k-

variate extension of the familiar binomial. In this section we define the multinomial and state those of its properties that bear directly on the problem of goodness-of-fit testing.

Given a series of n independent Bernoulli trials, each with success probability p, we know that the pdf for Y, the total number of successes, is

$$P(Y = y) = f_Y(y) = \binom{n}{y} p^y (1 - p)^{n-y} \qquad y = 0, 1, \ldots, n \qquad (9.2.1)$$

One of the obvious ways to generalize Equation 9.2.1 is to consider situations where at each trial k outcomes can occur, rather than just two. This means that Y will be allowed to take on any one of the values $y_1, y_2, \ldots, y_k$ with respective probabilities $p_1, p_2, \ldots, p_k$, the latter satisfying the constraint

$$\sum_{i=1}^{k} p_i = 1$$

Notice that if n such trials are observed, the resulting distribution of Y-values can be summarized by defining a new set of random variables, $X_1, X_2, \ldots, X_k$, where

$$X_i = \text{number of times that } Y = y_i \qquad i = 1, 2, \ldots, k$$

Of course, $\sum_{i=1}^{k} X_i = n$.

The vector $(X_1, X_2, \ldots, X_k)$ is a discrete multivariate random variable—its joint pdf is the multinomial we are seeking:

$$P(X_1 = x_1, X_2 = x_2, \ldots, X_k = x_k) = f_{X_1, X_2, \ldots, X_k}(x_1, x_2, \ldots, x_k)$$

$$= \frac{n!}{x_1! \, x_2! \cdots x_k!} \, p_1^{x_1} p_2^{x_2} \cdots p_k^{x_k} \qquad (9.2.2)$$

$$x_i = 0, \ldots, n; \quad i = 1, \ldots, k; \quad \sum_{i=1}^{k} x_i = n$$

It should be clear that Equation 9.2.2 can be obtained by appealing to the same arguments that gave rise to the binomial. For example, the combinatorial term in the multinomial is a direct extension of Theorem 2.9.2: the number of ways to arrange n items, of which x_1 are of one type, x_2 of a second type, $\ldots$, and x_k of a kth type, is $n!/(x_1! \, x_2! \cdots x_k!)$. When $k = 2$, and the two types are denoted simply as successes and failures, $n!/(x_1! \, x_2! \cdots x_k!)$ reduces to the familiar

$$\binom{n}{y} = \frac{n!}{y! \, (n - y)!}$$

Theorem 9.2.1 states a not-unexpected relationship between the binomial and the multinomial.

Theorem 9.2.1. Let the vector $(X_1, X_2, \ldots, X_k)$ be a multinomial random variable with parameters $n, p_1, \ldots, p_k$. Then the marginal pdf of X_i, $i = 1, 2, \ldots, k$, is the binomial with parameters n and p_i.

Proof. We will verify the theorem for $k = 3$. Let (X, Y, Z) have a *tri-nomial* distribution with parameters n, p_X, p_Y, and p_Z. What is to be shown is that

$$f_X(x) = \binom{n}{x} p_X^x (1 - p_X)^{n-x}$$

By definition,

$$f_X(x) = \sum_y \sum_z \frac{n!}{x! \, y! \, z!} \, p_X^x \, p_Y^y \, p_Z^z$$

$$y = 0, 1, \ldots, n - x; \quad z = 0, 1, \ldots, n - x; \quad y + z = n - x$$

$$= \sum_{y=0}^{n-x} \frac{n!}{x! \, y! \, (n - x - y)!} \, p_X^x \, p_Y^y (1 - p_X - p_Y)^{n-x-y} \tag{9.2.3}$$

Following the procedure of Example 3.8.4, we will first factor the "answer" out of the right-hand side of Equation 9.2.3 and then confirm that what remains sums to 1. The first step gives

$$f_X(x) = \frac{n!}{x! \, (n-x)!} \, p_X^x (1-p_X)^{n-x} \sum_{y=0}^{n-x} \frac{(n-x)!}{y! \, (n-x-y)!} \, p_Y^y \, \frac{(1-p_X-p_Y)^{n-x-y}}{(1-p_X)^{n-x}}$$

$$= \frac{n!}{x! \, (n-x)!} \, p_X^x (1-p_X)^{n-x} \sum_{y=0}^{n-x} \frac{(n-x)!}{y! \, (n-x-y)!}$$

$$\times \left(\frac{p_Y}{1 - p_X} \right)^y \left(1 - \frac{p_Y}{1 - p_X} \right)^{n-x-y}$$

Then, by inspection, note that the value of the sum appearing in the right-hand side of $f_X(x)$ is, indeed, 1, being the summation over all the values of a binomial pdf whose parameters are $n - x$ and $[p_Y/(1 - p_X)]$. This proves the theorem for $k = 3$.

It follows immediately from Theorem 9.2.1, in the general case, that $E(X_i) = np_i$ and $\text{Var}(X_i) = np_i(1 - p_i)$. Also it can be shown, although it is not a consequence of the theorem, that the maximum-likelihood estimates for the p_i's are the direct analogs of their binomial counterparts: namely, $\hat{p}_i = x_i/n$ (recall Question 5.8.1).

One final property of the multinomial deserves mention. Any random variable, discrete or continuous, can be "reduced" to a multinomial by partitioning its range into a set of k nonoverlapping intervals. For example, suppose that Y is a continuous random variable with pdf $f_Y(y)$ defined over the entire real line. Take as the k intervals the set, $(-\infty, a_1), [a_1, a_2), \ldots, [a_{k-1}, \infty)$, and let

$$p_i = \int_{a_{i-1}}^{a_i} f_Y(y) \, dy \qquad i = 1, 2, \ldots, k; \quad a_0 = -\infty; \quad a_k = \infty \tag{9.2.4}$$

Then, if n measurements are taken on Y, and if X_i is the number of Y's falling into the ith interval, the pdf for the vector $(X_1, X_2, \ldots, X_k)$ will be the multinomial with parameters n and the p_i's of Equation 9.2.4.

Question 9.2.1 An army enlistment officer categorizes potential recruits by IQ into three classes—class I: <90, class II: 90–110, and class III: >110. Given that the IQ distribution of the population from which the recruits are drawn is $N(100, (16)^2)$, compute the probability that of seven enlistees, two will belong to class I, four to class II, and one to class III.

Question 9.2.2 Suppose that five randomly selected students take the quantitative portion of the college boards, scores on which are normally distributed with a mean of 500 and a standard deviation of 100. What is the probability that two students earn scores below 400, two get scores between 400 and 650, inclusive, and one scores above 650?

Question 9.2.3 Recall the pipeline-missile problem described in Example 2.5.2. Suppose that six missiles are fired at the pipeline. What is the probability that two land within 20 feet to the left of the pipeline and four land within 20 feet to the right?

Question 9.2.4 Fifty observations are drawn from the pdf

$$f_Y(y) = 6y(1 - y) \qquad 0 < y < 1$$

Let X_i be the number of observations lying in the interval $((i - 1)/4, i/4)$, $i = 1, 2, 3, 4$.
(a) Write down a formula for $f_{X_1, X_2, X_3, X_4}(10, 15, 15, 10)$.
(b) Find Var (X_3).

Question 9.2.5 Suppose that a loaded die is tossed 12 times, where

$$p_i = P(i \text{ spots appear}) = ki \qquad i = 1, 2, \ldots, 6$$

Let X_i, $i = 1, 2, \ldots, 6$, denote the number of times the face with i spots appears. Find the probability that all the X_i's equal 2.

Question 9.2.6 Let $(x_1, x_2, \ldots, x_k)$ be the vector of sample observations representing a multinomial random variable with parameters n, p_1, p_2, $\ldots$, and p_k. Find the maximum-likelihood estimates for the p_i's.

Question 9.2.7 Let the vector of random variables (X_1, X_2, X_3) have the trinomial pdf with parameters n, p_1, p_2, and $p_3 = 1 - p_1 - p_2$. That is,

$$P(X_1 = x_1, X_2 = x_2, X_3 = x_3) = \frac{n!}{x_1! x_2! x_3!} p_1^{x_1} p_2^{x_2} p_3^{x_3}$$

$$x_i = 0, 1, \ldots, n; \ i = 1, 2, 3; \ x_1 + x_2 + x_3 = n$$

By definition, the moment-generating function for (X_1, X_2, X_3) is given by

$$M_{X_1, X_2, X_3}(t_1, t_2, t_3) = E(e^{t_1 X_1 + t_2 X_2 + t_3 X_3})$$

Show that

$$M_{X_1, X_2, X_3}(t_1, t_2, t_3) = (p_1 e^{t_1} + p_2 e^{t_2} + p_3 e^{t_3})^n$$

Question 9.2.8 If $M_{X_1, X_2, X_3}(t_1, t_2, t_3)$ is the moment-generating function for (X_1, X_2, X_3), then $M_{X_1, X_2, X_3}(t_1, 0, 0)$, $M_{X_1, X_2, X_3}(0, t_2, 0)$, and $M_{X_1, X_2, X_3}(0, 0, t_3)$ are the moment-generating functions for the marginal pdf's of X_1, X_2, and X_3, respectively. Use this fact, together with the result of Question 9.2.7, to verify the statement of Theorem 9.2.1.

Question 9.2.9 Four chips are allocated at random to three urns. Let X_i be the number of chips placed in the ith urn. Find $E(X_1^2 X_3)$.

9.3 GOODNESS-OF-FIT TESTS: ALL PARAMETERS KNOWN

The simplest version of a goodness-of-fit test arises when an experimenter is able to specify completely the probability model from which the sample data are alleged to have come. It might be supposed, for example, that the Y_i's are being generated by a Poisson pdf with parameter λ equal to 6.3, or by a normal distribution with $\mu = 500$ and $\sigma = 100$. For cases such as these, the hypotheses to be tested will be written

$$H_0: \quad f_Y(y) = f_0(y)$$

versus

$$H_1: \quad f_Y(y) \neq f_0(y)$$

where $f_Y(y)$ and $f_0(y)$ are the true and the presumed pdf's, respectively. In some situations it will prove more convenient to characterize the model in terms of the probabilities associated with the k nonoverlapping intervals described at the end of Section 9.2. Then the problem takes the form

$$H_0: \quad p_1 = p_{1_0}, p_2 = p_{2_0}, \dots, p_k = p_{k_0}$$

versus

$$H_1: \quad p_i \neq p_{i_0}, \quad \text{for at least one } i$$

The first statistic for testing either of these sets of hypotheses was proposed by Karl Pearson in 1900. Couched in the language of the multinomial, Pearson's method requires that the n observations be grouped into k intervals (or k "classes" if Y is discrete) and that $p_{1_0}, p_{2_0}, \dots, p_{k_0}$ [or $f_0(y)$] be specified. Theorem 9.3.1 defines Pearson's statistic, gives its asymptotic distribution, and locates its critical region.

Theorem 9.3.1. Let $(X_1, X_2, \dots, X_k)$ be a multinomial random variable with parameters $n, p_1, p_2, \dots, p_k$. Then:

(a) The random variable

$$C = \sum_{i=1}^{k} \frac{(X_i - np_i)^2}{np_i}$$

has approximately a χ^2 distribution with $k - 1$ degrees of freedom. (For the approximation to be adequate, the k classes should be defined so that np_i is greater than or equal to 5, for all i.)

(b) At the α level of significance, $H_0: p_1 = p_{1_0}, \ldots, p_k = p_{k_0}$ is rejected in favor of H_1: at least one $p_i \neq p_{i_0}$ if

$$c = \sum_{i=1}^{k} \frac{(x_i - np_{i_0})^2}{np_{i_0}} \geq \chi^2_{1-\alpha, k-1}$$

Proof. A formal proof lies beyond the scope of this text. We will present a heuristic argument for part (a) for the special case $k = 2$. That $c \geq \chi^2_{1-\alpha, k-1}$ is a reasonable critical region is evident by inspection. If agreement between the actual data and the presumed model were perfect, each x_i would equal its corresponding np_{i_0} (recall Theorem 9.2.1) and c would be 0 (and, of course, H_0 should be accepted). Conversely, as the discrepancies between the observed and expected frequencies proliferate, and c increases, the credibility of H_0 should surely diminish. On intuitive grounds, then, a test rejecting H_0 when c is large is eminently justifiable.

Now, returning to part (a), suppose $k = 2$. Then

$$C = \frac{(X_1 - np_1)^2}{np_1} + \frac{(X_2 - np_2)^2}{np_2}$$

$$= \frac{(X_1 - np_1)^2}{np_1} + \frac{[n - X_1 - n(1 - p_1)]^2}{n(1 - p_1)}$$

$$= \frac{(X_1 - np_1)^2(1 - p_1) + (-X_1 + np_1)^2 p_1}{np_1(1 - p_1)}$$

$$= \frac{(X_1 - np_1)^2}{np_1(1 - p_1)}$$

From Theorem 9.2.1, $E(X_1) = np_1$ and $\mathrm{Var}\,(X_1) = np_1(1 - p_1)$, the two implying that C can be written

$$C = \left[\frac{X_1 - E(X_1)}{\sqrt{\mathrm{Var}\,(X_1)}} \right]^2$$

By Theorem 4.3.1, then, C is the square of a variable that is asymptotically $N(0, 1)$, and the statement of part (a) follows (for $k = 2$) from Theorem 7.5.2. [A proof of the general statement can be accomplished by showing that the limit of the moment-generating function for C—as n goes to ∞—is the moment-generating function for a χ^2_{k-1} random variable. See (57) for details.]

Comment

Although Pearson formulated his statistic before any general theories of hypothesis testing had been developed, it has since been shown that C is, in fact, asymptotically equivalent to the generalized-likelihood-ratio test of H_0: $p_1 = p_{1_0}, p_2 = p_{2_0}, \ldots, p_k = p_{k_0}$.

Inhabiting many tropical waters is a small (<1 mm) crustacean, *Ceriodaphnia cornuta*, that occurs in two distinct morphological forms: one has a series of "horns" protruding from its exoskeleton, while the other is more rounded (Figure 9.3.1). Zaret (190) describes an experiment to test whether either variant is more conducive than the other to the survival of the species, in terms of its likelihood of being eaten.

Unhorned   Horned   **Figure 9.3.1**   Forms of *C. cornuta*.

A large number of *C. cornuta* were introduced into a holding tank in a 3-to-1 ratio—three of the unhorned variety were added to every one with horns. Also present in the tank was a natural predator of *C. cornuta*, a small (6-cm) fish, *Melaniris chagresi*. After approximately one hour, long enough for the predator to have completed its feeding, the fish was sacrificed and the contents of its stomach examined. Among the 44 crustacean casualties, the unhorned-to-horned ratio was 40 to 4. Can it be concluded from this that there is a *true* differential predation rate between the two polymorphs?

Here, the two *natural* classes for the response variable are "unhorned" and "horned," and under the null hypothesis that morphology has no effect on survival, it would follow that the probability of either form's being eaten should be proportional to the numbers of each kind available. If $p_1 = P$ (unhorned *C. cornuta* is eaten) and $p_2 = P$ (horned *C. cornuta* is eaten), the experimenter's objective reduces to a test of

$$H_0: \quad p_1 = \frac{3}{4}, \, p_2 = \frac{1}{4}$$

versus

$$H_1: \quad p_1 \neq \frac{3}{4}, \, p_2 \neq \frac{1}{4}$$

We will let $\alpha = 0.05$.

Since $k = 2$, the behavior of C will be approximated by a χ^2_1 distribution, for which the 0.05 critical value is 3.84 (see Figure 9.3.2). Substituting the values for the x_i's and p_{i_0}'s into the test statistic gives a c value of 5.93:

$$c = \frac{[40 - 44(3/4)]^2}{44(3/4)} + \frac{[4 - 44(1/4)]^2}{44(1/4)}$$

$$= 5.93$$

Our conclusion, then, is to *reject H_0*—it would appear that morphology *does* have an effect on *C. cornuta*'s chances of being eaten, the unhorned variety being significantly tastier!

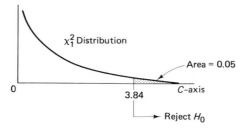

Figure 9.3.2 X_1^2 distribution.

The area under the X_1^2 Distribution curve shows Area = 0.05 at 3.84 on the C-axis, with a note to Reject H_0 beyond 3.84.

Comment

The final conclusion reached in this experiment was not exactly what the preceding analysis suggested. Using similar goodness-of-fit arguments, Zaret was able to show that the actual reason for the disparity in predation rates was not the presence or absence of horns but, rather, the enlarged eyespot characteristic of the latter. It is this feature that renders the otherwise nearly transparent *C. cornuta* more visible—and, as a result, more edible.

CASE STUDY 9.3.2

Examples of goodness-of-fit testing are quite common in genetics, where often the easiest way to support a theory about the inheritance of a particular trait is to perform a number of crosses and show (via a χ^2 test) that the appearance of the offspring is, in fact, what was predicted by the theory. The following study is a case in point.

The feathers of a frizzle chicken occur in three variations (or *phenotypes*)—extreme frizzle, mild frizzle, and normal. It has been suggested that two alleles (F and f) control frizzle and that F and f interact according to a phenomenon known as incomplete dominance. What the latter predicts is that progeny whose genetic complement is (F, F) will be extreme frizzles, those having an (F, f) genotype will be mild frizzles, while the (f, f) offspring will be normal.

If two Ff hybrids are crossed, and if the two alleles recombine randomly, the ratio of extreme frizzles to mild frizzles to normals should, of course, be 1 : 2 : 1. Table 9.3.1 shows the actual phenotypes produced in 93 such crossings (158).

TABLE 9.3.1 OFFSPRING OF Ff CROSSINGS

Phenotype	Observed Frequency
Extreme frizzle	23
Mild frizzle	50
Normal	20
	$\overline{93}$

Let $p_1 = P$ (offspring is extreme frizzle), $p_2 = P$ (offspring is mild frizzle), and $p_3 = P$ (offspring is normal). Then the hypotheses to be tested can be written

$$H_0: \quad p_1 = \frac{1}{4}, p_2 = \frac{1}{2}, p_3 = \frac{1}{4}$$

versus

$$H_1: \quad \text{at least one } p_i \neq p_{i_0}$$

If we elect 0.05 to be the level of significance, the critical value is 5.991, the 95th percentile of the χ_2^2 distribution. The value of c, though, is much less than that:

$$c = \frac{[23 - 93(1/4)]^2}{93(1/4)} + \frac{[50 - 93(1/2)]^2}{93(1/2)} + \frac{[20 - 93(1/4)]^2}{93(1/4)}$$

$$= 0.71$$

Thus, the assumptions that F and f recombine randomly and that the extent of frizzle depends on whether the progeny's genotype is (F, F), (F, f), or (f, f) are, together, supported by the data.

Question 9.3.1 Verify that the common belief in the propensity of babies to choose an inconvenient hour for birth has some basis in observation. A maternity hospital reported that out of one year's total of 2650 births, some 494 occurred between midnight and 4 A.M. (160). Use the goodness-of-fit test to show that these data are not what we should expect if births are assumed to occur uniformly in all time periods. Let $\alpha = 0.05$.

Question 9.3.2 One hundred samples of size 2 are drawn from an urn containing six red chips and four white chips. (Each selection is made without replacement.) Test the adequacy of the hypergeometric model if zero whites were obtained 35 times; one white, 55 times; and two whites, 10 times. Use the 0.10 decision rule.

Question 9.3.3 A sociologist interviews 100 families where the husband and wife, early in their marriage, decided to keep having children until they had their first girl. (All 100 couples did eventually have a female child.) Let X_i denote the total number of children in the ith family. The distribution of the x_i's is shown below.

Number of Children, x_i	Number of Families
1	55
2	19
3	12
4	8
5	3
6	3

Assume that p, the probability that any given child is a girl, is $\frac{1}{2}$. Do a goodness-of-fit test to see whether the distribution of the x_i's can be adequately described by a geometric pdf. Let $\alpha = 0.05$.

Question 9.3.4 Consider again the results of the sampling described in Question 9.3.2.

Number of White Chips, x_i	Number of Samples
0	35
1	55
2	10

Suppose, however, that we did not know whether the samples had been drawn with replacement or without replacement. Test whether sampling *with* replacement is a reasonable model. Let $\alpha = 0.05$.

Question 9.3.5 Records kept at an eastern racetrack showed the following distribution of winners as a function of their starting-post position. All 144 races were run with a full field of eight horses.

Starting post	1	2	3	4	5	6	7	8
Number of winners	32	21	19	20	16	11	14	11

Test an appropriate goodness-of-fit hypothesis. Let $\alpha = 0.05$.

Question 9.3.6 It was stated in Case Study 4.3.3 that the mean (μ) and the standard deviation (σ) of pregnancy durations are 266 days and 16 days, respectively. Accepting those as the true parameter values, test whether the additional assumption that pregnancy durations are normally distributed is supported by the following data.

70 PREGNANCY DURATIONS
(COUNTY GENERAL HOSPITAL, 1978)

251	264	234	283	226	244	269	241	276	274
263	243	254	276	241	232	260	248	284	253
265	235	259	279	256	256	254	256	250	269
240	261	263	262	259	230	268	284	259	261
268	268	264	271	263	259	294	259	263	278
267	293	247	244	250	266	286	263	274	253
281	286	266	249	255	233	245	266	265	264

Let $\alpha = 0.10$ be the level of significance. Use "220–229," "230–239," and so on, as classes.

Question 9.3.7 There have been a number of reports in the medical literature that season of birth and incidence of schizophrenia may be related, with a higher proportion of schizophrenics being born during the early months of the year. A recent study (65) following up on this hypothesis looked at 5139 persons born in England or Wales during the years 1921–1955 who were first admitted during the period 1970–1971 to a psychiatric ward in either England or Wales with a diagnosis of schizophrenia. Of these 5139, 1383 were born in the first quarter of a year. Based on census figures in the two countries, the expected number of persons (out of a random 5139) who would be born in the first quarter is 1292.1. Do an appropriate χ^2 test. Let $\alpha = 0.05$.

Question 9.3.8 Two traits that have been widely studied in tomato plants are *height* ("tall" versus "dwarf") and *leaf type* ("cut" versus "potato"). "Tall" and "cut" are dominant. When a homozygous "tall, cut" is crossed with a "dwarf, potato" the resulting progeny is called a dihybrid. (Its phenotype, of course, will be "tall" and "cut.") When dihybrids are crossed, the phenotypes "tall, cut," "tall, potato," "dwarf, cut," and "dwarf, potato" should appear in a $9 : 3 : 3 : 1$ ratio, provided the alleles governing the two traits segregate independently. In one experiment done with these two traits a total of 1611 progeny of dihybrid crosses were categorized by phenotype (158).

Phenotype	Frequency
Tall, cut	926
Tall, potato	288
Dwarf, cut	293
Dwarf, potato	104

Test the appropriateness of the $9 : 3 : 3 : 1$ model. Let $\alpha = 0.01$.

Question 9.3.9 Is it a reasonable hypothesis that the random sample of size 40 given below came from the beta pdf

$$f_Y(y) = 6y(1 - y) \qquad 0 < y < 1?$$

0.18	0.06	0.27	0.58	0.98
0.55	0.24	0.58	0.97	0.36
0.48	0.11	0.59	0.15	0.53
0.29	0.46	0.21	0.39	0.89
0.34	0.09	0.64	0.52	0.64
0.71	0.56	0.48	0.44	0.40
0.80	0.83	0.02	0.10	0.51
0.43	0.14	0.74	0.75	0.22

Start by dividing the data into five classes. Draw the corresponding histogram and the presumed model on the same graph. Carry out the hypothesis test at the $\alpha = 0.05$ level.

Question 9.3.10 Recall Example 2.5.1, where it was claimed that in a certain state the length of time, Y, that a person convicted of grand theft auto actually spends in jail is described by the pdf

$$f_Y(y) = \frac{1}{9} y^2 \qquad 0 \le y \le 3$$

As part of a general investigation into the inequities of judicial sentences, a district attorney reviews the records of 50 persons convicted of grand theft auto. Of those 50, 8 served less than one year in prison, 16 served between one and two years, and 26 between two and three years. Are these data compatible with the presumed $f_Y(y)$? Let $\alpha = 0.05$.

Question 9.3.11 Listed in the table are the lengths of the World Series for the 50 years from 1926 to 1975. Test whether these data are compatible with the model that each World Series game is an independent Bernoulli trial with

$$p = P(\text{AL wins}) = P(\text{NL wins}) = \frac{1}{2}$$

Number of Games	Number of Years
4	9
5	11
6	8
7	22
	50

Let $\alpha = 0.10$.

9.4 GOODNESS-OF-FIT TESTS: PARAMETERS UNKNOWN

More common than the sort of problems described in Section 9.3 are situations where the experimeter has reason to believe the response variable follows some particular *family* of pdf's—say, the gamma or the normal—but has little or no prior information to suggest what values should be assigned to the model's parameters. In cases such as these, we will carry out the goodness-of-fit test by first estimating all unknown parameters, preferably with the method of maximum likelihood, and then computing c_1, the obvious analog of Pearson's c:

$$c_1 = \sum_{i=1}^{k} \frac{(x_i - n\hat{p}_{i_0})^2}{n\hat{p}_{i_0}}$$

We pay a price, though, for having to rely on the data to fill in details about the presumed model: each parameter estimated reduces by one the number of degrees of freedom associated with the χ^2 distribution approximating the sampling distribution of C_1—and as the number of degrees of freedom decreases, so does the power of the test.

Theorem 9.4.1. Suppose that $f_0(y)$ is a pdf having r unknown parameters. To test $H_0: f_Y(y) = f_0(y)$ versus $H_1: f_Y(y) \neq f_0(y)$, reject the null hypothesis if

$$c_1 = \sum_{i=1}^{k} \frac{(x_i - n\hat{p}_{i_0})^2}{n\hat{p}_{i_0}} \geq \chi^2_{1-\alpha, k-1-r}$$

where $(n\hat{p}_{1_0}, \ldots, n\hat{p}_{k_0})$ are the *estimated* expected values under H_0 of the multinomial random variable $(X_1, \ldots, X_k)$. The $n\hat{p}_{i_0}$'s are computed by replacing the unknown parameters in $f_0(y)$ with their MLEs. (To ensure that C_1 is adequately described by the χ^2 distribution with $k - 1 - r$ degrees of freedom, the classes should be defined so that $n\hat{p}_{i_0} \geq 5$ for all i.)

Example 9.4.1

A sociologist is studying various aspects of the personal lives of preeminent nineteenth-century scholars. A total of 120 of the scholars in her sample had families consisting of two children. The distribution of the number of boys in those families is summarized in Table 9.4.1. Can it be concluded that Y, the number of boys in two-child families of preeminent scholars, is binomially distributed? Let $\alpha = 0.05$.

TABLE 9.4.1

Number of boys	0	1	2
Number of families	24	64	32

Here $f_0(y)$ has the *form*

$$f_0(y) = \binom{2}{y} p^y (1 - p)^{2-y} \qquad y = 0, 1, 2 \tag{9.4.1}$$

where $p = P$ (child is a boy) is an unknown parameter. The associated multinomial random variable referred to in Theorem 9.4.1 has three components:

$$X_1 = \text{number of families for which } y = 0$$
$$X_2 = \text{number of families for which } y = 1$$
$$X_3 = \text{number of families for which } y = 2$$

The expected values for the X_i's derive from Theorem 9.2.1:

$$E(X_1) = np_1 = 120 \cdot P(Y = 0) = 120 \cdot \binom{2}{0} p^0 (1 - p)^2$$

$$E(X_2) = np_2 = 120 \cdot P(Y = 1) = 120 \cdot \binom{2}{1} p^1 (1 - p)^1 \tag{9.4.2}$$

$$E(X_3) = np_3 = 120 \cdot P(Y = 2) = 120 \cdot \binom{2}{2} p^2 (1 - p)^0$$

Note that since the p in Equation 9.4.1 is unknown, the $E(X_i)$'s in Equation 9.4.2 cannot be determined. Instead, we need to estimate the expected values by first estimating p.

Recall from Example 5.4.2 that the maximum-likelihood estimate for the parameter in a binomial pdf is the ratio of the observed number of successes divided by the total number of trials:

$$\hat{p} = \text{MLE for } p = \frac{\text{observed number of successes}}{\text{total number of trials}}$$

For the data given in Table 9.4.1, the total number of trials is 240 (why?) and

$$\text{observed number of successes} = 24(0) + 64(1) + 32(2)$$
$$= 128$$

Therefore, $\hat{p} = 128/240 = 0.533$.

Substituting p into Equation 9.4.2 gives the $n\hat{p}_{i_0}$'s:

$$120 \cdot \hat{p}_{1_0} = 120 \cdot \binom{2}{0}(0.533)^0(1 - 0.533)^2$$
$$= 26.2$$

$$120 \cdot \hat{p}_{2_0} = 120 \cdot \binom{2}{1}(0.533)^1(1 - 0.533)^1$$
$$= 59.7$$

$$120 \cdot \hat{p}_{3_0} = 120 \cdot \binom{2}{2}(0.533)^2(1 - 0.533)^0$$
$$= 34.1$$

To assess the significance of the difference between the x_i's and their estimated expected values [(24, 64, 32) versus (26.2, 59.7, 34.1)] we compute c_1:

$$c_1 = \frac{(24 - 26.2)^2}{26.2} + \frac{(64 - 59.7)^2}{59.7} + \frac{(32 - 34.1)^2}{34.1}$$
$$= 0.62$$

By Theorem 9.4.1, we should reject the null hypothesis that Y is binomially distributed if

$$c_1 \geq \chi^2_{0.95,\, 3-1-1} = \chi^2_{0.95,\, 1} = 3.84 \tag{9.4.3}$$

Since Inequality 9.4.3 is *not* satisfied, our conclusion is to *accept* H_0.

CASE STUDY 9.4.1

Theoretical geographers often need to analyze spatial patterns; the context may be anything from the distribution of oak trees in a southern Appalachian forest to the ethnic mix populating a large Midwestern city. Knowing the nature of these patterns, and being able to characterize them mathematically, can help scientists understand them better.

Randomness, as might be expected, is one of the "patterns" most frequently encountered. But randomness is what occurs when the underlying random variable is Poisson (recall the discussion after Case Study 4.2.2). This means that investigating whether or not a particular pattern is random can be accomplished by doing a goodness-of-fit test with the model specified by the null hypothesis having the form $f_0(y) = e^{-\lambda}\lambda^y/y!$, $y = 0, 1, \ldots$.

West of Tokyo lies a large alluvial plain, dotted by a network of farming villages. Matui (100) analyzed the positioning of the 911 houses making up one of those villages. The area studied was a rectangle, 3 km by 4 km. A grid was superimposed over a map of the village, dividing its 12 square kilometers into 1200 plots, each 100 meters on a side. The numbers of houses located on each of those plots are displayed in the 30 × 40 matrix shown in Figure 9.4.1.

```
2 2 2 1 0 1 0 0 1 2 0 0 0 1 2 0 1 0 1 2 2 0 1 1 2 0 1 1 1 1 2 1 1 2 0 1 2 0 1 2 0 2
0 2 0 1 2 0 1 1 1 2 2 0 1 1 0 0 0 1 0 1 0 2 2 0 1 2 2 1 2 1 0 0 1 0 1 0 2 0 1 2
1 0 1 1 0 0 1 0 1 1 1 0 1 0 1 1 0 1 2 0 2 0 0 1 3 0 1 2 1 0 2 1 1 2 0 0 1 0 2 2
0 1 1 1 0 2 0 1 2 0 0 0 2 2 0 0 0 1 0 0 1 2 0 0 0 1 0 0 0 1 0 9 0 0 0 1 1 1 1 1
1 2 0 0 0 0 0 0 0 0 1 0 2 0 2 2 0 1 2 1 0 1 1 1 0 3 0 1 2 0 1 1 1 1 0 0 1 0 3 1
1 3 1 0 1 0 1 0 0 0 0 0 2 2 0 2 0 0 1 0 0 1 0 0 0 0 1 2 1 1 1 2 1 0 2 1 3 1 1 1
0 1 0 0 0 1 0 1 0 1 2 0 1 3 1 1 4 1 3 1 0 1 1 0 0 0 0 0 0 2 2 2 0 1 2 0 3 0 1
0 0 1 0 1 0 0 1 0 0 1 3 0 0 1 0 0 1 0 0 1 0 2 2 0 2 0 0 1 2 1 2 2 0 0 1 1 0 0 1
0 1 1 0 1 1 0 1 1 3 1 1 3 0 1 0 2 0 1 0 0 0 1 3 3 2 0 0 0 0 1 0 1 0 1 0 0 0 1 0
0 0 0 0 0 1 1 2 0 0 1 5 2 0 0 0 0 2 0 0 2 1 0 1 0 0 2 0 0 0 1 0 0 1 0 0 0 1 2 0
0 2 0 0 1 1 1 0 1 1 1 0 2 1 4 2 1 0 1 2 2 0 1 1 2 1 0 0 0 0 1 2 2 0 0 0 0 0 0 0
0 0 0 1 1 0 1 0 0 0 1 2 2 2 0 0 0 1 0 1 3 1 2 0 0 0 0 2 1 2 0 0 0 2 0 1 1 1
0 1 0 0 1 2 0 0 0 0 0 0 1 1 0 1 1 1 1 2 1 1 1 3 0 1 0 1 1 0 1 4 1 1 2 0 1 0 2
0 0 0 1 1 1 1 0 1 1 0 0 0 0 1 2 0 1 1 1 1 3 0 2 1 0 0 0 0 2 0 0 0 3 0 2 0 1 1 2
0 1 1 0 0 0 1 1 2 0 0 1 0 0 1 0 0 2 0 0 0 1 1 0 0 0 1 1 1 0 0 0 0 2 0 0 2 1 0 0
3 4 1 1 0 3 1 0 0 0 2 0 0 0 1 0 1 2 1 0 0 1 4 1 0 0 2 2 0 0 0 1 0 1 1 1 0 4 4 0
0 0 1 0 0 1 1 1 1 1 0 0 1 0 2 0 3 2 0 2 2 3 1 0 0 1 1 0 1 3 0 0 1 1 0 1 1 1 0
1 1 0 1 0 1 0 0 2 1 0 0 2 2 0 0 2 1 5 2 0 0 0 0 0 0 0 1 0 0 1 2 2 0 0 2 1 0 1
0 3 0 1 0 0 0 2 0 0 0 2 0 0 0 0 0 1 0 2 0 0 0 0 1 1 0 0 2 0 0 0 0 0 0 1 3 0 0 1
0 1 1 0 2 0 1 0 0 0 0 0 1 1 0 0 1 0 1 0 0 0 1 1 2 1 1 0 0 0 0 1 1 0 1 0 0 2 1 2
1 0 0 0 1 1 0 0 1 1 1 0 0 2 1 0 0 0 0 1 3 0 2 2 1 4 0 1 0 0 1 0 3 0 0 1 1 0 1 0
0 2 1 1 0 1 1 0 0 0 1 1 0 0 3 1 1 0 0 1 0 1 0 2 5 2 1 1 0 1 2 0 0 1 1 0 1 2 0 0
0 0 0 0 0 2 0 1 1 1 2 0 0 1 1 2 1 0 1 0 0 3 2 1 4 5 0 2 1 1 1 1 2 0 2 0 0 1 0 1
0 0 1 1 2 0 0 0 1 0 0 1 1 0 0 0 0 0 2 0 0 1 2 2 1 0 0 3 3 1 1 0 1 0 0 0 0 0 1 0
1 0 1 1 0 0 1 1 2 2 1 0 0 0 0 0 1 0 0 2 1 1 0 0 0 0 0 1 1 0 0 1 1 0 0 2 0 0 2
0 0 1 1 1 1 1 0 0 0 2 2 1 2 0 0 0 2 1 0 0 0 0 0 1 1 0 3 0 0 1 2 0 7 1 0 2 0 0 2
0 1 1 1 1 2 2 2 0 0 2 0 3 1 0 1 0 1 0 0 1 0 0 0 1 1 1 3 1 0 1 0 2 1 2 1 0 0 0 1
0 2 1 0 0 0 2 1 2 0 0 0 0 0 1 0 3 0 1 1 0 0 0 1 0 0 1 0 0 0 2 2 1 1 0 1 0 1 1 0
0 0 0 0 1 0 0 2 0 0 0 0 0 0 0 1 1 0 0 1 1 0 1 0 0 1 1 0 1 1 1 2 0 1 0 2 1 0 1 1
2 0 0 1 2 0 0 0 0 0 1 0 0 1 1 2 1 3 2 0 0 0 0 0 0 0 0 1 0 0 0 1 1 1 1 0 2 1 0
```

Figure 9.4.1 Village housing pattern

The first two columns of Table 9.4.2 summarize the results. Can we conclude at, say, the 0.01 level of significance that the houses are distributed randomly throughout the village—that is, is Y Poisson?

As it was in Example 9.4.1, our first step is to estimate the unknown parameter (λ) in $f_0(y)$. Recall Example 5.8.1: If Y is Poisson,

$$\hat{\lambda} = \text{MLE for } \lambda = \bar{y}$$

TABLE 9.4.2

Number of houses per plot, y	Observed frequency, x	$n\hat{p}_{i_0}$
0	584	561.6
1	398	427.2
2	168	162.0
3	35	40.8
4+	15	8.4
	1200	1200.0

From Table 9.4.2 and Figure 9.4.1,

$$\bar{y} = \frac{1}{1200} \left[0(584) + 1(398) + 2(168) + 3(35) + 4(9) + 5(4) \right.$$

$$+ 6(0) + 7(1) + 9(1) \big]$$

$$= 0.76$$

The model we are trying to fit to the data, then, is

$$f_0(y) = \frac{e^{-0.76}(0.76)^y}{y!} \qquad y = 0, 1, \ldots$$

The third column in Table 9.4.2 shows the estimated $E(X_i)$'s. For example, if X_1 denotes the number of plots having 0 houses, then $E(X_1) = 1200 \cdot P(Y = 0)$ and the *estimated* $E(X_1)$ is given by

$$n\hat{p}_{1_0} = 1200 \cdot f_0(0) = 1200 \cdot \frac{e^{-0.76}(0.76)^0}{0!}$$

$$= 561.6$$

The other numbers shown are derived similarly. Putting all this information together gives a test statistic of 9.12:

$$c_1 = \sum_{i=1}^{5} \frac{(x_i - n\hat{p}_{i_0})^2}{n\hat{p}_{i_0}}$$

$$= \frac{(584 - 561.1)^2}{561.1} + \cdots + \frac{(15 - 8.4)^2}{8.4}$$

$$= 9.12$$

Since $k - 1 - r = 5 - 1 - 1 = 3$, the appropriate critical value is 11.345, the 99th percentile of a χ_3^2 distribution. It follows by Theorem 9.4.1, then, that we should *accept* H_0, the housing pattern can be considered random (see Figure 9.4.2).

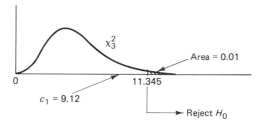

Figure 9.4.2 χ_3^2 distribution.

Comment

A more visually striking way to present the frequency information given in the Figure 9.4.1 matrix is to draw a *density contour map*. Adjacent plots having the same observed frequency are shaded a particular color. If the corners on the plots are then rounded, we get a "picture" like the one shown in Figure 9.4.3, a picture that looks more like a Pollock than a Poisson!

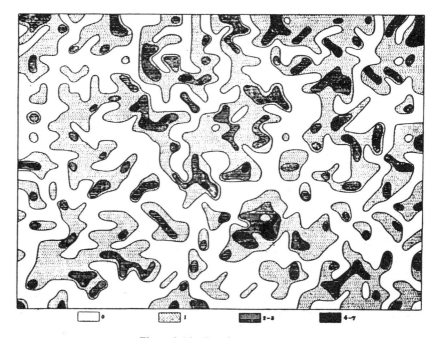

Figure 9.4.3 Density contour map.

CASE STUDY 9.4.2

Certainly the most frequent distributional assumption made in statistics, whether implicitly or explicitly, is that $Y \sim N(\mu, \sigma^2)$. As we have already seen, this was the starting point for all the t tests, χ^2 tests, and F tests presented in

Chapters 7 and 8. In practice, tests for normality are typically based on the two-parameter extension of Theorem 9.3.1, since both μ and σ^2 will be unknown.

Suppose that we wanted to test the normality of the Etruscan-skull data of Case Study 7.2.1 (as a first step, perhaps, in testing $H_0: \mu = \mu_0$ or $H_0: \sigma = \sigma_0$). Since Y is continuous, it will be necessary to group the observations into a set of nonoverlapping intervals and determine from H_0 the probability associated with each one. Table 9.4.3 shows one possible partitioning.

TABLE 9.4.3

Skull Width (mm)	Observed Frequency, x_i
125–129	1
130–134	4
135–139	10
140–144	33
145–149	24
150–154	9
155–159	3
	84

To compute the expected frequency corresponding to, say, the "140–144" class, we must first determine the probability of a single random observation's falling into that range. Using the continuity correction together with the usual Z transformation, and recalling that $\bar{y} = 143.8$ and $s = 6.0$, we get

$$P(140 < Y < 144) \doteq P(139.5 < Y < 144.5)$$

$$= P\left(\frac{139.5 - 143.8}{6.0} < Z < \frac{144.5 - 143.8}{6.0}\right)$$

$$= P(-0.72 < Z < 0.12) = 0.3120$$

It follows that the expected frequency for the "140–144" class is 26.2:

$$n\hat{p}_{i_0} = 84(0.3120) = 26.2$$

Table 9.4.4 lists the $\hat{p}_{i_0}$'s and $n\hat{p}_{i_0}$'s for the seven classes of Table 9.4.3.

TABLE 9.4.4

Skull Width	x_i	$\hat{p}_{i_0}$	$84\hat{p}_{i_0}$
≤ 129	1	0.0087	0.7
130–134	4	0.0519	4.4
135–139	10	0.1752	14.7
140–144	33	0.3120	26.2
145–149	24	0.2811	23.6
150–154	9	0.1336	11.2
155+	3	0.0375	3.2
	84	1.0000	84.0

Comment

Notice that the first and last classes have been made open-ended. This is a necessary adjustment whenever the model being fit has an infinite range. If this were not done, the $\hat{p}_{i_0}$'s would not sum to 1, a condition required of the multinomial distribution.

Table 9.4.5 shows the *final* set of classes, the first two and the last two intervals in Table 9.4.4 having been collapsed to comply with the "$n\hat{p}_{i_0} \geq 5$" condition. As indicated at the bottom of the last column, the calculated value for c_1 is 3.61.

TABLE 9.4.5

Skull Width (mm)	x_i	$\hat{p}_{i_0}$	$84\hat{p}_{i_0}$	$(x_i - 84\hat{p}_{i_0})^2/84\hat{p}_{i_0}$
≤ 134	5	0.0606	5.1	0.00
135–139	10	0.1752	14.7	1.50
140–144	33	0.3120	26.2	1.76
145–149	24	0.2811	23.6	0.01
150+	12	0.1711	14.4	0.34
	84	1.0000	84.0	3.61

Since $r = 2$ parameters were estimated and the final calculations were based on $k = 5$ classes, the number of degrees associated with C_1 is $5 - 2 - 1$, or 2. Had we elected to carry out the test at the $\alpha = 0.05$ level of significance, the corresponding critical value would be 5.991 $(= \chi^2_{0.95, 2})$, and our conclusion would be to *accept* the normality assumption.

Question 9.4.1 During the course of a season, a particular baseball player is involved in 200 games in which he has exactly four official at-bats. His distribution of hits during those games is summarized below.

Number of Hits	Number of Games
0	44
1	86
2	56
3	12
4	2

Test the assumption that X, the number of hits the player gets in four at-bats, behaves like a binomial random variable. Let 0.05 be the level of significance.

Question 9.4.2 Do a goodness-of-fit test on the war data described in Case Study 4.2.4.

Question 9.4.3 Carry out the details for a goodness-of-fit test on the tick data of Case Study 4.5.1. Use the 0.01 level of significance.

Question 9.4.4 As a way of studying the spread of a plant disease known as creeping rot, a field of cabbage plants was divided up into 270 *quadrats*, each quadrat containing the same number of plants. Below are listed the numbers of plants per quadrat showing signs of creeping rot infestation.

Number of Infected Plants/Quadrat	Number of Quadrats
0	38
1	57
2	68
3	47
4	23
5	9
6	10
7	7
8	3
9	4
10	2
11	1
12	1
13+	0

Can the number of plants infected with creeping rot per quadrat be described by a Poisson pdf? What might be a physical reason for the Poisson's not being appropriate in this situation? Which assumption of the Poisson appears to be violated?

Question 9.4.5 To raise money for a new rectory, the members of a church hold a raffle. A total of *n* tickets are sold (numbered 1 through *n*), out of which a total of 50 winners are to be drawn, presumably at random. Listed below are the 50 lucky numbers.

108	110	21	6	44
89	68	50	13	63
84	64	69	92	12
46	78	113	104	105
9	115	58	2	20
19	96	28	72	81
32	75	3	49	86
94	61	35	31	56
17	100	102	114	76
106	112	80	59	73

Set up a goodness-of-fit test that focuses on the randomness of the draw. Use the 0.05 level of significance.

Question 9.4.6 Do a goodness-of-fit test for normality on the traffic fatality data given in Case Study 2.5.1. Let 0.05 be the level of significance.

Too often inference is pictured as a static, one-shot process: a confidence interval is constructed or a hypothesis is tested, a conclusion is drawn, and the experimenter moves on to another problem, having disposed of days, maybe even months or years, of data with a simple "yes" or "no." In reality, though, interpreting data is more of a *dynamic* process, with the result of one test serving as input for a second, and *that*, in turn, suggesting still a third, and so on. The next example illustrates this idea by examining the relationship between parameter estimation, goodness-of-fit testing, and model building for a set of data related to the genetic phenomenon of *linkage*.

CASE STUDY 9.4.3

Corn can be either starchy (S) or sugary (s) and can have a green (G) base leaf or a white (g) one. Suppose the alleles for these two factors occur on separate chromosomes and are hence independent. Then each parent with alleles SsGg produces with equal likelihood gametes of the form (S, G), (S, g), (s, G), and (s, g). If two such hybrid parents are crossed, the phenotypes of the offspring will occur in the proportions suggested by Table 9.4.6. That is, the probability of an offspring of type (S, G) is $\frac{9}{16}$; type (S, g), $\frac{3}{16}$; type (s, G), $\frac{3}{16}$; and type (s, g), $\frac{1}{16}$.

TABLE 9.4.6

		Alleles of First Parent			
		SG	Sg	sG	sg
Alleles	SG	(S, G)	(S, G)	(S, G)	(S, G)
of Second	Sg	(S, G)	(S, g)	(S, G)	(S, g)
Parent	sG	(S, G)	(S, G)	(s, G)	(s, G)
	sg	(S, G)	(S, g)	(s, G)	(s, g)

Table 9.4.7 shows the $9:3:3:1$ set of ratios fit to a set of 3839 SsGg $\times$ SsGg crossings (18). With no parameters having been estimated, and given $k = 4$ classes, the critical value for, say, a 0.01 test would be $\chi^2_{0.99, 3}$, or 11.341. But the calculated value for c is 287.72, a number clearly indicating the inappropriateness of the initial model.

TABLE 9.4.7

Phenotype	x_i	p_{i_0}	$3839p_{i_0}$	$(x_i - 3839p_{i_0})^2/3839p_{i_0}$
Starchy green	1997	$\frac{9}{16}$	2159.44	12.22
Starchy white	906	$\frac{3}{16}$	719.81	48.16
Sugary green	904	$\frac{3}{16}$	719.81	47.13
Sugary white	32	$\frac{1}{16}$	239.94	180.21
	3839	1	3839.00	287.72

One possible explanation for the decided lack of fit between the data and the $9:3:3:1$ model is a type of genetic behavior known as *crossing-over*. Suppose a parent has one chromosome with alleles Sg and another with alleles sG. In the process of meiosis, the two members of the chromosome pair move close together, uncoil, and, in some cases, exchange parts—this latter being the phenomenon of crossing-over. It follows that there is a probability, p, of the chromosome pairs Sg and sG changing to SG and sg. Of course, whether or not a crossover occurs, either chromosome is equally likely to be present in the gamete. The probability, then, of any of the four possible allele combinations being present in the gamete can be determined from Theorem 2.6.1. For example,

$$P(\text{SG is in gamete}) = P(\text{SG is in gamete} \mid \text{crossover}) \cdot P(\text{crossover})$$
$$+ P(\text{SG is in gamete} \mid \text{no crossover})$$
$$\cdot P(\text{no crossover})$$
$$= \frac{1}{2} \cdot p + 0 \cdot (1 - p) = \frac{1}{2} p$$

Similarly,

$$P(\text{Sg is in gamete}) = \frac{1}{2}(1 - p)$$

$$P(\text{sG is in gamete}) = \frac{1}{2}(1 - p)$$

and

$$P(\text{sg is in gamete}) = \frac{1}{2} p$$

There is no reason to assume that p is the same for both parents, so the other parent would admit a similar array of gamete probabilities, but with p replaced by p'. The relative likelihood of a given phenotype can then be inferred from Table 9.4.6. For example,

$$P[(\text{s, g}) \text{ phenotype}] = P[(\text{s, g}) \text{ allele from first parent } \textit{and}$$
$$(\text{s, g}) \text{ allele from second parent}]$$
$$= \frac{1}{2} p \cdot \frac{1}{2} p' = \frac{1}{4} pp'$$

Similarly, the probability of an offspring with phenotype (s, G) will be $\frac{1}{4}(1 - pp')$, the sum of the probabilities of three different matings:

$$P[(\text{s, G}) \text{ phenotype}] = P[(\text{s, g}) \text{ allele} \times (\text{s, G}) \text{ allele}]$$
$$+ P[(\text{s, G}) \text{ allele} \times (\text{s, G}) \text{ allele}]$$
$$+ P[(\text{s, G}) \text{ allele} \times (\text{s, g}) \text{ allele}]$$
$$= \frac{1}{2} p \cdot \frac{1}{2}(1 - p') + \frac{1}{2}(1 - p) \cdot \frac{1}{2}(1 - p')$$
$$+ \frac{1}{2}(1 - p) \cdot \frac{1}{2} p'$$
$$= \frac{1}{4}(1 - pp')$$

By the same reasoning, $P[(S, g)$ phenotype$] = \frac{1}{4}(1 - pp')$ and $P[(S, G)$ phenotype$] = \frac{1}{4}(2 + pp')$.

Since each phenotype probability depends on the same parameter product, pp' can be replaced by a single parameter θ, where $0 < \theta < 1$. Table 9.4.8 summarizes the probability distribution for the four phenotypes. Note that if $p = p' = \frac{1}{2}$, making $\theta = \frac{1}{4}$, the probabilities of Table 9.4.8 reduce to the $9:3:3:1$ ratios of independent traits. If $p = p' = 1$, and $\theta = 1$, the simple $3:1$ dominant-recessive scheme arises.

TABLE 9.4.8

Phenotype	Probability
Starchy, green	$\frac{1}{4}(2 + \theta)$
Starchy, white	$\frac{1}{4}(1 - \theta)$
Sugary, green	$\frac{1}{4}(1 - \theta)$
Sugary, white	$\frac{1}{4}(\theta)$

Before the data of Table 9.4.7 can be tested against the theoretical construct of Table 9.4.8, the unknown θ needs to be estimated. Fortunately, its MLE is readily accessible. The likelihood function has the form,

$$L = [\tfrac{1}{4}(2 + \theta)]^{x_1} \cdot [\tfrac{1}{4}(1 - \theta)]^{x_2} \cdot [\tfrac{1}{4}(1 - \theta)]^{x_3} \cdot [\tfrac{1}{4}(\theta)]^{x_4}$$

Taking the ln of L and substituting for the x_i's gives

$$\ln L = 1997 \ln (2 + \theta) + (906 + 904) \ln (1 - \theta) + 32 \ln (\theta) - 3839 \ln (4)$$

Therefore,

$$\frac{d \ln L}{d\theta} = \frac{1997}{2 + \theta} - \frac{1810}{1 - \theta} + \frac{32}{\theta}$$

which, when set equal to 0, yields the maximum-likelihood estimate $\hat{\theta} = 0.0357$. Table 9.4.9 shows the initial data "refit" using the crossover model. Notice the improvement between the last column of Table 9.4.7 and the last column of Table 9.4.9.

TABLE 9.4.9

Phenotype	x_i	$\hat{p}_{i_0}$	$3839\hat{p}_{i_0}$	$(x_i - 3939\hat{p}_{i_0})^2/3839\hat{p}_{i_0}$
Starchy, green	1997	0.5089	1953.67	0.96
Starchy, white	906	0.2411	925.58	0.41
Sugary, green	904	0.2411	925.58	0.50
Sugary, white	32	0.0089	34.17	0.14
	3839	1.0000	3839.00	2.01

Since one parameter was estimated, the χ^2 statistic now has $4 - 1 - 1 = 2$ degrees of freedom. Also, with the 95th percentile of χ_2^2 being 5.991, and c_1 equal to 2.01, we *accept* H_0 at the 0.05 level, thus lending credence both to the notion of crossing-over and to the probability model we set up to characterize it.

Question 9.4.7 The Japanese morning glory (*Pharbitis nil*) admits a number of recessive traits, two of which are crumpled leaves and variegated leaves. The results of cross-breeding hybrid morning glories bearing a dominant and recessive gene in each trait were the 2176 progeny categorized in the table (81). Estimate the recombination parameter, θ, and fit these data using the methodology of Case Study 9.4.3.

OFFSPRING OF CROSS-BRED MORNING GLORIES

Phenotype	Observed Frequency
Normal	1450
Variegated only	184
Crumpled only	191
Variegated and crumpled	351

The sensitivity of the χ^2 procedure to parameter estimation was illustrated dramatically in Case Study 9.4.3. When the parameter θ was "estimated" to be $\frac{1}{4}$, yielding the no cross-over $\frac{9}{16}, \frac{3}{16}, \frac{3}{16}, \frac{1}{16}$ distribution, H_0 was resoundingly rejected. Yet with θ set equal to its MLE, the null hypothesis was accepted. A failsafe method of avoiding disagreements of this sort would be to assign the parameters those particular values that minimize the expression for c_1. Not surprisingly, $\hat{\theta}$'s obtained in such a way are known as *minimum χ^2 estimates*. The concluding remarks in this section relate the notion of minimum χ^2 estimation to the method of maximum likelihood and, ultimately, to the statement of Theorem 9.4.1. (For notational simplicity, we will restrict our attention to the case of a single parameter; the generalization to more than one parameter is straightforward.)

If, for all i, the multinomial p_i's are really $p_i(\theta)$'s, then C_1 is similarly $C_1(\theta)$, and for a fixed sample, $x_1, x_2, \ldots, x_n$,

$$C_1(\theta) = \sum_{i=1}^{k} \frac{[x_i - np_i(\theta)]^2}{np_i(\theta)}$$

Assuming appropriate differentiability conditions, we can find the minimum χ^2 estimate by setting $C_1'(\theta)$ equal to 0 and solving for θ. To begin,

$$C_1'(\theta) = -\sum_{i=1}^{k} \left\{ \frac{2[x_i - np_i(\theta)]p_i'(\theta)}{p_i(\theta)} + \frac{[x_i - np_i(\theta)]^2 p_i'(\theta)}{np_i^2(\theta)} \right\} \qquad (9.4.4)$$

In general, setting Equation 9.4.4 equal to 0 gives a decidedly difficult expression to solve. For large n, though, the second term being added is often negligible, so without unduly compromising matters, we can solve, instead, Equation 9.4.5:

$$\sum_{i=1}^{k} \frac{[x_i - np_i(\theta)]p_i'(\theta)}{p_i(\theta)} = 0 \qquad (9.4.5)$$

But this is just an old friend in a new suit. To see why, consider the likelihood equation,

$$L(\theta) = [p_1(\theta)]^{x_1}[p_2(\theta)]^{x_2} \cdots [p_k(\theta)]^{x_k}$$

Taking the ln of $L(\theta)$ and differentiating gives

$$\ln L(\theta) = \sum_{i=1}^{k} x_i \ln p_i(\theta)$$

and

$$\frac{d \ln L(\theta)}{d\theta} = \sum_{i=1}^{k} \frac{x_i p_i'(\theta)}{p_i(\theta)}$$

But since $\sum_{i=1}^{k} p_i(\theta) = 1$, it follows that $\sum_{i=1}^{k} p_i'(\theta) = 0$, and we can write

$$\sum_{i=1}^{k} \frac{np_i(\theta) \cdot p_i'(\theta)}{p_i(\theta)} = 0 \tag{9.4.6}$$

Now, combine Equation 9.4.6 with $[d \ln L(\theta)]/d\theta$ and compare the result with Equation 9.4.5:

$$\frac{d \ln L(\theta)}{d\theta} = \sum_{i=1}^{k} \frac{x_i p_i'(\theta)}{p_i(\theta)} - \sum_{i=1}^{k} \frac{np_i(\theta)p_i'(\theta)}{p_i(\theta)}$$

$$= \sum_{i=1}^{k} \frac{(x_i - np_i)p_i'(\theta)}{p_i(\theta)}$$

It follows that the MLE for θ is essentially the same as the minimum χ^2 estimate. Or, to put it another way, we now have some reassurance that the method of maximum likelihood gives estimates appropriate for goodness-of-fit testing, and that the statement of Theorem 9.4.1, calling for any unknown θ's to be replaced by their MLE's, is not unreasonable. (In Case Study 9.4.3 the minimum χ^2 estimate for θ is 0.035785; the MLE, 0.035712. The value of c_1 in the first case is 2.0153; in the second case, 2.0154.)

9.5 CONTINGENCY TABLES

We have seen that inferences both about parameters and about entire pdf's are two frequent objectives of statistical methodology. A third are inferences about *independence*: does knowing the value of a first random variable provide us with any insight into the probable behavior of a second? Examples of this latter type of problem are commonplace, even outside the scientific community. Perhaps the most familiar are the various studies purportedly linking the occurrence of cancer to such factors as smoking, industrial pollutants, saccharin, cyclamates, and panfried hamburgers.

Needless to say, questions of independence are by no means confined to health-related issues. Merchants ask if inventory requirements differ in downtown and suburban stores, educators speculate on the relationship between a city's busing efforts and students' performance on standardized exams, and not a few parents have wondered if juvenile delinquency is related to the use of Dr. Spock's baby book.

The notion of independence, of course, has already turned up in a number of different contexts, both probabilistic and statistical. To recast it in an inference setting requires nothing new, conceptually, and we can even use the same notation that was introduced in Chapter 2. Let $A_1, A_2, \ldots, A_r$ be a partition of the sample space S,

$$\bigcup_{i=1}^{r} A_i = S \qquad A_i \cap A_j = \varnothing, \quad i \neq j$$

Suppose that $B_1, B_2, \ldots, B_c$ is another partition of S. Then the question of whether criteria A and criteria B are independent can be written

$$H_0: \quad A_i \text{ and } B_j \text{ are independent for } 1 \leq i \leq r, \, 1 \leq j \leq c$$

versus

$$H_1: \quad A_i \text{ and } B_j \text{ are } \textit{not} \text{ independent for } 1 \leq i \leq r, \, 1 \leq j \leq c$$

The methodology for choosing between H_0 and H_1 follows closely the pattern established by Theorems 9.3.1 and 9.4.1. First, we select a random sample, $Y_1, Y_2, \ldots, Y_n$, from S, and then we define the random variable X_{ij} to be the number of observations belonging to the intersection $A_i \cap B_j$. Under H_0,

$$P[Y_k \in (A_i \cap B_j)] = P(Y_k \in A_i) \cdot P(Y_k \in B_j) = p_i q_j \qquad k = 1, 2, \ldots, n$$

where $\sum_{i=1}^{r} p_i = 1$ and $\sum_{j=1}^{c} q_j = 1$. Therefore, $E(X_{ij}) = n p_i q_j$, and, by analogy with C and C_1, the goodness-of-fit statistic

$$C_2 = \sum_{i=1}^{r} \sum_{j=1}^{c} \frac{(X_{ij} - n p_i q_j)^2}{n p_i q_j}$$

has an asymptotic χ^2 distribution. Typically, the p_i's and the q_j's will both be estimated, the former with MLE's

$$\hat{p}_i = \frac{1}{n} \sum_{j=1}^{c} x_{ij}$$

and the latter by

$$\hat{q}_j = \frac{1}{n} \sum_{i=1}^{r} x_{ij}$$

However, since

$$\sum_{i=1}^{r} \hat{p}_i = 1 \quad \text{and} \quad \sum_{j=1}^{c} \hat{q}_j = 1$$

only $(r - 1) + (c - 1)$ parameters need to be estimated directly. Since the Y's are being categorized into a total of $r \cdot c$ classes, it follows from Theorem 9.4.1 that the number of degrees of freedom associated with C_2 is

$$rc - 1 - (r - 1) - (c - 1) = rc - r - c + 1 = (r - 1)(c - 1)$$

Data for a test of independence are generally presented in tabular form, with rows representing the categories of one criteria and columns the categories of the other. Such displays are called *contingency tables*—a name probably due to Karl Pearson.

CASE STUDY 9.5.1

Market researchers often gather survey information by telephone. Ideally, the numbers dialed should be random, thus affording every subscriber an equal chance of being included in the sample. In practice, though, doing a survey that way would be very costly because of the sizable number of non-productive calls resulting from dialing invalid numbers. The obvious remedy is to sample from the telephone directory, but that raises another problem: unlisted numbers would not be included. Nationally, more than 15% of all telephone numbers are unlisted, and the percentage is much higher in certain locales (in Los Angeles, for example, more than 35% of all telephone numbers are unlisted). What this means is that the directory method is unreliable unless the factors being studied are independent of whether or not the telephone is listed.

Suppose that a political poll is being taken in California to assess the popularity of a proposition to reduce property taxes. For a directory sample to be unbiased in that context, the proportion of subscribers who own their own homes should not be significantly different for those with listed phones as compared to those with unlisted phones.

The responses (slightly modified) shown in Table 9.5.1 were obtained in a recent survey by Pacific Bell of 1000 subscribers (131). From the row totals we estimate p_1, the probability of a respondent's being a homeowner, as $\frac{774}{1000}$; p_2, the probability of his being a renter, as $\frac{226}{1000}$. Similarly, q_1, the probability of a respondent's having a listed phone, would be estimated by $\frac{800}{1000}$; and q_2, the probability of his having an unlisted phone, as $\frac{200}{1000}$. Under the null hypothesis asserting independence between home ownership and having a listed phone, the estimated probability of a respondent's being a homeowner with a listed

TABLE 9.5.1

	Listed	Unlisted	Total
Own	628	146	774
Rent	172	54	226
Total	800	200	1000

phone is $\hat{p}_1\hat{q}_1$, or (0.774) $(0.8) = 0.6192$. Thus, the expected *number* of such people is $1000(0.6192)$, or 619.2. Table 9.5.2 shows the similarly calculated expected frequencies (in parentheses) for each of the four data categories.

TABLE 9.5.2

	Listed	Unlisted	Total
Own	628 (619.2)	146 (154.8)	774
Rent	172 (180.8)	54 (45.2)	226
Total	800	200	1000

Here, the value of c_2 is 2.77:

$$\frac{(628-619.2)^2}{619.2} + \frac{(146-154.8)^2}{154.8} + \frac{(172-180.8)^2}{180.8} + \frac{(54-45.2)^2}{45.2} = 2.77$$

It follows that if we were to test

$$H_0: \text{ ownership and listing status are independent}$$

versus

$$H_1: \text{ ownership and listing status are dependent}$$

at the $\alpha = 0.05$ level, the null hypothesis would be accepted: with $(2-1)(2-1) = 1$ degree of freedom, the critical χ^2 value is $\chi^2_{0.95, 1} = 3.84$. (At the 10% level, though, the null hypothesis would have been rejected.)

Comment

The hazards in drawing inferences from telephone data are well documented. One of the most frequently cited "mistakes" was the *Literary Digest* presidential poll in 1936 that confidently predicted a Landon victory over Roosevelt.

CASE STUDY 9.5.2

The question of whether or not birth order is related to juvenile delinquency was examined in a large-scale study using a high school population. A total of 1154 girls attending public high school were given a questionnaire that measured the degree to which each had exhibited delinquent behavior, in terms of criminal acts, immoral conduct, and so on. After duly tabulating and interpreting the results, the researchers decided to categorize some 111 of the girls as "delinquent." Each girl in the initial sample was also asked to indicate her birth order, as being either (1) the oldest, (2) in between, (3) the youngest, or (4) an only child. Table 9.5.3 details the delinquency-by-birth-order breakdown (115).

TABLE 9.5.3

		Birth Order			
		Oldest	In Between	Youngest	Only Child
Delinquent?	Yes	24	29	35	23
	No	450	312	211	70
	Total	474	341	246	93
Percent "Yes"		5.1	8.5	14.2	24.7

We wish to test

$$H_0: \quad \text{birth order and delinquency are independent}$$

versus

$$H_1: \quad \text{birth order and delinquency are dependent}$$

Let $\alpha = 0.05$. The MLE for p_1, the probability of a girl's showing delinquent behavior, is

$$\hat{p}_1 = \frac{24 + 29 + 35 + 23}{1154} = \frac{111}{1154}$$

Similarly, the MLE for q_1, the probability of a girl's being an oldest child, is

$$\hat{q}_1 = \frac{24 + 450}{1154} = \frac{474}{1154}$$

Therefore, under H_0, the expected number of oldest girls exhibiting delinquent behavior would be

$$n\hat{p}_1\hat{q}_1 = 1154\left(\frac{111}{1154}\right)\left(\frac{474}{1154}\right) = 45.6$$

Table 9.5.4 shows the entire set of $n\hat{p}_i\hat{q}_j$ expected values, together with the original x_{ij}'s.

TABLE 9.5.4

	Oldest	In Between	Youngest	Only Child
Yes	24	29	35	23
	(45.6)	(32.8)	(23.7)	(8.9)
No	450	312	211	70
	(428.4)	(308.2)	(222.3)	(84.1)

At the 0.05 level, we should reject H_0 if $c_2 \geq \chi^2_{0.95, (2-1)(4-1)} = \chi^2_{0.95, 3} = 7.81$. But

$$c_2 = \frac{(24 - 45.6)^2}{45.6} + \frac{(29 - 32.8)^2}{32.8} + \cdots + \frac{(70 - 84.1)^2}{84.1}$$

$$= 42.5$$

a number far to the right of the critical value. Thus, we reject H_0 and conclude that birth order and delinquency are dependent. (This is not an unexpected result in light of the considerable differences in the "Percent 'Yes'" figures shown at the bottom of Table 9.5.3.

Question 9.5.1 A group of 57 elderly persons visiting a certain clinic on an outpatient basis were categorized as being either "compliers" or "noncompliers" according to whether or not they followed their doctor's orders in taking medication. Each of them was also classified by religious affiliation. The table summarizes their responses (171).

	Catholic	Protestant
Compliers	10	15
Noncompliers	7	25

Clearly, the sample proportion of Catholic compliers ($\frac{10}{17} = 0.59$) is considerably higher than the sample proportion of Protestant compliers ($\frac{15}{40} = 0.38$). Is this difference statistically significant? At the 0.05 level, test to see whether religion and compliance are independent.

Note: These data were collected to test a theory that the authoritarian nature of the Catholic church induces in its members a greater tolerance for taking orders than would be characteristic of members of Protestant faiths.

Question 9.5.2 In the United States, suicide rates tend to be much higher for men than for women—at all ages. That pattern may not extend to all professions, however. Death certificates obtained for the 3637 members of the American Chemical Society who died during the period from April 1948 to July 1967 revealed that 106 of the 3522 male deaths were suicides compared to 13 of the 115 female deaths (93). Do an appropriate test. State H_0 and H_1 explicitly and carry out the test at the $\alpha = 0.01$ level of significance.

Question 9.5.3 In a study (43) investigating the effect of rubella infections (German measles) on childbirth, a total of 578 pregnancies were classified in retrospect as having been either "normal" or "abnormal," the latter group including abortions, stillbirths, birth defects, and all infant deaths within two years. Altogether, there were 86 abnormal pregnancies. The second variable looked at was *when* the rubella infection occurred—during the first trimester or after the first trimester. It was found that 59 of the 86 abnormal births were among the 202 pregnancies complicated during the first trimester; the remaining 27 were born to mothers who contracted the virus after the first trimester. Is the risk of an abnormal birth dependent on when during the pregnancy the virus is contracted? Let $\alpha = 0.01$.

Question 9.5.4 A market research study has investigated the relationship between an adult's self-perception and his attitude toward small cars. A total of 299 persons living in a large metropolitan area were surveyed. On the basis of their responses to a questionnaire, each was "assigned" to one of three distinct personality types: (1) cautious conservative, (2) middle-of-the-roader, and (3) confident explorer. At the same time, each was solicited for his overall opinion of small cars. The results are displayed below (82).

		Self-Perception		
		Cautious Conservative	Middle-of-the-Roader	Confident Explorer
Opinion	Favorable	79	58	49
of Small	Neutral	10	8	9
Cars	Unfavorable	10	34	42

Test whether these two traits are independent. Use the 0.01 level of significance.

Question 9.5.5 High blood pressure is known to be one of the major contributors to coronary heart disease. A study was done to see whether or not there is a significant relationship between the blood pressures of children and those of their fathers (80). If such a relationship did exist, it might be possible to use one group to screen for high-risk individuals in the other group. The subjects were 92 eleventh graders, 47 males and 45 females, and their fathers. Blood pressures for both the children and the fathers were categorized as belonging to either the lower, middle, or upper third of their respective distributions.

		Child's blood pressure		
		Lower third	Middle third	Upper third
Father's	Lower third	14	11	8
blood	Middle third	11	11	9
pressure	Upper third	6	10	12

Test whether or not the blood pressures of children can be considered to be independent of the blood pressures of their fathers. Let $\alpha = 0.05$.

Question 9.5.6 In a recent UFO investigation, a total of 1276 "close encounters" were categorized according to (1) the location of the sighting, and (2) the nature of the sighting (79).

		Location of Sighting	
		In Spain	Not in Spain
Nature of sighting	Saucer seen on ground	53	705
	Saucer seen hovering	38	412
	Miscellaneous	9	59

Test whether "location of sighting" and "nature of sighting" are independent. Let $\alpha = 0.05$. Suppose that H_0 is accepted. Could that be considered evidence *in favor of* or *against* the existence of UFOs?

10

Regression

Francis Galton (1822–1911)

Galton had earned a Cambridge mathematics degree and completed two years of medical school when his father died, leaving him with a substantial inheritance. Free to travel, he became an explorer of some note, but when *Origin of the Species* was published in 1859, his interests began to shift from geography to statistics and anthropology (Charles Darwin was his cousin). It was Galton's work on fingerprints that made possible their use in human identification. He was knighted in 1909.

10.1 INTRODUCTION

One of the major objectives of all analytical inquiry is determining the relationships among the various components of the object or system under study. If these relationships are sufficiently understood, there is then a basis for intelligent action through prediction and control.

As an example, consider the formidable problem of relating the incidence of cancer to its many possible causes—diet, genetic makeup, pollution, and cigarette smoking, to name only a few. Or think of the Wall Stree financier, trying to predict stock prices from a myriad of potentially important factors involving market indices, economic climate, and corporation data.

This chapter discusses some of the basic techniques for measuring relationships between random variables. To be sure, in some situations many variables are involved, in which case the analysis becomes quite complex. Fortunately, most of the fundamental ideas emerge clearly for the case of only *two* variables, and we shall restrict our attention accordingly.

Section 10.2 introduces a measure of relationship between two random variables; the treatment is much in the spirit of Chapter 3 and, indeed, has only that chapter as a prerequisite. Section 10.3 discusses the historical development of the statistical theory of *linear* relationships. A computational technique for determining the "best" linear fit between two variables is given in Section 10.4. Sections 10.5 and 10.6 develop some inference procedures to answer questions raised in Sections 10.2 and 10.4. Then in Section 10.7 we give two convenient tests of the hypothesis that a pair of random variables are independent. The chapter concludes with some words of wisdom on statistical malpractice.

10.2 COVARIANCE AND CORRELATION

With the exception of the tests for independence described in the latter part of Chapter 9, our statistical efforts thus far have been directed at characterizing (with means, variances, pdf's, and so on) the behavior of either an individual random variable, X, or some function of a sample of random variables, $X_1, X_2, \ldots, X_n$, where the X_i's are independent and identically distributed. In this section we go off in a somewhat different direction and look for a *measure of relationship* between two random variables, X and Y, where X and Y are presumably *dependent* and not necessarily identically distributed. The *covariance of X and Y*, as given in Definition 10.2.1, is one such measure of relationship.

> **Definition 10.2.1.** Let X and Y be random variables with means μ_X and μ_Y. The *covariance of X and Y*, written Cov (X, Y), is given by
>
> $$\text{Cov } (X, Y) = E[(X - \mu_X)(Y - \mu_Y)]$$

Comment

Note that the concept of covariance generalizes the notion of variance, since Cov $(X, X) = $ Var (X).

A sometimes more convenient form for the covariance is given in Theorem 10.2.1. The proof is trivial, following directly from the distributive property of the expected value.

Theorem 10.2.1. For any random variables X and Y with means μ_X and μ_Y,

$$\text{Cov}(X, Y) = E(XY) - \mu_X \mu_Y$$

The statement of Theorem 10.2.2 should come as no real surprise. If X and Y vary independently, it follows that for a given $x - \mu_X$, $Y - \mu_Y$ will sometimes be positive and other times negative, thus precipitating a certain amount of "canceling" among the set of $(x - \mu_X)(y - \mu_Y)$ values. What the theorem states is that the canceling is, in fact, complete.

Theorem 10.2.2. If X and Y are independent,

$$\text{Cov}(X, Y) = 0$$

Proof. If X and Y are independent, $E(XY) = E(X) \cdot E(Y) = \mu_X \mu_Y$. The statement of the theorem follows immediately, then, from Theorem 10.2.1.

. The converse of Theorem 10.2.2 is not true, the following example being a case in point.

Example 10.2.1

Consider the sample space $S = \{(-2, 4), (-1, 1), (0, 0), (1, 1), (2, 4)\}$ with each point equally likely. Define the random variable X to be the first component of the sample point chosen, and Y the second. Therefore, $X(-2, 4) = -2$, $Y(-2, 4) = 4$, and so on. It is a simple matter here to show that X and Y are dependent, yet their covariance is 0. The former is true because

$$\frac{1}{5} = P(X = 1, Y = 1) \neq P(X = 1) \cdot P(Y = 1) = \frac{1}{5} \cdot \frac{2}{5} = \frac{2}{25}$$

To verify the latter, note that $E(XY) = [(-8) + (-1) + 0 + 1 + 8] \cdot \frac{1}{5} = 0$, $E(X) = 0$, and $E(Y) = [4 + 1 + 0 + 1 + 4] \cdot \frac{1}{5} = 2$. Therefore, $\text{Cov}(X, Y) = 0 - 0 \cdot 2 = 0$.

Question 10.2.1 Suppose that two dice are thrown. Let X be the number showing on the first die and let Y be the larger of the two numbers showing. Find $\text{Cov}(X, Y)$.

Question 10.2.2 Show that

$$\text{Cov}(aX + b, cY + d) = ac\,\text{Cov}(X, Y)$$

for any constants a, b, c, and d.

Question 10.2.3 Let U be a random variable uniformly distributed over $[0, 2\pi]$. Define $X = \cos U$ and $Y = \sin U$. Show that X and Y are dependent but that Cov $(X, Y) = 0$.

Question 10.2.4 Let X and Y be random variables with

$$f_{X, Y}(x, y) = \begin{cases} 1 & -y < x < y, \quad 0 < y < 1 \\ 0 & \text{elsewhere} \end{cases}$$

Show that Cov $(X, Y) = 0$ but that X and Y are dependent.

Theorem 3.10.3 gave a formula for the variance of a sum of independent random variables. With the notion of covariance, we can generalize that result to include *dependent* random variables.

> **Theorem 10.2.3.** Let $X_1, X_2, \ldots, X_n$ be any set of random variables. Let $X = X_1 + X_2 + \cdots + X_n$. Then
>
> $$\text{Var } (X) = \sum_{i=1}^{n} \text{Var } (X_i) + 2 \sum_{j<k} \text{Cov } (X_j, X_k)$$

Proof. The proof is a straightforward exercise in elementary algebra:

$$\text{Var } (X) = E(X^2) - [E(X)]^2$$

$$= E\left[\left(\sum_{i=1}^{n} X_i\right)^2\right] - \left[\sum_{i=1}^{n} E(X_i)\right]^2$$

$$= E\left(\sum_{i=1}^{n} X_i^2 + 2 \sum_{j<k} X_j X_k\right) - \left[\sum_{i=1}^{n} [E(X_i)]^2 + 2 \sum_{j<k} E(X_j)E(X_k)\right]$$

$$= \sum_{i=1}^{n} [E(X_i^2) - [E(X_i)]^2] + 2 \sum_{j<k} [E(X_j X_k) - E(X_j)E(X_k)]$$

$$= \sum_{i=1}^{n} \text{Var } (X_i) + 2 \sum_{j<k} \text{Cov } (X_j, X_k)$$

Example 10.2.2

We will use Theorem 10.2.3 to calculate the variance of a hypergeometric random variable. As a physical model, consider an urn containing N chips, r red and w white ($r + w = N$). A sample of n is to be selected *without* replacement. Let Y denote the number of red chips drawn. From Theorem 2.11.1,

$$P(Y = y) = \frac{\binom{r}{y}\binom{w}{n-y}}{\binom{N}{n}} \qquad \max (0, n - w) \leq y \leq \min (r, n)$$

Although the moments of Y could be gotten directly from the fundamental definitions, it is more instructive to introduce auxiliary random variables, as we did in the case of a binomial variable. Define

$$Y_i = \begin{cases} 1 & \text{if the } i\text{th chip is red} \\ 0 & \text{otherwise} \end{cases}$$

Then $Y = Y_1 + Y_2 + \cdots + Y_n$. It is easy to see that $E(Y_i) = r/N$ and $E(Y) = n(r/N) = np$, where $p = r/N$. Thus, $E(Y)$ is the same whether the sampling is done *with* replacement or *without* replacement.

For the variance, though, the mode of sampling makes a difference. The Y_i's are not independent; however, by symmetry, they *are* identically distributed. Also, any pair (Y_j, Y_k), $j \neq k$, has the same distribution as (Y_1, Y_2), again by symmetry. [There is no reason to suppose it is more (or less) likely to get chips on the third and fifth draws than on the first and second.] Since $Y_i^2 = Y_i$, $E(Y_i^2) = E(Y_i) = r/N$, and Var $(Y_i) = r/N - (r/N)^2 = (r/N)(1 - r/N) = p(1 - p)$. For any $j \neq k$, Cov $(Y_j, Y_k) = $ Cov $(Y_1, Y_2) = E(Y_1, Y_2) - E(Y_1)E(Y_2)$. But $E(Y_1 Y_2) = 1 \cdot P(Y_1 Y_2 = 1) = 1 \cdot P(\text{first chip is red and second chip is red}) = r/N \cdot (r - 1)/(N - 1)$, so

$$\text{Cov } (Y_j, Y_k) = \frac{r}{N} \cdot \frac{r - 1}{N - 1} - \frac{r}{N} \cdot \frac{r}{N} = \frac{r}{N}\left(\frac{r - 1}{N - 1} - \frac{r}{N}\right)$$

$$= -\frac{r}{N} \cdot \frac{N - r}{N} \cdot \frac{1}{N - 1}$$

Finally, applying Theorem 10.2.3, we get an expression for Var (Y):

$$\text{Var } (Y) = \sum_{i=1}^{n} \text{Var } (Y_i) + 2 \sum_{j<k} \text{Cov } (Y_j, Y_k)$$

$$= np(1 - p) - 2\binom{n}{2}p(1 - p) \cdot \frac{1}{N - 1}$$

$$= np(1 - p) \cdot \frac{N - n}{N - 1}$$

Recall that the variance for sampling *with* replacement is $np(1 - p)$, so the not-surprising result emerges that it is more efficient to estimate $p = r/N$ by sampling without replacement (particularly when n is relatively large compared to N).

For historical reasons, and because a measure of relationship should be dimensionless, the covariance is often normalized by dividing out the X and Y standard deviations. In this form, the quantity is called the *correlation coefficient*.

It is, of course, immediate that if X and Y are independent, $\rho(X, Y) = 0$. The converse is not true (recall Example 10.2.1, where $Y = X^2$). However, even though ρ is not a good indicator of functional relationships *in general*, it does recognize *linear* dependence, as Theorem 10.2.4 demonstrates.

Proof. Following the notation of Definition 10.2.2, let X^* and Y^* denote the normalized transforms of X and Y. Then

$$0 \le \text{Var}(X^* \pm Y^*) = \text{Var}(X^*) \pm 2\,\text{Cov}(X^*, Y^*) + \text{Var}(Y^*)$$

$$= 1 \pm 2\rho(X, Y) + 1$$

$$= 2[1 \pm \rho(X, Y)]$$

But $1 \pm \rho(X, Y) \ge 0$ implies that $|\rho(X, Y)| \le 1$, and part (a) of the theorem is proved.

Next, suppose that $\rho(X, Y) = 1$. Then $\text{Var}(X^* - Y^*) = 0$; however, a random variable with zero variance is constant, except possibly on a set of probability 0. From the constancy of $X^* - Y^*$, it readily follows that Y is a linear function of X. The case for $\rho(X, Y) = -1$ is similar.

The converse of part (b) is left as an exercise.

Question 10.2.5 Let X and Y have the joint pdf,

$$f_{X, Y}(x, y) = \begin{cases} \dfrac{x + 2y}{22} & \text{for } (x, y) = \{(1, 1), (1, 3), (2, 1), (2, 3)\} \\ 0 & \text{elsewhere} \end{cases}$$

Find $\text{Cov}(X, Y)$ and $\rho(X, Y)$.

Question 10.2.6 Suppose that X and Y have the joint pdf,

$$f_{X, Y}(x, y) = x + y \qquad 0 < x < 1, 0 < y < 1$$

Find $\rho(X, Y)$.

Question 10.2.7 Find the covariance for X and Y defined by the joint pdf

$$f_{X,Y}(x, y) = \begin{cases} 8xy & 0 \leq x \leq y \leq 1 \\ 0 & \text{otherwise} \end{cases}$$

Also find $\rho(X, Y)$.

Question 10.2.8 Suppose that X and Y are discrete random variables with the following joint pdf:

(x, y)	$f_{X,Y}(x, y)$
$(1, 2)$	$\frac{1}{2}$
$(1, 3)$	$\frac{1}{4}$
$(2, 1)$	$\frac{1}{8}$
$(2, 4)$	$\frac{1}{8}$

Find the correlation coefficient between X and Y.

Question 10.2.9 Prove that $\rho(a + bX, c + dY) = \rho(X, Y)$ for constants a, b, c, and d. Note that this result allows for a change of scale to one convenient for computation.

Question 10.2.10 Let the random variable X take on the values $1, 2, \ldots, n$, each with probability $1/n$. Define Y to be X^2. Find $\rho(X, Y)$ and $\lim_{n \to \infty} \rho(X, Y)$.

Question 10.2.11 For random variables X and Y, show that

$$\text{Cov} (X + Y, X - Y) = \text{Var} (X) - \text{Var} (Y)$$

Question 10.2.12 Suppose that $\text{Cov} (X, Y) = 0$. Prove that

$$\rho(X + Y, X - Y) = \frac{\text{Var} (X) - \text{Var} (Y)}{\text{Var} (X) + \text{Var} (Y)}$$

We conclude this section with an estimation problem whose full significance will become apparent in Section 10.3. Suppose that the correlation coefficient between X and Y is unknown but we have some relevant information about it in the form of n measurements made on (X, Y)—specifically, (X_1, Y_1), (X_2, Y_2), $\ldots$, (X_n, Y_n). How can we use that information to *estimate* $\rho(X, Y)$?

Since the correlation coefficient can be written in terms of various theoretical moments,

$$\rho(X, Y) = \frac{E(XY) - E(X)E(Y)}{\sqrt{\text{Var} (X)} \sqrt{\text{Var} (Y)}}$$

it would seem reasonable to estimate each component of $\rho(X, Y)$ with its corresponding *sample* moment. That is, we will take $\bar{X}$ and $\bar{Y}$ to be estimators of $E(X)$ and $E(Y)$, replace $E(XY)$ with

$$\frac{1}{n} \sum_{i=1}^{n} X_i Y_i$$

and substitute

$$\frac{1}{n}\sum_{i=1}^{n}(X_i - \bar{X})^2 \quad \text{and} \quad \frac{1}{n}\sum_{i=1}^{n}(Y_i - \bar{Y})^2$$

for Var (X) and Var (Y). Putting these together gives R, the *sample correlation coefficient*:

$$R = \frac{\dfrac{1}{n}\sum_{i=1}^{n}X_i Y_i - \bar{X}\bar{Y}}{\sqrt{\dfrac{1}{n}\sum_{i=1}^{n}(X_i - \bar{X})^2}\sqrt{\dfrac{1}{n}\sum_{i=1}^{n}(Y_i - \bar{Y})^2}} \tag{10.2.1}$$

For computational purposes, it is better to express R in a slightly different way:

$$R = \frac{n\sum_{i=1}^{n}X_i Y_i - \left(\sum_{i=1}^{n}X_i\right)\left(\sum_{i=1}^{n}Y_i\right)}{\sqrt{n\sum_{i=1}^{n}X_i^2 - \left(\sum_{i=1}^{n}X_i\right)^2}\sqrt{n\sum_{i=1}^{n}Y_i^2 - \left(\sum_{i=1}^{n}Y_i\right)^2}} \tag{10.2.2}$$

Question 10.2.13 Derive Equation 10.2.2 from Equation 10.2.1.

CASE STUDY 10.2.1

On Wall Street, thousands of highly trained analysts use everything from crystal balls to multimillion-dollar computers to predict the price of stocks, bonds, and other financial commodities. While the actual number of variables involved in these predictions is enormous, the kind of reasoning that goes on can be illustrated by looking at just a few simple indicators of general stock market behavior. Two of the most familiar of these are the Dow Jones average and Standard and Poor's stock price index. Since both of these purport to measure the same quantity, "the state of the market," they should be expected to be highly correlated.

The Dow Jones Composite Average is derived from the daily quotations of some 65 stocks listed on the New York Stock Exchange. The Standard and Poor's "500" Composite Price Index is an average of daily prices from a large sample of stocks, where the index is expressed relative to base values calculated in 1941–1943. Table 10.2.1 gives for each of these the week's end closing average or index for the months of January, February, and March 1978. Also quoted are the prices for one stock (Eastman Kodak Common) that is used in determining the Dow Jones average and one stock (Eaton Corporation Common) that is not (29). (Our "expectation" would be that the correlation will be highest for the Standard and Poor's index, next highest for the Eastman Kodak prices, and lowest for the Eaton Corporation figures).

TABLE 10.2.1 END OF WEEK CLOSING PRICE, JANUARY–MARCH 1978

Week	Dow Jones Composite	Standard and Poor's Composite	Eastman Kodak Common	Eaton Common
1	276.61	91.62	49.75	36.00
2	271.26	89.69	48.50	33.87
3	272.47	89.89	48.87	34.75
4	268.35	88.58	45.75	33.75
5	271.44	89.62	45.12	34.12
6	272.35	90.08	45.25	34.87
7	263.86	87.96	43.37	34.37
8	265.07	88.49	43.75	34.00
9	262.15	87.45	41.62	33.50
10	265.38	88.88	43.50	33.87
11	269.41	90.20	43.25	34.00
12	267.01	89.36	42.12	34.25
13	266.94	89.21	42.25	34.75

Let Standard and Poor's index be the X-variable and Dow Jones the Y. Then

$$\sum_{i=1}^{13} x_i = 1161.03 \qquad \sum_{i=1}^{13} x_i^2 = 103{,}705.57$$

$$\sum_{i=1}^{13} y_i = 3492.3 \qquad \sum_{i=1}^{13} y_i^2 = 938{,}367.32$$

$$\sum_{i=1}^{13} x_i y_i = 311{,}946.5$$

and, from Equation 10.2.2,

$$r = \frac{13(311{,}946.5) - (1161.03)(3492.3)}{\sqrt{13(103{,}705.57) - (1161.03)^2}\ \sqrt{13(938{,}367.32) - (3492.3)^2}}$$

$$= 0.93$$

Similarly, the sample correlation coefficient between Eastman Kodak Common and the Dow Jones average comes to 0.83, and between Eaton Corporation and Dow Jones, 0.69.

Comment

Despite the statement of Theorem 10.2.4, the interpretation of r is far from obvious. What does it mean to say that one pair of variables has a sample correlation coefficient of, say, 0.93, and another, 0.69? How much stronger is one relationship than the other? Does it follow that either or both sets of variables are dependent? We will defer such questions to later sections. Here we are simply illustrating a computing formula.

Question 10.2.14 There is a theory espoused by some baseball fans that the number of home runs a team hits is markedly affected by the altitude of the club's home park, the rationale being that the air is thinner at the higher altitudes and the balls should go further. Below are the altitudes of the 12 American League ballparks and the number of home runs each of those teams hit during the 1972 season (164).

Club	Altitude (ft)	Number of Home Runs
Cleveland	660	138
Milwaukee	635	81
Detroit	585	135
New York	55	90
Boston	21	120
Baltimore	20	84
Minnesota	815	106
Kansas City	750	57
Chicago	595	109
Texas	435	74
California	340	61
Oakland	25	120

Compute the sample correlation coefficient for these data.

Question 10.2.15 The table below gives the mean temperature for 20 successive days and the average butterfat content in the milk of ten cows sampled on those days (128).

Date	Temperature	Percent Butterfat
Apr. 3	64	4.65
Apr. 4	65	4.58
Apr. 5	65	4.67
Apr. 6	64	4.60
Apr. 7	61	4.83
Apr. 8	55	4.55
Apr. 9	39	5.14
Apr. 10	41	4.71
Apr. 11	46	4.69
Apr. 12	59	4.65
Apr. 13	56	4.36
Apr. 14	56	4.82
Apr. 15	62	4.65
Apr. 16	37	4.66
Apr. 17	37	4.95
Apr. 18	45	4.60
Apr. 19	57	4.68
Apr. 20	58	4.65
Apr. 21	60	4.60
Apr. 22	55	4.46

Graph these data and compute their sample correlation coefficient.

Question 10.2.16 By late 1971, all cigarette packs had to be labeled with the words, "Warning: The Surgeon General Has Determined That Smoking Is Dangerous to Your Health." The case against smoking rested heavily on statistical, rather than laboratory, evidence. Extensive surveys of smokers and nonsmokers had revealed the former to have much higher risks of dying from a variety of causes, including heart disease. The table below shows (1) the annual cigarette consumption and (2) the mortality rate due to coronary heart disease (CHD) for 21 countries (109).

Year	Country	Cigarette Consumption per Adult per Year	CHD Mortality per 100,000 (Ages 35–64)
1962	United States	3900	256.9
1962	Canada	3350	211.6
1962	Australia	3220	238.1
1962	New Zealand	3220	211.8
1963	United Kingdom	2790	194.1
1962	Switzerland	2780	124.5
1962	Ireland	2770	187.3
1962	Iceland	2290	110.5
1962	Finland	2160	233.1
1963	West Germany	1890	150.3
1962	Netherlands	1810	124.7
1962	Greece	1800	41.2
1962	Austria	1770	182.1
1962	Belgium	1700	118.1
1962	Mexico	1680	31.9
1963	Italy	1510	114.3
1961	Denmark	1500	144.9
1962	France	1410	59.7
1962	Sweden	1270	126.9
1961	Spain	1200	43.9
1962	Norway	1090	136.3

Graph these data and compute their correlation coefficient.

Question 10.2.17 When two closely related species are crossed, the progeny will tend to have physical traits that lie somewhere "between" the corresponding traits of the parents. But what about behavioral traits? Are they similarly "mixed"? And does a hybrid whose physical traits favor one particular parent tend to have behavioral patterns that resemble those of the same parent? One attempt at answering these questions was an experiment done with mallard and pintail ducks. A total of 11 males were studied; all were second-generation crosses. A rating scale was devised that measured the extent to which the plumage of each of the ducks resembled the plumage of the first generation's parents. A score of 0 indicated that the hybrid had the same appearance (phenotype) as a pure mallard; a score of 20 meant that the hybrid looked like a pintail. Similarly, certain behavioral traits were quantified and a second scale was constructed that ranged from 0 (completely mallardlike) to 15 (completely pintaillike). The table shows the results (154). Plot these data and compute their sample correlation coefficient.

Male	Plumage Index, x	Behavioral Index, y
R	7	3
S	13	10
D	14	11
F	6	5
W	14	15
K	15	15
U	4	7
O	8	10
V	7	4
J	9	9
L	14	11

10.3 SIMPLE LINEAR REGRESSION: A HISTORICAL APPROACH

Our objective, as stated at the outset of this chapter, is to examine the relationship between two random variables. What we have accomplished in the way of preliminaries is the definition of the correlation coefficient, a quantity related to the notion of independence. We now need to look at the question of functional relationships in more detail and, in particular, at the role that $\rho(X, Y)$ plays in their interpretation.

First, a basic question: Given two presumably dependent random variables, how should we go about studying their joint behavior? One approach would be to consider the conditional pdf of Y given x as a function of x—that is, make a graph of $f_{Y|x}(y)$ versus x. While such graphs would show the relationship between X and Y, their use and interpretation are difficult. A better solution is to replace $f_{Y|x}$ with one of its numerical descriptors—say $E(Y|x)$—and graph $E(Y|x)$ versus x.

Example 10.3.1

Let X and Y have the following joint density:

$$f_{X,Y}(x, y) = y^2 e^{-y(x+1)} \qquad x \geq 0, y \geq 0$$

The marginal pdf for X is readily shown to be

$$f_X(x) = \frac{2}{(x+1)^3}.$$

Therefore, by Definition 3.7.1 the conditional pdf of Y given x is

$$f_{Y|x}(y) = \frac{y^2 e^{-y}(x+1)}{\dfrac{2}{(x+1)^3}} = \frac{1}{2}(x+1)^3 y^2 e^{-y(x+1)}$$

Then

$$E(Y \mid x) = \int_0^\infty y \cdot \frac{1}{2} (x + 1)^3 y^2 e^{-y(x+1)} \, dy$$

$$= \frac{1}{2(x + 1)} \int_0^\infty u^3 e^{-u} \, du$$

the latter expression resulting from the substitution $u = y(x + 1)$. But

$$\int_0^\infty u^3 e^{-u} \, du = \Gamma(4) = 3! = 6$$

so the conditional expectation reduces to

$$E(Y \mid x) = \frac{3}{x + 1}$$

(see Figure 10.3.1).

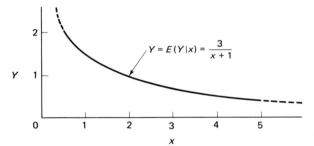

Figure 10.3.1   A graph of $E(Y \mid x)$ versus x.

In more general terminology, the search for a relationship between $E(Y \mid x)$ and x is known as the *regression problem* (and the function $h(x) = E(Y \mid x)$ is referred to as the *regression curve of Y on x*). Experimentally, problems of this sort arise when a researcher records a set of (x_i, y_i) pairs, plots the points on a scatter diagram (Figure 10.3.2), and then asks for an equation that adequately describes the functional relationship between the two variables. What the equation is seeking to approximate, of course, is $E(Y \mid x)$ versus x.

For the remainder of this chapter we will confine much of our attention to one particular regression problem, the special case where $E(Y \mid x)$ is a *linear* function of

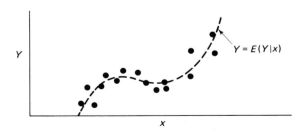

Figure 10.3.2   Scatter diagram.

x: $E(Y \mid x) = a + bx$. Although this is admittedly a very simple data model, it nevertheless occurs quite often in the real world. Furthermore, a number of phenomena whose relationships are not initially linear can be suitably transformed in such a way that the results for linear relationships remain valid. We present several examples of this sort in Section 10.4. Finally, many relationships that are not linear in their entirety are still, to a good approximation, linear over some interval of interest. The linear model, then, while simple, is not unrealistic.

Francis Galton, the renowned British biologist and scientist, perhaps more than any other person was responsible for launching *regression analysis* as a worthwhile field of statistical inquiry. Galton was a redoubtable data analyst, whose keen insight enabled him to intuit much of the basic mathematical structure that we now associate with the regression problem. One of his more famous endeavors was an examination of the relationship between parents' heights and their adult children's heights. Table 10.3.1 shows one of his sets of data (52). Galton used as a measure of parent height the average of the paternal height and 1.08 times the maternal height; this quantity was called the *mid-parent height.*

From cross-tabulations of this sort, Galton made a number of critically important discoveries bearing on the general problem of regression. If X denotes a mid-parent height and Y a child's height, then:

1. The marginal pdf's of both X and Y are normal.
2. $E(Y \mid x)$ is a linear function of x (see Figure 10.3.3).
3. Var $(Y \mid x)$ is constant in x (this property has the marvelous name, *homoscedasticity*).

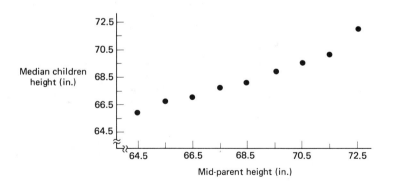

Figure 10.3.3   Children's heights versus parent's heights.

Sir Francis performed still one other feat of empirical legerdemain on Table 10.3.1. He replaced any four adjacent figures in the table by their average, which he positioned at the center of the four figures. Taking this average as the value of the joint density function at that point, he proceeded to draw contour lines—smooth

TABLE 10.3.1 NUMBER OF ADULT CHILDREN OF VARIOUS STATURES BORN OF 205 MID-PARENTS OF VARIOUS STATURES (ALL FEMALE HEIGHTS HAVE BEEN MULTIPLIED BY 1.08)

Height of the Mid-Parents (inches)	Heights of the Adult Children														Total Number of:		Medians or Values of M
	Below 62.2	62.2	63.2	64.2	65.2	66.2	67.2	68.2	69.2	70.2	71.2	72.2	73.2	Above	Adult Children	Mid-Parents	
Above 72.5	—	—	—	—	—	—	—	—	—	—	—	1	3	—	4	5	—
72.5	—	—	—	—	—	—	—	1	2	1	2	7	2	4	19	6	72.2
71.5	—	—	—	1	3	4	3	5	10	4	9	2	2	—	43	11	69.9
70.5	1	—	1	1	1	3	3	12	18	14	7	4	3	—	68	22	69.5
69.5	—	1	1	16	4	17	27	20	33	25	20	11	4	5	183	41	68.9
68.5	1	—	7	11	16	25	31	34	48	21	18	4	3	—	219	49	68.2
67.5	—	3	5	14	15	36	38	28	38	19	11	—	—	—	211	33	67.6
66.5	—	3	3	5	2	17	17	14	13	4	—	—	—	—	78	20	67.2
65.5	1	—	9	5	7	11	11	7	7	5	2	1	—	—	66	12	67.2
64.5	1	1	4	4	1	5	5	—	2	—	—	—	—	—	23	5	66.7
Below 64.5	1	—	2	4	1	2	2	1	1	—	—	—	—	—	14	1	65.8
Totals	5	7	32	59	48	117	138	120	167	99	64	41	17	14	928	205	
Medians	—	—	66.3	67.8	67.9	67.7	67.9	68.3	68.5	69.0	69.0	70.0	—	—			

curves connecting points of like value. From these he drew a truly remarkable conclusion:

4. The set of points in the plane for which $f_{X,Y}(x, y) = c$ is an ellipse for each constant $c > 0$. Furthermore, these ellipses are concentric.[1] [See Figure 10.3.4, reproduced from (122). Also, see (189) for a fuller discussion of the smoothing procedure.]

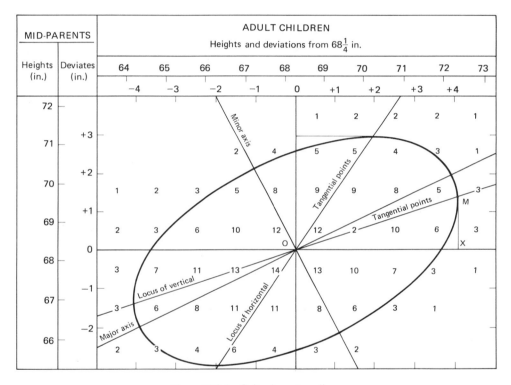

Figure 10.3.4 Galton's contour lines.

Galton (51) gave the following recollection of the birth of property 4:

At length, one morning, while waiting at a roadside station near Ramsgate for a train, and poring over the diagram in my notebook, it struck me that the lines of equal frequency ran in concentric ellipses. The cases were too few for my certainty, but my eye, being accustomed to such things, satisfied me that I was approaching the solution. More careful drawings strongly corroborated the first impression.

[1] Of this discovery, Pearson said, "That Galton should have evolved all this from his observations is to my mind one of the most noteworthy scientific discoveries arising from pure analysis of observations" (122).

Galton was curious as to the nature of the probability density functions that would possess these four properties, and he posed the question to J. D. H. Dickson, Tutor at St. Peter's College, Cambridge. Dickson readily supplied the answer, which we will consider in Section 10.6.

To conclude this section, we give a theorem showing the prominent role the correlation coefficient plays in linear regression.

> **Theorem 10.3.1.** If X and Y are random variables such that $E(Y \mid x) = a + bx$ for all values x taken on by X, then
>
> $$E(Y \mid x) = \mu_Y + \frac{\rho(X, Y)\sigma_Y}{\sigma_X} (x - \mu_X)$$

Proof. The proof for continuous variables is given; for the discrete case, replace the integrals by summations.

First, we note that

$$\int_{-\infty}^{\infty} E(Y \mid x) f_X(x) \, dx = \mu_Y \tag{10.3.1}$$

To see this, write the left-hand side as

$$\int_{-\infty}^{\infty} E(Y \mid x) f_X(x) \, dx = \int_{-\infty}^{\infty} \left[\int_{-\infty}^{\infty} y \, \frac{f_{X, Y}(x, y)}{f_X(x)} \, dy \right] \cdot f_X(x) \, dx$$

$$= \int_{-\infty}^{\infty} y \left[\int_{-\infty}^{\infty} f_{X, Y}(x, y) \, dx \right] dy$$

with the latter expression arising from a change in the order of integration. But

$$\int_{-\infty}^{\infty} y \left[\int_{-\infty}^{\infty} f_{X, Y}(x, y) \, dx \right] dy = \int_{-\infty}^{\infty} y f_Y(y) \, dy = \mu_Y$$

(The result in Equation 10.3.1 is not in the least surprising; think of what it says intuitively.)

Next, we multiply the equation $E(Y \mid x) = a + bx$ by $f_X(x)$ and integrate (with respect to x):

$$\int_{-\infty}^{\infty} E(Y \mid x) f_X(x) \, dx = \int_{-\infty}^{\infty} a f_X(x) \, dx + \int_{-\infty}^{\infty} b x f_X(x) \, dx$$

in which case

$$\mu_Y = a + b\mu_X \quad \text{or} \quad a = \mu_Y - b\mu_X$$

Therefore,

$$E(Y \mid x) = \mu_Y + b(x - \mu_X)$$

Now, transpose μ_Y and multiply by $(x - \mu_X)f_X(x)$:

$$[E(Y \mid x) - \mu_Y](x - \mu_X)f_X(x) = b(x - \mu_X)^2 f_X(x)$$

An equivalent expression (the reader should verify this) is

$$\int_{-\infty}^{\infty} (y - \mu_Y)(x - \mu_X)f_{X,Y}(x, y) \, dy = b(x - \mu_X)^2 f_X(x) \qquad (10.3.2)$$

If x is integrated out, Equation 10.3.2 becomes

$$\text{Cov}(X, Y) = b\sigma_X^2$$

or

$$b = \frac{\text{Cov}(X, Y)}{\sigma_X^2} = \frac{\rho(X, Y)\sigma_Y}{\sigma_X}$$

thus completing the proof.

If x is considered to be the independent variable and $E(Y \mid x)$ the dependent variable, then

$$E(Y \mid x) = \mu_Y + \frac{\rho(X, Y)\sigma_Y}{\sigma_X}(x - \mu_X) \qquad (10.3.3)$$

is the equation of a straight line through the point (μ_X, μ_Y). Equations of this sort are known as *regression lines*.

Comment

The term "regression line" derives from a consequence of Equation 10.3.3. Suppose we make the simplifying assumption that $\mu_X = \mu_Y = \mu$ and $\sigma_X = \sigma_Y$. Then Equation 10.3.3 becomes

$$E(Y \mid x) - \mu = \rho(X, Y)(x - \mu)$$

But recall that $|\rho(X, Y)| \le 1$—and, in this case, $0 < \rho(X, Y) < 1$. Here the positive sign of $\rho(X, Y)$ tells us that, on the average, tall parents have tall children. However, $\rho(X, Y) < 1$ means (again, *on the average*) that the children's heights are closer to the mean than the parents': a group of parents with mid-height, say, one inch above μ will have children with average height above μ but by less than one inch. Galton called this phenomenon "regression to mediocrity."

Question 10.3.1 Suppose that X and Y are two discrete random variables with joint pdf

$$f_{X,Y}(x, y) = \begin{cases} \dfrac{x + y}{15} & x = 1, 2, 3; \quad y = 0, 1 \\ 0 & \text{elsewhere} \end{cases}$$

Graph $f_{Y \mid x}(y)$ versus x for $x = 1, 2,$ and 3.

Question 10.3.2   Let X and Y have the joint pdf

$$f_{X, Y}(x, y) = \begin{cases} x + y & 0 < x < 1, \quad 0 < y < 1 \\ 0 & \text{elsewhere} \end{cases}$$

Plot $E(Y \mid x)$ versus x. Is it linear?

Question 10.3.3   Let X and Y have the pdf

$$f_{X, Y}(x, y) = \begin{cases} 2 & 0 < x < y, \quad 0 < y < 1 \\ 0 & \text{elsewhere} \end{cases}$$

Graph the function $Y = E(Y \mid x)$.

Question 10.3.4   Let X and Y have the trinomial pdf

$$f_{X, Y}(x, y) = \frac{n!}{x! \, y! \, (n - x - y)!} \, p_X^x \, p_Y^y (1 - p_X - p_Y)^{n - x - y}$$

(see Section 9.2). Find the regression curve of Y on X.

Question 10.3.5   Find the regression line of Y on X, where

$$f_{X, Y}(x, y) = \frac{y}{(x + 1)^4} \, e^{-y/(x + 1)} \qquad 0 < x, 0 < y$$

Verify that the coefficients agree with the statement of Theorem 10.3.1.

Question 10.3.6   Let X and Y have the bivariate pdf

$$f_{X, Y}(x, y) = 24xy \qquad 0 < x, \quad 0 < y, \quad x + y < 1$$

Find the regression line of Y on X.

Question 10.3.7   Suppose that the regression of Y on X is linear, as is the regression of X on Y. Verify that $\rho(X, Y)$ is the geometric mean of the slopes of the two regression lines. (Recall that the geometric mean of two numbers a and b is $\sqrt{ab}$.)

Question 10.3.8   Let X and Y be discrete random variables. The *conditional variance of Y given that $X = x$* is defined to be

$$\text{Var}\,(Y \mid x) = \sum_{\text{all } y} [y - E(Y \mid x)]^2 f_{Y \mid x}(y)$$

(An analogous statement holds when X and Y are continuous.) Show that an equivalent way of expressing $\text{Var}\,(Y \mid x)$ is

$$\text{Var}\,(Y \mid x) = E(Y^2 \mid x) - [E(Y \mid x)]^2$$

Question 10.3.9   Find $\text{Var}\,(Y \mid x)$ for $x = 1$ and $x = 2$ for the probability model given in Question 10.2.8.

Question 10.3.10   Let X and Y have the joint pdf,

$$f_{X, Y}(x, y) = \begin{cases} 2 & 0 \leq x < y < 1 \\ 0 & \text{elsewhere} \end{cases}$$

Find $\text{Var}\,(Y \mid x)$.

10.4 THE METHOD OF LEAST SQUARES

Theorem 10.3.1 addresses a theoretical problem: If $E(Y \mid x)$ is linear, it will necessarily be true that

$$E(Y \mid x) = \mu_Y + \frac{\rho(X, Y)\sigma_Y}{\sigma_X} \cdot (x - \mu_X)$$

Equivalently, the slope of the linear relationship is

$$\frac{\rho(X, Y)\sigma_Y}{\sigma_X}$$

and the y-intercept can be written

$$\mu_Y - \frac{\rho(X, Y)\sigma_Y}{\sigma_X} \mu_X$$

Section 10.4 explores a related practical problem: Given a set of n observations $(X_1, Y_1), (X_2, Y_2), \ldots, (X_n, Y_n)$, and a presumed linear relationship between X and Y, how should we estimate

$$\frac{\rho(X, Y)\sigma_Y}{\sigma_X} \quad \text{and} \quad \mu_Y - \frac{\rho(X, Y)\sigma_Y}{\sigma_X} \mu_X \quad ?$$

We begin by phrasing the question in a purely geometrical, nonprobabilistic context: *Given a set of n points $(x_1, y_1), (x_2, y_2), \ldots, (x_n, y_n)$ and a positive integer m, which polynomial of degree m is "closest" to the given points?* Even though we are dropping any reference here to random variables, we are still subsuming a variety of problems. How we fit polynomials to data is the same regardless of whether the points are treated as constants or as random variables. What we can infer from those fitted polynomials, though, does depend on the assumptions imposed on the n points. [Sections 10.5 and 10.6 look at the cases where the (x_i, y_i)'s are (x_i, Y_i)'s and (X_i, Y_i)'s.]

Suppose that the desired polynomial $p(x)$ is written

$$\sum_{k=0}^{m} a_k x^k$$

where $a_0, a_1, \ldots, a_m$ are to be determined. The *method of least squares* chooses as "solutions" those coefficients minimizing the sum of the squares of the vertical distances from the data points to the presumed polynomial. That is, the polynomial we will term "best" is the one whose coefficients minimize the function L, where

$$L = \sum_{i=1}^{n} [y_i - p(x_i)]^2$$

For the most part, our attention in this section will focus on the linear polynomial $p(x) = a + bx$. But we begin with a discussion of the special case where $p(x)$ is linear but $a = 0$.

Example 10.4.1

Suppose that the polynomial to be fitted to a set of n points is $p(x) = bx$. It is a simple matter to show that the value of the slope satisfying the least-squares criterion can be written as the ratio of two sums:

$$b = \frac{\sum_{i=1}^{n} x_i y_i}{\sum_{i=1}^{n} x_i^2} \tag{10.4.1}$$

Let L denote the sum of the squared vertical deviations of the (x_i, y_i)'s from $p(x)$:

$$L = \sum_{i=1}^{n} (y_i - bx_i)^2 \tag{10.4.2}$$

Remember that the x_i's and y_i's here are constants: the only variable on the right-hand side of Equation 10.4.2 is b. To find the value of b that minimizes L (and satisfies the least-squares criterion), we need to solve the equation $dL/db = 0$. But

$$\frac{dL}{db} = 2 \sum_{i=1}^{n} (y_i - bx_i)(-x_i)$$

$$= -2 \sum_{i=1}^{n} (x_i y_i - bx_i^2) \tag{10.4.3}$$

Setting Equation 10.4.3 equal to 0 gives

$$0 = -2 \sum_{i=1}^{n} (x_i y_i - bx_i^2)$$

or

$$b \sum_{i=1}^{n} x_i^2 = \sum_{i=1}^{n} x_i y_i$$

from which Equation 10.4.1 follows.

CASE STUDY 10.4.1

One of the most startling and profound scientific revelations of the twentieth century was the discovery in 1929 by the American astronomer Edwin Hubble that the universe is expanding. Hubble's announcement shattered forever the ancient belief that the heavens are basically in a state of cosmic equilibrium: quite the contrary, galaxies are receding from each other at fantastic velocities. If v is a galaxy's recession velocity (relative to any other galaxy) and d, its distance (from that other galaxy), Hubble's law states that

$$v = Hd$$

where H is known as Hubble's constant.

Table 10.4.1 lists distance and velocity measurements made on 11 galactic clusters (22). Estimate Hubble's constant.

TABLE 10.4.1

Cluster	Distance (millions of light years)	Velocity (thousands of miles/sec)
Virgo	22	0.75
Pegasus	68	2.4
Perseus	108	3.2
Coma Berenices	137	4.7
Ursa Major No. 1	255	9.3
Leo	315	12.0
Corona Borealis	390	13.4
Gemini	405	14.4
Bootes	685	24.5
Ursa Major No. 2	700	26.0
Hydra	1100	38.0

According to Equation 10.4.1,

$$H = b = \frac{\sum_{i=1}^{11} d_i v_i}{\sum_{i=1}^{11} d_i^2}$$

But

$$\sum_{i=1}^{11} d_i v_i = (22)(0.75) + \cdots + (1100)(38.0)$$

$$= 95,161.2$$

and

$$\sum_{i=1}^{11} d_i^2 = (22)^2 + \cdots + (1100)^2$$

$$= 2,685,141$$

Therefore,

$$H = \frac{95,161.2}{2,685,141}$$

$$= 0.03544 \tag{10.4.4}$$

Figure 10.4.1 shows the equation $v = 0.03544d$ superimposed on the data of Table 10.4.1.

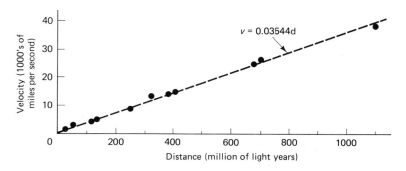

$v = 0.03544d$

Distance (million of light years)

Figure 10.4.1 Hubble's law.

Comment

Hubble's constant is an extremely significant number to cosmologists: its reciprocal is an estimate of the age of the universe. According to currently held theories, the universe began with the so-called Big Bang. The galactic recession that Hubble's law documents is nothing but a consequence of that initial conflagration. Since, in general, distance (d) equals rate (v) times time (t) and Hubble showed that $v = Hd$, we can write

$$t = \frac{d}{v}$$

$$= \frac{d}{Hd}$$

$$= \frac{1}{H}$$

But t in this context is the age of the universe—or, at the very least, the amount of time elapsed since the Big Bang. To express t in, say, *years*, we need to reexpress the units of H. Since the v_i's and d_i's in Table 10.4.1 are given in thousands of miles per second and millions of light years, respectively, the units of H are

$$\frac{\text{thousand miles}}{(\text{sec})(\text{millions of light years})}$$

But a million light years is 5.87850×10^{18} miles, so

$$H = 0.03544$$

$$= \frac{0.03544 \text{ thousand miles}}{(\text{sec})(5.87850 \times 10^{18} \text{ miles})}$$

Therefore,

$$t = \text{age of the universe (in sec)}$$

$$= \frac{1}{H} = \frac{(\text{sec})(5.87850 \times 10^{18} \text{ miles})}{35.44 \text{ miles}}$$

$$= 1.6587 \times 10^{17} \text{ seconds}$$

or approximately 5.2 billion years. (Astronomers today put the age of the universe closer to 15 billion years. The reason for the discrepancy is that techniques for measuring interstellar distances have been greatly refined since the 1920s when Hubble worked with the data of Table 10.4.1. If the most current distance/velocity measurements are fit with an equation of the form $v = Hd$, the resulting value of H is found to be about a third of Hubble's original figure, a revision that effectively multiplies by three the estimated age of the universe.)

Hubble's application notwithstanding, situations that call for the fitting of $y = bx$ are decidedly uncommon. Much more typical are data sets where the y-intercept is not necessarily zero. Theorem 10.4.1 summarizes the conclusions of the method of least squares as it applies to polynomials of the form $p(x) = a + bx$.

Theorem 10.4.1. Given n points $(x_1, y_1), \ldots, (x_n, y_n)$, the straight line $y = a + bx$ minimizing

$$L = \sum_{i=1}^{n} [y_i - (a + bx_i)]^2$$

has slope

$$b = \frac{n \sum_{i=1}^{n} x_i y_i - \left(\sum_{i=1}^{n} x_i\right)\left(\sum_{i=1}^{n} y_i\right)}{n \left(\sum_{i=1}^{n} x_i^2\right) - \left(\sum_{i=1}^{n} x_i\right)^2}$$

and y-intercept

$$a = \frac{\sum_{i=1}^{n} y_i - b \sum_{i=1}^{n} x_i}{n}$$

Proof. The proof is accomplished by the usual device of taking the partial derivatives of L with respect to a and with respect to b, setting the resulting expressions equal to 0, and solving. By the first step, we get

$$\frac{\partial L}{\partial a} = \sum_{i=1}^{n} (-2)[y_i - (a + bx_i)]$$

and

$$\frac{\partial L}{\partial b} = \sum_{i=1}^{n} (-2)x_i[y_i - (a + bx_i)]$$

Now, set the right-hand sides of $\partial L/\partial a$ and $\partial L/\partial b$ equal to 0 and simplify. This gives

$$na + \left(\sum_{i=1}^{n} x_i\right) b = \sum_{i=1}^{n} y_i$$

and

$$\left(\sum_{i=1}^{n} x_i\right) a + \left(\sum_{i=1}^{n} x_i^2\right) b = \sum_{i=1}^{n} x_i y_i$$

An application of Cramer's rule gives the solution for b stated in the theorem. The expression for a follows immediately.

CASE STUDY   10.4.2

A manufacturer of air conditioning units was having assembly problems due to the failure of a connecting rod to meet finished-weight specifications. Too many rods were being completely finished, then rejected as overweight. To reduce this cost, the manufacturer sought to estimate the relationship between the weight of the finished rod and that of the rough casting. Castings likely to produce rods that were too heavy could then be discarded before undergoing the final tooling process.

A total of 25 (x_i, y_i) pairs were measured, x_i being the weight of the ith rough casting and y_i the weight of the finished rod (129). From Table 10.4.2 we find

$$\sum_{i=1}^{25} x_i = 66.075 \qquad \sum_{i=1}^{25} x_i^2 = 174.673$$

$$\sum_{i=1}^{25} y_i = 50.12 \qquad \sum_{i=1}^{25} y_i^2 = 100.499$$

$$\sum_{i=1}^{25} x_i y_i = 132.491$$

Therefore,

$$b = \frac{25(132.491) - (66.075)(50.12)}{25(174.673) - (66.075)^2} = 0.648$$

and

$$a = \frac{50.12 - 0.648(66.075)}{25} = 0.292$$

making the least-squares line

$$y = 0.292 + 0.648x \tag{10.4.5}$$

TABLE 10.4.2 ROUGH AND FINISHED ROD WEIGHTS

Rod Number	Rough Weight	Finished Weight	Rod Number	Rough Weight	Finished Weight
1	2.745	2.080	14	2.635	1.990
2	2.700	2.045	15	2.630	1.990
3	2.690	2.050	16	2.625	1.995
4	2.680	2.005	17	2.625	1.985
5	2.675	2.035	18	2.620	1.970
6	2.670	2.035	19	2.615	1.985
7	2.665	2.020	20	2.615	1.990
8	2.660	2.005	21	2.615	1.995
9	2.655	2.010	22	2.610	1.990
10	2.655	2.000	23	2.590	1.975
11	2.650	2.000	24	2.590	1.995
12	2.650	2.005	25	2.565	1.955
13	2.645	2.015			

Figure 10.4.2 is a graph of Equation 10.4.5 superimposed over the data of Table 10.4.2.

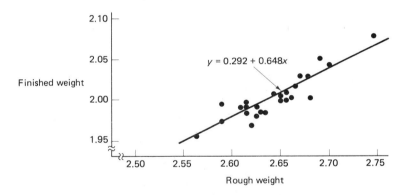

Figure 10.4.2 Regression line for rough and finished rod weights.

It follows that if the weight of a rough casting is, say, 2.71, the estimate of its finished weight would be 2.05 (see Figure 10.4.2):

$$\text{estimated weight} = E(Y \mid 2.71)$$
$$= 0.292 + 0.648(2.71)$$
$$= 2.05$$

If an expected value of 2.05 is judged to be too heavy, then the rough casting should be discarded.

Question 10.4.1 Crickets make their chirping sound by sliding one wing cover very rapidly back and forth over the other. Biologists have long been aware that there is a linear relationship between *temperature* and the frequency with which a cricket chirps, although the slope and *y*-intercept of the relationship varies from species to species. Listed below are 15 frequency-temperature observations recorded for the striped ground cricket, *Nemobius fasciatus fasciatus* (125). Plot these data and find the equation of the least-squares line, $y = a + bx$. Suppose a cricket of this species is observed to chirp 18 times per second. What would be the estimated temperature?

Observation Number	Chirps per Second, x_i	Temperature, y_i (°F)
1	20.0	88.6
2	16.0	71.6
3	19.8	93.3
4	18.4	84.3
5	17.1	80.6
6	15.5	75.2
7	14.7	69.7
8	17.1	82.0
9	15.4	69.4
10	16.2	83.3
11	15.0	79.6
12	17.2	82.6
13	16.0	80.6
14	17.0	83.5
15	14.4	76.3

Question 10.4.2 The aging of whisky in charred oak barrels brings about a number of chemical changes that enhance its taste and darken its color. Shown below is the change in a whisky's proof as a function of the number of years it is stored (148).

Age, x (years)	Proof, y
0	104.6
0.5	104.1
1	104.4
2	105.0
3	106.0
4	106.8
5	107.7
6	108.7
7	110.6
8	112.1

(*Note:* The proof initially decreases because of dilution by moisture in the staves of the barrels.) Graph these data and draw in the least-squares line.

Question 10.4.3 To understand better the plight of people returning to professions after periods of inactivity, a survey was conducted at some 67 randomly selected hospitals throughout the United States: The administrators were asked about their willingness to hire medical technologists who had been away from the field for a certain number of years. The results are summarized below (134).

Years of Inactivity	Percent of Hospitals Willing to Hire
$\frac{1}{2}$	100
$1\frac{1}{2}$	94
4	75
8	44
13	28
18	17

Graph these data and fit them with a straight line.

Question 10.4.4 Listed below are the average weights of varsity football players at the University of Texas for selected years from the turn of the century to the middle 1960s. [Data slightly modified from (98).]

Year	Average Weight (lb)
1905	164
1919	163
1932	181
1945	192
1955	194
1965	199

Graph these data and fit them with a straight line. If the linear model were to remain valid for this phenomenon in the years ahead (a highly unlikely state of affairs!), what would the average University of Texas football player weigh in the year 3000? What does this tell you about (1) extrapolating linear models, and/or (2) scheduling the University of Texas to play your school on Homecoming in the year 3000?

Question 10.4.5 Verify that the coefficients a and b of the least-squares straight line are solutions of the matrix equation

$$\begin{pmatrix} n & \sum x_i \\ \sum x_i & \sum x_i^2 \end{pmatrix}\begin{pmatrix} a \\ b \end{pmatrix} = \begin{pmatrix} \sum y_i \\ \sum x_i y_i \end{pmatrix}$$

Question 10.4.6 Suppose that we wish to fit a least-squares *parabola*, $y = a + bx + cx^2$, to a set of n points, $(x_1, y_1), (x_2, y_2), \ldots, (x_n, y_n)$. Show that the coefficients a, b, and c solve the matrix equation

$$\begin{pmatrix} n & \sum x_i & \sum x_i^2 \\ \sum x_i & \sum x_i^2 & \sum x_i^3 \\ \sum x_i^2 & \sum x_i^3 & \sum x_i^4 \end{pmatrix}\begin{pmatrix} a \\ b \\ c \end{pmatrix} = \begin{pmatrix} \sum y_i \\ \sum x_i y_i \\ \sum x_i^2 y_i \end{pmatrix}$$

Question 10.4.7 Derive the matrix equation giving the coefficients of the least-squares polynomial of degree k.

Question 10.4.8 Set up (but do not solve) the equations necessary for determining the least-squares estimates for a trigonometric regression,

$$y = a + bx + c \sin x + d \cos x$$

Assume that a random sample $(x_1, y_1), \ldots, (x_n, y_n)$ has been observed.

An experimenter collects a set of n (x_i, y_i)'s. What should be the first step in their statistical analysis, the computation of a and b? No! *The first step in any single-variable regression problem is to plot the data.* This is critically important because the a and b coming out of Theorem 10.4.1 are meaningless if the xy-relationship is not linear to begin with. Consider, for example, the set of six points listed in Table 10.4.3. Substituting $\sum_{i=1}^{6} x_i = 21.0$, $\sum_{i=1}^{6} y_i = 22.0$, $\sum_{i=1}^{6} x_i^2 = 94.5$, and $\sum_{i=1}^{6} x_i y_i = 77.0$ into Theorem 10.4.1 gives

$$b = \frac{6(77.0) - (21.0)(22.0)}{6(94.5) - (21.0)^2} = 0$$

and

$$a = \frac{22.0 - (0)(21.0)}{6} = 3.67$$

TABLE 10.4.3

x_i	y_i
5.5	4.0
1.0	1.0
3.0	6.0
1.5	4.0
4.0	6.0
6.0	1.0

Is $y = 3.67 + 0 \cdot x$ the best straight line through the data? Yes. Is it a meaningful mathematical description of the xy-relationship? No. Figure 10.4.3 shows the equation $y = 3.67 + 0 \cdot x$ superimposed over the six given points. Quite clearly, no straight line would adequately characterize the pattern suggested by this particular set of data: the relationship is *curvilinear* rather than linear.

What should be done, then, if a scatter diagram shows the data to be distinctly nonlinear? In practice, two variations of this problem are likely to be encountered. Sometimes the very nature of the phenomenon being studied suggests a priori what form the regression function should take. For example, suppose that y is a radiation

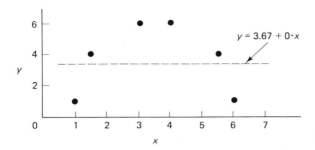

Figure 10.4.3   A curvilinear relationship.

level and x denotes time. The kinetics of radioactive decay suggest that x and y are likely to be related by a function of the form $y = ae^{bx}$, rather than $y = a + bx$. On other occasions, though, we have no contextual clues hinting at the form that $y = p(x)$ should take: all we know is limited to whatever can be inferred from the scatter diagram itself.

For both these situations, it is often possible to "transform" x and y into a new set of variables—x' and y'—that *are* linearly related. Theorem 10.4.1 can then be applied to the (x_i', y_i')'s. Once those data have been fit (with a straight line), transforming back to the original (x_i, y_i)'s (and the original curve) is usually easy.

To introduce this idea, let's consider the simplest kind of nonlinear problem, situations where the experimenter *knows* the general form of $y = p(x)$. Suppose, for example, that the phenomenon is such that the xy-relationship is theoretically described by a function of the form

$$y = ae^{bx} \qquad (10.4.6)$$

Taking the natural log of both sides of Equation 10.4.6 gives

$$\ln y = \ln a + bx$$

Let $y' = \ln y$, $x' = x$, $a' = \ln a$, and $b' = b$. Then $\ln y = \ln a + bx$ can be written

$$y' = a' + b'x'$$

showing that y' is linear with x'. That being the case, we can use the formulas given in Theorem 10.4.1 to find a' and b' (based on x_i' and y_i', $i = 1, 2, \ldots, n$). Then, once a' and b' have been determined, we can deduce a and b and write the regression function in its original (nonlinear) form, $y = ae^{bx}$.

CASE STUDY   10.4.3

Radioactive gold (^{195}Au–aurothiomalate) has an affinity for inflamed tissue and is sometimes used as a tracer to help diagnose arthritis. The data in Table 10.4.4 (54) come from an experiment investigating the length of time, and in what concentrations, ^{195}Au–aurothiomalate is retained in a person's serum. Listed are the serum gold concentrations found in 10 blood samples taken from patients given an initial dose of 50 mg. Follow-up readings were made at various times, ranging from one to seven days after injection. In each case, the retention is expressed in terms of the patient's day-zero serum gold concentration.

Sec. 10.4   The Method of Least Squares

TABLE 10.4.4

Days after injection, x_i	Serum gold, y_i (% of day-zero conc.)
1	94.5
1	86.4
2	71
2	80.5
2	81.4
3	67.4
5	49.3
6	46.8
6	42.3
7	36.6

Figure 10.4.4 shows a graph of the data. As would be expected given the nature of the data, the xy-relationship does not look particularly linear. The physics of the situation would suggest that we fit, instead, a model of the form $y = ae^{bx}$, or, equivalently, $\ln y = \ln a + bx$. Table 10.4.5 shows the original data as well as the transformed values, $y_i' = \ln y_i$ and $x_i' = x_i$.

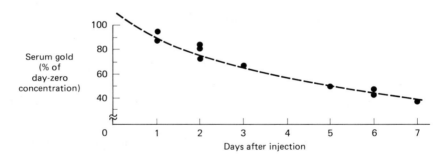

Figure 10.4.4 Serum gold retention regression.

TABLE 10.4.5

x_i	y_i	$x_i' (= x_i)$	$y_i' (= \ln y_i)$
1	94.5	1	4.54860
1	86.4	1	4.45899
2	71	2	4.26268
2	80.5	2	4.38826
2	81.4	2	4.39938
3	67.4	3	4.21065
5	49.3	5	3.89792
6	46.8	6	3.84588
6	42.3	6	3.74479
7	36.6	7	3.60005

Ignoring the (x_i, y_i)'s and working instead with the (x_i', y_i')'s, we find $\sum_{i=1}^{10} x_i' = 35$, $\sum_{i=1}^{10} y_i' = 41.35720$, $\sum_{i=1}^{10} x_i'^2 = 169$, and $\sum_{i=1}^{10} x_i' y_i' = 137.97415$. Substituting into Theorem 10.4.1 gives

$$b' = \frac{10(137.97415) - (35)(41.35720)}{10(169) - (35)^2} = -0.14572$$

and

$$a' = \frac{41.35720 - (-0.14572)(35)}{10} = 4.64574$$

implying that the best straight line describing the relationship between x' and y' is

$$y' = 4.64574 - 0.14572x' \tag{10.4.7}$$

The final step is to rewrite Equation 10.4.7 in terms of the original data (and the original a and b). Since $y = ae^{bx}$ is equivalent to $\ln y = \ln a + bx$, it follows that

$$b = b' = -0.14572 = -0.146$$

and

$$a = e^{a'} = e^{4.64574} = 104.141$$

Therefore, the equation of the form $y = ae^{bx}$ that best fits the data of Table 10.4.5 is

$$y = 104.141e^{-0.146x}$$

Question 10.4.9 Estimate the effective half-life of ^{195}Au–aurothiomalate. (How long does it take for half the gold to disappear from a person's serum?)

Question 10.4.10 One of the factors thought to contribute to the incidence of skin cancer is ultraviolet (UV) radiation coming from the sun. It is well known that the amount of UV radiation a person receives is a function of the shielding thickness of the earth's ozone layer, which, in turn, depends on the person's latitude. Listed below are the malignant skin cancer (melanoma) rates for white males determined for nine areas throughout the United States during the three-year period from 1969 to 1971 (39). The location of each area is given in "degrees north latitude."

Location number	Degrees north latitude	Melanoma rate (per 100,000)
1	32.8	9.0
2	33.9	5.9
3	34.1	6.6
4	37.9	5.8
5	40.0	5.5
6	40.8	3.0
7	41.7	3.4
8	42.2	3.1
9	45.0	3.8

Fit these data with an exponential model, $y = ae^{bx}$. Let x denote "degrees north latitude" and y, "melanoma rate." Plot the data and sketch in the regression line.

Another common nonlinear equation easy to "linearize" with an appropriate transformation is

$$y = ax^b \tag{10.4.8}$$

Models of this type have many biological and engineering applications. Taking the common log of both sides of Equation 10.4.8 gives

$$\log y = \log a + b \cdot \log x$$

which shows that *log x is linear with log y*. Or, written in the notation we introduced in the previous example,

$$y' = a' + b'x' \tag{10.4.9}$$

where $y' = \log y$, $a' = \log a$, $b' = b$, and $x' = \log x$. (Fitting a "log-log" relationship like Equation 10.4.9 is described more fully in Case Study 10.4.4.)

In many areas of both the physical as well as the social sciences, the *growth* of a phenomenon can be an important variable. In biology, it might be the doubling time of a *Drosophila* population; in economics, the rise of a financial institution; in political science, the gradual acceptance of a government policy. Prominent among the many "growth" models that are used to describe situations of this sort is the *logistic equation*,

$$y = \frac{L}{1 + e^{a+bx}} \tag{10.4.10}$$

where a, b, and L are constants. For various values of a and b, Equation 10.4.10 generates a variety of S-shaped curves. Like the models in Equations 10.4.6 and 10.4.8, though, a logistic regression can be linearized with a suitable transformation. Start by taking the reciprocal of Equation 10.4.10:

$$\frac{1}{y} = \frac{1 + e^{a+bx}}{L}$$

Therefore,

$$\frac{L}{y} = 1 + e^{a+bx}$$

and

$$\frac{L-y}{y} = e^{a+bx}$$

Equivalently,

$$\ln\left(\frac{L-y}{y}\right) = a + bx$$

which is now in the "standard" form, $y' = a' + b'x'$, where $y' = \ln((L-y)/y)$, $a' = a$, $b' = b$, and $x' = x$. (A set of data for which Equation 10.4.10 is an appropriate model is the focus of Question 10.4.13.)

One final example of an *a priori* model is worth mentioning. In reliability studies, engineers often collect data on the proportion of items that fail (y) as a function of how long those items have been "on test" (x). In certain contexts, the relationship between x and y is of the form

$$y = 1 - e^{-x^b/a} \qquad (10.4.11)$$

Here, once again, a little algebra can produce a linear equivalent of Equation 10.4.11. If $y = 1 - e^{-x^b/a}$, then

$$y - 1 = -e^{-x^b/a}$$

$$1 - y = e^{-x^b/a}$$

$$\ln(1-y) = -\frac{x^b}{a}$$

$$\ln\left(\frac{1}{1-y}\right) = \frac{x^b}{a}$$

and finally,

$$\ln \ln \left(\frac{1}{1-y}\right) = -\ln a + b \cdot \ln x \qquad (10.4.12)$$

Setting $y' = \ln \ln (1/(1-y))$, $a' = -\ln a$, $b' = b$, and $x' = \ln x$ reduces Equation 10.4.11 to the familiar $y' = a' + b'x'$.

Question 10.4.11 Suppose that a set of n (x_i, y_i)'s are measured on a phenomenon whose theoretical xy-relationship is of the form $y = ax^b$. On what kind of graph paper would the (x_i, y_i)'s show a linear relationship?

Question 10.4.12 Suppose that theoretical considerations suggest that two variables, x and y, have a relationship described by the function $y = x/(a + bx)$. Derive the linearizing transformation.

Question 10.4.13 The growth of the American intercontinental ballistic missile force during the 1960s is summarized in the table below (7).

Year, x	Number of ICBM's, y
1960	18
1961	63
1962	294
1963	424
1964	834
1965	854
1966	904
1967	1054
1968	1054
1969	1054

Fit these data with a logistic model. Let $L = 1060$. Graph the original data and sketch in the curve, $y = 1060/(1 + e^{a + bx})$.

For most researchers, working with a phenomenon for which an *a priori* regression model is readily apparent is the exception rather than the rule. Much more common than the kind of situation described in Case Study 10.4.3 are problems where all that is known about the xy-relationship is what appears on the graph of the original data. There are no theoretical principles to appeal to or past experience to rely on. Finding an appropriate $y = f(x)$ under these conditions is essentially an exercise in trial and error. Our prospects, though, are not as bleak as they might seem. In the shape of the scatter diagram are clues that can help us make intelligent guesses.

Suppose that an experimenter plots the observed (x_i, y_i)'s and gets a curvilinear pattern similar to one of the three appearing in Figure 10.4.5. A good model to try would be

$$y = ax^b \tag{10.4.13}$$

If the scatter diagram looks more like either Figure 10.4.6a or b, a better choice would be $y = ae^{bx}$. Other patterns are associated with still other functions. [For a more complete discussion of how the appearances of scattered diagrams suggest various nonlinear models, see (112).]

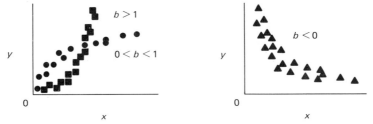

Figure 10.4.5 Regressions of the form $y = ax^b$.

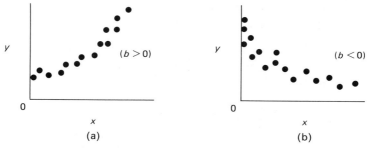

Figure 10.4.6 Regressions of the form $y = ae^{bx}$.

CASE STUDY 10.4.4

Among mammals, the relationship between the age at which an animal develops locomotion and the age at which it first begins to play has been widely studied. Listed in Table 10.4.6 are typical "onset" times for locomotion and for play in 11 different species (36). Graphed, the data show a pattern for which $y = ax^b$ would be a suitable model (see Figure 10.4.7).

TABLE 10.4.6

Species	Locomotion begins, x_i (days)	Play begins, y_i (days)
Homo sapiens	360	90
Gorilla gorilla	165	105
Felis catus	21	21
Canis familiaris	23	26
Rattus norvegicus	11	14
Turdus merula	18	28
Macaca mulatta	18	21
Pan troglodytes	150	105
Saimiri sciurens	45	68
Cercocebus alb.	45	75
Tamiasciureus hud.	18	46

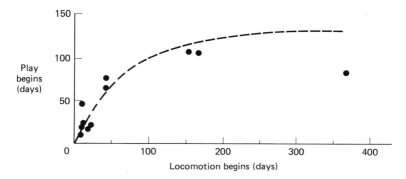

Figure 10.4.7 Regression of play onset versus locomotion onset.

Since $y = ax^b$, we can write $\log y = \log a + b \log x$. Columns 2 and 4 of Table 10.4.7 list $x_i' (= \log x_i)$ and $y_i' (= \log y_i)$.

TABLE 10.4.7

x_i	x_i'	y_i	y_i'
360	2.55630	90	1.95424
165	2.21748	105	2.02119
21	1.32222	21	1.32222
23	1.36173	26	1.41497
11	1.04139	14	1.14613
18	1.25527	28	1.44716
18	1.25527	21	1.32222
150	2.17609	105	2.02119
45	1.65321	68	1.83251
45	1.65321	75	1.87506
18	1.25527	46	1.66276

For the transformed data, we find

$$\sum_{i=1}^{11} x_i' = 17.74744 \qquad \sum_{i=1}^{11} y_i' = 18.01965$$

$$\sum_{i=1}^{11} x_i'^2 = 31.06764 \qquad \sum_{i=1}^{11} x_i' y_i' = 30.43743$$

so

$$b' = \frac{11(30.43743) - (17.74744)(18.01965)}{11(31.06764) - (17.74744)^2}$$

$$= 0.56 = b$$

and

$$a' = \frac{18.01965 - (0.56)(17.74744)}{11}$$

$$= 0.73364$$

But

$$a' = \log a$$

which implies that

$$a = 10^{0.73364}$$

$$= 5.42$$

(The equation $y = 5.42 \cdot x^{0.56}$ is shown as the dashed curve in Figure 10.4.7.)

Question 10.4.14 Over the years, many efforts have been made to demonstrate that the human brain is appreciably different in structure from the brains of lower-order primates. In point of fact, such differences in gross anatomy are disconcertingly difficult to discern. Listed below are the average areas of the striate cortex (x) and the prestriate cortex (y) found for humans and for three species of chimpanzees (120).

	Area	
Primate	Striate Cortex, x (mm^2)	Prestriate Cortex, y (mm^2)
Homo	2613	7838
Pongo	1876	2864
Cercopithecus	933	1334
Galago	78.9	40.8

Fit these data with an equation of the form

$$y = ax^b$$

Question 10.4.15 Listed below are the average prices of used cars in 1957 (20). Figures are given for cars ranging from 1 year old to 10 years old.

Age	Average Price
1	$2651
2	1943
3	1494
4	1087
5	765
6	538
7	484
8	290
9	226
10	204

Let "age" be the x variable and "average price," the y variable. Fit these data with a regression function of the form $y = ae^{bx}$. Graph the data and sketch in the least-squares curve.

Question 10.4.16 Mistletoe is a plant that grows parasitically in the upper branches of large trees. Left unchecked, it can seriously stunt a tree's growth. Recently an experiment was done to test a theory that older trees are less susceptible to mistletoe growth than younger trees. A number of shoots were cut from 3-, 4-, 9-, 15-, and 40-year-old Ponderosa pines. These were then side-grafted to 3-year-old nursery stock and planted in a preserve. Each tree was "inoculated" with mistletoe seeds. Five years later, a count was made of the number of mistletoe plants in each stand of trees. (A stand consisted of approximately ten trees; there were three stands of each of the four youngest age groups and two stands of the oldest.) The results are shown on page 468 (138).

	Age of Trees, x (years)				
	3	4	9	15	40
Number	28	10	15	6	1
of Mistletoe	33	36	22	14	1
Plants, y	22	24	10	9	

(a) Graph the original data. Also, graph x versus $\ln y$.

(b) Fit the data with an exponential model, $y = ae^{bx}$. Sketch in the least-squares curve on the graph of the original data.

10.5 AN INTRODUCTION TO THE LINEAR MODEL

In the geometric view of Section 10.4, distributions for the x- and y-variables were ignored. The remainder of this chapter examines linear relationships where one or both of the measurements do have an assigned density. This section is devoted to the former—where x is not a random variable but Y is. Imagine, for instance, a clinician administering various *fixed* dosages (x) of a blood coagulant to hemophiliacs and then measuring the subjects' resulting clotting times, the *random* variable Y.

Our objective here is to propose a reasonable set of assumptions for Y and explore some of the inference procedures that will inevitably follow. In practice, the (x, Y)-regression model most often used is

$$Y = a + bx + \epsilon \qquad \epsilon \sim N(0, \sigma^2)$$

where the quantity ϵ represents "random error." Equivalently,

$$Y \sim N(a + bx, \sigma^2)$$

Writing ϵ explicitly into the model simply emphasizes the notion that Y differs from its mean by an error term.

Comment

A more general form of the preceding model is

$$Y = \sum_{j=1}^{k} b_j g_j(x) + \epsilon \qquad \epsilon \sim N(0, \sigma^2) \tag{10.5.1}$$

for suitably chosen $g_j(x)$. Even though the $g_j(x)$ may be nonlinear, Equation 10.5.1 is still referred to as a *linear model* because

$$E(Y) = \sum_{j=1}^{k} b_j g_j(x)$$

is a linear function of the parameters $b_1, b_2, \ldots, b_k$. A further direction of generalization is the extension to more than one variable—that is, letting Y

depend on the *set* of nonrandom variables $x_1, x_2, \ldots, x_m$. A comprehensive treatment of these more general models can be found in (59) or (150), to mention but two of a large number of books on the subject.

Placing a distribution on Y does not change our basic interest in estimating the coefficients a and b. What it does do, though, is let us address other issues. With a pdf assumed for Y, we can find maximum-likelihood estimators for a and b (rather than least-squares estimators) and use the distributions of those MLEs to set up inference procedures.

Theorem 10.5.1. Let $Y_1, \ldots, Y_n$ be independent random variables, $Y_i \sim N(a + bx_i, \sigma^2)$, $i = 1, \ldots, n$, with parameters a, b, and σ^2 all being unknown. The corresponding maximum-likelihood estimators are given by

$$\hat{A} = \bar{Y} - \hat{B}\bar{x}$$

$$\hat{B} = \frac{n \sum_{i=1}^{n} x_i Y_i - \left(\sum_{i=1}^{n} x_i\right)\left(\sum_{i=1}^{n} Y_i\right)}{n \sum_{i=1}^{n} x_i^2 - \left(\sum_{i=1}^{n} x_i\right)^2} = \frac{\sum_{i=1}^{n} (x_i - \bar{x})(Y_i - \bar{Y})}{\sum_{i=1}^{n} (x_i - \bar{x})^2}$$

and

$$\hat{\sigma^2} = \frac{1}{n} \sum_{i=1}^{n} (Y_i - \hat{Y}_i)^2$$

where $\hat{Y}_i = \hat{A} + \hat{B}x_i$, $i = 1, \ldots, n$.

Proof. Since $Y_i \sim N(a + bx_i, \sigma^2)$, the likelihood function is

$$L = \frac{1}{(2\pi\sigma^2)^{n/2}} \exp\left[-\frac{1}{2\sigma^2} \sum_{i=1}^{n} (y_i - a - bx_i)^2 \right]$$

or

$$-2 \ln L = n \ln (2\pi) + n \ln (\sigma^2) + \frac{1}{\sigma^2} \sum_{i=1}^{n} (y_i - a - bx_i)^2$$

As frequently happens, the maximum of L occurs when the partials of $\ln L$ (or $-2 \ln L$) with respect to a, b, and σ^2 all vanish. Here, the relevant equations are

$$\frac{2}{\sigma^2} \sum_{i=1}^{n} (y_i - a - bx_i)(-1) = 0$$

$$\frac{2}{\sigma^2} \sum_{i=1}^{n} (y_i - a - bx_i)(-x_i) = 0$$

$$\frac{n}{\sigma^2} - \frac{1}{(\sigma^2)^2} \sum_{i=1}^{n} (y_i - a - bx_i)^2 = 0$$

The first two equations depend only on a and b, and the resulting solutions for $\hat{A}$ and $\hat{B}$ have the same form as those found in the proof of Theorem 10.4.1. Substituting the solutions for the first two equations into the third gives the claimed $\widehat{\sigma^2}$.

Theorems 10.5.2 and 10.5.3 establish the distributions of $\hat{A}$, $\hat{B}$, and $\widehat{\sigma^2}$ as well as an important independence result. These are the key properties from which hypothesis tests and confidence intervals for a and b will follow almost immediately.

Theorem 10.5.2.

(a) $\hat{A}$ and $\hat{B}$ are both normally distributed.

(b) $\hat{A}$ and $\hat{B}$ are both unbiased: $E(\hat{A}) = a$ and $E(\hat{B}) = b$.

(c) $\text{Var}(\hat{B}) = \dfrac{\sigma^2}{\sum\limits_{i=1}^{n}(x_i - \bar{x})^2}$.

(d) $\text{Var}(\hat{A}) = \dfrac{\sigma^2 \sum\limits_{i=1}^{n} x_i^2}{n \sum\limits_{i=1}^{n}(x_i - \bar{x})^2}$.

Proof. (See Question 10.5.3.)

Theorem 10.5.3.

(a) The random variables $\hat{B}$, $\bar{Y}$, and $\widehat{\sigma^2}$ are mutually independent.

(b) $\dfrac{n\widehat{\sigma^2}}{\sigma^2} \sim \chi^2_{n-2}$.

Proof. (See Appendix 10.1)

Corollary. $[n/(n-2)]\widehat{\sigma^2}$ is an unbiased estimator of σ^2.

Proof. The corollary follows readily from the basic properties of the expected value:

$$E\left(\frac{n}{n-2}\widehat{\sigma^2}\right) = \frac{\sigma^2}{n-2} E\left(\frac{n\widehat{\sigma^2}}{\sigma^2}\right) = \frac{\sigma^2}{n-2} \cdot (n-2) = \sigma^2$$

The next-to-last equality is a result of the fact that $n\widehat{\sigma^2}/\sigma^2 \sim \chi^2_{n-2}$, and the expected value of a chi-square random variable with $n-2$ degrees of freedom is $n-2$ (see Question 7.5.2).

We now have the theoretical background necessary to develop inference procedures for a and b. Our focus will be on the latter because hypothesis tests and confidence intervals for the slope are much more common than their analogs for the y-intercept.

Theorem 10.5.4. The random variable

$$\frac{\sqrt{n-2}\,(\hat{B}-b)\sqrt{\sum_{i=1}^{n}(x_i-\bar{x})^2}}{\sqrt{n\widehat{\sigma^2}}} \sim T_{n-2}$$

Proof. The statistic T is the quotient of the standard normal

$$\frac{\hat{B}-b}{\sqrt{\dfrac{\sigma^2}{\sum_{i=1}^{n}(x_i-\bar{x})^2}}}$$

by the square root of the χ^2_{n-2} variable, $n\widehat{\sigma^2}/\sigma^2$, divided by its degrees of freedom, $n-2$. Since Theorem 10.5.3 establishes independence of the numerator and denominator, the conclusion is immediate from Definition 7.6.1.

Theorem 10.5.5. A $100(1-\alpha)\%$ confidence interval for b is given by

$$\left(\hat{b} - t_{\alpha/2,\,n-2}\sqrt{\frac{n\widehat{\sigma^2}}{(n-2)\sum_{i=1}^{n}(x_i-\bar{x})^2}}, \; \hat{b} + t_{\alpha/2,\,n-2}\sqrt{\frac{n\widehat{\sigma^2}}{(n-2)\sum_{i=1}^{n}(x_i-\bar{x})^2}} \right)$$

Proof. The proof is left as an exercise.

CASE STUDY   10.5.1

An airline's freight revenues should be a function of the amount of freight carried. If the function is approximately linear in the usual form, $y = a + bx$, we would interpret b to be the marginal revenue—that is, b gives the additional revenue resulting from a unit increase in the amount of cargo consigned. Table 10.5.1 gives the amount of freight carried (measured in ton-miles) and the corresponding revenues earned by the 10 largest airline freight companies in 1973 (87). For these data,

$$\sum_{i=1}^{10} x_i = 4{,}322 \qquad \sum_{i=1}^{10} y_i = 901 \qquad \sum_{i=1}^{10} x_i^2 = 2{,}408{,}810$$

$$\sum_{i=1}^{10} y_i^2 = 103{,}195 \qquad \sum_{i=1}^{10} x_i y_i = 494{,}774$$

TABLE 10.5.1 AIRLINE FREIGHT AND REVENUES

Airline	Freight Ton-Miles, x_i (millions)	Freight Revenues, y_i (millions)
Pan American	860	188
Flying Tiger	681	120
United	645	135
American	529	114
TWA	475	98
Seaboard	359	53
Northwest	246	52
Eastern	207	56
Delta	176	56
Continental	144	29

Given these values, we calculate $\hat{a} = 5.902760$ and $\hat{b} = 0.194811$. A further computation yields

$$n\widehat{\sigma^2} = 1489.2624 \qquad \text{and} \qquad \sum_{i=1}^{10} (x_i - \bar{x})^2 = 540841.6$$

Interpreting $\hat{b}$ as marginal revenue tells an airline that an increase of 1 million freight ton-miles will result in an additional \$194,811 in revenue. If an airline is willing to accept the validity of the linear model, it might be interested in a confidence interval for b. According to Theorem 10.5.5 (and noting that $t_{0.025, 8} = 2.306$), a 95% confidence interval for b extends from 0.152 to 0.238:

$$\left(0.194811 - (2.306)\sqrt{\frac{1489.2624}{(8)(540841.6)}}, \quad 0.194811 + (2.306)\sqrt{\frac{1489.2624}{8(540841.6)}} \right)$$

$$= (0.194811 - 0.042782, 0.194811 + 0.042782) \doteq (0.152, 0.238)$$

It should be noted that the parameter b in a linear model replaces the correlation coefficient as a measure of linear dependence: To reject the null hypothesis H_0: $b = 0$ is to claim a statistically significant dependence of Y on x. The test procedure is derived from the T statistic of Theorem 10.5.4. We will state the result but omit our usual GLRT justification.

Theorem 10.5.6.   To test H_0: $b = 0$ versus H_1: $b \neq 0$ at the α level of significance, reject the null hypothesis if

$$\frac{\sqrt{n-2}\, \hat{b} \sqrt{\sum_{i=1}^{n} (x_i - \bar{x})^2}}{\sqrt{n\widehat{\sigma^2}}}$$

is either (a) $\leq -t_{\alpha/2, n-2}$ or (b) $\geq t_{\alpha/2, n-2}$.

Comment

The usual modifications for one-sided tests also apply here. In some applications it would be clear that either $b = 0$ or $b < 0$. An example of this occurs if $x =$ amount of a commodity offered for sale and $Y =$ unit price of the commodity. Example 10.5.1 discusses a situation where it would be appropriate to test $H_0: b = 0$ versus $H_1: b > 0$.

Example 10.5.1

Case Study 1.2.4 suggested a possible linear relationship between radioactive contamination (the x-variable) and cancer mortality (the y-variable). Figure 1.2.2 offered strong visual evidence for such a dependence. As always, however, we must be careful not to mistake random fluctuations for a real phenomenon. Since no reputable medical experts feel that radiation exposure of this sort would ever *decrease* cancer mortality, it makes sense to test $H_0: b = 0$ versus $H_1: b > 0$. The data from Table 1.2.5 yield the following sums:

$$\sum_{i=1}^{9} x_i = 41.56 \qquad \sum_{i=1}^{9} y_i = 1{,}416.1$$

$$\sum_{i=1}^{9} x_i^2 = 289.4222 \qquad \sum_{i=1}^{9} y_i^2 = 232{,}498.97$$

$$\sum_{i=1}^{9} x_i y_i = 7439.37$$

We then calculate $\sum_{i=1}^{9} (x_i - \bar{x})^2 = 97.507356$, $\hat{b} = 9.231456$, and $9\widehat{\sigma^2} = 1373.9464$. Thus the statistic T (with $b = 0$) has the value

$$\frac{\sqrt{7}(9.231456)\sqrt{97.507356}}{\sqrt{1373.9464}} = 6.51$$

We should reject the null hypothesis only if the test statistic exceeds $t_{\alpha, n-2}$. Here, at the 5% level, $t_{0.05, 7} = 1.895$. Our conclusion, then, is clear-cut: Reject H_0.

Question 10.5.1 Show that an equivalent form for the MLE for B is

$$\hat{B} = \frac{\displaystyle\sum_{i=1}^{n} (x_i - \bar{x}) Y_i}{\displaystyle\sum_{i=1}^{n} (x_i - \bar{x})^2}$$

Question 10.5.2 Prove the following two identities. The second is especially useful for computational purposes.

$$n\widehat{\sigma^2} = \sum_{i=1}^{n} (Y_i - \bar{Y})^2 - \hat{B} \sum_{i=1}^{n} (x_i - \bar{x})(Y_i - \bar{Y})$$

$$= \sum_{i=1}^{n} Y_i^2 - \frac{\left(\sum_{i=1}^{n} Y_i\right)^2}{n} - \hat{B}\left[\sum_{i=1}^{n} x_i Y_i - \frac{\left(\sum_{i=1}^{n} x_i\right)\left(\sum_{i=1}^{n} Y_i\right)}{n}\right]$$

$$= \sum_{i=1}^{n} Y_i^2 - \hat{A} \sum_{i=1}^{n} Y_i - \hat{B} \sum_{i=1}^{n} x_i Y_i$$

Question 10.5.3 Let (x_i, Y_i), $i = 1, 2, \ldots, n$, be a random sample where $Y_i \sim N(a + bx_i, \sigma^2)$.
(a) Prove that $E(\bar{Y}) = a + b\bar{x}$.
(b) Prove that $E(\hat{B}) = b$. (*Hint:* Write $\hat{B}$ in the form given in Question 10.5.1.)
(c) Prove that

$$\text{Var}(\hat{B}) = \frac{\sigma^2}{\displaystyle\sum_{i=1}^{n} (x_i - \bar{x})^2}$$

(*Hint:* Write $\hat{B}$ in the form given in Question 10.5.1.)
(d) Prove that $E(\hat{A}) = a$. [*Hint:* Use parts (a) and (b).]
(e) Prove that

$$\text{Cov}(\hat{B}, Y_j) = \frac{x_j - \bar{x}}{\displaystyle\sum_{i=1}^{n} (x_i - \bar{x})^2}\, \sigma^2$$

(f) Prove that $\text{Cov}(\hat{B}, \bar{Y}) = 0$.
(g) Prove that

$$\text{Var}(\hat{A}) = \frac{\sigma^2 \displaystyle\sum_{i=1}^{n} x_i^2}{n \displaystyle\sum_{i=1}^{n} (x_i - \bar{x})^2}$$

[*Hint:* Use parts (c) and (f).]

Question 10.5.4 We wish to estimate the coefficients of the linear model. As usual, n x-values are to be selected. Suppose that the values for x must lie in the interval $[-5, 5]$ and, for convenience, n is even. How should the x's be chosen so as to minimize Var $(\hat{B})$?

Question 10.5.5 Suppose that half the x-values referred to in Question 10.5.4 were set at -4 and half at $+4$. What would be the relative efficiency of estimating b with that sampling scheme as opposed to the optimal strategy?

Question 10.5.6 Suppose in Case Study 10.5.1 that σ were known to be 13.64.

    (a) Use the $N(0, 1)$ statistic

$$\frac{(\hat{B} - b)}{\sigma} \sqrt{\sum_{i=1}^{n} (x_i - \bar{x})^2}$$

    to construct a 95% confidence interval for b.

    (b) Use the $N(0, 1)$ statistic

$$\frac{(\hat{A} - a)}{\sigma} \sqrt{\frac{n \sum_{i=1}^{n} (x_i - \bar{x})^2}{\sum_{i=1}^{n} x_i^2}}$$

    to construct a 95% confidence interval for a.

Question 10.5.7 Find a 95% confidence interval for the slope of the chirping frequency-temperature data given in Question 10.4.1.

Question 10.5.8 Insect flight ability can be measured in a laboratory by attaching the insect to a nearly frictionless rotating arm by means of a very thin wire. The "tethered" insect then flies in circles until exhausted. The nonstop distance flown can easily be calculated from the number of revolutions made by the arm. Shown below are measurements of this sort made on *Culex tarsalis* mosquitoes of four different ages. The response variable is the average (tethered) distance flown until exhaustion for 40 females of the species (140).

Age, x_i (weeks)	Distance Flown, y_i (thousands of meters)
1	12.6
2	11.6
3	6.8
4	9.2

Fit a straight line to these data and test that the slope is 0. Use a two-sided alternative and the 0.05 level of significance.

Question 10.5.9 For the six airlines with the largest 1973 passenger revenues, the table (87) gives the number of passengers carried (x) and the corresponding passenger revenues (Y). Since passenger revenues depend heavily on routes, class of travel, and distances, there could fail to be significance between x and Y. Examine this possibility by testing $H_0 : b = 0$ versus $H_1 : b > 0$ at the 0.05 level.

Airline	Number of Passengers	Passenger Revenues
United	30,250	1,654
Eastern	26,201	1,130
Delta	24,604	1,021
American	21,163	1,296
TWA	14,148	1,054
Pan American	10,409	1,049

Question 10.5.10   Refer to the smoking–CHD mortality data of Question 10.2.16.
  (a) If x denotes a country's cigarette consumption and y its CHD mortality rate, fit a straight line to the data of the form $y = a + bx$.
  (b) Test $H_0: b = 0$ versus $H_1: b > 0$. (Why should the alternative be one-sided here?) Let $\alpha = 0.05$.
  (c) Construct a 95% confidence interval for b.

Question 10.5.11   Set up the test statistic that would be used in making inferences about the y-intercept, a, in the linear model, $y = a + bx$. (Assume that σ^2 is unknown.) What distribution would the statistic have?

Question 10.5.12   Determining small quantities of calcium in the presence of magnesium is a difficult problem for the analytical chemist. Direct precipitation is not feasible. One of the procedures proposed involves the use of alcohol as a solvent. The data below present the results of applying the alcohol method to ten mixtures containing known quantities of CaO. The second column gives, in each instance, the amount of CaO recovered (68).

CaO Present, x (mg)	CaO Recovered, y (mg)
4.0	3.7
8.0	7.8
12.5	12.1
16.0	15.6
20.0	19.8
25.0	24.5
31.0	31.1
36.0	35.5
40.0	39.4
40.0	39.5

Perhaps the most immediate question to be answered is whether the discrepancies between what was present and what was found are (1) random, or (2) intrinsic to the procedure. One way to examine those options is by expressing the amount recovered (y) as a linear function of the amount present (x)—that is, $y = a + bx$—and then testing the null hypotheses that $a = 0$ and $b = 1$.
  (a) At the 5% level, test $H_0: a = 0$ versus $H_1: a \neq 0$.
  (b) At the 5% level, test $H_0: b = 1$ versus $H_1: b \neq 1$. Use the statistic of Theorem 10.5.4.

Question 10.5.13   Let $(x_1, Y_1), \ldots, (x_n, Y_n)$, with $Y_i \sim N(a + bx_i, \sigma^2)$, be a random sample and $(u_1, V_1), \ldots, (u_m, V_m)$, with $V_i \sim N(c + du_i, \sigma^2)$, a second random sample independent of the first. Let $\widehat{\sigma_1^2}$ be the estimator of σ^2 from the first sample and $\widehat{\sigma_2^2}$ the estimator from the second. Find a statistic to test $H_0: b = d$ versus $H_1: b \neq d$.

Question 10.5.14   A second method for recovering calcium from magnesium (see Question 10.5.12) has been found that requires fewer precipitations than the alcohol method. In a preliminary study of this second technique, the following data were obtained.

CoO Present, x (mg)	CaO Recovered, y (mg)
4.0	3.9
8.0	8.1
12.5	12.4
16.0	16.0
20.0	19.8
25.0	25.0
31.0	31.1
36.0	35.8
40.0	40.1
40.0	40.1

Fit these data with a regression equation of the form $y = c + dx$. At the 5% level of significance, test $H_0: b = d$ versus $H_1: b \neq d$. Use the statistic from Question 10.5.13. If H_0 is accepted, it seems clear the new method should be adopted because of its simplicity. What if H_0 is rejected?

As useful as inferences about a and b are inferences about $E(Y \mid x) = a + bx$, the mean of Y for a fixed value of x. For example, if x represents the quantity of goods manufactured and Y the cost for producing that amount, we can well imagine a manager's interest in the average cost for a run of a certain size. A reasonable choice of a point estimator for $E(Y \mid x)$ is $\hat{Y} = \hat{A} + \hat{B}x$, and it proves to be unbiased:

$$E(\hat{Y}) = E(\hat{A} + \hat{B}x) = E(\hat{A}) + E(\hat{B})x = a + bx$$

The last equality, of course, is a consequence of the unbiasedness of $\hat{A}$ and $\hat{B}$.

Before we can use $\hat{Y}$ in any kind of confidence interval, we must also know its variance. Fortunately, the necessary computation is relatively straightforward:

$$\text{Var}(\hat{Y}) = \text{Var}(\hat{A} + \hat{B}x) = \text{Var}(\bar{Y} - \hat{B}\bar{x} + \hat{B}x)$$

$$= \text{Var}[\bar{Y} + \hat{B}(x - \bar{x})]$$

$$= \text{Var}(\bar{Y}) + (x - \bar{x})^2 \text{Var}(\hat{B}) \qquad \text{(Why?)}$$

$$= \frac{1}{n}\sigma^2 + \frac{(x - \bar{x})^2}{\sum_{i=1}^{n}(x_i - \bar{x})^2}\sigma^2$$

$$= \sigma^2 \left[\frac{1}{n} + \frac{(x - \bar{x})^2}{\sum_{i=1}^{n}(x_i - \bar{x})^2} \right]$$

Out of this last expression comes the not-surprising result that the variance of the estimated average response increases as the value of x becomes more extreme. That is, we are better able to predict the behavior of $E(Y \mid x)$ for an x-value close to $\bar{x}$ than we are for x-values that are either very small or very large (relative to $\bar{x}$).

By Theorem 10.5.3, $\hat{Y}$ is independent of $n\widehat{\sigma^2}$, so the statistic

$$\frac{\dfrac{\hat{Y} - (a + bx)}{\sigma\sqrt{\dfrac{1}{n} + \dfrac{(x - \bar{x})^2}{\displaystyle\sum_{i=1}^{n}(x_i - \bar{x})^2}}}}{\dfrac{1}{\sigma}\dfrac{1}{\sqrt{n-2}}\sqrt{n\widehat{\sigma^2}}} = \frac{\sqrt{n-2}\,[\hat{Y} - (a + bx)]}{\sqrt{n\widehat{\sigma^2}\left[\dfrac{1}{n} + \dfrac{(x - \bar{x})^2}{\displaystyle\sum_{i=1}^{n}(x_i - \bar{x})^2}\right]}}$$

has a Student t distribution with $n - 2$ degrees of freedom. Knowing that, we can readily obtain an expression for a confidence interval for $E(Y \mid x)$.

Theorem 10.5.7. A $100(1 - \alpha)\%$ confidence interval for $E(Y \mid x) = a + bx$, the expected value of Y at the fixed value x, is given by

$$(\hat{y} - w, \hat{y} + w)$$

where

$$w = t_{\alpha/2,\,n-2}\,\frac{1}{\sqrt{n-2}}\sqrt{n\widehat{\sigma^2}\left[\frac{1}{n} + \frac{(x - \bar{x})^2}{\displaystyle\sum_{i=1}^{n}(x_i - \bar{x})^2}\right]}$$

A variation of the preceding theme is the so-called "prediction problem," where we wish to estimate the particular outcome of a random variable, rather than the parameters of its density or its conditional moments. Recall Case Study 10.5.1 and imagine a certain airline wanting to predict the revenues it would accrue by carrying a given amount of freight—say, x. To make such an inference we will need to consider both the original random sample, $(x_1, Y_1), \ldots, (x_n, Y_n)$, and an additional observation, (x, Y), with Y being independent of the Y_i's. A *prediction interval* is one that contains Y with a desired probability.

The appropriate statistic is again $\hat{Y}$, but now $\hat{Y} - Y$ will be the quantity of interest, rather than $\hat{Y} - (a + bx)$. First, note that

$$E(\hat{Y} - Y) = E(\hat{Y}) - E(Y) = (a + bx) - (a + bx) = 0$$

Also,

$$\mathrm{Var}\,(\hat{Y} - Y) = \mathrm{Var}\,(\hat{Y}) + \mathrm{Var}\,(Y)$$

$$= \sigma^2\left[\frac{1}{n} + \frac{(x - \bar{x})^2}{\displaystyle\sum_{i=1}^{n}(x_i - \bar{x})^2}\right] + \sigma^2$$

$$= \sigma^2\left[1 + \frac{1}{n} + \frac{(x - \bar{x})^2}{\displaystyle\sum_{i=1}^{n}(x_i - \bar{x})^2}\right]$$

Given the latter expression, it becomes a simple matter to recast the confidence-interval formula for $E(Y \mid x)$ into a prediction interval for Y.

Theorem 10.5.8. A $100(1 - \alpha)\%$ prediction interval for Y at a fixed value x is given by

$$(\hat{y} - w, \hat{y} + w)$$

where

$$w = t_{\alpha/2,\, n-2} \frac{1}{\sqrt{n-2}} \sqrt{n\hat{\sigma}^2 \left[1 + \frac{1}{n} + \frac{(x - \bar{x})^2}{\sum\limits_{i=1}^{n} (x_i - \bar{x})^2} \right]}$$

Example 10.5.2

In Case Study 10.5.1, let us suppose that American Airlines is considering increasing its capacity for freight to 600 million freight ton-miles. It would not be unreasonable for them to want some intelligent estimate of their possible revenues. A 95% prediction interval would be one way to phrase what they want to know.

The relevant x-value is 600, so the corresponding estimated revenue is $\hat{y} = \hat{a} + 600\hat{b} = 5.902760 + 600(0.194811) = 122.78936$ (millions of dollars). This latter number is the center of the prediction interval. The interval's radius, or half-length, is given by

$$(2.306) \frac{1}{\sqrt{8}} \sqrt{(1489.2624) \left[1 + \frac{1}{10} + \frac{(600 - 432.2)^2}{540841.6} \right]} = 33.770515$$

Thus they could anticipate revenues somewhere between \$89.02 million and \$156.56 million:

$$(122.78936 - 33.770515,\ 122.78936 + 33.770515) = (89.02,\ 156.56)$$

Question 10.5.15 Construct a 95% confidence interval for $E(Y \mid 600)$ using the airline data of Case Study 10.5.1. Suppose a Senate subcommittee were investigating allegations of profit gouging in the airline freight industry. Would the committee members be more interested in a 95% confidence interval for $E(Y \mid 600)$ or a 95% prediction interval for Y given that $x = 600$? Explain.

Question 10.5.16 Construct a 95% confidence interval for the expected CHD mortality rate in a country where the cigarette consumption is 2500 per adult per year (see Question 10.5.10).

Question 10.5.17 The tables on page 480 give the monthly sales and the monthly sales expenses for a small manufacturing firm in Norman, Oklahoma (108).

Month	Sales (thousands)	Sales Expenses (thousands)
4/78	$187.1	$25.4
5/78	179.5	22.8
6/78	157.0	20.6
7/78	197.0	21.8
8/78	239.4	32.4
9/78	217.8	24.4
10/78	227.1	29.3
11/78	233.4	27.9
12/78	242.0	27.8
1/79	251.9	34.2
2/79	190.0	29.2
3/79	295.8	30.0

(a) Find the least-squares straight line describing sales expenses (y) as a function of sales (x).

(b) Test $H_0: b = 0$ versus $H_1: b > 0$ at the 0.05 level of significance.

(c) Find a 95% confidence interval for b.

(d) Suppose that the company expects monthly sales to become essentially constant at $200,000 per month. Give a 95% confidence interval for the average sales expenses.

(e) The company is considering what the costs would be if sales were boosted to $300,000 in a certain month. Find the corresponding 95% prediction interval for the sales expenses.

(f) Use the statistic from Question 10.5.11 to find a 95% confidence interval for the y-intercept, a. Note that, here, a represents the fixed or overhead costs for sales. The large width of a's interval emphasizes the uncertainty in estimating for values of x that are numerically distant from the x-values actually sampled.

Question 10.5.18 One of the obvious applications of regression analysis is the situation where two variables, x and y, are related, but one of them—say, y—is difficult to measure. We can finesse such a problem by simply measuring the x-variable and estimating y via the regression function. It may be extremely difficult, for example, to measure the volume of an irregularly shaped object but very simple to weigh it. The table below shows the weight in kilograms and the volume in cubic decimeters of 18 children between the ages of 5 and 8 (14).

Weight, x	Volume, y	Weight, x	Volume, y
17.1	16.7	15.8	15.2
10.5	10.4	15.1	14.8
13.8	13.5	12.1	11.9
15.7	15.7	18.4	18.3
11.9	11.6	17.1	16.7
10.4	10.2	16.7	16.6
15.0	14.5	16.5	15.9
16.0	15.8	15.1	15.1
17.8	17.6	15.1	14.5

(a) Graph the data and find the least-squares line, $y = \hat{a} + \hat{b}x$.

(b) Construct a 95% confidence interval for $E(Y \mid 14.0)$.

(c) Construct a 95% prediction interval for the volume of a child weighing 14.0 kilograms.

Question 10.5.19 Prove that the variance of $\hat{Y}$ can also be written

$$\text{Var}(\hat{Y}) = \frac{\sigma^2 \sum_{i=1}^{n}(x_i - x)^2}{n \sum_{i=1}^{n}(x_i - \bar{x})^2}$$

Question 10.5.20 Show that

$$\sum_{i=1}^{n}(Y_i - \bar{Y})^2 = \sum_{i=1}^{n}(Y_i - \hat{Y}_i)^2 + \sum_{i=1}^{n}(\hat{Y}_i - \bar{Y})^2$$

for any set of points (x_i, Y_i), $i = 1, 2, \ldots, n$.

Question 10.5.21 The expression

$$\frac{\sum_{i=1}^{n}(\hat{Y}_i - \bar{Y})^2}{\sum_{i=1}^{n}(Y_i - \bar{Y})^2}$$

is called the *coefficient of determination*. Express the coefficient of determination as a function of the sample correlation coefficient (ignore the fact that x is not a random variable here). What does the coefficient of determination represent? (See Question 10.5.20.)

10.6 THE BIVARIATE NORMAL DENSITY

As already mentioned, Galton, having empirically discovered properties 1 through 4 of Section 10.3, enlisted the aid of J. D. H. Dickson in finding a joint density function with these properties. Dickson did so—and with alacrity—the desired distribution being well known and its derivation not particularly difficult. In fact, Pearson (122) commented, "Why Galton did not at once write down the equation to his surface . . . has always been a puzzle to me," and then reasoned, "The explanation of Galton's action possibly lies in the fact that Galton was very modest and throughout his life underrated his own mathematical powers." Taking encouragement from this last remark, let *us* try to derive the appropriate density function $f_{X,Y}(x, y)$.

The first clue to consider is property 4, the elliptical nature of the contour lines. For simplicity, suppose we assume these ellipses have one axis on the line $y = x$ and are centered at the origin. Then $f_{X,Y}(x, y)$ has the form $g(x^2 - 2vxy + y^2)$. The notation of the cross-term constant as $-2v$ is suggested by the recognition that completing the square is often a useful tool when dealing with quadratics. Since the marginal distributions are to be normal, an exponential form seems reasonable—for example,

$$f_{X,Y}(x, y) = Ke^{-(1/2)c(x^2 - 2vxy + y^2)} \qquad (10.6.1)$$

where $c > 0$ is arbitrary and K must ensure that the integral of $f_{X,Y}(x, y)$ is 1.

The integral can be calculated by completing the square in y and writing

$$f_{X,Y}(x, y) = Ke^{-(1/2)c(1 - v^2)x^2} \cdot e^{-(1/2)c(y - vx)^2}$$

Since the first exponent must be negative, we need to insist that $|v| < 1$. Then

$$\int_{-\infty}^{\infty} \int_{-\infty}^{\infty} e^{-(1/2)c(1 - v^2)x^2} \cdot e^{-(1/2)c(y - vx)^2} \, dy \, dx$$

$$= \int_{-\infty}^{\infty} e^{-(1/2)c(1 - v^2)x^2} \left(\int_{-\infty}^{\infty} e^{-(1/2)c(y - vx)^2} \, dy \right) dx$$

$$= \int_{-\infty}^{\infty} e^{-(1/2)c(1 - v^2)x^2} \left(\frac{\sqrt{2\pi}}{\sqrt{c}} \right) dx$$

$$= \frac{\sqrt{2\pi}}{\sqrt{c}} \frac{\sqrt{2\pi}}{\sqrt{c}\sqrt{1 - v^2}}$$

$$= \frac{2\pi}{c\sqrt{1 - v^2}}$$

It follows that we should take

$$K = \frac{c\sqrt{1 - v^2}}{2\pi}$$

The constant c can be any positive value, but a convenient form proves to be $c = 1/(1 - v^2)$. Substitution of these values for K and c into Equation 10.6.1 gives

$$f_{X,Y}(x, y) = \frac{1}{2\pi\sqrt{1 - v^2}} e^{-(1/2)[1/(1 - v^2)](x^2 - 2vxy + y^2)}$$

$$= \frac{1}{2\pi\sqrt{1 - v^2}} e^{-x^2} \cdot e^{-(1/2)[1/(1 - v^2)](y - vx)^2} \qquad (10.6.2)$$

Recall that our choice of the form of $f_{X,Y}(x, y)$ was predicated on a wish for the marginal pdf's to be normal. A simple integration shows that to be the case:

$$f_X(x) = \int_{-\infty}^{\infty} f_{X,Y}(x, y) \, dy$$

$$= \frac{1}{2\pi\sqrt{1 - v^2}} e^{-(1/2)x^2} \int_{-\infty}^{\infty} e^{-(1/2)[1/(1 - v^2)](y - vx)^2} \, dy$$

$$= \frac{1}{2\pi\sqrt{1 - v^2}} e^{-(1/2)x^2} \cdot \sqrt{2\pi}\sqrt{1 - v^2}$$

$$= \frac{1}{\sqrt{2\pi}} e^{-(1/2)x^2}$$

Since $f_{X,Y}(x, y)$ is symmetric in x and y, $f_Y(y)$ is also $N(0, 1)$.

Next we show that the constant v appearing in Equations 10.6.1 and 10.6.2 is really $\rho(X, Y)$. Since $E(X) = E(Y) = 0$,

$$\rho(X, Y) = E(XY) = \int_{-\infty}^{\infty} \int_{-\infty}^{\infty} xy f_{X, Y}(x, y) \, dx \, dy$$

$$= \frac{1}{\sqrt{2\pi}} \int_{-\infty}^{\infty} xe^{-(1/2)x^2} \left(\frac{1}{\sqrt{2\pi} \sqrt{1 - v^2}} \int_{-\infty}^{\infty} ye^{-(1/2)[1/(1 - v^2)](y - vx)^2} \, dy \right) dx$$

$$= \frac{1}{\sqrt{2\pi}} \int_{-\infty}^{\infty} xe^{-(1/2)x^2} \cdot vx \, dx \qquad \text{(why?)}$$

$$= v \frac{1}{\sqrt{2\pi}} \int_{-\infty}^{\infty} x^2 e^{-(1/2)x^2} \, dx = v \, \text{Var}(X) = v$$

[To avoid unnecessary notation, we will write $\rho(X, Y)$ as simply ρ.]

There is a two-variable version of Theorem 3.5.1 allowing a generalization of Equation 10.6.2 to the case where x is replaced by $(x - \mu_X)/\sigma_X$ and y by $(y - \mu_Y)/\sigma_Y$. To accomplish this, the original density must be multiplied by the derivative of both the X-transformation and the Y-transformation—that is, by $(1/\sigma_X \sigma_Y)$. Definition 10.6.1 gives the resulting extension of Equation 10.6.2.

Definition 10.6.1. Let X and Y be random variables with joint pdf

$$f_{X, Y}(x, y) = \frac{1}{2\pi\sigma_X \sigma_Y \sqrt{1 - \rho^2}}$$

$$\cdot \exp \left\{ -\frac{1}{2} \left(\frac{1}{1 - \rho^2} \right) \left[\frac{(x - \mu_X)^2}{\sigma_X^2} - 2\rho \frac{x - \mu_X}{\sigma_X} \cdot \frac{y - \mu_Y}{\sigma_Y} + \frac{(y - \mu_Y)^2}{\sigma_Y^2} \right] \right\}$$

for all x and y. Then X and Y are said to have the *bivariate normal distribution*.

Comment

Knowledge of the bivariate normal dates back to the beginning of the nineteenth century: Laplace, for example, wrote about it as early as 1810. Until the time of Galton, though, its applications were primarily confined to astronomy.

Comment

For bivariate normal densities, $\rho(X, Y) = 0$ imples that X and Y are independent, a result not true in general (recall Example 10.2.1).

Comment

There is a natural generalization from the bivariate normal to the *multivariate normal*, the latter having the form

$$f_{\bar{X}}(\hat{x}) = \frac{1}{(2\pi)^{n/2} \sqrt{|A^{-1}|}} \exp \left[-\frac{1}{2} (\hat{x} - \hat{\mu})^T A(\hat{x} - \hat{\mu}) \right]$$

where $\hat{x}$ and $\hat{\mu}$ are n-dimensional column vectors, A is an $n \times n$ symmetric positive-definite matrix, and T denotes the transpose operation. For further details, see (28) or (59).

Theorem 10.6.1 summarizes what has already been established for the bivariate normal and confirms that it possesses the four properties discovered by Galton.

> **Theorem 10.6.1.** Suppose that X and Y are random variables having the bivariate normal distribution given in Definition 10.6.1. Then
>
> (a) $X \sim N(\mu_X, \sigma_X^2)$ and $Y \sim N(\mu_Y, \sigma_Y^2)$.
> (b) $\rho(X, Y) = v = \rho$.
> (c) $E(Y \mid x) = \mu_Y = \dfrac{\rho \sigma_Y}{\sigma_X}(x - \mu_X)$.
> (d) $\mathrm{Var}\,(Y \mid x) = (1 - \rho^2)\sigma_Y^2$.

Proof. We have already established (a) and (b). Properties (c) and (d) will be examined for the special case $\mu_X = \mu_Y = 0$ and $\sigma_X = \sigma_Y = 1$. The extension to arbitrary μ_X, μ_Y, σ_X, and σ_Y is straightforward.

First, note that

$$f_{Y|x}(y) = \frac{f_{X,Y}(x, y)}{f_X(x)}$$

$$= \frac{\dfrac{1}{2\pi\sqrt{1 - \rho^2}}\, e^{-(1/2)x^2} e^{-(1/2)[1/(1-\rho^2)](y - \rho x)^2}}{\dfrac{1}{\sqrt{2\pi}}\, e^{-(1/2)x^2}}$$

$$= \frac{1}{\sqrt{2\pi}\sqrt{1 - \rho^2}}\, e^{-(1/2)[1/(1-\rho^2)](y - \rho x)^2} \tag{10.6.3}$$

But, by inspection, Equation 10.6.3 is the pdf for an $N(\rho x, 1 - \rho^2)$ random variable. Therefore, $E(Y \mid x) = \rho x$ and $\mathrm{Var}\,(Y \mid x) = 1 - \rho^2$. Replacing Y by $(Y - \mu_Y)/\sigma_Y$ and x by $(x - \mu_X)/\sigma_X$ gives the desired results.

Seeing for ourselves the bivariate normal emerge as the pdf satisfying Galton's four conditions, we can perhaps appreciate the exuberance of Sir Francis, who upon receiving Dickson's response (basically Theorem 10.6.1), wrote (52):

> The problem may not be difficult to an accomplished mathematician, but I certainly never felt such a glow of loyalty and respect towards the sovereignty and wide sway of mathematical analysis as when his answer arrived, confirming by purely mathematical reasoning my various and laborious statistical conclusions with far more minuteness than I had dared to hope, because the data ran somewhat roughly and I had to smooth them with tender caution.

Question 10.6.1 Suppose that X and Y have a bivariate normal pdf with $\mu_X = 3$, $\mu_Y = 6$, $\sigma_X^2 = 4$, $\sigma_Y^2 = 10$, and $\rho = \frac{1}{2}$. Find $P(5 < Y < 6\frac{1}{2})$ and $P(5 < Y < 6\frac{1}{2} \mid x = 2)$.

Question 10.6.2 Suppose that X and Y have a bivariate normal distribution with Var (X) = Var (Y).
(a) Show that X and $Y - \rho X$ are independent.
(b) Show that $X + Y$ and $X - Y$ are independent. (*Hint*: See Question 10.2.11.)

Question 10.6.3 Suppose that X and Y have a bivariate normal distribution.
(a) Prove that $cX + dY$ has a normal distribution for any constants c and d.
(b) Find $E(cX + dY)$ and Var $(cX + dY)$ in terms of μ_X, μ_Y, σ_X, σ_Y, and $\rho(X, Y)$.

Question 10.6.4 Suppose that the random variables X and Y have a bivariate normal pdf with $\mu_X = 56$, $\mu_Y = 11$, $\sigma_X^2 = 1.2$, $\sigma_Y^2 = 2.6$, and $\rho = 0.6$. Compute $P(10 < Y < 10.5 \mid x = 55)$. Suppose that $n = 4$ values were to be observed with x fixed at 55. Find $P(10.5 < \bar{Y} < 11 \mid x = 55)$.

Question 10.6.5 If the joint pdf of the random variables X and Y is

$$f_{X,Y}(x, y) = ke^{-(2/3)[(1/4)x^2 - (1/2)xy + y^2]}$$

find $E(X)$, $E(Y)$, Var (X), Var (Y), $\rho(X, Y)$, and k.

Question 10.6.6 Give conditions on $a > 0$, $b > 0$, and u so that

$$f_{X,Y}(x, y) = ke^{-(ax^2 - 2uxy + by^2)}$$

is the bivariate normal density of random variables X and Y each having expected value zero. Also, find Var (X), Var (Y), and $\rho(X, Y)$.

10.7 TESTING $H_0: \rho(X,Y) = 0$ WHEN X AND Y ARE BIVARIATE NORMAL

Situations arise where it is desirable to test the independence of two random variables. We are not at this point able to formulate such a test in any general context, but for the special case where X and Y are bivariate normal, independence is equivalent to $\rho(X, Y)$'s being 0, and the latter can be examined without too much difficulty. Not surprisingly, the test statistic appropriate for hypotheses involving ρ, the *true* correlation coefficient, is R, the *sample* correlation coefficient (recall Equation 10.2.2). Among its other properties, R is the maximum-likelihood estimator for ρ.

> **Theorem 10.7.1.** Given that $f_{X,Y}(x, y)$ is a bivariate normal pdf, the maximum-likelihood estimators for μ_X, μ_Y, σ_X^2, σ_Y^2, and $\rho(X, Y)$, assuming that all are unknown, are $\bar{X}$, $\bar{Y}$, $(1/n) \sum_{i=1}^{n} (X_i - \bar{X})^2$, $(1/n) \sum_{i=1}^{n} (Y_i - \bar{Y})^2$, and R, respectively.

Proof. The proof is accomplished by the usual method of taking partial derivatives of the likelihood function with respect to each of the five parameters, setting those equal to 0, and solving simultaneously. The details are best left to the energetic reader.

Although we have already made a case for using R to test H_0: $\rho(X, Y) = 0$, it turns out to be more convenient to use a function of R as a test statistic. The derivation of that function's density is beyond the scope of this text, but we can state the result and illustrate it with an example.

Theorem 10.7.2. Under the null hypothesis that $\rho(X, Y) = 0$, the statistic

$$\frac{\sqrt{n-2}\,R}{\sqrt{1-R^2}}$$

has a Student t distribution with $n - 2$ degrees of freedom.

CASE STUDY 10.7.1

Not long ago, Alvin Toffler wrote a best-seller called *Future Shock* in which he speculated about the cultural, medical, and moral effects of change. One question that was raised involved the extent to which change should be considered a health hazard—if, indeed, it should be considered a health hazard at all. A partial answer was provided in a recent study done on a group of patients hospitalized for various chronic illnesses, some serious and some not. Each patient was asked to fill out a Schedule of Recent Experience (SRE) questionnaire. The questions related to 42 different life-change situations (a new job, another child, and so forth). A composite score was computed that summarized the average degree of change each patient had experienced in the preceding two years. Higher values on the SRE questionnaire reflected greater life-style changes. At the same time, each patient's health condition (the one he was being hospitalized for) was graded according to the Seriousness of Illness Rating Scale (SIRS). Values on this scale ranged from a low of 21 for dandruff to a high of 1020 for cancer.

The patients in the sample had a total of 17 different conditions that were considered chronic. In some cases more than one patient had the same condition. Table 10.7.1 and Figure 10.7.1 show the SIRS values associated with the various illnesses as well as the average SRE scores recorded for the patients who had each one (186).

The scatter diagram for Table 10.7.1 suggests a possible relationship between SRE and SIRS, but the linearity is not overwhelming. Consequently, before fitting a least-squares line through these 17 points it would be a good idea to test formally the null hypothesis that X and Y are independent.

TABLE 10.7.1 LIFE-STYLE CHANGES (SRE) AND HEALTH EVALUATIONS (SIRS)

Admitting Diagnosis	Average SRE	SIRS
Dandruff	26	21
Varicose veins	130	173
Psoriasis	317	174
Eczema	231	204
Anemia	325	312
Hyperthyroidism	816	393
Gallstones	563	454
Arthritis	312	468
Peptic ulcer	603	500
High blood pressure	405	520
Diabetes	599	621
Emphysema	357	636
Alcoholism	688	688
Cirrhosis	443	733
Schizophrenia	609	776
Heart failure	772	824
Cancer	777	1020

Let ρ denote the true correlation coefficient between X and Y. The hypotheses to be tested, then, are

$$H_0: \quad \rho = 0$$

versus

$$H_1: \quad \rho \neq 0$$

To accept H_0 is to conclude that X and Y are independent (assuming the (x_i, y_i) values in Table 10.7.1 are a random sample from a bivariate normal distribution).

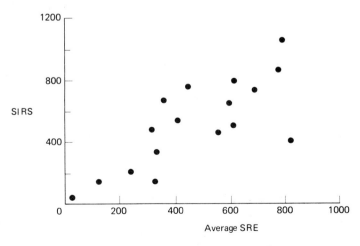

Figure 10.7.1 Health problems (SIRS) and social change (SRE).

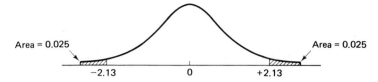

Figure 10.7.2 Student t distribution with 15 degrees of freedom.

We will reject the null hypothesis only if the sample correlation coefficient, r, is either too much less than 0 or too much greater than 0. Let α be 0.05. Then the pdf of the test statistic is the Student t distribution with $n - 2 = 15$ degrees of freedom, and H_0 should be rejected if

$$\frac{\sqrt{n - 2}\, r}{\sqrt{1 - r^2}}$$

is either (1) less than or equal to -2.13, or (2) greater than or equal to $+2.13$ (see Figure 10.7.2). Since

$$\sum_{i=1}^{17} x_i = 7973 \qquad \sum_{i=1}^{17} x_i^2 = 4,611,291$$

$$\sum_{i=1}^{17} y_i = 8517 \qquad \sum_{i=1}^{17} y_i^2 = 5.421,917$$

$$\sum_{i=1}^{17} x_i y_i = 4,759,470$$

it follows that

$$r = \frac{17(4,759,470) - (7973)(8517)}{\sqrt{17(4,611,291) - (7973)^2}\,\sqrt{17(5,421,917) - (8517)^2}}$$

$$= 0.76$$

Therefore,

$$\frac{\sqrt{n - 2}\, r}{\sqrt{1 - r^2}} = \frac{\sqrt{15}\,(0.76)}{\sqrt{1 - (0.76)^2}} = 4.53$$

Since the observed t ratio exceeds the upper critical value ($+2.13$), our conclusion is to *reject* the null hypothesis.

Comment

An alternate approach to testing H_0: $\rho = 0$ was given by Fisher (44). He showed that the statistic

$$\frac{1}{2} \ln \frac{1 + R}{1 - R}$$

is asymptotically normal with mean $\frac{1}{2} \ln [(1 + \rho)/(1 - \rho)]$ and variance approximately $1/(n - 3)$. Fisher's formulation makes it relatively easy to determine the power of a correlation test—a computation that would be much more difficult if the inference had to be based on $\sqrt{n - 2} \, R/\sqrt{1 - R^2}$.

Question 10.7.1 Test whether X and $\ln Y$ are independent for the data given in Question 10.4.9. Let X denote "degrees north latitude" and Y "melanoma rate (per 100,000)". Let $\alpha = 0.05$.

Question 10.7.2 Refer to the baseball data of Question 10.2.14. Test the independence of altitude and frequency of home runs. Let $\alpha = 0.05$.

Question 10.7.3 Test $H_0: \rho(X, Y) = 0$ versus $H_1: \rho(X, Y) > 0$ for the smoking–CHD mortality rate data of Question 10.2.16. Let $\alpha = 0.05$. Does your answer agree with part (b) of Question 10.5.10?

Question 10.7.4 Use the statistic

$$\frac{1}{2} \ln \frac{1 + R}{1 - R}$$

to get an expression for an approximate $100(1 - \alpha)\%$ confidence interval for ρ. Use your result on the data of Question 10.2.17 to construct a 90% confidence interval for the correlation coefficient between the behavioral index and the plumage index of mallard–pintail hybrids.

10.8 EPILOGUE AND CAVEAT

We have seen the usefulness of the correlation coefficient and its intimate relationship with the problem of linear regression. Regression and correlation have emerged, essentially unbidden, in a number of natural phenomena. Yet no discussion of the merits of these concepts would be complete without the warning that *correlation does not imply causality*. An excellent amplification of this dictum can be found in George Bernard Shaw's *The Doctor's Dilemma* (155):

> [C]omparisons which are really comparisons between two social classes with different standards of nutrition and education are palmed off as comparisons between the results of a certain medical treatment and its neglect. Thus it is easy to prove that the wearing of tall hats and the carrying of umbrellas enlarges the chest, prolongs life, and confers comparative immunity from disease; for the statistics show that the classes which use these articles are bigger, healthier, and live longer than the class which never dreams of possessing such things. It does not take much perspicacity to see that what really makes this difference is not the tall hat and the umbrella, but the wealth and nourishment of which they are evidence, and that a gold watch or membership of a club in Pall Mall might be proved in the same way to have the like sovereign virtues. A university degree, a daily bath, the owning of thirty pairs of trousers, a knowledge of Wagner's music, a pew in church, anything, in short, that implies more means and better nurture than the mass of laborers enjoy, can be statistically palmed off as a magic-spell conferring all sorts of privileges.

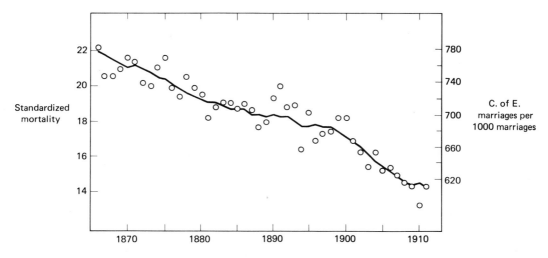

Figure 10.8.1 Correlation between standardized mortality per 1000 persons in England and Wales (circles) and the proportion of Church of England marriages per 1000 of all marriages (line), 1866–1911 ($r = 0.9512$).

In 1926, Yule (188) gave one of the first cohesive mathematical treatments of meaningless correlations (the technical term is *spurious* correlations). He began his exposition with the presentation of data showing a very high correlation between Church of England marriages and mortality (see Figure 10.8.1). He then commented:

> Now I suppose it is possible, given a little ingenuity and goodwill, to rationalize very nearly anything. And I can imagine some enthusiast arguing that the fall in the proportion of Church of England marriages is simply due to the Spread of Scientific Thinking since 1866, and the fall in mortality is also clearly to be ascribed to the Progress of Science; hence both variables are largely or mainly influenced by a common factor and consequently ought to be highly correlated. But most people would, I think, agree with me that the correlation is simply sheer nonsense; that it has no meaning whatever; that it is absurd to suppose that the two variables in question are in any sort of way, however indirect, causally related to one another.
>
> And yet, if we apply the ordinary test of significance in the ordinary way, the result suggests that the correlation is certainly "significant"—that it lies far outside the probable limits of fluctuations of sampling.

The reader should not suppose the problem of correlation and causality has disappeared; it seems an intrinsic part of discovery and understanding. Only recently a study announced a high correlation between coffee drinking and coronary heart disease. Some newspaper reports of the study portrayed the fragrant essence of the roasted berries of *Coffea arabica* as a menace to public health. Later papers [see (71)] suggested that the risk comes from such factors as cigarette smoking and sucrose intake, both highly correlated with coffee drinking.

There are also misuses associated with the theory of regression. Too often complex problems have been forced into the Procrustean bed of the linear model.

This is particularly obvious in the use of linear extrapolation, which recalls Mark Twain's oft-quoted satire in *Life on the Mississippi* (23): "In the space of 176 years the Lower Mississippi has shortened itself 252 miles. That is an average of a trifle more than one mile and a third per year. Therefore any calm person, who is not blind or idiotic, can see that in 742 years from now the lower Mississippi will be one mile and three quarters long, and Cairo, Ill., and New Orleans will have joined their streets together."

APPENDIX 10.1 A PROOF OF THEOREM 10.5.3

The strategy for the proof is to express $n\widehat{\sigma}^2$ in terms of the squares of normal random variables and then apply Fisher's lemma (see Appendix 7.2). The random variables to be used are $(\hat{B} - b)$, $W_i = Y_i - a - bx_i$, $i = 1, 2, \ldots, n$, and

$$\bar{W} = \frac{1}{n} \sum_{i=1}^{n} W_i = \bar{Y} - a - b\bar{x}$$

Let us first note that $\hat{B} - b$ can be written as a linear combination of the W_i's. The demonstration requires just a little algebra, beginning with the definition of $\hat{B}$:

$$\hat{B} - b = \frac{\sum_{i=1}^{n} (x_i - \bar{x})(Y_i - \bar{Y})}{\sum_{i=1}^{n} (x_i - \bar{x})^2} - b$$

$$= \frac{\sum_{i=1}^{n} (x_i - \bar{x})(Y_i - \bar{Y}) - b\sum_{i=1}^{n} (x_i - \bar{x})^2}{\sum_{i=1}^{n} (x_i - \bar{x})^2}$$

$$= \frac{\sum_{i=1}^{n} (x_i - \bar{x})(Y_i - \bar{Y}) - a\sum_{i=1}^{n} (x_i - \bar{x}) - b\sum_{i=1}^{n} (x_i - \bar{x})x_i}{\sum_{i=1}^{n} (x_i - \bar{x})^2}$$

$$= \frac{\sum_{i=1}^{n} (x_i - \bar{x})[(Y_i - a - bx_i) - (\bar{Y} - a - b\bar{x})]}{\sum_{i=1}^{n} (x_i - \bar{x})^2}$$

$$= \frac{\sum_{i=1}^{n} (x_i - \bar{x})(W_i - \bar{W})}{\sum_{i=1}^{n} (x_i - \bar{x})^2}$$

Further,

$$Y_i - \hat{Y}_i = Y_i - \hat{A} - \hat{B}x_i = Y_i - \bar{Y} + \hat{B}\bar{x} - \hat{B}x_i$$

$$= (Y_i - a - bx_i) - (\bar{Y} - a - b\bar{x}) - (x_i - \bar{x})(\hat{B} - b)$$

$$= (W_i - \bar{W}) - (x_i - \bar{x})(\hat{B} - b)$$

Thus

$$n\widehat{\sigma^2} = \sum_{i=1}^{n} (Y_i - \hat{Y}_i)^2 = \sum_{i=1}^{n} [(W_i - \bar{W}) - (x_i - \bar{x})(\hat{B} - b)]^2$$

$$= \sum_{i=1}^{n} (W_i - \bar{W})^2 + \sum_{i=1}^{n} (x_i - \bar{x})^2(\hat{B} - b)^2 - 2\sum_{i=1}^{n} (W_i - \bar{W})(x_i - \bar{x})(\hat{B} - b)$$

Now we use the expression for $(\hat{B} - b)$ established at the beginning of the proof to obtain the equality

$$-2\sum_{i=1}^{n} (W_i - \bar{W})(x_i - \bar{x})(\hat{B} - b) = -2(\hat{B} - b)\sum_{i=1}^{n} (x_i - \bar{x})(W_i - \bar{W})$$

$$= -2(\hat{B} - b)(\hat{B} - b)\sum_{i=1}^{n} (x_i - \bar{x})^2$$

$$= -2\sum_{i=1}^{n} (x_i - \bar{x})^2(\hat{B} - b)^2$$

Substituting this last version of the cross-product term into the expression for $n\widehat{\sigma^2}$ gives

$$n\widehat{\sigma^2} = \sum_{i=1}^{n} (W_i - \bar{W})^2 - \sum_{i=1}^{n} (x_i - \bar{x})^2(\hat{B} - b)^2$$

$$= \sum_{i=1}^{n} W_i^2 - n\bar{W}^2 - \sum_{i=1}^{n} (x_i - \bar{x})^2(\hat{B} - b)^2$$

Now, choose an $n \times n$ orthogonal matrix, M, whose first two rows are

$$\frac{x_1 - \bar{x}}{\sqrt{\sum\limits_{i=1}^{n} (x_i - \bar{x})^2}} \quad \cdots \quad \frac{x_n - \bar{x}}{\sqrt{\sum\limits_{i=1}^{n} (x_i - \bar{x})^2}}$$

and

$$\frac{1}{\sqrt{n}} \quad \cdots \quad \frac{1}{\sqrt{n}}$$

Define the random variables $Z_1, \ldots, Z_n$ through the transformation

$$\begin{pmatrix} Z_1 \\ \vdots \\ Z_n \end{pmatrix} = M \begin{pmatrix} W_1 \\ \vdots \\ W_n \end{pmatrix}$$

By Fisher's lemma, the Z_i's are independent, $N(0, \sigma^2)$, and

$$\sum_{i=1}^{n} Z_i^2 = \sum_{i=1}^{n} W_i^2$$

Also, $Z_1^2 = \sum_{i=1}^{n} (x_i - \bar{x})^2 (\hat{B} - b)^2$ and $Z_2^2 = n\bar{W}^2$, so

$$n\widehat{\sigma^2} = \sum_{i=1}^{n} W_i^2 - Z_1^2 - Z_2^2 = \sum_{i=3}^{n} Z_i^2$$

From this follows the independence of $n\widehat{\sigma^2}$, $\hat{B}$, and $\bar{Y}$. Finally, observe that

$$\frac{n\widehat{\sigma^2}}{\sigma^2} = \sum_{i=3}^{n} \left(\frac{Z_i}{\sigma}\right)^2$$

and the latter has a chi-square distribution with $n - 2$ degrees of freedom.

11

The Analysis of Variance

Ronald A. Fisher

No aphorism is more frequently repeated in connection with field trials, than that we must ask Nature few questions or, ideally, one question, at a time. The writer is convinced that this view is wholly mistaken. Nature, he suggests, will best respond to a logical and carefully thought-out questionnaire; indeed, if we ask her a single question, she will often refuse to answer until some other topic has been discussed.

11.1 INTRODUCTION

In this chapter we take up an important extension of the two-sample location problem introduced in Chapter 8. The *completely randomized, one-factor design* is a conceptually similar *k*-sample location problem, but one that requires a substantially different sort of analysis than its prototype. Here, the appropriate test statistic turns out to be a ratio of variance estimates, the sampling behavior of which is described by an *F* distribution rather than a Student *t*. The name attached to this procedure, in deference to the form of its test statistic, is the *analysis of variance* (or "ANOVA," for short). A very flexible method, the analysis of variance finds many other applications, a particularly important one being the *randomized block design*, which is the subject of Chapter 12.

Comment

Credit for much of the early development of the analysis of variance goes to Sir Ronald A. Fisher. Shortly after the end of World War I, Fisher resigned a public school teaching position that he was none too happy with and accepted a post at the Rothamsted Statistical Laboratory, a facility heavily involved in agricultural research. There he suddenly found himself entangled in problems where differences in the response variable (crop yields, for example) were constantly in danger of being obscured by the high level of uncontrollable heterogeneity in the experimental environment (different soil qualities, drainage gradients, and so on). Quickly seeing that traditional techniques were hopelessly inadequate under these conditions, Fisher set out to look for alternatives and in just a few years succeeded in fashioning an entirely new statistical methodology, a panoply of data-collecting principles and mathematical tools that is today known as *experimental design*. The centerpiece of Fisher's creation—what makes it all work—is the analysis of variance.

Suppose an experimenter wishes to compare the average effects elicited by k different levels of some given factor, where k is greater than or equal to 2. The factor, for example, might be "stop-smoking" therapies and the levels, three specific methods. Or the factor might be crowdedness as it relates to aggression in captive monkeys, with the levels being five different monkey-per-square-foot densities in five separate enclosures. Still another example might be an engineering study comparing the effectiveness of four kinds of catalytic converters in reducing the concentrations of harmful emissions in automobile exhaust. Whatever the circumstances, data from a completely randomized, one-factor design will consist of k independent random samples of sizes $n_1, n_2, \ldots,$ and n_k, the total sample size being denoted $n\left(= \sum_{j=1}^{k} n_j \right)$.

We will let Y_{ij} represent the *i*th observation recorded for the *j*th level. Table 11.1.1 shows some additional notation.

The dot notation of Table 11.1.1 is standard in analysis-of-variance problems. The presence of a dot in lieu of a subscript indicates that that particular subscript

TABLE 11.1.1

	Factor Level			
	1	2	$\cdots$	k
	Y_{11}	Y_{12}		Y_{1k}
	Y_{21}	Y_{22}		
	$\vdots$	$\vdots$	$\cdots$	$\vdots$
	$Y_{n_1 1}$	$Y_{n_2 2}$		$Y_{n_k k}$
Sample sizes	n_1	n_2	$\cdots$	n_k
Sample totals	$T_{.1}$	$T_{.2}$		$T_{.k}$
Sample means	$\bar{Y}_{.1}$	$\bar{Y}_{.2}$		$\bar{Y}_{.k}$
True means	μ_1	μ_2	$\cdots$	μ_k

has been summed over. Thus the response total for the jth sample is written

$$T_{.j} = \sum_{i=1}^{n_j} Y_{ij}$$

and the corresponding sample mean,

$$\bar{Y}_{.j} = \frac{1}{n_j} \sum_{i=1}^{n_j} Y_{ij} = \frac{T_{.j}}{n_j}$$

By the same convention, $T_{..}$ and $\bar{Y}_{..}$ will denote the overall total and overall mean, respectively:

$$T_{..} = \sum_{j=1}^{k} \sum_{i=1}^{n_j} Y_{ij} = \sum_{j=1}^{k} T_{.j}$$

$$\bar{Y}_{..} = \frac{1}{n} \sum_{j=1}^{k} \sum_{i=1}^{n_j} Y_{ij} = \frac{1}{n} \sum_{j=1}^{k} n_j \bar{Y}_{.j} = \frac{1}{n} \sum_{j=1}^{k} T_{.j}$$

Appearing at the bottom of Table 11.1.1 are a set of "true means," $\mu_1, \mu_2, \ldots, \mu_k$. Each μ_j is an unknown location parameter reflecting the true average response characteristic of level j. Depending on the physical circumstances of the problem, our objective will be either to estimate the μ_j's or test their equality. The latter is perhaps the more common, and the test takes the form

$$H_0: \quad \mu_1 = \mu_2 = \cdots = \mu_k$$

versus

$$H_1: \quad \text{not all the } \mu_j\text{'s are equal}$$

In the next several sections we will propose a variance-ratio statistic for testing H_0, investigate its sampling behavior under both H_0 and H_1, and introduce a set of computing formulas to simplify its evaluation. We will also explore the possibility of testing more specific *subhypotheses* about the μ_j's—for instance, $H_0^1: \mu_2 = \mu_4$ or $H_0^2: (\mu_2 + \mu_5)/2 = \mu_3$.

11.2 THE F TEST

To derive a procedure for testing $H_0: \mu_1 = \mu_2 = \cdots = \mu_k$ we could once again invoke the generalized-likelihood-ratio criterion, compute $\lambda = L(\hat{\omega})/L(\hat{\Omega})$, and begin the search for a monotonic function of λ having a known distribution. But since we have already seen several examples of formal GLRT calculations in Chapters 7 and 8, the benefits of doing still another would be minimal. Instead, we will work backward: the test statistic will be stated at the outset and the "derivation" will be confined to an investigation of some of its properties.

The data structure for a completely randomized, one-factor design was outlined in Section 11.1. To that basic setup we now add a *distribution* assumption—that the Y_{ij}'s are independent and normally distributed with mean $\mu_j, j = 1, 2, \ldots, k$, and variance σ^2 (constant for all j). With our previous notation, this could have been written

$$Y_{ij} \sim N(\mu_j, \sigma^2) \tag{11.2.1}$$

but in analysis-of-variance problems—as in regression problems—distribution assumptions are more typically expressed in terms of *model equations*. In these equations the response variable is represented as the sum of one or more fixed components and one or more random components. Here, one possible model equation would be

$$Y_{ij} = \mu_j + \epsilon_{i(j)} \tag{11.2.2}$$

where $\epsilon_{i(j)}$ denotes the "noise" associated with Y_{ij}—that is, the amount by which Y_{ij} differs from its expected value. Of course, from Equation 11.2.1 it follows that $\epsilon_{i(j)} \sim N(0, \sigma^2)$.

While Equation 11.2.2 is a perfectly acceptable way to represent the data for a completely randomized, one-factor design, another parameterization is more commonly used. Equations 11.2.3 and 11.2.4 define μ, the overall average effect of the k levels and τ_j, the *differential effect* of level j (relative to μ):

$$\tau_j = \mu_j - \mu \tag{11.2.3}$$

and

$$\mu = \frac{1}{n} \sum_{j=1}^{k} n_j \mu_j \tag{11.2.4}$$

In terms of μ and τ_j,

$$Y_{ij} = \mu + \tau_j + \epsilon_{i(j)} \tag{11.2.5}$$

The equivalence of Equations 11.2.2 and 11.2.5 should be obvious: testing $H_0: \mu_1 = \mu_2 = \cdots = \mu_k$ is the same as testing $H_0: \tau_1 = \tau_2 = \cdots = \tau_k = 0$.

Example 11.2.1

Suppose that a completely randomized, one-factor design involves three levels whose true means are $\mu_1 = 2$, $\mu_2 = 7$, and $\mu_3 = 5$. Furthermore, suppose that

TABLE 11.2.1

Sample 1 ($\mu_1 = 2$)	Sample 2 ($\mu_2 = 7$)	Sample 3 ($\mu_3 = 5$)
Y_{11}	Y_{12}	Y_{13}
Y_{21}	Y_{22}	Y_{23}
Y_{31}		
Y_{41}		

four observations are taken from level 1 and two each from levels 2 and 3 (see Table 11.2.1). With the second parameterization, the model equation becomes

$$Y_{ij} = \mu + \tau_j + \epsilon_{i(j)} \qquad i = 1, 2, \ldots, n_j; \quad j = 1, 2, 3$$

and the values of the parameters would be

$$\mu = \frac{1}{8}[4(2) + 2(7) + 2(5)] = 4.0$$

$$\tau_1 = 2 - 4.0 = -2.0$$

$$\tau_2 = 7 - 4.0 = +3.0$$

$$\tau_3 = 5 - 4.0 = +1.0$$

Comment

Keep in mind that in any real problem neither μ nor the τ_j's are ever known; however, they can be estimated using the method of least squares. In this context the least-squares function, L, of Section 10.4 becomes

$$L = L(\mu, \tau_1, \tau_2, \ldots, \tau_k) = \sum_{j=1}^{k} \sum_{i=1}^{n_j} (y_{ij} - \mu - \tau_j)^2$$

Taking derivatives of L with respect to the unknown parameters, and setting those derivatives equal to 0, gives a set of $k + 1$ equations:

$$\frac{\partial L}{\partial \mu} = n\mu + \sum_{j=1}^{k} n_j \tau_j - \sum_{j=1}^{k} \sum_{i=1}^{n_j} y_{ij} = 0$$

$$\frac{\partial L}{\partial \tau_j} = n_j \mu + n_j \tau_j - \sum_{i=1}^{n_j} y_{ij} = 0 \qquad j = 1, 2, \ldots, k \tag{11.2.6}$$

Unique solutions to Equations 11.2.6 can be obtained only by adding a $(k + 2)$nd equation, the side condition that

$$\sum_{j=1}^{k} n_j \tau_j = 0$$

The resulting estimates, which should come as no surprise, are

$$\hat{\mu} = \bar{y}_{..} \quad \text{and} \quad \hat{\tau}_j = \bar{y}_{.j} - \bar{y}_{..} \qquad j = 1, 2, \ldots, k \tag{11.2.7}$$

Our development of the analysis of the variance for testing $H_0: \tau_1 = \tau_2 = \cdots = \tau_k = 0$ will proceed along three lines:

    I. Definition of the test statistic.

    II. Derivation of the distribution of the test statistic under H_0.

    III. Proof that the test statistic is sensitive to departures from H_0.

Taken together, the conclusions of parts I, II, and III will demonstrate the feasibility of using variance ratios to test the equality of means. The important question of the test statistic's optimality, though—specifically, its relationship to the generalized-likelihood-ratio criterion—will be deferred to the exercises.

Part I

We begin with an identity:

$$Y_{ij} = \bar{Y}_{..} + (\bar{Y}_{.j} - \bar{Y}_{..}) + (Y_{ij} - \bar{Y}_{.j})$$

or, equivalently,

$$(Y_{ij} - \bar{Y}_{..}) = (\bar{Y}_{.j} - \bar{Y}_{..}) + (Y_{ij} - \bar{Y}_{.j}) \tag{11.2.8}$$

which should be recognized as simply Equation 11.2.5, with the parameters μ and τ_j having been replaced by their least-squares estimators and $\epsilon_{i(j)}$ by $Y_{ij} - \bar{Y}_{.j}$. Since Equation 11.2.8 holds for *any* i and j, it must be true that

$$\sum_{j=1}^{k} \sum_{i=1}^{n_j} (Y_{ij} - \bar{Y}_{..})^2 = \sum_{j=1}^{k} \sum_{i=1}^{n_j} [(\bar{Y}_{.j} - \bar{Y}_{..}) + (Y_{ij} - \bar{Y}_{.j})]^2 \tag{11.2.9}$$

Expanding the right-hand side of Equation 11.2.9 gives

$$\sum_{j=1}^{k} \sum_{i=1}^{n_j} (\bar{Y}_{.j} - \bar{Y}_{..})^2 + \sum_{j=1}^{k} \sum_{i=1}^{n_j} (Y_{ij} - \bar{Y}_{.j})^2$$

since the cross-product term vanishes:

$$\sum_{j=1}^{k} \sum_{i=1}^{n_j} (\bar{Y}_{.j} - \bar{Y}_{..})(Y_{ij} - \bar{Y}_{.j}) = \sum_{j=1}^{k} (\bar{Y}_{.j} - \bar{Y}_{..}) \sum_{i=1}^{n_j} (Y_{ij} - \bar{Y}_{.j})$$

$$= \sum_{j=1}^{k} (\bar{Y}_{.j} - \bar{Y}_{..})(0) = 0$$

Therefore,

$$\sum_{j=1}^{k} \sum_{i=1}^{n_j} (Y_{ij} - \bar{Y}_{..})^2 = \sum_{j=1}^{k} \sum_{i=1}^{n_j} (\bar{Y}_{.j} - \bar{Y}_{..})^2 + \sum_{j=1}^{k} \sum_{i=1}^{n_j} (Y_{ij} - \bar{Y}_{.j})^2$$

or, more conveniently,

$$Q = Q_1 + Q_2 \tag{11.2.10}$$

It would be well to pause at this point and reflect on what these algebraic manipulations have accomplished. First, consider the term on the left-hand side of Equation 11.2.10. We will call

$$Q = \sum_{j=1}^{k} \sum_{i=1}^{n_j} (Y_{ij} - \bar{Y}_{..})^2$$

the *total sum of squares* and denote it SS_{total}. What Q measures is the total variability in the data—that is, it quantifies the extent to which the Y_{ij}'s are not all equal. (If they *were* all the same, each would equal $\bar{Y}_{..}$ and Q would be 0.)

On the right-hand side of Equation 11.2.10,

$$Q_1 = \sum_{j=1}^{k} \sum_{i=1}^{n_j} (\bar{Y}_{.j} - \bar{Y}_{..})^2$$

is an estimate of the extent to which the τ_j's are not all equal: accordingly, Q_1 will be called the *treatment sum of squares* and abbreviated $SS_{\text{treatments}}$.

Finally,

$$Q_2 = \sum_{j=1}^{k} \sum_{i=1}^{n_j} (Y_{ij} - \bar{Y}_{.j})^2$$

measures the variability *within* each level. Deviations of Y_{ij} from $\bar{Y}_{.j}$ estimate *experimental error*, the combined effect on the response variable of factors other than level j. Written SS_{error}, Q_2 is termed the *error sum of squares*.

What Equation 11.2.10, has accomplished, then, is a *partitioning* of the overall variability among the Y_{ij}'s (as measured by Q) into two components—the first, Q_1, measures that portion of the total variability due to (or "caused by") the fact that the τ_j's are not all 0; the second, Q_2, measures the variability due to factors other than the one being controlled. Conceptually, Q_1 and Q_2 play roles analogous to the numerator and denominator in the two-sample t statistic: Q_2 (like $S_p \sqrt{(1/n) + (1/m)}$) reflects the inherent variability in the data and serves as a yardstick against which the magnitude of the treatment effect, as measured by Q_1, can be compared. (In the two-sample problem, of course, the magnitude of the treatment effect is gauged by $\bar{X} - \bar{Y}$.)

It might seem reasonable at this point to define the quotient Q_1/Q_2 to be our test statistic, but for reasons part II will make clear, both the numerator and denominator first have to be *scaled*—Q_1 by $k - 1$ and Q_2 by $n - k$. The proposed statistic, then, or F ratio, for testing

$$H_0: \quad \tau_1 = \tau_2 = \cdots = \tau_k$$

versus

$$H_1: \quad \text{not all the } \tau_j\text{'s} = 0$$

will be

$$F = \frac{Q_1/(k-1)}{Q_2/(n-k)} = \frac{SS_{\text{treatments}}/(k-1)}{SS_{\text{error}}/(n-k)}$$

Part II

In part II we will find the distributions of Q_1 and Q_2 and, from them, of the F ratio, under the null hypothesis that the population means are all equal. First, consider Q_2. Let S_j^2 denote the sample variance for the jth level. In the notation of Table 11.1.1,

$$S_j^2 = \frac{1}{n_j - 1} \sum_{i=1}^{n_j} (Y_{ij} - \bar{Y}_{.j})^2$$

It follows by inspection that S_j^2 and Q_2 are directly related:

$$Q_2 = \sum_{j=1}^{k} (n_j - 1)S_j^2 \tag{11.2.11}$$

Of course, for each j,

$$\frac{(n_j - 1)S_j^2}{\sigma^2} \sim \chi_{n_j-1}^2$$

(see Theorem 7.5.3). Furthermore, the k variance ratios are all independent, since the k samples themselves are independent; therefore, by the additive property of chi square variables as stated in Theorem 7.5.1,

$$\frac{Q_2}{\sigma^2} \sim \chi_{n-k}^2$$

where, as before, $n = \sum_{j=1}^{k} n_j$.

Finding the distribution of Q_1/σ^2 presents a few more problems than Q_2/σ^2 did, but the basic approach is much the same. We begin by writing Q_1 as the difference of two sums of squares:

$$Q_1 = \sum_{j=1}^{k} \sum_{i=1}^{n_j} (\bar{Y}_{.j} - \bar{Y}_{..})^2 = \sum_{j=1}^{k} n_j(\bar{Y}_{.j} - \bar{Y}_{..})^2$$

$$= \sum_{j=1}^{k} n_j[(\bar{Y}_{.j} - \mu) - (\bar{Y}_{..} - \mu)]^2$$

$$= \sum_{j=1}^{k} n_j[(\bar{Y}_{.j} - \mu)^2 + (\bar{Y}_{..} - \mu)^2 - 2(\bar{Y}_{.j} - \mu)(\bar{Y}_{..} - \mu)]$$

$$= \sum_{j=1}^{k} n_j(\bar{Y}_{.j} - \mu)^2 - n(\bar{Y}_{..} - \mu)^2$$

Therefore,

$$\frac{Q_1}{\sigma^2} = \sum_{j=1}^{k} \left(\frac{\bar{Y}_{.j} - \mu}{\sigma/\sqrt{n_j}}\right)^2 - \left(\frac{\bar{Y}_{..} - \mu}{\sigma/\sqrt{n}}\right)^2$$

Under the assumption that H_0 is true, both $E(\bar{Y}_{.j})$ and $E(\bar{Y}_{..})$ equal μ, implying that

$$\frac{\bar{Y}_{.j} - \mu}{\sigma/\sqrt{n_j}} \sim N(0, 1) \qquad \text{for all } j$$

and

$$\frac{\bar{Y}_{..} - \mu}{\sigma/\sqrt{n}} \sim N(0, 1)$$

Now, by an application of Fisher's lemma, similar to that in Appendices 7.2 and 10.1, it can be shown that $Q_1/\sigma^2 \sim \chi^2_{k-1}$ and that Q_1 and $\bar{Y}_{..}$ are independent. The details of the argument will be omitted.

The random variables Q_1 and Q_2 are also independent. Since $\bar{Y}_{..}$ is a function of the $\bar{Y}_{.j}$'s, Q_1 is a function of those same variables. Furthermore, Q_2 is a weighted sum of the sample variances, $S_1^2, S_2^2, \ldots, S_k^2$, where each S_i^2 is independent of $\bar{Y}_{.j}$, $i \neq j$, the two having been derived from independent random samples. Finally, recall that S_j^2 is independent of $\bar{Y}_{.j}$ by Theorem A7.2.1. From these facts the independence of Q_1 and Q_2 readily follows.

The upshot of these distribution and independence arguments is that, by Theorem 7.6.1, the proposed F ratio has an F distribution (when H_0 is true). That is,

$$F = \frac{\dfrac{Q_1}{k-1}}{\dfrac{Q_2}{n-k}} = \frac{\dfrac{Q_1}{(k-1)\sigma^2}}{\dfrac{Q_2}{(n-k)\sigma^2}} \sim \frac{\dfrac{\chi^2_{k-1}}{k-1}}{\dfrac{\chi^2_{n-k}}{n-k}} \sim F_{k-1,\, n-k}$$

Part III

It remains to be demonstrated that the proposed F ratio will be sensitive to departures from H_0—that is, that the probability of F's lying in the critical region (which has yet to be defined) given that H_1 is true will be greater than α, the probability of its lying in the critical region when H_0 is true. To demonstrate this, we need to compute the expected values of $Q_1/(k-1)$ and $Q_2/(n-k)$ when H_1 is true. The denominator of the F ratio will be considered first. From Equation 11.2.11,

$$E(Q_2) = \sum_{j=1}^{k} (n_j - 1)E(S_j^2)$$

But for any j, and regardless of whether H_0 or H_1 is true, S_j^2 is an unbiased estimator for σ^2. Thus

$$E(Q_2) = (n - k)\sigma^2$$

or, equivalently,

$$E\left(\frac{Q_2}{n-k}\right) = \sigma^2 \tag{11.2.12}$$

The derivation of $E[Q_1/(k-1)]$ is somewhat more involved. First, we simplify the expression for $E(Q_1)$ from a double sum to a single sum and collect terms:

$$E(Q_1) = E\left[\sum_{j=1}^{k}\sum_{i=1}^{n_j}(\bar{Y}_{.j} - \bar{Y}_{..})^2\right] = E\left[\sum_{j=1}^{k} n_j(\bar{Y}_{.j} - \bar{Y}_{..})^2\right]$$

$$= E\left[\sum_{j=1}^{k} n_j(\bar{Y}_{.j}^2 - 2\bar{Y}_{.j}\bar{Y}_{..} + \bar{Y}_{..}^2)\right]$$

$$= E\left[\sum_{j=1}^{k} n_j\bar{Y}_{.j}^2 - n\bar{Y}_{..}^2\right] \qquad \text{(Why?)}$$

Then, after an application of Theorem 3.10.1 to rewrite $E(\bar{Y}_{.j}^2)$ and $E(\bar{Y}_{..}^2)$,

$$E(Q_1) = \sum_{j=1}^{k} n_j\{\text{Var }(\bar{Y}_{.j}) + [E(\bar{Y}_{.j})]^2\} - n\{\text{Var }(\bar{Y}_{..}) + [E(\bar{Y}_{..})]^2\}$$

Of course,

$$\text{Var }(\bar{Y}_{.j}) = \frac{\sigma^2}{n_j}$$

and, regardless of whether H_0 or H_1 is true, Var $(\bar{Y}_{..})$ equals σ^2/n:

$$\text{Var }(\bar{Y}_{..}) = \text{Var }\left(\frac{1}{n}\sum_{j=1}^{k} n_j\bar{Y}_{.j}\right) = \frac{1}{n^2}\sum_{j=1}^{k} n_j^2 \text{ Var }(\bar{Y}_{.j}) = \frac{1}{n^2}\sum_{j=1}^{k} n_j\sigma^2 = \frac{\sigma^2}{n}$$

Similarly, $E(\bar{Y}_{.j}) = \mu_j$ and $E(\bar{Y}_{..}) = \mu$, no matter what the values of the τ_j's might be. Substituting all these results into the last expression for $E(Q_1)$ gives

$$E(Q_1) = \sum_{j=1}^{k} n_j\left(\frac{\sigma^2}{n_j} + \mu_j^2\right) - n\left(\frac{\sigma^2}{n} + \mu^2\right)$$

$$= (k-1)\sigma^2 + \sum_{j=1}^{k} n_j(\mu_j - \mu)^2$$

$$= (k-1)\sigma^2 + \sum_{j=1}^{k} n_j\tau_j^2$$

Notice that when H_0 is true, $\sum_{j=1}^{k} n_j\tau_j^2 = 0$ and

$$E\left(\frac{Q_1}{k-1}\right) = \sigma^2 \qquad (11.2.13)$$

When H_1 is true, though, $\sum_{j=1}^{k} n_j\tau_j^2 > 0$ and

$$E\left(\frac{Q_1}{k-1}\right) = \sigma^2 + \frac{1}{k-1}\sum_{j=1}^{k} n_j\tau_j^2 \qquad (11.2.14)$$

a number *greater* than σ^2.

Equations 11.2.12, 11.2.13, and 11.2.14 dictate the nature of the critical region. When H_0 is true the expected values of $Q_1/(k-1)$ and $Q_2/(n-k)$ are both σ^2, meaning that the F ratio should be close to 1. But when H_1 is true,

$$E\left(\frac{Q_1}{k-1}\right) > E\left(\frac{Q_2}{n-k}\right)$$

and the F ratio will tend to be larger than 1. It follows that the critical region should be in the right-hand tail of the $F_{k-1,\,n-k}$ distribution and that H_0 should be rejected at the α level of significance if the observed F ratio is greater than or equal to $F_{1-\alpha,\,k-1,\,n-k}$ (see Figure 11.2.1). Case Study 11.2.1 illustrates the technique.

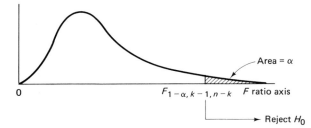

Figure 11.2.1 F distribution with $k-1$ and $n-k$ degrees of freedom.

CASE STUDY   11.2.1

A certain fraction of antibiotics injected into the bloodstream are "bound" to serum proteins. This is a phenomenon of considerable pharmacological importance, because as the extent of the binding increases, the systemic uptake of the drug decreases. This, in turn, bears directly on the effectiveness the medication will ultimately have.

In one recent study, determinations were made of the binding percentages characteristic of five widely used antibiotics. Bovine serum was used in each instance, but the results are comparable to what could be expected in human serum (193) (see Table 11.2.2).

TABLE 11.2.2

	Penicillin G	Tetra-cycline	Strepto-mycin	Erythro-mycin	Chloram-phenicol
	29.6	27.3	5.8	21.6	29.2
	24.3	32.6	6.2	17.4	32.8
	28.5	30.8	11.0	18.3	25.0
	32.0	34.8	8.3	19.0	24.2
$t_{.j}$	114.4	125.5	31.3	76.3	111.2
$\bar{y}_{.j}$	28.6	31.4	7.8	19.1	27.8

In terms of the μ and τ_j parameterization, the model equation for these data would be written

$$Y_{ij} = \mu + \tau_j + \epsilon_{i(j)} \qquad j = 1, 2, \ldots, 5; \, i = 1, 2, 3, 4$$

with τ_j being the differential effect of the jth antibiotic. We will assume that $\epsilon_{i(j)} \sim N(0, \sigma^2)$. A preliminary hypothesis that might be of interest is whether or not the true binding percentages $(\mu_j, j = 1, 2, 3, 4, 5)$ for the five antibiotics are all equal. This requires testing

$$H_0: \quad \tau_1 = \tau_2 = \cdots = \tau_5 = 0$$

versus

$$H_1: \quad \text{not all the } \tau_j\text{'s} = 0$$

For these data, $k = 5$ and $n = 20$, so the F test calls for H_0 to be rejected at the α level of significance if

$$\frac{(20 - 5) \sum\limits_{j=1}^{5} 4(\bar{y}_{.j} - \bar{y}_{..})^2}{(5 - 1) \sum\limits_{j=1}^{5} \sum\limits_{i=1}^{4} (y_{ij} - \bar{y}_{.j})^2} \geq F_{1-\alpha, \, 4, \, 15}$$

If α is chosen to be 0.05, $F_{0.95, \, 4, \, 15} = 3.06$ (see Figure 11.2.2).

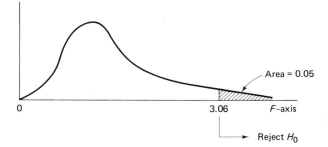

Area = 0.05

0   3.06   F–axis

Reject H_0

Figure 11.2.2 F distribution with 4 and 15 degrees of freedom.

From Table 11.2.2,

$$\bar{y}_{..} = \frac{1}{20} \sum_{j=1}^{5} t_{.j} = \frac{1}{20}(458.7) = 22.94$$

making the treatment sum of squares 371.19:

$$\sum_{j=1}^{5} (\bar{y}_{.j} - \bar{y}_{..})^2 = \sum_{j=1}^{5} (\bar{y}_{.j} - 22.94)^2 = 371.19$$

Similarly,

$$\sum_{j=1}^{5} \sum_{i=1}^{4} (y_{ij} - \bar{y}_{.j})^2 = 135.83$$

Substituting these two sums into the formula for the test statistic gives

$$\frac{15(4)(371.19)}{4(135.83)} = 40.99$$

Because 40.99 exceeds the 0.05 critical value (3.06), we reject the null hypothesis that the true binding percentages are all equal, a conclusion hardly unexpected in light of the substantial between-level differences readily apparent in Table 11.2.2.

Question 11.2.1 In 1965 a silver shortage in the United States prompted Congress to authorize the minting of silverless dimes and quarters. They also recommended that the silver content of half-dollars be reduced from 90% to 40%. Historically, fluctuations in the amount of rare metals found in coins are not uncommon. The data in the table concern a possible shift in silver content of a Byzantine coin minted on two separate occasions during the reign of Manuel I (1143–1180).

Early Coinage (%)	Late Coinage (%)
5.9	5.3
6.8	5.6
6.4	5.5
7.0	5.1
6.6	6.2
7.7	5.8
7.2	5.8
6.9	
6.2	

The 16 coins examined (nine belonging to the earlier minting) were part of a large hoard discovered not too long ago in Cyprus. The chemical analysis consisted of first dissolving chips from a particular specimen in a 50% nitric acid solution and then titrating that solution with sodium chloride until all the silver chloride had precipitated out. By weighing the precipitate, the percentage of silver in the coin could be determined (70). Do an analysis of variance on the data given in the table. Let 0.05 be the level of significance.

Question 11.2.2 There is a possibility that a teacher's opinion of the academic abilities of her students may be translated into actions and attitudes that help produce those expectations (that is, the opinion may be a self-fulfilling prophecy). Not too long ago an experiment was set up (137) to investigate that hypothesis. The data given here, while fictitious, are based on the findings of that study. Children in the first grade at a certain school were all given a standard IQ test, but the children's teachers were told it was a special test for predicting whether a child would show sudden spurts of intellectual growth in the near future. The experimenters then divided the children into three groups *at random* but informed the teachers that, according to the test, the children in group I probably would not demonstrate any pronounced intellectual growth for the next year, those in group II would develop at a moderate rate, while those in group III could be expected to advance considerably during the coming year. A year later the children were again given a standard IQ test. The differences in the two IQ scores (second test—first test) for each child are shown below.

IQ INCREASES

Group I	Group II	Group III
3	10	20
2	4	10
6	11	19
10	14	15
11	5	9
5	3	18

Follow the method described in Case Study 11.2.1 and do an analysis of variance on these data. Use the 0.05 level of significance. Do the data support the contention that teachers' attitudes can affect IQ?

Question 11.2.3 Carry out the details to show that

$$\frac{1}{\sigma^2} \left\{ \sum_{j=1}^{k} n_j [(\bar{Y}_{.j} - \mu)^2 + (\bar{Y}_{..} - \mu)^2 - 2(\bar{Y}_{.j} - \mu)(\bar{Y}_{..} - \mu)] \right\}$$
$$= \sum_{j=1}^{k} \left(\frac{\bar{Y}_{.j} - \mu}{\sigma/\sqrt{n_j}} \right)^2 - \left(\frac{\bar{Y}_{..} - \mu}{\sigma/\sqrt{N}} \right)^2$$

(see part II of the derivation of the F test).

The analysis of variance was introduced in Section 11.1 as a k-sample *extension* of the two-sample t test. The two procedures overlap, though, when k is equal to 2, which raises an obvious question: which procedure is better for testing H_0: $\mu_X = \mu_Y$? The answer, as Example 11.2.2 shows, is "neither": despite their different test statistics, the two procedures are equivalent—if one rejects H_0, so will the other. (This should have been expected, of course, since both the F test and the t test are GLRTs.)

Example 11.2.2

Let $X_1, X_2, \ldots, X_n \sim N(\mu_X, \sigma^2)$ and $Y_1, Y_2, \ldots, Y_m \sim N(\mu_Y, \sigma^2)$ be two independent random samples. If these data were subjected to the analysis of variance, the test statistic

$$\frac{\frac{Q_1}{k-1}}{\frac{Q_2}{n-k}} = \frac{\frac{Q_1}{1}}{\frac{Q_2}{n+m-2}}$$

would have the F distribution with 1 and $n + m - 2$ degrees of freedom. Now, consider the corresponding t test. By Theorem 8.2.2, $H_0: \mu_X = \mu_Y$ would be rejected in favor of $H_1: \mu_X \neq \mu_Y$ if either $t \leq -t_{\alpha/2, n+m-2}$ or $t \geq +t_{\alpha/2, n+m-2}$, or, equivalently, if $t^2 \geq t^2_{\alpha/2, n+m-2}$. But

$$
T^2 = \left(\frac{\bar{X} - \bar{Y}}{S_p \sqrt{\dfrac{1}{n} + \dfrac{1}{m}}} \right)^2 = \frac{\left[\dfrac{\bar{X} - \bar{Y}}{\sigma \sqrt{\dfrac{1}{n} + \dfrac{1}{m}}} \right]^2}{\dfrac{(n + m - 2)S_p^2}{\sigma^2(n + m - 2)}}
$$

$$
\sim \frac{[N(0, 1)]^2}{\dfrac{\chi^2_{n+m-2}}{n + m - 2}} \sim \frac{\dfrac{\chi^2_1}{1}}{\dfrac{\chi^2_{n+m-2}}{n + m - 2}} \sim F_{1, n+m-2}
$$

(recall Theorem 7.6.1). Therefore, the t test will reject H_0 if and only if the F test does.

Question 11.2.4 Verify the conclusion of Example 11.2.2 by doing a t test on the data of Question 11.2.1. Show that the observed F ratio is the square of the observed t ratio and that the F critical value is the square of the t critical value.

Question 11.2.5 Do an analysis of variance on the Mark Twain–Quintus Curtius Snodgrass data of Case Study 8.2.1. Verify that the observed F ratio is the square of the observed t ratio (3.86).

To parallel the treatment given the t test in Chapters 7 and 8, we conclude this section with a few remarks concerning the *distribution* of the F ratio when H_1 is true (recall Appendices 7.1 and 8.2). Knowing our test statistic's distribution under the alternative hypothesis is important: it makes possible, for example, the construction of a power function. But first we require two preliminary definitions.

Definition 11.2.1. Let $Y_i \sim N(\mu_i, 1)$ for $i = 1, 2, \ldots, t$. Then

$$
Y = \sum_{i=1}^{t} Y_i^2
$$

is said to have the *noncentral* χ^2 distribution with t degrees of freedom and noncentrality parameter γ, where

$$
\gamma = \frac{1}{2} \sum_{i=1}^{t} \mu_i^2
$$

Definition 11.2.2. Let V_1 be a noncentral χ^2 random variable with r_1 degrees of freedom and noncentrality parameter γ and let V_2 be an independent (central) χ^2 variable with r_2 degrees of freedom. Then the ratio

$$\frac{V_1/r_1}{V_2/r_2}$$

is said to have a *noncentral F distribution* with r_1 and r_2 degrees of freedom and noncentrality parameter γ.

Substituting $\mu_j - \tau_j$ for μ in part II's expression for Q_1/σ^2 gives

$$\frac{Q_1}{\sigma^2} = \sum_{j=1}^{k} \left(\frac{\bar{Y}_{.j} - \mu}{\frac{\sigma}{\sqrt{n_j}}} \right)^2 - \left(\frac{\bar{Y}_{..} - \mu}{\frac{\sigma}{\sqrt{n}}} \right)^2$$

$$= \sum_{j=1}^{k} \left(\frac{\bar{Y}_{.j} - \mu_j + \tau_j}{\frac{\sigma}{\sqrt{n_j}}} \right)^2 - \left(\frac{\bar{Y}_{..} - \mu}{\frac{\sigma}{\sqrt{n}}} \right)^2$$

Observe that when H_1 is true, and $E(\bar{Y}_{.j}) = \mu_j$,

$$\frac{\bar{Y}_{.j} - \mu_j + \tau_j}{\frac{\sigma}{\sqrt{n_j}}} \sim N\left(\frac{\tau_j \sqrt{n_j}}{\sigma}, 1 \right)$$

Unaffected by which hypothesis is true, the distribution of the second component of Q_1/σ^2 remains unchanged:

$$\frac{\bar{Y}_{..} - \mu}{\frac{\sigma}{\sqrt{n}}} \sim N(0, 1)$$

By Definition 11.2.1, then, Q_1/σ^2 has a noncentral χ^2 distribution with $k - 1$ degrees of freedom and noncentrality parameter γ, where

$$\gamma = \frac{1}{2\sigma^2} \sum_{j=1}^{k} n_j \tau_j^2$$

Finally, since Q_2/σ^2 is always χ^2_{n-k}, it follows immediately from Definition 11.2.2 that, when H_1 is true,

$$F = \frac{\dfrac{Q_1}{k-1}}{\dfrac{Q_2}{n-k}}$$

has a noncentral F distribution with $k - 1$ and $n - k$ degrees of freedom and noncentrality parameter γ.

Comment

As H_1 gets further away from H_0, as measured by γ, the noncentral F will shift more and more to the right of the central F. Accordingly, the power of the F test will increase. That is,

$$P(F \geq F_{1-\alpha,\, k-1,\, n-k}) \to 1 \qquad \text{as } \gamma \to \infty$$

The pdf for the noncentral F is not very tractable, but its integral has been evaluated by numerical approximation, and this has allowed the power function for the F test to be tabulated [see, for instance, (99)].

Question 11.2.6 Suppose an experimenter has taken three independent measurements on each of five treatment levels and intends to use the analysis of variance to test

$$H_0: \quad \tau_1 = \tau_2 = \tau_3 = \tau_4 = \tau_5 \; (=0)$$

versus

$$H_1: \quad \text{not all the } \tau_j\text{'s are } 0$$

where the τ_j's are the differential effects of the treatment levels. Two of the possible alternatives in H_1 are

$$H_1^*: \quad \tau_1 = -1,\, \tau_2 = 2,\, \tau_3 = 0,\, \tau_4 = 1,\, \tau_5 = -2$$

and

$$H_1^{**}: \quad \tau_1 = -3,\, \tau_2 = 2,\, \tau_3 = 1,\, \tau_4 = 0,\, \tau_5 = 0$$

Against which alternative will the F test have greater power? Explain. Is H_1^{***}: $\tau_1 = 2,\, \tau_2 = 1,\, \tau_3 = 1,\, \tau_4 = -3,\, \tau_5 = 0$ an "admissible" member of H_1?

Question 11.2.7 If the random variable Y has a noncentral χ^2 distribution with r degrees of freedom and noncentrality parameter γ (see Definition 11.2.1), then its moment-generating function is given by

$$M_Y(t) = \frac{1}{(1-2t)^{r/2}}\, e^{2t\gamma/(1-2t)} \qquad t < \tfrac{1}{2}; \quad 0 < \gamma$$

[for a derivation, see (73)]. Find $E(Y)$ and Var (Y).

Question 11.2.8 Suppose $Y_1,\, Y_2,\, \ldots,\, Y_n$ are independent noncentral χ^2 random variables having $r_1,\, r_2,\, \ldots,\, r_n$ degrees of freedom, respectively, and with noncentrality parameters, $\gamma_1,\, \gamma_2,\, \ldots,\, \gamma_n$. Find the distribution of $W = Y_1 + Y_2 + \cdots + Y_n$.

11.3 COMPUTING FORMULAS FOR THE F TEST

In practice, there is a much easier way to calculate the F ratio than indicated in Section 11.2. The expressions for SS_{total} and $SS_{\text{treatments}}$ given in Theorem 11.3.1 are in forms particularly easy to evaluate with an electronic calculator.

Theorem 11.3.1. Let $c = T_{..}^2/n$. Then

$$SS_{total} = \sum_{j=1}^{k} \sum_{i=1}^{n_j} Y_{ij}^2 - c$$

and

$$SS_{treatments} = \sum_{j=1}^{k} \frac{T_{.j}^2}{n_j} - c$$

Proof. By definition,

$$SS_{total} = \sum_{j=1}^{k} \sum_{i=1}^{n_j} (Y_{ij} - \bar{Y}_{..})^2 = \sum_{j=1}^{k} \sum_{i=1}^{n_j} (Y_{ij}^2 - 2Y_{ij}\bar{Y}_{..} + \bar{Y}_{..}^2)$$

$$= \sum_{j=1}^{k} \sum_{i=1}^{n_j} Y_{ij}^2 - 2\bar{Y}_{..} \sum_{j=1}^{k} \sum_{i=1}^{n_j} Y_{ij} + n\bar{Y}_{..}^2$$

But

$$\sum_{j=1}^{k} \sum_{i=1}^{n_j} Y_{ij} = n\bar{Y}_{..}$$

and the first part of the theorem follows:

$$SS_{total} = \sum_{j=1}^{k} \sum_{i=1}^{n_j} Y_{ij}^2 - n\bar{Y}_{..}^2 = \sum_{j=1}^{k} \sum_{i=1}^{n_j} Y_{ij}^2 - c$$

Similarly,

$$SS_{treatments} = \sum_{j=1}^{k} \sum_{i=1}^{n_j} (\bar{Y}_{.j} - \bar{Y}_{..})^2 = \sum_{j=1}^{k} n_j(\bar{Y}_{.j} - \bar{Y}_{..})^2$$

$$= \sum_{j=1}^{k} n_j(\bar{Y}_{.j} - 2\bar{Y}_{.j}\bar{Y}_{..} + \bar{Y}_{..}^2)$$

$$= \sum_{j=1}^{k} \frac{T_{.j}^2}{n_j} - 2\bar{Y}_{..} \sum_{j=1}^{k} n_j\bar{Y}_{.j} + n\bar{Y}_{..}^2$$

and, since $\sum_{j=1}^{k} n_j\bar{Y}_{.j} = n\bar{Y}_{..}$,

$$SS_{treatments} = \sum_{j=1}^{k} \frac{T_{.j}^2}{n_j} - n\bar{Y}_{..}^2 = \sum_{j=1}^{k} \frac{T_{.j}^2}{n_j} - c$$

The error sum of squares can be gotten either directly or by subtraction. The latter approach follows immediately from Equation 11.2.10:

$$SS_{error} = SS_{total} - SS_{treatments}$$

It will be left as an exercise to prove that Equation 11.3.1 gives the error sum of squares directly:

$$SS_{error} = \sum_{j=1}^{k} \left(\sum_{i=1}^{n_j} Y_{ij}^2 - \frac{T_{.j}^2}{n_j} \right) \tag{11.3.1}$$

TABLE 11.3.1

Source	df	SS	MS	EMS	F
Treatments	$k-1$	$\displaystyle\sum_{j=1}^{k} \frac{T_{\cdot j}^2}{n_j} - c$	$\dfrac{SS_{tr}}{k-1}$	$\sigma^2 + \dfrac{1}{k-1}\displaystyle\sum_{j=1}^{k} n_j \tau_j^2$	$\dfrac{MS_{tr}}{MS_{er}}$
Error	$n-k$	$SS_{tot} - SS_{tr}$	$\dfrac{SS_{er}}{n-k}$	σ^2	
Total	$n-1$	$\displaystyle\sum_{j=1}^{k}\sum_{i=1}^{n_j} Y_{ij}^2 - c$			

Calculations for an F test are typically presented in an *analysis-of-variance* (or *ANOVA*) *table* (see Table 11.3.1). The first column of an ANOVA table lists each of the *sources of variation*, as singled out by the model equation. For the completely randomized, one-factor design there are three: treatments, error, and total. Since each source of variation can be used to provide an estimate of σ^2, each has an associated number of degrees of freedom. These are listed in the second column. The third column, denoted SS, records the sum of squares corresponding to each source of variation. Next is the mean-square (MS) column: entries in this fourth column are each source of variation's sum of squares divided by its degrees of freedom. No mean square is calculated for "total." Still next is the EMS, or expected-mean-square, column. For each source of variation (other than "total"), $EMS = E(SS/df)$. Finally, the sixth column displays whatever F ratios are appropriate for the hypotheses being tested. Of course, for the completely randomized, one-factor design there is only one:

$$\frac{SS_{treatments}/(k-1)}{SS_{error}/(n-k)} = \frac{MS_{treatments}}{MS_{error}}$$

Often a variety of symbols appear adjacent to the final F ratios to signify their statistical significance. By convention, a single asterisk means that the corresponding H_0 can be rejected at the $\alpha = 0.05$ level of significance. Two asterisks indicate that H_0 can be rejected at $\alpha = 0.01$. A blank, or an NS ("not significant"), means that H_0 is accepted at $\alpha = 0.05$.

Comment

The EMS column is not always included in ANOVA tables, particularly when the experimental designs are relatively straightforward, as is the case here and in Chapter 12. In more complicated situations, though, the EMS column is absolutely essential in determining how the analysis is to proceed.

Example 11.3.1

Consider again the antibiotic serum binding data of Table 11.2.2. For the 20 observations listed,

$$\sum_{j=1}^{5}\sum_{i=1}^{4} y_{ij} = 458.7$$

and

$$\sum_{j=1}^{5}\sum_{i=1}^{4} y_{ij}^2 = 12{,}136.93$$

This makes the correction factor, c, 10,520.28 and the total sum of squares 1616.65:

$$c = \frac{(458.7)^2}{20} = 10,520.28$$

$$SS_{total} = 12,136.93 - 10,520.28 = 1616.65$$

Squaring the treatment totals, dividing each by n_j, and subtracting c gives the treatment sum of squares:

$$SS_{treatments} = \frac{(114.4)^2}{4} + \frac{(125.5)^2}{4} + \cdots + \frac{(111.2)^2}{4} - 10,520.28$$

$$= 1480.83$$

Table 11.3.2 is the completed ANOVA table.

TABLE 11.3.2 ANOVA COMPUTATIONS

Source	df	SS	MS	EMS	F	
Antibiotics	4	1480.83	370.21	$\sigma^2 + \sum\limits_{j=1}^{5} \tau_j^2$	40.9	**
Error	15	135.82	9.05	σ^2		
Total	19	1616.65				

Note that the entry appearing in the last column is the same value that we got for the observed F ratio in Section 11.2. The two asterisks signify that the test statistic exceeds $F_{0.99, 4, 15} = 3.06$ and that $H_0: \tau_1 = \tau_2 = \cdots = \tau_5 = 0$ should be rejected at the $\alpha = 0.01$ level of significance.

Question 11.3.1 Each of five varieties of corn is planted in three plots in a large field. The respective yields, in bushels per acre, are indicated below.

Var. 1	Var. 2	Var. 3	Var. 4	Var. 5
46.2	49.2	60.3	48.9	52.5
51.9	58.6	58.7	51.4	54.0
48.7	57.4	60.4	44.6	49.3

Test whether the differences among the average yields are statistically significant. Show the ANOVA table. Let 0.05 be the level of significance.

Question 11.3.2 A car manufacturer is trying to decide which of three catalytic converters to install on its new models. All three meet the government emission standards but they may not all give the same mileage. Ten new cars, all alike, are selected for the test. Converter A is installed on four of the cars, while converters B and C are each put on three. The cars are then driven over a specially engineered track that simulates city driving. The mileage estimates are listed below.

Converter A	Converter B	Converter C
21.6	22.8	23.9
23.2	25.6	24.6
20.5	24.7	23.8
21.7		

Test whether all three converters are equally economical. Let $\alpha = 0.05$.

Question 11.3.3 Derive the computing formula for the error sum of squares given in Equation 11.3.1,

$$SS_{error} = \sum_{j=1}^{k} \left(\sum_{i=1}^{n_j} Y_{ij}^2 - \frac{T_{.j}^2}{n_j} \right)$$

Then use the computing formula to double-check the value gotten for SS_{error} in Question 11.3.2.

Question 11.3.4 Suppose k independent random samples, each of size r, are taken from an $N(\mu_j, \sigma^2)$ pdf, $j = 1, 2, \ldots, k$, where the μ_j's and σ^2 are all unknown. We wish to set up the generalized-likelihood-ratio test for

$$H_0: \quad \mu_1 = \mu_2 = \cdots = \mu_k \ (=\mu)$$

versus

$$H_1: \quad \text{not all the } \mu_j\text{'s are equal}$$

(a) Write out $L(\omega)$ and $L(\Omega)$.

(b) Show that the MLE's for μ and σ^2 under ω are

$$\hat{\mu} = \frac{1}{rk} \sum_{j=1}^{k} \sum_{i=1}^{r} y_{ij} = \bar{y}_{..}$$

and

$$\hat{\sigma}^2 = \frac{1}{rk} \sum_{j=1}^{k} \sum_{i=1}^{r} (y_{ij} - \bar{y}_{..})^2$$

(c) Show that the MLE's for μ_j and σ^2 under Ω are

$$\hat{\mu}_j = \frac{1}{r} \sum_{i=1}^{r} y_{ij} = \bar{y}_{.j} \qquad j = 1, 2, \ldots, k$$

and

$$\hat{\sigma}^2 = \frac{1}{rk} \sum_{j=1}^{k} \sum_{i=1}^{r} (y_{ij} - \bar{y}_{.j})^2$$

(d) Show that the generalized likelihood ratio reduces to

$$\lambda = \frac{L(\hat{\omega})}{L(\hat{\Omega})} = \left[\frac{\sum_{j=1}^{k} \sum_{i=1}^{r} (y_{ij} - \bar{y}_{.j})^2}{\sum_{j=1}^{k} \sum_{i=1}^{r} (y_{ij} - \bar{y}_{..})^2} \right]^{rk/2}$$

Question 11.3.5 Refer to Question 11.3.4. Express $\lambda^{2/rk}$ as a function of the treatment sum of squares and the error sum of squares. Show that rejecting $H_0: \mu_1 = \mu_2 = \cdots = \mu_k$ whenever λ is too small (recall Definition 6.4.2) is equivalent to rejecting H_0 when the observed F ratio,

$$F = \frac{Q_1/(k-1)}{Q_2/k(r-1)}$$

is too large.

11.4 TESTING SUBHYPOTHESES: ORTHOGONAL CONTRASTS

The F test as described in Sections 11.2 and 11.3 is the usual beginning in an analysis-of-variance problem, but it is seldom the ending. Few researchers, after working months—or even years—to design and carry out an experiment, will be satisfied with a statistician whose conclusion is simply "Yes, the means are all equal," or "No, they are not." In most situations there are other, more specific, inferences an experimenter would like to make. Consider, for example, the serum binding data of Case Study 11.2.1. Although a total of five antibiotics were tested, it might be that a researcher is especially interested in one particular illness for which only the first two drugs are recommended. It would make sense under those circumstances to single out penicillin G and tetracycline and compare *their* binding percentages apart from all the others: that is, test the hypothesis $H_0': \mu_1 = \mu_2$. Clearly, the ability to do this—break down a broad null hypothesis into smaller, more relevant *subhypotheses*—would add to the analysis of variance a much needed measure of flexibility.

In general, there are two ways to test subhypotheses, the choice depending, strangely enough, on *when* the H_0' is first specified. If, on the basis of physical considerations, economic factors, past experience, and so on, a particular subhypothesis can be formulated *before any data are taken*, then H_0' can best be tested using an *orthogonal contrast*. On the other hand, if the researcher wishes to do the experiment first and let the results suggest a suitable subhypothesis, the appropriate analysis is any of the so-called *multiple-comparison* techniques. In this section we discuss the construction of orthogonal contrasts; the multiple-comparison problem will be taken up in Section 11.5.

> **Definition 11.4.1.** Let $\mu_1, \mu_2, \ldots, \mu_k$ denote the true means of the k factor levels being sampled. A linear combination, C, of the μ_j's is said to be a *contrast* if the sum of its coefficients is zero. That is, C is a contrast if
>
> $$C = \sum_{j=1}^{k} c_j \mu_j \quad \text{and} \quad \sum_{j=1}^{k} c_j = 0$$

Contrasts are important because of their relationship to hypothesis tests. Consider the situation just described—testing H_0': $\mu_1 = \mu_2$, or, in a form clearly equivalent, H_0': $\mu_1 - \mu_2 = 0$. Note that in this second format, H_0' is a statement about a contrast—specifically, the contrast C, where

$$C = \mu_1 - \mu_2 = (1)\mu_1 - (1)\mu_2 + (0)\mu_3 + (0)\mu_4 + (0)\mu_5$$

As another example, suppose that there was a valid pharmacological reason for comparing the average level of serum binding for the first two antibiotics to the average level for the last three. Written as a subhypothesis, the statement of no difference would be

$$H_0': \frac{\mu_1 + \mu_2}{2} = \frac{\mu_3 + \mu_4 + \mu_5}{3}$$

As a contrast, it becomes

$$C = \frac{1}{2}\mu_1 + \frac{1}{2}\mu_2 - \frac{1}{3}\mu_3 - \frac{1}{3}\mu_4 - \frac{1}{3}\mu_5$$

In both these cases, the numerical value of the contrast will be 0 if H_0' is true. This suggests that the choice between H_0' and H_1' can be accomplished by first estimating C and then determining, via a significance test, whether that estimate is too far from 0. The details are presented in the next several paragraphs.

We begin by considering some of the mathematical properties of contrasts and their estimates. Since $\bar{Y}_{.j}$ is always an unbiased estimator for μ_j, it seems reasonable to estimate C, a linear combination of population means, with $\hat{C}$, a linear combination of *sample* means:

$$\hat{C} = \sum_{j=1}^{k} c_j \bar{Y}_{.j}$$

(The coefficients appearing in $\hat{C}$, of course, are the same as those that defined C.) It follows that

$$E(\hat{C}) = \sum_{j=1}^{k} c_j E(\bar{Y}_{.j}) = C$$

and

$$\mathrm{Var}\,(\hat{C}) = \sum_{j=1}^{k} c_j^2 \, \mathrm{Var}\,(\bar{Y}_{.j}) = \sigma^2 \sum_{j=1}^{k} \frac{c_j^2}{n_j}$$

Comment

Replacing the unknown error variance, σ^2, by its estimate from the ANOVA table—$\hat{\sigma}^2 = MS_{\mathrm{error}}$—gives a formula for the estimated variance of the estimated contrast:

$$S_{\hat{C}}^2 = \widehat{\mathrm{Var}(\hat{C})} = MS_{\mathrm{error}} \sum_{j=1}^{k} \frac{c_j^2}{n_j}$$

The sampling behavior of $\hat{C}$ is easily derived. By Theorem 7.3.1, the normality of the Y_{ij}'s ensures that $\hat{C}$ is also normal, and by the usual Z transformation, the ratio

$$\frac{\hat{C} - E(\hat{C})}{\sqrt{\text{Var}(\hat{C})}} = \frac{\hat{C} - C}{\sqrt{\text{Var}(\hat{C})}}$$

is a *standard* normal. Therefore,

$$\left[\frac{\hat{C} - C}{\sqrt{\text{Var}(\hat{C})}}\right]^2 \sim X_1^2 \qquad (11.4.1)$$

Of course, if $H_0: \mu_1 = \mu_2 = \cdots = \mu_k$ is true, C is 0, and Equation 11.4.1 reduces to

$$\frac{\hat{C}^2}{\sigma^2 \sum\limits_{j=1}^{k} \dfrac{c_j^2}{n_j}} \sim \chi_1^2 \qquad (11.4.2)$$

One final property of contrasts plays a role in the significance testing of subhypotheses. Two contrasts,

$$C_1 = \sum_{j=1}^{k} c_{1j}\mu_j \qquad \text{and} \qquad C_2 = \sum_{j=1}^{k} c_{2j}\mu_j$$

are said to be *orthogonal* if

$$\sum_{j=1}^{k} \frac{c_{1j}c_{2j}}{n_j} = 0$$

Similarly, a set of q contrasts, $\{C_i\}_{i=1}^{q}$, are said to be *mutually orthogonal* if

$$\sum_{j=1}^{k} \frac{c_{sj}c_{tj}}{n_j} = 0 \qquad \text{for all } s \neq t$$

(The same definitions apply, of course, to *estimated* contrasts.)

Definition 11.4.2 and Theorems 11.4.1 and 11.4.2, both stated here without proof, summarize the relationship between contrasts and the analysis of variance. In short, the treatment sum of squares can be partitioned into $k - 1$ "contrast" sums of squares, provided the contrasts are mutually orthogonal. These $k - 1$ sums of squares can then be used to form $k - 1$ F ratios—and to test $k - 1$ subhypotheses.

Definition 11.4.2. Let $C_i = \sum\limits_{j=1}^{k} c_{ij}\mu_j$ be any contrast. The sum of squares associated with C_i is given by

$$SS_{C_i} = \frac{\hat{C}_i^2}{\sum\limits_{j=1}^{k} \dfrac{c_{ij}^2}{n_j}}$$

where $\hat{C}_i = \sum\limits_{j=1}^{k} c_{ij}\bar{Y}_{.j}$.

Theorem 11.4.1. Let $\left\{C_i = \sum_{j=1}^{k} c_{ij}\mu_j\right\}_{i=1}^{k-1}$ be a set of $k-1$ mutually orthogonal contrasts. Let $\left\{\hat{C}_i = \sum_{j=1}^{k} c_{ij}\bar{Y}_{.j}\right\}_{i=1}^{k-1}$ be their estimators. Then

$$SS_{\text{treatments}} = \sum_{j=1}^{k} \sum_{i=1}^{n_j} (\bar{Y}_{.j} - \bar{Y}_{..})^2$$

$$= SS_{C_1} + SS_{C_2} + \cdots + SS_{C_{k-1}}$$

Theorem 11.4.2. Let C be a contrast having the same coefficients as the subhypothesis H_0': $c_1\mu_1 + c_2\mu_2 + \cdots + c_k\mu_k = 0$, where $\sum_{i=1}^{k} c_i = 0$. As before, let

$$SS_{\text{error}} = \sum_{j=1}^{k} \sum_{i=1}^{n_j} (Y_{ij} - \bar{Y}_{.j})^2 \quad \text{and} \quad n = \sum_{j=1}^{k} n_j$$

Then, if H_0' is true,

$$F = \frac{\dfrac{SS_C}{1}}{\dfrac{SS_{\text{error}}}{n-k}} \sim F_{1,\,n-k}$$

Also, if H_0' is not true, the F ratio will have a noncentral F distribution with $\gamma > 0$, implying that H_0' should be rejected at level α if $F \geq F_{1-\alpha,\,1,\,n-k}$.

Comment

Theorem 11.4.1 is not meant to imply that only mutually orthogonal contrasts can, or should, be tested. It is simply a statement of a partitioning relationship that exists between $SS_{\text{treatments}}$ and the sum of squares for mutually orthogonal C_i's. In any given experiment, the contrasts that should be singled out are those the experimenter has some prior reason to test.

CASE STUDY 11.4.1

As a rule, infants are not able to walk by themselves until they are almost 14 months old. A recent study, however, investigated the possibility of reducing that time through the use of special "walking" exercises (191). A total of 23 infants were included in the experiment—all were one-week-old white males. They were randomly divided into four groups, and for seven weeks each group followed a different training program. Group A received special walking and placing exercises for 12 minutes each day. Group B also had daily 12-minute exercise periods but were not given the special walking and placing exercises. Groups C and D received no special instruction. The progress of groups A, B, and C was checked every week; the progress of group D was checked only once, at the end of the study.

After seven weeks the formal training ended and the parents were told they could continue with whatever procedure they desired. Listed in Table 11.4.1 are the ages (in months) at which each of the 23 children first walked alone. Table 11.4.2 shows the analysis-of-variance computations and the 2.14 F ratio. Based on 3 and 19 degrees of freedom, the $\alpha = 0.05$ critical value is 3.13, so $H_0: \mu_A = \mu_B = \mu_C = \mu_D$ is accepted.

TABLE 11.4.1 AGE WHEN INFANTS FIRST WALKED ALONE (MONTHS)

	Group A	Group B	Group C	Group D
	9.00	11.00	11.50	13.25
	9.50	10.00	12.00	11.50
	9.75	10.00	9.00	12.00
	10.00	11.75	11.50	13.50
	13.00	10.50	13.25	11.50
	9.50	15.00	13.00	
$t_{.j}$	60.75	68.25	70.25	61.75
$\bar{y}_{.j}$	10.12	11.38	11.71	12.35

TABLE 11.4.2 ANOVA COMPUTATIONS

Source	df	SS	MS	EMS	F	
Exercises	3	14.77	4.92	$\sigma^2 + \dfrac{1}{3}\sum\limits_{j=1}^{4} n_j \tau_j^2$	2.14	NS
Error	19	43.70	2.30	σ^2		
Total	22	58.47				

Comment

At this point the analysis would usually end; with the overall H_0 being accepted, there is no pressing need to look at contrasts. We will continue with the subhypothesis procedures, however, to illustrate the application of Theorem 11.4.2.

Recall that groups A and B spent equal amounts of time exercising but followed different regimens. Consequently, a test of $H_0': \mu_A = \mu_B$ versus $H_1': \mu_A \neq \mu_B$ would be an obvious way to assess the effectiveness of the special walking and placing exercises. The associated contrast, of course, would be $C_1 = \mu_A - \mu_B$. Similarly, a test of $H_0'': \mu_C = \mu_D$ ($C_2 = \mu_C - \mu_D$) would provide an evaluation of the psychological effect of periodic progress checks.

From Definition 11.4.2 and the data in Table 11.4.1,

$$SS_{C_1} = \frac{\left[1\left(\dfrac{60.75}{6}\right) - 1\left(\dfrac{68.25}{6}\right) \right]^2}{\dfrac{1^2}{6} + \dfrac{(-1)^2}{6}} = 4.68$$

and

$$SS_{C_2} = \frac{\left[1\left(\dfrac{70.25}{6}\right) - 1\left(\dfrac{61.75}{5}\right) \right]^2}{\dfrac{1^2}{6} + \dfrac{(-1)^2}{5}} = 1.12$$

TABLE 11.4.3 SUBHYPOTHESIS COMPUTATIONS

Subhypothesis	Contrast	SS	F	
$H_0': \mu_A = \mu_B$	$C_1 = \mu_A - \mu_B$	4.68	2.03	NS
$H_0'': \mu_C = \mu_D$	$C_2 = \mu_C - \mu_D$	1.12	0.49	NS

Dividing these sums of squares by the error mean square (2.30) gives F ratios of $4.68/2.30 = 2.03$ and $1.12/2.30 = 0.49$, neither of which is significant at the $\alpha = 0.05$ level ($F_{0.95, 1, 19} = 4.38$) (see Table 11.4.3).

Question 11.4.1 The cathode warm-up time (in seconds) was determined for three different types of X-ray tubes using 15 observations on each type. The results are listed below.

WARM-UP TIMES (SEC)

	Tube Type				
A		B		C	
19	27	20	24	16	14
23	31	20	25	26	18
26	25	32	29	15	19
18	22	27	31	18	21
20	23	40	24	19	17
20	27	24	25	17	19
18	29	22	32	19	18
35		18		18	

Do an analysis of variance on these data and test the hypothesis that the three tube types require the same average warm-up time. Include a pair of orthogonal contrasts in your ANOVA table. Define one of the contrasts so it tests $H_0': \mu_A = \mu_C$. What does the other contrast test? Check to see that the sums of squares associated with your two contrasts verify the statement of Theorem 11.4.1.

Question 11.4.2 Verify that $C_3 = \frac{11}{12}\mu_A + \frac{11}{12}\mu_B - \mu_C - \frac{5}{6}\mu_D$ is mutually orthogonal to the C_1 and C_2 of Case Study 11.4.1. Find SS_{C_3} and illustrate the statement of Theorem 11.4.1.

Question 11.4.3 Refer to the IQ data of Question 11.2.2. Suppose that the experimenter had reason to believe, before collecting any data, that students in groups II and III would show, on the average, the same increase in IQ, but that students in group I would respond differently. Set up and test the appropriate contrast. Let $\alpha = 0.05$.

Question 11.4.4 Test the hypothesis that the average of the true yields for the first three varieties of corn described in Question 11.3.1 is the same as the average for the last two. Let $\alpha = 0.05$.

Question 11.4.5 One of the assumptions underlying the analysis of variance is that the errors in the responses are normally distributed (recall Equation 11.2.2). Investigate the reasonableness of this assumption in the case of the cathode warm-up data of Question 11.4.1. First, use Equation 11.2.7 to find the estimates, $\hat{\mu}$ and $\hat{\tau}_j$, $j = 1, 2, 3$. Then calculate the set of *residuals*, $y_{ij} - \hat{\mu} - \hat{\tau}_j$, $i = 1, 2, \ldots, 15; j = 1, 2, 3$. Make a histogram of the residuals. Does the shape of the histogram lend credibility to the normality assumption? What sort of formal analysis would be appropriate here?

Question 11.4.6 For many years sodium nitrite has been used as a curing agent for bacon, and until recently it was thought to be perfectly harmless. But now it appears that during frying, sodium nitrite induces the formation of nitrosopyrrolidine (NPy), a substance suspected of being a carcinogen. In one study focusing on this problem, measurements were made of the amount of NPy (in ppb) recovered after the frying of three slices of four commercially available brands of bacon (152). Do the analysis

NPy RECOVERED FROM BACON (PPB)

	Brand		
A	B	C	D
20	75	15	25
40	25	30	30
18	21	21	31

of variance for the data in the table and partition the treatment sum of squares into a complete set of three mutually orthogonal contrasts. Let the first contrast test H'_0: $\mu_A = \mu_B$ and the second, H''_0: $(\mu_A + \mu_B)/2 = (\mu_C + \mu_D)/2$. Do all tests at the 0.05 level of significance.

11.5 MULTIPLE COMPARISONS: TUKEY'S METHOD

Imagine a consumer researcher working for, say, Procter & Gamble, who faces the task of comparing six new laundry detergents in a whiteness test. We have seen that the routine way to begin such a comparison would be with the analysis of variance, testing $H_0: \mu_1 = \mu_2 = \cdots = \mu_6$. What comes next, though, depends on how much was known, or expected, about the detergents *before* the data were collected. Specifically, there may be bits of peripheral information about some or all of the six products to suggest the formation of one or more contrasts. If there *are*, then the methodology of Section 11.4 applies. For example, detergents 1 and 4 may be much cheaper than all the others, in which case it would make sense to test the sub-hypothesis $H_0': \mu_1 = \mu_4$. But suppose there is *no* prior information or expectation about any of the six—and, therefore, no reason to specify in advance any particular subhypotheses. How, then, should we proceed?

Perhaps the most tempting course of action would be to wait until the results were in and then begin testing the extremes. If, for instance, detergent 1 rated out to be the best of the six, it might seem reasonable, as a way of "confirming" its superiority, to set up a contrast for testing

$$H_0': \mu_1 = \frac{\mu_2 + \mu_3 + \mu_4 + \mu_5 + \mu_6}{5}$$

or, perhaps, a set of *five* contrasts to test $H_0^1: \mu_1 = \mu_2$, $H_0^2: \mu_1 = \mu_3$, ..., $H_0^5: \mu_1 = \mu_6$. Using the contrast approach here, though, raises some disturbing questions. The problem is that by singling out as the groups to compare those whose $\bar{y}_{.j}$'s were markedly different in the first place, we are prejudicing the outcome. This is obvious if we look at the probability of committing a Type I error for (1) two groups chosen at random, and (2) the two extreme groups. Let i and j denote two arbitrary factor levels chosen in advance. Let c^* denote the smallest value of $\hat{C} = \bar{Y}_{.i} - \bar{Y}_{.j}$ for which $H_0': \mu_i = \mu_j$ can be rejected at level α. Then, by definition,

$$P(\bar{Y}_{.i} - \bar{Y}_{.j} \geq c^* \mid H_0: \mu_1 = \mu_2 = \cdots = \mu_k) = \alpha$$

But, clearly,

$$P(\bar{Y}_{max} - \bar{Y}_{min} \geq c^* \mid H_0) > P(\bar{Y}_{.i} - \bar{Y}_{.j} \geq c^* \mid H_0)$$

—an inequality that should serve as fair warning against proceeding in this fashion.

An alternative approach, perhaps, would be to use the contrast technique on *all* comparisons of a certain kind—say, all possible $\binom{6}{2} = 15$ pairwise hypothesis tests: $H_0^1: \mu_1 = \mu_2$, $H_0^2: \mu_1 = \mu_3$, ..., $H_0^{15}: \mu_5 = \mu_6$. While this adroitly sidesteps the selection problem of the first method, it brings up some disturbing questions of its own about the meaning of α. We know that for any one given test, α is the probability of incorrectly rejecting a true null hypothesis. But suppose that not one but m tests are done, each at level α. The probability of committing *at least one* Type I error among these m is much larger than the nominal α associated with each one. If the tests are independent,

P (at least one Type I error is made in m tests)

$$= 1 - P \text{ (no Type I errors are made in } m \text{ tests)}$$

$$= 1 - (1 - \alpha)^m$$

Let α' denote the "overall" Type I error for an experiment—that is, the probability of rejecting at least one true H_0. Table 11.5.1 shows, for the independent case, the rapid deterioration of α' as a function of m. Thus, if an experimenter did 15 independent tests, each at the $\alpha = 0.05$ level, his chances of committing at least one Type I error would not be 0.05, or anything even close to 0.05, but, rather, 0.54.

TABLE 11.5.1 VALUES OF α'

m	$\alpha = 0.10$	$\alpha = 0.05$	$\alpha = 0.01$
1	0.10	0.05	0.01
2	0.19	0.10	0.02
5	0.41	0.23	0.05
10	0.65	0.40	0.10
15	0.80	0.54	0.14
20	0.88	0.64	0.18

Whether or not the behavior of α' is worth worrying about (not all statisticians think it is), it does suggest an interesting question: is there any way to test a large, maybe even unspecified, number of subhypotheses and still keep α' fixed (and small)? Or, what is equivalent, is there any valid way to formulate and test hypotheses *after* seeing the data? The answer is "Yes"; in fact, there are quite a few ways. The *multiple-comparison problem*, as this has come to be known, has received a good deal of attention from mathematical statisticians in recent years. In this section, we develop the simplest formulation of one of the earliest multiple-comparison procedures, a method due to Tukey.

Definition 11.5.1 gives a background result. The derivation of Tukey's method hinges on a distribution we have not seen yet, the *studentized range*.

> **Definition 11.5.1.** Let $Y_1, Y_2, \ldots, Y_k$ be k independent $N(\mu, \sigma^2)$ random variables, and let R be their *range*:
>
> $$R = \max_i Y_i - \min_i Y_i$$
>
> Let S^2 be an estimator of σ^2 having v degrees of freedom and let S^2 and the Y_i's be independent. (In the completely randomized, one-factor design, S^2 will be MS_{error} and v will be $n - k$.) The *studentized range*, $Q_{k, v}$, is taken to be the ratio
>
> $$Q_{k, v} = \frac{R}{S}$$

Table A.5 in the Appendix gives values of $Q_{\alpha, k, v}$, the $[100(1 - \alpha)]$th percentile of $Q_{k, v}$, for $\alpha = 0.05$ and 0.01 and for various k and v.

Theorem 11.5.1 outlines Tukey's procedure. Notice the restrictions: n_j must be the same for all j, and only pairwise tests are included. (Although more difficult to prove, there are other, more general, multiple-comparison procedures that apply to *any* contrast and do not require equal sample sizes [see, for instance, (146) or (104)].)

Proof. Let $W_t = \bar{Y}_{.t} - \mu_t$; then $W_t \sim N(0, \sigma^2/r)$. From the definition of the studentized range,

$$\frac{\max\limits_t W_t - \min\limits_t W_t}{\sqrt{\dfrac{MS_{error}}{r}}} \sim Q_{k, rk-k}$$

which implies that

$$P\left(\frac{\max\limits_t W_t - \min\limits_t W_t}{\sqrt{\dfrac{MS_{error}}{r}}} < Q_{\alpha, k, rk-k}\right) = 1 - \alpha$$

or, equivalently,

$$P(\max\limits_t W_t - \min\limits_t W_t < D\sqrt{MS_{error}}) = 1 - \alpha \qquad (11.5.1)$$

where $D = Q_{\alpha, k, rk-k}/\sqrt{r}$. But if Equation 11.5.1 is true, it must also be true that

$$P(|W_i - W_j| < D\sqrt{MS_{error}}) = 1 - \alpha \qquad \text{for } all \ i \text{ and } j \qquad (11.5.2)$$

Rewriting Equation 11.5.2 gives the statement of the theorem:

$$P(-D\sqrt{MS_{error}} < W_i - W_j < D\sqrt{MS_{error}}) = 1 - \alpha \qquad \text{for } all \ i \text{ and } j$$

or, after simplifying,

$$P(\bar{Y}_{.i} - \bar{Y}_{.j} - D\sqrt{MS_{error}} < \mu_i - \mu_j < \bar{Y}_{.i} - \bar{Y}_{.j} + D\sqrt{MS_{error}})$$
$$= 1 - \alpha \qquad \text{for } all \ i \text{ and } j$$

Example 11.5.1

Table 11.5.2 lists the average serum binding percentages for the $k = 5$ antibiotics of Case Study 11.2.1. Each average was based on $n_j = r = 4$ replicates.

TABLE 11.5.2 AVERAGE SERUM BINDING PERCENTAGES

	Penicillin G	Tetracycline	Streptomycin	Erythromycin	Chloramphenicol
$\bar{y}_{.j}$	28.6	31.4	7.8	19.1	27.8

Suppose we want to make all $\binom{5}{2} = 10$ pairwise comparisons while maintaining an overall Type I error, α', of 0.05. Recall that when the analysis of variance was done on these data, it was found that $MS_{error} = 9.05$ (see Table 11.3.2). Also, from Table A.5 in the Appendix, $Q_{0.05,\,5,\,15} = 4.37$. This makes $D = 4.37/\sqrt{4} = 2.185$ and $D\sqrt{MS_{error}} = 6.58$.

For each of the ten pairwise subhypothesis tests, H_0': $\mu_i = \mu_j$ versus H_1': $\mu_i \neq \mu_j$, Table 11.5.3 lists the corresponding estimated contrast, $\bar{y}_{.i} - \bar{y}_{.j}$, and its 95% "Tukey" confidence interval. As the last column indicates, seven of the subhypotheses are rejected (those whose Tukey intervals do not contain 0) and three are accepted.

TABLE 11.5.3 ANALYSIS OF PAIRWISE SUBHYPOTHESIS TESTS

Pairwise Difference	$\bar{y}_{.i} - \bar{y}_{.j}$	Tukey Interval		Conclusion
$\mu_1 - \mu_2$	−2.8	(−9.38,	3.78)	NS
$\mu_1 - \mu_3$	20.8	(14.22,	27.38)	Reject
$\mu_1 - \mu_4$	9.5	(2.92,	16.08)	Reject
$\mu_1 - \mu_5$	0.8	(−5.78,	7.38)	NS
$\mu_2 - \mu_3$	23.6	(17.02,	30.18)	Reject
$\mu_2 - \mu_4$	12.3	(5.72,	18.88)	Reject
$\mu_2 - \mu_5$	3.6	(−2.98,	10.18)	NS
$\mu_3 - \mu_4$	−11.3	(−17.88,	−4.72)	Reject
$\mu_3 - \mu_5$	−20.0	(−26.58,	−13.42)	Reject
$\mu_4 - \mu_5$	−8.7	(−15.28,	−2.12)	Reject

Question 11.5.1 Use Tukey's method to make all the pairwise comparisons for the IQ data of Question 11.2.2.

Question 11.5.2 Construct 95% Tukey confidence intervals for all $\binom{5}{2} = 10$ pairwise differences, $\mu_i - \mu_j$, for the data of Question 11.3.1. Summarize the results by plotting the five sample averages on a horizontal axis and drawing straight lines under varieties whose average yields are not significantly different.

Question 11.5.3 Intravenous infusion fluids produced by three different pharmaceutical companies (Cutter, Abbott, and McGaw) were tested for their concentrations of particulate contaminants. Six samples were inspected from each company. The figures listed in the table are, for each sample, the number of particles per liter greater than 5 microns in diameter (169).

NUMBER OF CONTAMINANT
PARTICLES IN INFUSION
FLUIDS MADE BY THREE
FIRMS

Cutter	Abbott	McGaw
255	105	577
264	288	515
342	98	214
331	275	413
234	221	401
217	240	260

Do the analysis of variance to test $H_0: \mu_C = \mu_A = \mu_M$ and then test each of the three pairwise subhypotheses by constructing 95% Tukey confidence intervals.

11.6 DATA TRANSFORMATIONS

The three assumptions required by the analysis of variance have already been mentioned: the $\epsilon_{i(j)}$'s must be independent, normally distributed (for each j), and have the same variance. In practice, the three are not equally difficult to satisfy, nor do their violations have equal consequences for the validity of the F test. Independence is certainly a critical property for the $\epsilon_{i(j)}$'s to have, but randomizing the order in which the observations are taken tends to eliminate systematic bias and thereby to satisfy the assumption, at least to a good approximation. Normality is a much more difficult property to induce or even to verify (recall Section 9.4); fortunately, departures from normality, unless extreme, do not seriously compromise the probabilistic integrity of the F test (more will be said about this in Chapter 13).

If the final assumption is violated, though, and all the Y_{ij}'s do *not* have the same variance, the effect on certain inference procedures—for example, the construction of confidence intervals for individual means—can be more unsettling. However, it is possible in some situations to "stabilize" the level-to-level variances by a suitable *data transformation*. In particular, if the variance of the observations in the jth level is a known function of the mean for that same level, we can appeal to Theorem 11.6.1.

Theorem 11.6.1. Suppose that $Y_{ij} \sim f_Y(y_{ij}; \mu_j)$, $i = 1, 2, \ldots, n_j$; $j = 1, 2, \ldots, k$, where Var $(Y_{ij}) = g(\mu_j)$. Then the transformed variables $Z_{ij} = A(Y_{ij})$ will have a nearly constant variance for all j if

$$A(Y_{ij}) = c_1 \int \frac{1}{\sqrt{g(y_{ij})}} \, dy_{ij} + c_2$$

where c_1 and c_2 are arbitrary constants.

Proof. We wish to find a transformation, A, which when applied to the Y_{ij}'s will generate a new set of variables having a constant variance. That is, we seek $A(Y_{ij}) = Z_{ij}$, where Var $(Z_{ij}) = c_1^2$, a constant. By Taylor's theorem,

$$Z_{ij} \doteq A(\mu_j) + (Y_{ij} - \mu_j)A'(\mu_j)$$

Of course, $E(Z_{ij}) = A(\mu_j)$, since $E(Y_{ij} - \mu_j) = 0$. Also,

$$\text{Var } (Z_{ij}) = E[Z_{ij} - E(Z_{ij})]^2$$
$$= E[(Y_{ij} - \mu_j)A'(\mu_j)]^2$$
$$= [A'(\mu_j)]^2 \text{ Var } (Y_{ij}) = [A'(\mu_j)]^2 g(\mu_j)$$

Solving for $A'(\mu_j)$ gives

$$A'(\mu_j) = \frac{\sqrt{\text{Var } (Z_{ij})}}{\sqrt{g(\mu_j)}} = \frac{c_1}{\sqrt{g(\mu_j)}}$$

For Y_{ij} in the neighborhood of μ_j, it follows that

$$A(Y_{ij}) = c_1 \int \frac{1}{\sqrt{g(y_{ij})}} \, dy_{ij} + c_2$$

and the theorem is proved.

Examples 11.6.1 and 11.6.2 show the origin of two of the more commonly used data transformations, the square root and the arc sine. The former is appropriate if the given measurements are Poisson, the latter if the Y_{ij}'s are binomial.

Example 11.6.1

Suppose that the response variable for the jth level in a completely randomized, one-factor design is Poisson with parameter μ_j,

$$Y_{ij} \sim f_Y(y_{ij}; \mu_j) = \frac{e^{-\mu_j} \mu_j^{y_{ij}}}{y_{ij}!}$$

For this particular pdf, the variance is *equal* to the mean (recall Theorem 4.2.2):

$$\text{Var } (Y_{ij}) = E(Y_{ij}) = \mu_j = g(\mu_j)$$

Therefore, from Theorem 11.6.1,

$$A(Y_{ij}) = c_1 \int \frac{1}{\sqrt{y_{ij}}} \, dy_{ij} + c_2 = 2c_1\sqrt{y_{ij}} + c_2$$

or, letting $c_1 = \frac{1}{2}$ and $c_2 = 0$ to make the transformation as simple as possible,

$$A(Y_{ij}) = Z_{ij} = \sqrt{Y_{ij}} \qquad (11.6.1)$$

Equation 11.6.1 implies that if the data are known in advance to be Poisson, each of the observations should be replaced by its square root before we proceed with the analysis of variance.

Example 11.6.2

Suppose that Y_{ij} is binomial with parameters n and p_j,

$$Y_{ij} \sim f_Y(y_{ij}; np_j) = \binom{n}{y_{ij}} p_j^{y_{ij}} (1 - p_j)^{n - y_{ij}}$$

Then, since $E(Y_{ij}) = np_j = \mu_j$,

$$\text{Var } (Y_{ij}) = np_j(1 - p_j) = \mu_j \left(1 - \frac{\mu_j}{n}\right) = g(\mu_j)$$

In this case, the stabilizing transformation involves the inverse sine:

$$A(Y_{ij}) = c_1 \int \frac{1}{\sqrt{y_{ij}(1 - y_{ij}/n)}} \, dy_{ij} + c_2$$

$$= c_1 2\sqrt{n} \arcsin \left(\frac{y_{ij}}{n}\right)^{1/2} + c_2$$

or, what is equivalent,

$$A(Y_{ij}) = \arcsin \left(\frac{y_{ij}}{n}\right)^{1/2}$$

Question 11.6.1 A commercial film processor is experimenting with two kinds of fully automatic color developers. Six sheets of exposed film are put through each developer. The number of flaws on each negative visible with the naked eye is then counted.

NUMBER OF VISIBLE FLAWS

Developer A	Developer B
1	8
4	6
5	4
6	9
3	11
7	10

Assume the number of flaws on a given negative is a Poisson random variable. Make an appropriate data transformation and do the indicated analysis of variance.

Question 11.6.2 An experimenter wants to do an analysis of variance on a set of data involving five treatment groups, each with three replicates. She has computed $\bar{y}_{.j}$ and s_j for each group and gotten the results listed in the table.

SUMMARY OF DATA FOR ANALYSIS OF VARIANCE

Treatment Group

	1	2	3	4	5
$\bar{y}_{.j}$	9.0	4.0	16.0	9.0	1.0
s_j	3.0	2.0	4.0	3.0	1.0

What should the experimenter do before computing the various sums of squares necessary to carry out the F test? Be as quantitative as possible.

Question 11.6.3 Three air-to-surface missile launchers are tested for their accuracy. The same gun crew fires four rounds with each launcher, each round consisting of 20 missiles. A " hit " is scored if the missile lands within 10 yards of the target. The table below gives the number of hits registered in each round.

NUMBER OF HITS PER ROUND

Launcher A	Launcher B	Launcher C
13	15	9
11	16	11
10	18	10
14	17	8

Compare the accuracy of these three launchers by using the analysis of variance, after making a suitable data transformation. Let $\alpha = 0.05$.

12

Randomized Block Designs

Pictured is a 5-by-5 Latin square laid out at Bettgelert Forest in 1929 for studying the effect of exposure on Sitka spruce, Norway spruce, Japanese larch, European larch, *Pinus contorta*, and beech. (A Latin square is a special type of experimental design that extends to two dimensions the important notion of "blocking" that is introduced in this chapter. Latin squares are particularly useful in agricultural research.)

12.1 INTRODUCTION

In any experiment, reducing the magnitude of the experimental error is a highly desirable objective: the smaller σ^2 is, the better will be our chances of rejecting a false null hypothesis. Basically, there are two ways to reduce experimental error. The nonstatistical approach is simply to refine the experimental technique—use better equipment, minimize subjective error, and so on. The statistical method, which can often produce results much more dramatic, is by collecting the data in "blocks," in what is referred to as a *randomized block design*. Such designs are the subject of this chapter.

Historically, it was Fisher who first advanced the notion of blocking. He saw it initially as a statistical defense against the obfuscating effects of soil heterogeneity in agricultural experiments. Suppose, for example, a researcher wishes to compare the yields of four different varieties of corn. Figure 12.1.1a shows the simplest experimental layout: variety A is planted in the leftmost portion of the field, variety B is planted next to A, and so on. Even to a city slicker the statistical hazards involved in using this design should be obvious. Suppose, for example, there was a soil *gradient* in the field, with the best soil being in the westernmost part (where variety A was planted). Then if variety A achieved the highest yield, we would not know whether to attribute its success to its inherent quality or to its location (or to some combination of both).

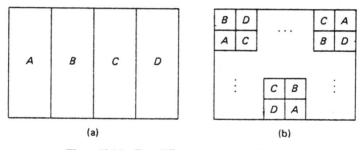

Figure 12.1.1 Two different experimental designs.

A more sensible *modus operandi* is pictured in Figure 12.1.1b. There the field is divided up into a number of smaller "blocks," each block being still further parceled into four "plots." All four varieties are planted in each block, one to a plot, with the plot assignments being chosen at random. Notice that the geographical contiguity of the four plots within a given block insures that the environmental conditions from plot to plot will be relatively uniform and will not lead to any biasing of the observed yields. What the analysis of variance will then do is "pool" from block to block the *within*-block information concerning the treatment differences while bypassing the *between*-block differences—that is, the heterogeneity in the experimental environment. As a result, the treatment comparisons can be made with greater precision. Analytically, where the total sum of squares was partitioned into *two* components in a completely randomized, one-factor design, it will be split into *three* separate sums in a randomized block design: one for treatments, another for blocks, and a third for experimental error.

It did not take long for scientists to realize that the benefits of blocking could be extended well beyond the confines of agricultural experimentation. In medical research, blocks are often made up of subjects of the same age, sex, and overall physical condition. A common practice in animal studies is to form blocks out of littermates. Industrial experiments often require that "time" be a blocking criterion: measurements taken by personnel on the day shift might be considered one block and those taken by the night shift a second block. In some sense the ultimate form of blocking, although one not always physically possible, is to apply the entire set of treatment levels to each subject, thus making each subject his own block.

Section 12.2 begins with a development of the analysis of variance for the general randomized block design, where t factor levels are administered within each of b blocks. As before, the hypotheses to be tested are $H_0: \mu_1 = \mu_2 \cdots = \mu_t$ versus H_1: not all the μ_j's are equal. Section 12.2 concludes with a set of case studies illustrating the blocking concept in a variety of different settings.

12.2 THE F TEST FOR A RANDOMIZED BLOCK DESIGN

As indicated in Section 12.1, the typical data structure for a randomized block design is a matrix with b rows and t columns, the former representing the b blocks, the latter the t levels of the treatment. We will let γ_{ij}, $i = 1, 2, \ldots, b$; $j = 1, 2, \ldots, t$, denote the true average response associated with the application of level j to block i. Similarly, ω_i and μ_j will denote the true average (separate) responses for block i and level j, respectively. The model equation for Y_{ij}, the observation recorded for the ith block and jth level, is a straightforward extension of Equation 11.2.5:

$$Y_{ij} = \mu + \beta_i + \tau_j + \epsilon_{ij} \qquad i = 1, \ldots, b; \quad j = 1, \ldots, t \qquad (12.2.1)$$

where $\quad \mu = \dfrac{1}{bt} \displaystyle\sum_{i=1}^{b} \sum_{j=1}^{t} \gamma_{ij} = $ overall mean

$\beta_i = \omega_i - \mu = $ differential effect of the ith block (relative to μ)

$\tau_j = \mu_j - \mu = $ differential effect of the jth treatment (relative to μ)

$\epsilon_{ij} = $ experimental error

The constraints imposed on the parameters of Equation 12.2.1 will vary somewhat from problem to problem. The block effect and treatment effect, for example, might be either *fixed* or *random*, the choice depending on how the levels for the b blocks and t treatments were selected. If the levels for a factor are in any way preselected or unique, then that factor is said to be *fixed*; otherwise, if the levels included in the experiment are just an arbitrary sample of all possible levels, the factor is said to be *random*. In a typical randomized block design, the treatment effect will be fixed and the block effect random. The τ_j's, then, will be a set of constants for which

$$\sum_{j=1}^{t} \tau_j = 0$$

The β_i's, on the other hand, will be random variables—specifically, $\beta_i \sim N(0, \sigma_B^2)$, $i = 1, 2, \ldots, b$, where σ_B^2 is unknown. The assumption about the ϵ_{ij}'s will be the same as it was in the completely randomized, one-factor design:

$$\epsilon_{ij} \sim N(0, \sigma^2) \qquad i = 1, \ldots, b; j = 1, \ldots, t$$

(More will be said about the distinction between fixed and random effects in the case studies.)

Example 12.2.1 parallels Example 11.2.1 in showing what the components of Equation 12.2.1 mean numerically.

Example 12.2.1

Suppose the (unknown) γ_{ij}'s for a randomized block design with three treatment levels and two bocks are the entries appearing in Table 12.2.1. (It is assumed here that both "treatments" and "blocks" are fixed).

TABLE 12.2.1 A SET OF γ_{ij}'S

		Treatment Level		
		1	2	3
Block	1	14	10	6
	2	15	15	6

Then

$$\mu = \frac{1}{2(3)} (14 + 10 + \cdots + 6) = \frac{66}{6} = 11.0$$

$$\mu_1 = \frac{29}{2} = 14.5 \qquad \mu_2 = \frac{25}{2} = 12.5 \qquad \mu_3 = \frac{12}{2} = 6.0$$

$$\omega_1 = \frac{30}{3} = 10.0 \qquad \omega_2 = \frac{36}{3} = 12.0$$

It follows that τ_1, the differential effect for treatment level 1, is $14.5 - 11.0 = 3.5$. Similarly, $\tau_2 = 12.5 - 11.0 = 1.5$, and $\tau_3 = 6.0 - 11.0 = -5.0$. The differential effects of the blocks are -1.0 and $+1.0$, respectively: $\beta_1 = 10.0 - 11.0 = -1.0$ and $\beta_2 = 12.0 - 11.0 = +1.0$. Notice that

$$\sum_{j=1}^{3} \tau_j = 0 = \sum_{i=1}^{2} \beta_i \qquad (12.2.2)$$

Question 12.2.1 Given the information in Table 12.2.1, what would ϵ_{11} equal if y_{11} equals 13.3?

Question 12.2.2 Differentiate the least-squares function

$$L = \sum_{i=1}^{b} \sum_{j=1}^{t} (y_{ij} - \mu - \beta_i - \tau_j)^2$$

with respect to all $bt + 1$ parameters and use the two constraints indicated in Equation 12.2.2 to show that $\hat{\mu} = \bar{y}_{..}$, $\hat{\beta}_i = \bar{y}_{i.} - \bar{y}_{..}$, and $\hat{\tau}_j = \bar{y}_{.j} - \bar{y}_{..}$.

Following the procedure established for the completely randomized, one-factor design, the first step in deriving the F test for $H_0: \tau_1 = \tau_2 = \cdots = \tau_t = 0$ is to partition the total sum of squares, $\sum_{i=1}^{b} \sum_{j=1}^{t} (Y_{ij} - \bar{Y}_{..})^2$, into components related to the terms in the model equation. The appropriate identity to start with (see Equation 11.2.8) is

$$(Y_{ij} - \bar{Y}_{..}) = (\bar{Y}_{i.} - \bar{Y}_{..}) + (\bar{Y}_{.j} - \bar{Y}_{..}) + (Y_{ij} - \bar{Y}_{i.} - \bar{Y}_{.j} + \bar{Y}_{..}) \qquad (12.2.3)$$

Squaring both sides of Equation 12.2.3 and summing over i and j gives

$$\sum_{i=1}^{b} \sum_{j=1}^{t} (Y_{ij} - \bar{Y}_{..})^2 = \sum_{i=1}^{b} \sum_{j=1}^{t} (\bar{Y}_{i.} - \bar{Y}_{..})^2 + \sum_{i=1}^{b} \sum_{j=1}^{t} (\bar{Y}_{.j} - \bar{Y}_{..})^2$$

$$+ \sum_{i=1}^{b} \sum_{j=1}^{t} (Y_{ij} - \bar{Y}_{i.} - \bar{Y}_{.j} + \bar{Y}_{..})^2 \qquad (12.2.4)$$

or

$$Q = Q_1 + Q_2 + Q_3$$

(It will be left as an exercise to verify that the three cross-product terms are identically 0.)

Three of the four terms in Equation 12.2.4 have obvious interpretations: Q measures the total variability in the data, Q_1 measures the block effect, and Q_2 the treatment effect. To understand Q_3, notice that

$$Y_{ij} - \bar{Y}_{i.} - \bar{Y}_{.j} + \bar{Y}_{..} = Y_{ij} - (\bar{Y}_{i.} - \bar{Y}_{..}) - (\bar{Y}_{.j} - \bar{Y}_{..}) - \bar{Y}_{..}$$

$\qquad$ = observed response − estimated differential block effect
$\qquad$ − estimated differential treatment effect − estimated overall effect

$\qquad$ = experimental error

Therefore, the decomposition achieved in Equation 12.2.4 can be written

$$SS_{total} = SS_{blocks} + SS_{treatments} + SS_{error}$$

Theorem 12.2.1 gives the distributions for Q_2 and Q_3 under H_0 and their expected values under both H_0 and H_1. The proof closely parallels what was done in parts II and III of Section 11.2 and will not be given. (A similar result holds for Q_1 but will not be needed.)

Theorem 12.2.1.

(a) When $H_0: \tau_1 = \tau_2 = \cdots = \tau_t = 0$ is true,

$$\frac{Q_2}{\sigma^2} \sim \chi_{t-1}^2$$

and $E[Q_2/(t-1)] = \sigma^2$. When H_1: not all the τ_j's $= 0$ is true, Q_2/σ^2 has a noncentral χ^2 distribution and

$$E\left(\frac{Q_2}{t-1}\right) = \sigma^2 + \frac{b}{t-1} \sum_{j=1}^{t} \tau_j^2$$

(b) Regardless of what is true about the β_i's or the τ_j's,

$$\frac{Q_3}{\sigma^2} \sim \chi_{(b-1)(t-1)}^2$$

and $E\{Q_3/[(b-1)(t-1)]\} = \sigma^2$.

It follows from Theorem 12.2.1, Theorem 7.6.1, and Definition 11.2.2 that

$$\frac{\dfrac{Q_2}{(t-1)\sigma^2}}{\dfrac{Q_3}{(b-1)(t-1)\sigma^2}} = \frac{MS_{\text{treatments}}}{MS_{\text{error}}} \sim \frac{\dfrac{\chi_{t-1}^2}{t-1}}{\dfrac{\chi_{(b-1)(t-1)}^2}{(b-1)(t-1)}}$$

will have a (central) F distribution with $t-1$ and $(b-1)(t-1)$ degrees of freedom when $H_0: \tau_1 = \tau_2 = \cdots = \tau_t = 0$ is true, and a noncentral F distribution when H_1: not all the τ_j's $= 0$ is true. Therefore, the null hypothesis that the means for the t treatment levels are all equal should be rejected at the α level of significance if

$$\frac{MS_{\text{treatments}}}{MS_{\text{error}}} \geq F_{1-\alpha,\, t-1,\, (b-1)(t-1)} \tag{12.2.5}$$

While Equation 12.2.5 is sufficient to carry out the analysis of variance, the F ratio can be gotten much easier with the help of computing formulas for $SS_{\text{treatments}}$ and SS_{error}. Theorem 12.2.2 is the analog of Theorem 11.3.1. Its proof is a straightforward exercise in summation algebra and will not be given.

Theorem 12.2.2. Let $c = T_{..}^2/bt$. Then

$$SS_{\text{total}} = \sum_{i=1}^{b} \sum_{j=1}^{t} Y_{ij}^2 - c$$

$$SS_{\text{blocks}} = \sum_{i=1}^{b} \frac{T_{i.}^2}{t} - c$$

$$SS_{\text{treatments}} = \sum_{j=1}^{t} \frac{T_{.j}^2}{b} - c$$

and

$$SS_{\text{error}} = SS_{\text{total}} - SS_{\text{blocks}} - SS_{\text{treatments}}$$

The next several examples illustrate the analysis of variance calculations for a randomized block design. Notice the different blocking criteria being utilized. Case Study 12.2.1 employs the most direct sort of blocking: each subject, having been measured three times, functions as its own block. In Case Study 12.2.2 a pretest is the blocking mechanism, while in Case Study 12.2.3, time, in the form of months, delineates one block from another.

CASE STUDY   12.2.1

Hypnotic age regression is a procedure whereby an adult subject is instructed under hypnosis to return to an earlier chronological age. Although this process can recapture *behavioral* patterns of the earlier age, it is not known whether it can reinstate *perceptual* patterns. A recent study addressed itself to that point by using the Ponzo illusion pictured in Figure 12.2.1.

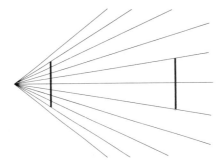

Figure 12.2.1   The Ponzo illusion.

When the two vertical lines are equal in length, most people will perceive the right-most one as being shorter. Furthermore, studies have shown that the strength of the illusion is a function of age: it becomes more and more pronounced from the childhood years through adolescence; then, in the late teens, it levels off.

The Ponzo illusion was shown to eight college students. These same students were then regressed under hypnosis to age nine and to age five. At each age, their perceptions of the illusion were measured. If hypnosis can, in fact, recapture perceptual patterns, the illusion strengths at these three ages should be significantly different.

The Ponzo illusion was drawn on a set of 21 cards. On each card the leftmost line was 4 inches long; the rightmost line varied from card to card, ranging from $2\frac{1}{2}$ to 5 inches in increments of $\frac{1}{8}$ inch. The cards were shown to each subject in random order. In each instance the subject was instructed to point to the longer line. If the subject perceived the rightmost line to be longer when it was actually $4\frac{6}{8}$ inches but shorter when it was $4\frac{5}{8}$ inches, the strength of the illusion was defined to be $\frac{5}{8} + \frac{1}{2}(\frac{6}{8} - \frac{5}{8}) = 0.69$ inch. The results for the eight subjects are listed in Table 12.2.2.

TABLE 12.2.2 ILLUSION STRENGTHS

Subject	Awake	Regressed to Age Nine	Regressed to Age Five	$t_{i.}$
1	0.81	0.69	0.56	2.06
2	0.44	0.31	0.44	1.19
3	0.44	0.44	0.44	1.32
4	0.56	0.44	0.44	1.44
5	0.19	0.19	0.31	0.69
6	0.94	0.44	0.69	2.07
7	0.44	0.44	0.44	1.32
8	0.06	0.19	0.19	0.44
$t_{.j}$	3.88	3.14	3.51	10.53

Here the "subjects" would be considered a random effect (since the eight being used were presumably in no way special, but just an arbitrary sample from the population of all college students) and the "regressed ages" a fixed effect (the three treatment levels—awake, age nine, and age five—were preselected because of their particular relevance to the expected strength of the illusion). The model equation would be

$$Y_{ij} = \mu + \beta_i + \tau_j + \epsilon_{ij} \qquad i = 1, \ldots, 8; \quad j = 1, 2, 3$$

From Theorem 12.2.2,

$$c = \frac{(10.53)^2}{8(3)} = 4.6200$$

Also, $\sum_{i=1}^{8} \sum_{j=1}^{3} y_{ij}^2 = 5.5889$. Therefore,

$$SS_{\text{total}} = 5.5889 - 4.6200 = 0.9689$$

$$SS_{\text{treatments}} = \frac{(3.88)^2}{8} + \frac{(3.14)^2}{8} + \frac{(3.51)^2}{8} - 4.6200 = 0.0343$$

$$SS_{\text{blocks}} = \frac{(2.06)^2}{3} + \cdots + \frac{(0.44)^2}{3} - 4.6200 = 0.7709$$

and, by subtraction,

$$SS_{\text{error}} = 0.9689 - 0.0343 - 0.7709 = 0.1637$$

The analysis of variance is detailed in Table 12.2.3. For any randomized block design, the degrees of freedom for blocks, treatments, error, and total will be $b - 1$, $t - 1$, $(b - 1)(t - 1)$, and $bt - 1$, respectively.

TABLE 12.2.3 ANOVA COMPUTATIONS

Source	df	SS	MS	EMS	F
Subjects	7	0.7709	0.1101		
Ages	2	0.0343	0.0172	$\sigma^2 + \dfrac{8}{3-1} \sum\limits_{i=1}^{3} \tau_j^2$	1.47
Error	14	0.1637	0.0117	σ^2	
Total	23	0.9689			

If we elect to test $H_0: \mu_1 = \mu_2 = \mu_3$ (or $H_0: \tau_1 = \tau_2 = \tau_3 = 0$) at the $\alpha = 0.05$ level of significance, our conclusion is to *accept* the null hypothesis (see Figure 12.2.2).

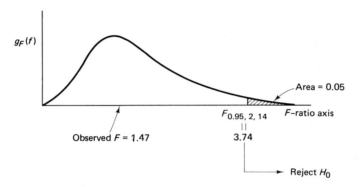

Figure 12.2.2 *F* distribution with 2 and 14 degrees of freedom.

CASE STUDY 12.2.2

Acrophobia is a fear of heights. It can be treated in a number of different ways. Using contact desensitization, a therapist demonstrates some task that would be difficult for someone with acrophobia to do, such as looking over a ledge or standing on a ladder. Then he guides the subject through the very same maneuver, always keeping in physical contact. Another method of treatment is demonstration participation. Here the therapist tries to *talk* the subject through the task; no physical contact is made. A third technique, live modeling, requires the subject simply to *watch* the task being done—he does not attempt it himself.

These three techniques were compared in a recent study involving 15 volunteers, all of whom had a history of severe acrophobia (133). It was realized at the outset, though, that the affliction was much more incapacitating in some subjects than in others, and that this heterogeneity might compromise the therapy comparison. Accordingly, the experiment began with each subject being given the Height Avoidance Test (HAT), a series of 44 tasks related to ladder climbing. A subject received a "point" for each task successfully completed. On the basis of their final scores the 15 volunteers were divided into five blocks (A, B, C, D, and E), each of size three. The subjects in Block A had the *lowest* scores (that is, the most severe acrophobia), those in block B the second lowest scores, and so on.

Each of the three therapies was then assigned at random to one of the three subjects in each block. When the counseling sessions were over, the subjects retook the HAT. Table 12.2.4 lists the *changes* in their scores (score after therapy—score before therapy). Our objective will be to test the equality of the therapies at the $\alpha = 0.01$ level of significance.

TABLE 12.2.4 HAT SCORE CHANGES

Block	Contact Desensitization	Therapy Demonstration Participation	Live Modeling	$t_{i.}$
A	8	2	−2	8
B	11	1	0	12
C	9	12	6	27
D	16	11	2	29
E	24	19	11	54
$t_{.j}$	68	45	17	130

Since $c = (130)^2/15 = 1126.7$ and $\sum_{i=1}^{5} \sum_{j=1}^{3} y_{ij}^2 = 1894$, it follows that

$$SS_{\text{total}} = 1894 - 1126.7 = 767.3$$

$$SS_{\text{blocks}} = \frac{(8)^2}{3} + \cdots + \frac{(54)^2}{3} - 1126.7 = 438.0$$

$$SS_{\text{treatments}} = \frac{(68)^2}{5} + \frac{(45)^2}{5} + \frac{(17)^2}{5} - 1126.7 = 260.9$$

giving an error sum of squares of 68.4:

$$SS_{\text{error}} = 767.3 - 438.0 - 260.9 = 68.4$$

The analysis of variance is summarized in Table 12.2.5. As the two asterisks signify, $H_0: \tau_1 = \tau_2 = \tau_3 = 0$ can be rejected at the 0.01 level: $F_{0.99, 2, 8} = 8.65$.

TABLE 12.2.5 ANOVA COMPUTATIONS

Source	df	SS	MS	EMS	F
Blocks	4	438.0	109.5		
Therapies	2	260.9	130.4	$\sigma^2 + \frac{5}{3-1} \sum_{j=1}^{3} \tau_j^2$	15.2 **
Error	8	68.4	8.6	σ^2	
Total	14	767.3			

The Tukey multiple-comparison technique of Section 11.5 can also be applied to a randomized block design. In the notation of Theorem 11.5.1, the values here for k and r would be 3 and 5, respectively. The definition of D is slightly different, however, since the associated studentized range is no longer $Q_{k, rk-k}$ but rather $Q_{k, (b-1)(t-1)}$, a change reflecting the difference in the number of degrees of freedom available for estimating σ^2. Table 12.2.6 shows the three pairwise therapy differences and their 95% Tukey intervals. The value used for D was 1.81:

$$D = \frac{Q_{0.05, 3, 8}}{\sqrt{5}} = \frac{4.04}{2.24} = 1.81$$

TABLE 12.2.6 THERAPY DIFFERENCES AND 95% TUKEY INTERVALS

Pairwise Difference	$\bar{y}_{.i} - \bar{y}_{.j}$	Tukey Interval	Conclusion
$\mu_1 - \mu_2$	4.6	$(-0.7, 9.9)$	NS
$\mu_1 - \mu_3$	10.2	$(4.9, 15.5)$	Reject
$\mu_2 - \mu_3$	5.6	$(0.3, 10.9)$	Reject

Thus we would conclude, *after the fact*, at the 0.05 level that the therapeutic values of contact desensitization and demonstration participation are not significantly different but that both contact desensitization and demonstration participation are significantly better than live modeling.

CASE STUDY 12.2.3

Case Study 1.2.3 described an investigation designed to "measure" the so-called Transylvania effect—the possible influence of the full moon on human behavior. Table 12.2.7 reproduces those data (admission rates to the emergency room of a mental health clinic) and gives the appropriate row and column totals.

Table 12.2.8 summarizes the ANOVA calculations. For 2 and 22 degrees of freedom the 0.05 critical value is 3.44. Thus, our conclusion at the 0.05 level would to be accept $H_0: \mu_1 = \mu_2 = \mu_3$; a lunar "effect" has not been demonstrated.

Testing the general H_0 is not the only appropriate way to analyze these data, though. An a priori subhypothesis is clearly suggested by the circumstances of the problem—specifically, it would make sense to test whether the admission rate during the full moon is different from the *average* of the rates for "before" and "after." What needs to be considered, then, is $H_0': \mu_2 = (\mu_1 + \mu_3)/2$.

TABLE 12.2.7 MENTAL HEALTH CLINIC EMERGENCY ROOM ADMISSION RATES

| | Admission Rates (patients/day) | | | |
Month	(1) Before Full Moon	(2) During Full Moon	(3) After Full Moon	$t_{i.}$
Aug.	6.4	5.0	5.8	17.2
Sept.	7.1	13.0	9.2	29.3
Oct.	6.5	14.0	7.9	28.4
Nov.	8.6	12.0	7.7	28.3
Dec.	8.1	6.0	11.0	25.1
Jan.	10.4	9.0	12.9	32.3
Feb.	11.5	13.0	13.5	38.0
Mar.	13.8	16.0	13.1	42.9
Apr.	15.4	25.0	15.8	56.2
May	15.7	13.0	13.3	42.0
June	11.7	14.0	12.8	38.5
July	15.8	20.0	14.5	50.3
$t_{.j}$	131.0	160.0	137.5	

TABLE 12.2.8 ANOVA COMPUTATIONS

Source	df	SS	MS	EMS	F
Months	11	451.08	41.01		
Lunar cycles	2	38.59	19.30	$\sigma^2 + \dfrac{12}{3-1} \displaystyle\sum_{j=1}^{3} \tau_j^2$	3.22
Error	22	132.08	6.00	σ^2	
Total	35	621.75			

Following the procedure outlined in Section 11.4, the contrast associated with H_0' is

$$C = -\frac{1}{2}\mu_1 + \mu_2 - \frac{1}{2}\mu_3$$

while its estimate is

$$\hat{C} = -\frac{1}{2}(10.92) + 1(13.3) - \frac{1}{2}(11.46)$$

$$= 2.11$$

From Definition 11.4.2, the sum of squares associated with C is 35.62:

$$SS_C = \frac{(2.11)^2}{\dfrac{1/4}{12} + \dfrac{1}{12} + \dfrac{1/4}{12}} = 35.62$$

Dividing SS_C by the mean square for error gives an F ratio with 1 and 22 degrees of freedom:

$$\frac{35.62/1}{132.08/22} = 5.93$$

But, for the 0.05 level, $F_{0.95, 1, 22} = 4.30$. Thus, contrary to our acceptance of H_0, we would *reject* H_0' and conclude that the admission rate *is* significantly higher during a full moon than it is for the rest of the month.

Comment

It is always more than a little disconcerting when two valid statistical techniques applied to the same data lead to opposite conclusions. That such apparent contradictions occur, though, should not be unexpected. Different methods of analysis simply utilize the data in different ways. Disagreements from time to time are inevitable.

Question 12.2.3 In recent years a number of research projects in extrasensory perception have examined the possibility that hypnosis may be helpful in bringing out ESP in persons who did not think they had any. The obvious way to test such a hypothesis is with a self-paired design: the ESP ability of a subject when he is awake is compared to his ability when hypnotized. In one study of this sort, 15 college students were each asked to guess the identity of 200 Zener cards (see Case Study 4.3.1). The same "sender"—that is, the person concentrating on the card—was used for each trial. For 100 of the trials both the student and the sender were awake; for the other 100 both were hypnotized. If chance were the only factor involved, the expected number of correct identifications in each set of 100 trials would be 20. The observed average numbers of correct guesses for subjects awake and subjects hypnotized were 18.9 and 21.7, respectively (19).

NUMBER OF CORRECT RESPONSES (OUT OF 100) IN ESP EXPERIMENT

Student	Sender and Student in Waking State	Sender and Student in Hypnotic State
1	18	25
2	19	20
3	16	26
4	21	26
5	16	20
6	20	23
7	20	14
8	14	18
9	11	18
10	22	20
11	19	22
12	29	27
13	16	19
14	27	27
15	15	21
$\bar{y}_{.j}$	18.9	21.7

Use the analysis of variance to determine whether that difference is statistically significant at the 0.05 level.

Question 12.2.4 Rat poison is normally made by mixing the active chemical ingredients with ordinary cornmeal. In many urban areas, though, rats can find food that they prefer to cornmeal, so that the poison is left untouched. One solution is to make the cornmeal more palatable by adding food supplements such as peanut butter or meat. This is effective, but the cost is very high and the food supplements tend to spoil rather quickly. The purpose of the experiment described here was to see whether *artificial* food supplements might be similarly enticing. The study took place in Milwaukee and consisted of five two-week surveys. For each survey, approximately 3200 baits were placed around garbage-storage areas. About 800 of the baits were plain cornmeal; a second 800 were cornmeal mixed with artificial butter-vanilla flavoring; a third set of 800 were cornmeal mixed with artificial roast beef flavor; the remaining baits were cornmeal mixed with an artificial bread flavor. The different kinds of baits were all placed fairly close to one another so that a rat would have equal access to all four. After two weeks the test sites were inspected and the number of baits that were gone was recorded. Then a different set of locations in the same general area was selected and the experiment was repeated for another two weeks. Altogether the study was replicated five times and lasted ten weeks (77). The results are summarized in the table.

Survey Number	Percentage of Baits Accepted			
	Plain	Butter-Vanilla	Roast Beef	Bread
1	13.8	11.7	14.0	12.6
2	12.9	16.7	15.5	13.8
3	25.9	29.8	27.8	25.0
4	18.0	23.1	23.0	16.9
5	15.2	20.2	19.0	13.7

Do an analysis of variance on these data. Is there anything about these percentages to suggest that a series of rat-poison surveys should *not* be done as a completely randomized design? Explain.

Question 12.2.5 Show that the three cross-product terms that arise when Equation 12.2.3 is squared are all identically zero, thereby verifying Equation 12.2.4.

Question 12.2.6 In general, does a "large" value for SS_{blocks} (relative to the magnitude of SS_{error}) indicate that an experiment should or should not have been done as a randomized block design? Explain.

Question 12.2.7 Derive the computing formulas given in Theorem 12.2.2 for SS_{total}, SS_{blocks}, and $SS_{treatments}$.

Question 12.2.8 Heart rates were monitored (9) for six tree shrews (*Tupaia glis*) during three different stages of sleep: LSWS (light slow-wave sleep), DSWS (deep slow-wave sleep), and REM (rapid-eye-movement sleep).

HEART RATES (BEATS/5 SECONDS)

Tree Shrew	LSWS	DSWS	REM
1	14.1	11.7	15.7
2	26.0	21.1	21.5
3	20.9	19.7	18.3
4	19.0	18.2	17.0
5	26.1	23.2	22.5
6	20.5	20.7	18.9

(a) Do the analysis of variance to test the equality of the heart rates during these three phases of sleep. Let $\alpha = 0.05$.

(b) Because of the marked physiological difference between REM sleep and LSWS and DSWS sleep, it was decided before the data were collected to test the REM rate against the average of the other two. Test the appropriate subhypothesis with a contrast. Use the 0.05 level of significance. Also, find a second contrast orthogonal to the first and verify that the sum of the sum of squares for the two contrasts equals $SS_{treatments}$.

Question 12.2.9 Refer to the rat-poison data of Question 12.2.4. Partition the treatment sum of squares into three orthogonal contrasts. Let one contrast test the hypothesis that the true acceptance percentage for the plain cornmeal is equal to the true acceptance percentage for the cornmeal with artificial roast beef flavoring. Let a second contrast compare the effectiveness of the "butter–vanilla" and "bread" baits. What does the third contrast test? Do all testing at the $\alpha = 0.10$ level of significance.

Question 12.2.10 A comparison was made of the efficiency of four different unit-dose injection systems. A group of pharmacists and nurses were the "blocks." For each system, they were to remove the unit from its outer package, assemble it, and simulate an injection. In addition to the standard system of using a disposable syringe and needle to draw the medication from a vial, the other systems tested were Vari-Ject (CIBA Pharmaceutical), Unimatic (Squibb), and Tubex (Wyeth). Listed in the table are the average times (in seconds) needed to implement each of the systems (139).

AVERAGE TIMES (SEC) FOR IMPLEMENTING INJECTION SYSTEMS

Subject	Standard	Vari-Ject	Unimatic	Tubex	$t_{i.}$
1	35.6	17.3	24.4	25.0	102.3
2	31.3	16.4	22.4	26.0	96.1
3	36.2	18.1	22.8	25.3	102.4
4	31.1	17.8	21.0	24.0	93.9
5	39.4	18.8	23.3	24.2	105.7
6	34.7	17.0	21.8	26.2	99.7
7	34.1	14.5	23.0	24.0	95.6
8	36.5	17.9	24.1	20.9	99.4
9	32.2	14.6	23.5	23.5	93.8
10	40.7	16.4	31.3	36.9	125.3
$t_{.j}$	351.8	168.8	237.6	256.0	1014.2

State the model equation for these data and indicate what assumptions are being made about its components. Do the analysis of variance to test $H_0: \tau_1 = \cdots = \tau_4 = 0$ versus H_1: not all the τ_j's $= 0$ at the $\alpha = 0.05$ level of significance. Also, use Tukey's method to test all six pairwise differences. *Note:*

$$\sum_{i=1}^{10} \sum_{j=1}^{4} y_{ij}^2 = 27,771.14$$

12.3 THE PAIRED t TEST

An important special case of a randomized block design arises when the number of treatment levels is reduced to $t = 2$. The simplicity of the data structure then allows the F test to be replaced by the more familiar one-sample t test. Specifically, if X_i and Y_i denote the responses in the ith block to treatment level 1 and treatment level 2, respectively, we will define D_i to be the within-block difference—that is, $D_i = X_i - Y_i$. If there are b blocks, the sample mean and the sample variance of the D_i's have formulas analogous to their Chapter 7 counterparts:

$$\bar{D} = \frac{1}{b} \sum_{i=1}^{b} D_i$$

and

$$S_D^2 = \frac{1}{b-1} \sum_{i=1}^{b} (D_i - \bar{D})^2 = \frac{b \sum_{i=1}^{b} D_i^2 - \left(\sum_{i=1}^{b} D_i \right)^2}{b(b-1)}$$

Let μ_X and μ_Y denote the true means of levels 1 and 2, and let $\mu_D = \mu_X - \mu_Y$. Testing the null hypothesis $H_0: \mu_X = \mu_Y$ is then the same as testing $H_0: \mu_D = 0$. If it can be assumed that the D_i's are an independent random sample from an $N(\mu_D, \sigma^2)$ distribution, we can look to Theorem 7.7.1 to provide an appropriate critical region.

Comment

Testing $H_0: \mu_D = 0$ is a one-sample problem analytically but a two-sample problem conceptually. However, it is not analogous to the data structure presented in Chapter 8, because here the two samples are *dependent*, X_i and Y_i being measured on members of the same block. To emphasize these distinctions, the application of Theorem 7.7.1 in this context is referred to as a *paired t test*.

CASE STUDY   12.3.1

Blood coagulates only after a complex chain of chemical reactions has taken place. The final step is the transformation of fibrinogen, a protein found in the plasma, into fibrin, which forms the clot. The fibrinogen-fibrin reaction is triggered by another protein, thrombin, which itself is formed under the influence of still other proteins, including one called prothrombin.

A person's blood-clotting ability is typically expressed in terms of a "prothrombin time," which is defined to be the interval between the initiation of the prothrombin-thrombin reaction and the formation of the final clot. Factors affecting the length of a person's prothrombin time have profound medical significance. Recently a study was undertaken to determine what effect, if any, *aspirin* has on this particular function. Twelve adult males participated (187). Their prothrombin times were measured *before* and *three hours after* each was given two aspirin tablets (650 mg) (see Table 12.3.1).

TABLE 12.3.1 BLOOD-CLOTTING STUDY

Subject	Before Aspirin, x_i	After Aspirin, y_i	$d_i = x_i - y_i$
1	12.3	12.0	0.3
2	12.0	12.3	−0.3
3	12.0	12.5	−0.5
4	13.0	12.0	1.0
5	13.0	13.0	0
6	12.5	12.5	0
7	11.3	10.3	1.0
8	11.8	11.3	0.5
9	11.5	11.5	0
10	11.0	11.5	−0.5
11	11.0	11.0	0
12	11.3	11.5	−0.2
			1.3

The columns above the table are headed "Prothrombin Times (sec)".

If μ_X and μ_Y denote the true average prothrombin times *before* and *after* aspirin, respectively, and if $\mu_D = \mu_X - \mu_Y$, the hypotheses to be tested are

$$H_0: \quad \mu_D = 0$$

versus

$$H_1: \quad \mu_D \neq 0$$

We will let 0.05 be the level of significance.

From Table 12.3.1,

$$\sum_{i=1}^{12} d_i = 1.3 \quad \text{and} \quad \sum_{i=1}^{12} d_i^2 = 2.97$$

Therefore,

$$\bar{d} = \frac{1}{12}(1.3) = 0.108$$

and

$$s_D^2 = \frac{12(2.97) - (1.3)^2}{12(11)} = 0.257$$

Since $b = 12$, the critical values for the test statistic will be the 2.5th and 97.5th percentiles of the Student t distribution with 11 degrees of freedom: ± 2.20. The appropriate decision rule, then, from Theorem 7.7.1, would be

$$\text{Reject } H_0: \mu_D = 0 \text{ if } \frac{\bar{d}}{s_D/\sqrt{12}} \text{ is either } \begin{cases} \leq -2.20 \\ \text{or} \\ \geq +2.20 \end{cases}$$

In this case, the t ratio is

$$\frac{\bar{d}}{s_D/\sqrt{12}} = \frac{0.108}{\sqrt{0.257}/\sqrt{12}} = 0.74$$

and our conclusion is to *accept* H_0—aspirin has no demonstrable effect on prothrombin time.

Question 12.3.1 Triticale is a hybrid cereal grain derived from wheat and rye. Only recently cultivated, its nutritional properties are largely unknown. Not long ago an experiment was set up to compare the protein content in two different varieties of triticale, 6TA–204 and Rosner. Both were grown at four widely separated sites in central California. The table below shows the average protein content (expressed as a dry-weight percentage) measured for the eight crops (141).

PERCENT PROTEIN IN TWO TRITICALE VARIETIES

Location	6TA–204	Rosner
El Centro	16.2	16.6
Five Points	15.6	16.7
Davis	16.9	16.8
Tulelake	15.9	16.7

Test whether the protein yields for the 6TA–204 and Rosner varieties of triticale are significantly different. Use the 0.05 level of significance.

Question 12.3.2 A paint manufacturer is experimenting with a new latex derivative that may improve the chalkiness of its house paint. The problem is that it might also affect the tint, thereby requiring the company to rework its blending procedures. As a test, a quality control engineer draws a sample from each of seven batches of Osage Orange. Each sample is then split into two aliquots, with the latex derivative being added to one of the two. Both samples are then examined with a spectroscope. The data are shown below in standardized lumen units (if the tint of the aliquot were exactly right, it would register a reading of 1.00).

	Spectroscopy readings	
Batch	Osage Orange	Osage Orange + Additive
1	1.10	1.06
2	1.05	1.02
3	1.08	1.17
4	0.98	1.21
5	1.01	1.01
6	0.96	1.23
7	1.02	1.19

Analyze these data using (a) a paired t test and (b) a randomized block design. Show the equivalence of the results (see Question 12.3.5). Let 0.05 be the level of significance.

Question 12.3.3 Let $D_1, D_2, \ldots, D_b$ be within-block differences as defined in Section 12.3. Assume that $D_i \sim N(\mu_D, \sigma_D^2)$, for $i = 1, 2, \ldots, b$. Derive a formula for a $100(1 - \alpha)\%$ confidence interval for μ_D. Use your formula, together with the data of Case Study 12.3.1, to construct a 95% confidence interval for the true average prothrombin time difference ("before aspirin"—"after aspirin").

Question 12.3.4 Tracheobronchial clearance can be measured by having a subject inhale a radioactive aerosol and then metering at some later time the radiation level in his lungs. In one such experiment, seven pairs of monozygotic twins inhaled aerosols of radioactive Teflon particles. One member of each twin pair lived in a rural area; the other was a city-dweller. The objective of the study was to see what effect environment had on tracheobronchial clearance. The table lists the percentages of the radioactivity retained in the lungs of each subject one hour after the initial inhalation (17).

RADIOACTIVITY RETAINED
IN THE LUNGS

	Retention (%)	
Twin Pair	Rural	Urban
1	10.1	28.1
2	51.8	36.2
3	33.5	40.7
4	32.8	38.8
5	69.0	71.0
6	38.9	47.0
7	54.6	57.0

Define the appropriate μ_D for these data and test $H_0: \mu_D = 0$ versus $H_1: \mu_D \neq 0$ using a paired t test. Let $\alpha = 0.01$.

Question 12.3.5 Show that the paired t test is equivalent to the F test in a randomized block design when the number of treatment levels is two. (*Hint:* Consider the distribution of $T^2 = b\bar{D}^2/S_D^2$ and follow the approach outlined in Example 11.2.2.)

13

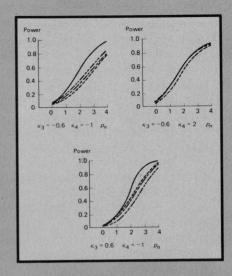

Nonparametric Statistics

The figures show a comparison of the power functions of the one-sample t test (solid line) and the sign test (dashed lines) for three different sets of hypotheses, various degrees of nonnormality, a sample size of 10, and a level of significance of 0.05. (The parameter ρ_n measures the shift from H_0 to H_1; κ_3 and κ_4 measure the extent of nonnormality in the sampled population.)

13.1 INTRODUCTION

Behind every confidence interval and hypothesis test we have looked at thus far has been a very specific set of assumptions about the origin of the data from which the inference was to be made. For example, in doing a two-sample Z test for proportions—H_0: $p_X = p_Y$ versus H_1: $p_X \neq p_Y$—it is assumed that n independent and identically distributed Bernoulli random variables make up the X-sample and m independent and identically distributed Bernoullis the Y-sample. Of course, the most common assumption made in data analysis is that each set of observations is a random sample from a *normal distribution*. This is the condition that was specified in earlier chapters when the inference procedure being discussed involved either the t, the χ^2, or the F distribution.

The need to make such assumptions raises an obvious question: what is affected when what we assume to be true about the data is not? Certainly the test statistic is the same and the computational formulas do not change. The answer, of course, lies with the sampling distribution of the test statistic. In the case of a one-sample t test, $T = (\bar{Y} - \mu_0)/(S/\sqrt{n})$ will no longer behave exactly like a Student t variable with $n - 1$ degrees of freedom if the Y_i's are not $N(\mu_0, \sigma^2)$. As a result, the *actual* probability of committing a Type I error will not necessarily equal the *nominal* probability. That is,

$$\int_{-\infty}^{-t_{\alpha/2,\,n-1}} f_T(t \mid H_0)\, dt + \int_{t_{\alpha/2,\,n-1}}^{\infty} f_T(t \mid H_0)\, dt \neq \alpha \qquad (13.1.1)$$

where $f_T(t \mid H_0)$ is the (unknown) sampling distribution of $T = (\bar{Y} - \mu_0)/(S/\sqrt{n})$ under the assumption that $\mu = \mu_0$ and $f_Y(y)$ is something other than

$$(1/\sqrt{2\pi}\,\sigma)e^{-(1/2)[(y - \mu_0)/\sigma]^2}$$

Statisticians have sought to overcome the problems implicit in Equation 13.1.1 in two very different ways. On the one hand, efforts have been made to show that even though T will not have an *exact* Student t distribution when the Y_i's are not normal, $f_T(t \mid H_0)$ will still, in many cases, be similar enough to that distribution so that the actual α is, for all intents and purposes, equal to the nominal α. On the whole, these efforts have been quite successful: unless the Y_i's are markedly non-normal (and n is very small), the distribution of T will, indeed, have a shape much like that of a Student t, implying that in many cases the assumption of normality is not all that critical after all. This is summarized by saying that the *t test is robust with respect to departures from normality*.

Example 13.1.1

Questions of robustness tend to be prohibitively difficult to handle in any formal mathematical way. The usual method of attack is through Monte Carlo methods. For example, a study of the robustness of the two-sample t test might begin by generating two independent random samples of size 5 from an *exponential* distribution with λ equal to 1. From the resulting x_i's and y_i's,

$\bar{x}$, $\bar{y}$, and s_p are calculated and substituted into the usual two-sample t statistic: $t = (\bar{x} - \bar{y})/(s_p \sqrt{\frac{1}{5} + \frac{1}{5}})$. The resulting value is then compared to what would have been the two critical values, $\pm t_{\alpha/2,\,8}$, and it is recorded whether or not the observed t fell into the rejection region. The same process is then repeated over and over again, each time starting with a different set of random samples. After, say, a thousand such trials, we compute α', the actual proportion of observed t's falling outside the interval $(-t_{\alpha/2,\,8}, +t_{\alpha/2,\,8})$. If α' is numerically close to α, we will conclude that the two-sample t test is robust against that particular kind of nonnormality.

Table 13.1.1 summarizes the results of several Monte Carlo studies focusing on the two-sample t test (12). The notation $N(\mu, \sigma^2)(n)$ means that n random observations were drawn from a normal pdf with mean μ and variance σ^2; $U(\mu, \sigma^2)(n)$ represents a random sample of size n drawn from a uniform pdf with mean μ and variance σ^2. In each instance a total of 1000 sets of samples were generated. The nominal level of significance was $\alpha = 0.05$.

As Table 13.1.1 shows, certain departures from the t-test assumptions have little effect on α—for example, when both random samples are *uniform* variables. Other violations, though, take more of a toll—see, for example, line 2, where five $N(0, 4)$ x_i's and fifteen $N(0, 1)$ y_i's combined to give an α' of 0.160, more than three times the nominal 0.05.

TABLE 13.1.1 PROPORTION OF t RATIOS FALLING IN THE $\alpha = 0.05$ CRITICAL REGION (1000 TRIALS)

Samples Drawn		Proportion of t Ratios in Critical Region
X	Y	
$N(0, 1)(5)$	$N(0, 4)(15)$	0.064
$N(0, 4)(5)$	$N(0, 1)(15)$	0.160
$U(0, 1)(5)$	$U(0, 1)(5)$	0.051
$U(0, 1)(5)$	$U(0, 4)(5)$	0.071

A second way of dealing with Equation 13.1.1 is to replace the t test with procedures having test statistics whose distribution under H_0 remains the same, regardless of how the population sampled may change. Inference procedures having this sort of latitude are called *nonparametric*, or more appropriately, *distribution free*.

The number of nonparametric procedures proposed since the early 1940s has been enormous and continues to grow. We do not presume to give any sort of survey of those procedures here; for that, the reader is referred to more specialized texts, such as (55) or (74). Our more limited objective is to introduce some of the basic methodology of nonparametric statistics in the context of problems whose "parametric" solutions we have already seen. In so doing we will state, and in some cases derive, a nonparametric solution to the paired-data problem, the one-sample location problem, and both the experimental design models of Chapters 11 and 12.

13.2 THE SIGN TEST

Among the simplest of all nonparametric procedures is the *sign test*. It finds a number of different applications: if a set of Y_i's come from a *symmetric* (but not necessarily normal) pdf, a sign test can be used for the one-sample location problem, $H_0: \tilde{\mu} = \tilde{\mu}_0$ versus $H_1: \tilde{\mu} \neq \tilde{\mu}_0$, where $\tilde{\mu}$ denotes the median of the population being sampled. The same test is also appropriate in a paired-data situation where the normality of the X_i's and Y_i's is questionable.

Consider the latter situation: $(X_1, Y_1), (X_2, Y_2), \ldots, (X_n, Y_n)$ is a random sample of paired observations. Let $p = P(Y_i > X_i)$. We wish to test whether or not the X's are shifted in location relative to the Y's. In terms of p, the hypotheses reduce to

$$H_0: \quad p = \tfrac{1}{2} \quad \text{(no shift in location)}$$

versus

$$H_1: \quad p \neq \tfrac{1}{2} \quad \text{(shift in location)}$$

Let each (X_i, Y_i) pair be replaced by either a plus sign or a minus sign: a plus sign if $Y_i > X_i$ and a minus sign if $Y_i < X_i$. [It will be assumed that both $f_X(x)$ and $f_Y(y)$ are continuous, so the event of a tie has probability zero.] It follows that if H_0 is true, each (X_i, Y_i) pair constitutes a Bernoulli trial, and Y_+, the number of plus signs, will have a binomial pdf with parameters n and $\tfrac{1}{2}$. The appropriate decision rule, then, is a direct extension of the argument given in Section 6.2: H_0 should be rejected if y_+ is either less than or equal to Y_+^* or greater than or equal to Y_+^{**}, where

$$P(Y_+ \leq Y_+^*) = P(Y_+ \geq Y_+^{**}) = \frac{\alpha}{2} \tag{13.2.1}$$

Comment

For small n, the discreteness of the binomial pdf might make it impossible to find a Y_+^* and a Y_+^{**} satisfying Equation 13.2.1 exactly. For large n, that becomes less of a problem: we simply appeal to the DeMoivre–Laplace limit theorem and approximate, say, Y_+^{**} by solving

$$z_{1-\alpha/2} = \frac{Y_+^{**} - \tfrac{1}{2} \cdot n}{\sqrt{n/4}}$$

CASE STUDY   13.2.1

Children with severe learning problems often have electroencephalograms and behavior patterns similar to those of children with petit-mal, a mild form of epilepsy. This led to speculation that drugs helpful in treating petit-mal might also be useful as "learning facilitators" (157). To test that hypothesis, ten children, ranging in age from 8 to 14 and all having a history of learning and behavioral problems, were recruited to participate in a six-week study. For three of those weeks a child was given a placebo; for the other three weeks, ethosuximide, a widely prescribed anticonvulsant. After each three-week period the children were given several parts of the standard Wechsler IQ test. Because a child might be expected to do better on the test the second time he took it, the order in which the placebo and the ethosuximide were administered was randomized: some children were given the placebo for the first three weeks while others began the study by taking the ethosuximide.

Table 13.2.1 shows the two verbal IQ scores recorded for each subject. The last entry in the fourth column indicates the value of the test statistic, Y_+.

TABLE 13.2.1 IQ SCORES IN ETHOSUXIMIDE EXPERIMENT

Child	IQ After Placebo, x_i	IQ After Ethosuximide, y_i	$y_i > x_i$?
1	97	113	+
2	106	113	+
3	106	101	−
4	95	119	+
5	102	111	+
6	111	122	+
7	115	121	+
8	104	106	+
9	90	110	+
10	96	126	+
			$y_+ = 9$

Suppose that we wish to test

$$H_0: \quad p = \frac{1}{2}$$

versus

$$H_1: \quad p \neq \frac{1}{2}$$

at the $\alpha = 0.10$ level of significance, where $p = P(Y_i > X_i)$. The two-sided alternative is used here to allow for the possibility that ethosuximide might actually have an adverse effect on IQ.

For the binomial pdf with parameters $n = 10$ and $p = \frac{1}{2}$,

$$P(Y_+ \leq 2) = \sum_{y=0}^{2} \binom{10}{y}\left(\frac{1}{2}\right)^y\left(\frac{1}{2}\right)^{10-y} \doteq 0.05$$

and

$$P(Y_+ \geq 8) = \sum_{y=8}^{10} \binom{10}{y}\left(\frac{1}{2}\right)^y\left(\frac{1}{2}\right)^{10-y} \doteq 0.05$$

Thus, to maintain a Type I error probability of approximately 0.10, we should reject H_0 if y_+ is either less than or equal to 2 or greater than or equal to 8. In point of fact, the latter was true ($y_+ = 9$), so our conclusion is to *reject* the null hypothesis: it would appear that ethosuximide *does* have an effect on IQ, and a beneficial one at that.

Question 13.2.1 One reason cited for the mental deterioration so often seen in the very elderly is the reduction in cerebral blood flow that accompanies the aging process. Addressing itself to this notion, a study was recently done in a rest home to see whether cyclandelate, a vasodilator, might be able to stimulate the cerebral circulation and thereby slow down the rate of mental deterioration (3). The drug was given to 11 subjects on a daily basis. To measure its physiological effect, radioactive tracers were used to determine each subject's mean circulation time (MCT) at the start of the experiment and four months later when the study was discontinued. (The MCT is the length of time it takes blood to travel from the carotid artery to the jugular vein.) The results are shown in the table.

CEREBRAL CIRCULATION EXPERIMENT

Subject	Mean Circulation Time (sec)	
	Before, x_i	After, y_i
J.B.	15	13
M.B.	12	8
A.B.	12	12.5
M.B.	14	12
J.L.	13	12
S.M.	13	12.5
M.M.	13	12.5
S.McA.	12	14
A.McL.	12.5	12
F.S.	12	11
P.W.	12.5	10

Test the appropriate hypothesis with a sign test. Use a one-sided alternative.

Question 13.2.2 Recall Case Study 13.2.1. What could be concluded if the "after ethosuximide" IQs were significantly higher than the "after placebo" IQs but if all 10 children had been given the placebo first?

Question 13.2.3 Synovial fluid is the clear, viscid secretion that lubricates joints and tendons. For some ailments, its hydrogen-ion concentration (pH) has diagnostic importance. In healthy adults the average pH for synovial fluid is 7.39. Listed below are synovial pH values measured for fluids drawn from the knees of 44 patients with various arthritic conditions (167).

SYNOVIAL FLUID pH

7.02	7.26	7.31	7.14	7.45	7.32	7.21	7.36	7.36
7.35	7.25	7.24	7.20	7.39	7.40	7.33	7.09	6.60
7.32	7.35	7.34	7.41	7.28	6.99	7.28	7.32	7.29
7.33	7.38	7.32	7.77	7.34	7.10	7.35	6.95	7.31
7.15	7.20	7.34	7.12	7.22	7.30	7.24	7.35	

Is an abnormal synovial pH a symptom of arthritis? Do a sign test on

$$H_0: \quad \tilde{\mu} = 7.39$$

versus

$$H_1: \quad \tilde{\mu} \neq 7.39$$

where $\tilde{\mu}$ denotes the synovial fluid pH typical of adults with arthritis. Let $\alpha = 0.01$.

Question 13.2.4 Let $X_1, X_2, \ldots, X_{22}$ be a random sample of normally distributed random variables with an unknown mean μ and a known variance of 6.0. We wish to test

$$H_0: \quad \mu = 10$$

versus

$$H_1: \quad \mu > 10$$

Construct a large sample sign test having a Type I error probability of 0.05. What will the power of the test be if $\mu = 11$?

Question 13.2.5 Suppose that $n = 7$ paired observations, (X_i, Y_i), are recorded, $i = 1, 2, \ldots,$ 7. Let $p = P(Y_i > X_i)$. Write out the entire probability distribution for Y_+, the number of positive differences among the set of $Y_i - X_i$'s, $i = 1, 2, \ldots, 7$, assuming that $p = \frac{1}{2}$. What α levels are possible for testing $H_0: p = \frac{1}{2}$ versus $H_1: p > \frac{1}{2}$?

Question 13.2.6 Analyze the Shoshoni rectangle data (Case Study 1.2.2) with a sign test. Let $\alpha = 0.05$.

Question 13.2.7 In a marketing research test, 28 adult males were asked to shave one side of their face with one brand of razor blade and the other side with a second brand. They were to use the blades for seven days and then decide which was giving the smoother shave. Suppose that 19 of the subjects preferred blade A. Use a sign test to determine whether it can be claimed, at the 0.05 level, that the two blades are significantly different.

Question 13.2.8 Suppose that a random sample of size 36, $X_1, X_2, \ldots, X_{36}$, is drawn from a uniform pdf defined over the interval $(0, \theta)$, where θ is unknown. Set up a large-sample sign test for deciding whether or not the 25th percentile of the X distribution is equal to 6. Let $\alpha = 0.05$. With what probability will your procedure commit a Type II error if 7 is the true 25th percentile?

13.3 THE WILCOXON SIGNED RANK TEST

Although the sign test is a bona fide nonparametric procedure, its extreme simplicity makes it somewhat atypical. The *Wilcoxon signed rank test* introduced in this section is more representative of nonparametric procedures as a whole. Like the sign test, it can be adapted to several different data structures. Here, we will pose it within the framework of a one-sample test for location, where it becomes an alternative to the one-sample t test. The procedure carries over immediately to the paired-data problem and with only minor modifications can also serve as a two-sample test for location and a two-sample test for dispersion (provided the two populations have equal locations). Historically, the Wilcoxon signed rank test was one of the first nonparametric procedures to be developed (it dates back to 1945), and it remains one of the most widely used.

Let $Y_1, Y_2, \ldots, Y_n$ be a random sample of size n from a pdf $f_Y(y)$ that is both continuous and symmetric. Let $\tilde{\mu}$ be the median of $f_Y(y)$. We wish to test

$$H_0: \quad \tilde{\mu} = \tilde{\mu}_0$$

versus

$$H_1: \quad \tilde{\mu} \neq \tilde{\mu}_0$$

where $\tilde{\mu}_0$ is some prespecified value for $\tilde{\mu}$.

The Wilcoxon statistic is based on the magnitudes, and directions, of the deviations of the Y_i's from $\tilde{\mu}_0$. Let $|Y_1 - \tilde{\mu}_0|, |Y_2 - \tilde{\mu}_0|, \ldots, |Y_n - \tilde{\mu}_0|$ denote the set of absolute deviations of the Y_i's from $\tilde{\mu}_0$. These can be ordered from smallest to largest, and we will define R_i to be the *rank* of $|Y_i - \tilde{\mu}_0|$ in the set $\{|Y_j - \tilde{\mu}_0|\}_{j=1}^n$. (The smallest $|Y_j - \tilde{\mu}_0|$ is assigned a rank of 1, the second smallest, a rank of 2, and so on up to n.)

Associated with each R_i will be a sign indicator, Z_i, where

$$Z_i = \begin{cases} 0 & \text{if } Y_i - \tilde{\mu}_0 < 0 \\ 1 & \text{if } Y_i - \tilde{\mu}_0 > 0 \end{cases}$$

We will define the Wilcoxon signed rank statistic, W, to be the linear combination,

$$W = \sum_{i=1}^n Z_i R_i$$

To illustrate this terminology, consider the case where $n = 3$ and $y_1 = 6.0$, $y_2 = 4.9$, and $y_3 = 11.2$. Suppose that the problem is to test

$$H_0: \quad \tilde{\mu} = 10.0$$

versus

$$H_1: \quad \tilde{\mu} \neq 10.0$$

Note that $|y_1 - \tilde{\mu}_0| = 4.0$, $|y_2 - \tilde{\mu}_0| = 5.1$, and $|y_3 - \tilde{\mu}_0| = 1.2$. Since $1.2 < 4.0 < 5.1$, it follows that $r_1 = 2$, $r_2 = 3$, and $r_3 = 1$. Also, $z_1 = 0$, $z_2 = 0$, and $z_3 = 1$. Combining the r_i's and the z_i's we have that

$$w = \sum_{i=1}^n z_i r_i$$
$$= (0)(2) + (0)(3) + (1)(1)$$
$$= 1$$

Comment

Notice that W is based on the *ranks* of the deviations from $\tilde{\mu}_0$ and not on the deviations themselves. For this example, the value of W would remain unchanged if y_2 were 4.9, 3.6, or $-10,000$. In each case, r_2 would be 3 and Z_2 would be 0. If the test statistic *did* depend on the magnitude of the deviations, it would have been necessary to specify a particular distribution for $f_Y(y)$, and the resulting procedure would no longer be nonparametric.

It should be clear that W ranges from 0 (all deviations negative) to $\sum_{i=1}^{n} i = [n(n + 1)]/2$ (all deviations positive). Intuitively, if H_0 were true we would expect the Y_i's to be distributed at random above and below $\tilde{\mu}_0$, meaning that W should tend to take on values close to $[n(n + 1)]/4$. The next theorem enables us to determine how close W has to get to either 0 or $[n(n + 1)]/2$ before H_0 can be rejected.

Theorem 13.3.1. For $\{Y_i\}_{i=1}^{n}$, $\{R_i\}_{i=1}^{n}$, and $\{Z_i\}_{i=1}^{n}$ defined as above, the probability distribution of $W = \sum_{i=1}^{n} Z_i R_i$, when $H_0: \tilde{\mu} = \tilde{\mu}_0$ is true, is given by

$$P(W = w) = f_W(w) = \left(\frac{1}{2^n}\right) \cdot c(w)$$

where $c(w)$ is the coefficient of e^{wt} in the expansion of

$$\prod_{i=1}^{n} (1 + e^{it})$$

Proof. The statement and proof of Theorem 13.3.1 are typical of many nonparametric results. Closed-form expressions for sampling distributions are seldom possible: the combinatorial nature of nonparametric test statistics lends itself more readily to a generating-function format.

To begin, note that if H_0 is true, the distribution of $W = \sum_{i=1}^{n} Z_i R_i$ is equivalent to the distribution of $U = \sum_{i=1}^{n} U_i$, where

$$U_i = \begin{cases} 0 & \text{with probability } \frac{1}{2} \\ i & \text{with probability } \frac{1}{2} \end{cases}$$

Therefore, W and U have the same moment-generating function. Also, since the Z_i's are independent, so are the U_i's, and from Theorem 3.12.3,

$$M_U(t) = M_W(t)$$

$$= \prod_{i=1}^{n} M_{U_i}(t)$$

$$= \prod_{i=1}^{n} E(e^{U_i t})$$

$$= \prod_{i=1}^{n} \left(\frac{1}{2} e^{0t} + \frac{1}{2} e^{it}\right)$$

$$= \left(\frac{1}{2^n}\right) \prod_{i=1}^{n} (1 + e^{it}) \tag{13.3.1}$$

Sec. 13.3   The Wilcoxon Signed Rank Test   **557**

Now, consider the *structure* of $f_W(w)$, the pdf for the Wilcoxon statistic. In forming W, we can prefix R_1 by either a plus sign or a zero; similarly for R_2, $R_3, \ldots$, and R_n. It follows that since each R_i can take on two different values, the total number of ways to make signed rank sums is 2^n. Under H_0, of course, each of these arrangements is equally likely so, in general terms, the probability distribution of W has the form

$$P(W = w) = f_W(w) = \frac{c(w)}{2^n} \qquad (13.3.2)$$

where $c(w)$ is the number of ways to assign $+$'s and "zeros" to the first n integers so that $\sum_{i=1}^{n} Z_i R_i$ has the value w.

The conclusion of Theorem 13.3.1 follows immediately by comparing the form of $f_W(w)$ to Equation 13.3.1 and to the general expression for a moment-generating function. By definition,

$$M_W(t) = E(e^{Wt}) = \sum_{w=1}^{n(n+1)/2} e^{wt} f_W(w)$$

but from Equations 13.3.1 and 13.3.2 we can write

$$\sum_{w=1}^{n(n+1)/2} e^{wt} f_W(w) = \left(\frac{1}{2^n}\right) \prod_{i=1}^{n} (1 + e^{it}) = \sum_{w=1}^{n(n+1)/2} e^{wt} \cdot \frac{c(w)}{2^n}$$

We prove the theorem by recognizing that $c(w)$ is the coefficient of e^{wt} in the expansion of $\prod_{i=1}^{n} (1 + e^{it})$.

It may help to clarify these ideas by considering a numerical example. Suppose $n = 4$. Then, by Equation 13.3.1, the moment-generating function for W becomes

$$M_W(t) = \left(\frac{1 + e^t}{2}\right)\left(\frac{1 + e^{2t}}{2}\right)\left(\frac{1 + e^{3t}}{2}\right)\left(\frac{1 + e^{4t}}{2}\right)$$

$$= \left(\frac{1}{16}\right)\{1 + e^t + e^{2t} + 2e^{3t} + 2e^{4t} + 2e^{5t} + 2e^{6t} + 2e^{7t} + e^{8t} + e^{9t} + e^{10t}\}$$

Thus, the probability that W equals, say, 2 is $\frac{1}{16}$ (since the coefficient of e^{2t} is 1); the probability that W equals 7 is $\frac{2}{16}$, and so on. The first two columns of Table 13.3.1 show the complete probability distribution of W, as given by the expansion of $M_W(t)$. The last column enumerates the particular assignments of $+$'s and 0's that generate each value of W.

Cumulative tail area probabilities,

$$P(W \le w_1^*) = \sum_{w=0}^{w_1^*} f_W(w) \qquad \text{and} \qquad P(W \ge w_2^*) = \sum_{w=w_2^*}^{n(n+1)/2} f_W(w)$$

are listed in Table A.6 of the Appendix for sample sizes ranging from $n = 4$ to $n = 12$. Knowing these probabilities, it is easy to construct decision rules for either

TABLE 13.3.1 PROBABILITY DISTRIBUTION OF W

w	$f_W(w) = P(W = w)$	r_i 1	2	3	4
0	$\frac{1}{16}$	0	0	0	0
1	$\frac{1}{16}$	+	0	0	0
2	$\frac{1}{16}$	0	+	0	0
3	$\frac{2}{16}$	$\begin{cases}+ \\ 0\end{cases}$	$\begin{matrix}+ \\ 0\end{matrix}$	$\begin{matrix}0 \\ +\end{matrix}$	$\begin{matrix}0 \\ 0\end{matrix}$
4	$\frac{2}{16}$	$\begin{cases}+ \\ 0\end{cases}$	$\begin{matrix}0 \\ 0\end{matrix}$	$\begin{matrix}+ \\ 0\end{matrix}$	$\begin{matrix}0 \\ +\end{matrix}$
5	$\frac{2}{16}$	$\begin{cases}+ \\ 0\end{cases}$	$\begin{matrix}0 \\ +\end{matrix}$	$\begin{matrix}0 \\ +\end{matrix}$	$\begin{matrix}+ \\ 0\end{matrix}$
6	$\frac{2}{16}$	$\begin{cases}+ \\ 0\end{cases}$	$\begin{matrix}+ \\ +\end{matrix}$	$\begin{matrix}+ \\ 0\end{matrix}$	$\begin{matrix}0 \\ +\end{matrix}$
7	$\frac{2}{16}$	$\begin{cases}+ \\ 0\end{cases}$	$\begin{matrix}+ \\ 0\end{matrix}$	$\begin{matrix}0 \\ +\end{matrix}$	$\begin{matrix}+ \\ +\end{matrix}$
8	$\frac{1}{16}$	+	0	+	+
9	$\frac{1}{16}$	0	+	+	+
10	$\frac{1}{16}$	+	+	+	+
	1				

one-sided or two-sided alternatives. For example, suppose n is 7 and we wish to test

$$H_0: \quad \tilde{\mu} = \tilde{\mu}_0$$

versus

$$H_1: \quad \tilde{\mu} \neq \tilde{\mu}_0$$

at the $\alpha = 0.05$ level of significance. The critical region would be the set of w-values less than or equal to 2 or greater than or equal to 26: $C = \{w: w \leq 2 \text{ or } w \geq 26\}$. This follows immediately from Table A.6, since

$$\sum_{w \in C} f_W(w) = 0.023 + 0.023 \doteq 0.05$$

CASE STUDY   13.3.1

Swell sharks (*Cephaloscyllium ventriosum*) are small, reef-dwelling sharks that inhabit the California coastal waters south of Monterey Bay. There is a second population of these fish living nearby in the vicinity of Catalina Island, but it has been hypothesized (61) that the two populations never mix. Separating Santa Catalina from the mainland is a deep basin, which, according to the "separation" hypothesis, is an inpenetrable barrier for these particular fish.

One way to test this theory would be compare the morphology of sharks caught in the two regions. If there were no mixing, we would expect a certain number of differences to have evolved. Table 13.3.2 lists the total length (*TL*),

TABLE 13.3.2 MEASUREMENTS MADE ON TEN SHARKS CAUGHT NEAR SANTA CATALINA

Total Length (mm)	Height of First Dorsal Fin (mm)	TL/HDI
906	68	13.32
875	67	13.06
771	55	14.02
700	59	11.86
869	64	13.58
895	65	13.77
662	49	13.51
750	52	14.42
794	55	14.44
787	51	15.43

the height of the first dorsal fin (HDI), and the ratio TL/HDI for ten male swell sharks caught near Santa Catalina.

It has been estimated on the basis of past data that the median TL/HDI ratio for male swell sharks caught *off the coast* is 14.60. Is this figure consistent with the data of Table 13.3.2? In more formal terms, if $\tilde{\mu}$ denotes the true median TL/HDI ratio for the Santa Catalina population, can we reject H_0: $\tilde{\mu} = 14.60$, and thereby lend support to the separation theory?

Table 13.3.3 gives the values of TL/HDI ($=Y_i$), $Y_i - \tilde{\mu}_0 = Y_i - 14.60$, $|Y_i - 14.60|$, R_i, Z_i, and $R_i Z_i$ for the ten Santa Catalina sharks.

TABLE 13.3.3 COMPUTATIONS FOR WILCOXON SIGNED RANK TEST

TL/HDI ($=y_i$)	$y_i - 14.60$	$|y_i - 14.60|$	r_i	z_i	$r_i z_i$
13.32	−1.28	1.28	8	0	0
13.06	−1.54	1.54	9	0	0
14.02	−0.58	0.58	3	0	0
11.86	−2.74	2.74	10	0	0
13.58	−1.02	1.02	6	0	0
13.77	−0.83	0.83	4.5	0	0
13.51	−1.09	1.09	7	0	0
14.42	−0.18	0.18	2	0	0
14.44	−0.16	0.16	1	0	0
15.43	+0.83	0.83	4.5	1	4.5

Comment

When two or more numbers being ranked are equal, each is assigned the *average* of the ranks they would otherwise have received. In the data of Table 13.3.3 there are two sharks whose $|y_i - 14.60|$ value is 0.83. Had these two deviations been just slightly different, they would have been given ranks 4 and 5, but since they are equal, each is assigned a rank of 4.5 [$=(4 + 5)/2$].

By summing the last column we see that $w = 4.5$. According to Table A.6 in the Appendix the $\alpha = 0.05$ decision rule for testing

$$H_0: \quad \tilde{\mu} = 14.60$$

versus

$$H_1: \quad \tilde{\mu} \neq 14.60$$

requires that H_0 be rejected if w is either less than or equal to 8 or greater than or equal to 47. (Why is the alternative hypothesis two-sided here?) (*Note:* The *exact* level of significance associated with $C = \{w: w \leq 8 \text{ or } w \geq 47\}$ is $0.024 + 0.024 = 0.048$.) Thus we should *reject* H_0, since the observed w was less than 8. These particular data, then, would support the separation hypothesis.

Question 13.3.1 The average energy expenditures for eight elderly women were estimated on the basis of information received from a battery-powered heart rate monitor that each subject wore. Two overall averages were calculated for each woman, one for the summer months and one for the winter months (144), as shown in the table.

AVERAGE DAILY ENERGY EXPENDITURES (KCAL)

Subject	Summer, x_i	Winter, y_i
1	1458	1424
2	1353	1501
3	2209	1495
4	1804	1739
5	1912	2031
6	1366	934
7	1598	1401
8	1406	1339

Let μ_D denote the location difference between the summer and winter energy-expenditure populations. Compute $y_i - x_i$, $i = 1, 2, \ldots, 8$, and use the Wilcoxon signed rank procedure to test

$$H_0: \quad \mu_D = 0$$

versus

$$H_1: \quad \mu_D \neq 0$$

Let $\alpha = 0.15$.

Question 13.3.2 Use the expansion of

$$\prod_{i=1}^{n} (1 + e^{it})$$

to find the pdf of W when $n = 5$. What α levels are available for testing $H_0: \tilde{\mu} = \tilde{\mu}_0$ versus $H_1: \tilde{\mu} > \tilde{\mu}_0$?

The usefulness of Table A.6 in the Appendix is clearly limited to tests where the sample size is small, less than or equal to 12. To accommodate a larger n, we can define a normalized test statistic, W', where

$$W' = \frac{W - E(W)}{\sqrt{\text{Var}(W)}}$$

and $E(W)$ and Var (W) are the expected value and variance of W *when H_0 is true* (see Theorem 13.3.2). It can be shown that as n gets large, the distribution of W' converges to the standard normal. Furthermore, for n even as small as 13, $f_{W'}(w')$ and the $N(0, 1)$ are remarkably similar. Therefore, to test

$$H_0: \quad \tilde{\mu} = \tilde{\mu}_0$$

versus

$$H_1: \quad \tilde{\mu} \neq \tilde{\mu}_0$$

at, say, the $\alpha = 0.05$ level of significance, we would reject H_0 if w' was either less than or equal to -1.96 or greater than or equal to $+1.96$.

Theorem 13.3.2. When H_0 is true, the mean and standard deviation of the Wilcoxon signed rank statistic, W, are given by

$$E(W) = \frac{n(n + 1)}{4}$$

and

$$\text{Var}(W) = \frac{n(n + 1)(2n + 1)}{24}$$

Also, for $n > 12$, the distribution of

$$W' = \frac{W - [n(n + 1)]/4}{\sqrt{[n(n + 1)(2n + 1)]/24}}$$

can be adequately approximated by the standard normal.

Proof. We will derive $E(W)$ and Var (W); for a proof of the asymptotic normality, see (73). Recall that W has the same distribution as $U = \sum_{i=1}^{n} U_i$, where

$$U_i = \begin{cases} 0 & \text{with probability } \frac{1}{2} \\ i & \text{with probability } \frac{1}{2} \end{cases}$$

Therefore,

$$E(W) = E\left(\sum_{i=1}^{n} U_i\right) = \sum_{i=1}^{n} E(U_i)$$

$$= \sum_{i=1}^{n} \left(0 \cdot \frac{1}{2} + i \cdot \frac{1}{2}\right) = \sum_{i=1}^{n} \frac{i}{2}$$

$$= \frac{n(n + 1)}{4}$$

Similarly,

$$\text{Var } (W) = \text{Var } (U) = \sum_{i=1}^{n} \text{Var } (U_i)$$

since the U_i's are independent. But

$$\text{Var } (U_i) = E(U_i^2) - [E(U_i)]^2$$

$$= \frac{i^2}{2} - \left(\frac{i}{2}\right)^2 = \frac{i^2}{4}$$

making

$$\text{Var } (W) = \sum_{i=1}^{n} \frac{i^2}{4} = \left(\frac{1}{4}\right)\left[\frac{n(n+1)(2n+1)}{6}\right]$$

$$= \frac{n(n+1)(2n+1)}{24}$$

CASE STUDY   13.3.2

Methadone is a drug widely used in the treatment of heroin addiction; another is cyclazocine. Recently a study was done (130) to evaluate the effectiveness of the latter in reducing a person's psychological dependence on heroin. The subjects were 14 males, all chronic heroin addicts. Each was asked a battery of questions that compared his feelings when he was using heroin to his feelings when he was "clean." The resultant Q-scores ranged from a possible minimum of 11 to a possible maximum of 55, as shown in Table 13.3.4. (From the way the questions were worded, higher scores represented *less* psychological dependence.)

TABLE 13.3.4 *Q*-SCORES
OF HEROIN ADDICTS AFTER
CYCLAZOCINE THERAPY

51	43
53	45
43	27
36	21
55	26
55	22
39	43

The median score for addicts *not* treated with cyclazocine is known from past experience to be 28. Can we conclude on the basis of the data in Table 13.3.4 that cyclazocine is an effective treatment?

Since high Q-scores represent *less* dependence on heroin (and assuming cyclazocine would not tend to worsen an addict's condition), the alternative hypothesis should be one-sided *to the right*:

$$H_0: \quad \tilde{\mu} = 28$$

versus

$$H_1: \quad \tilde{\mu} > 28$$

We will set α equal to 0.05.

Table 13.3.5 shows the computations necessary to find W—in this case, 95.0. With n being larger than 12, $W' = [W - E(W)]/\sqrt{\text{Var}(W)}$ has approximately a standard normal distribution, and the 0.05 decision rule becomes

$$\text{Reject } H_0: \tilde{\mu} = 28 \text{ if } w' \geq 1.64$$

TABLE 13.3.5 COMPUTATIONS TO FIND W

Q-Score, y_i	$(y_i - 28)$	$\|y_i - 28\|$	r_i	z_i	$r_i z_i$
51	+23	23	11	1	11
53	+25	25	12	1	12
43	+15	15	8	1	8
36	+8	8	5	1	5
55	+27	27	13.5	1	13.5
55	+27	27	13.5	1	13.5
39	+11	11	6	1	6
43	+15	15	8	1	8
45	+17	17	10	1	10
27	−1	1	1	0	0
21	−7	7	4	0	0
26	−2	2	2	0	0
22	−6	6	3	0	0
43	+15	15	8	1	8
					95.0

By Theorem 13.3.2, $E(W) = [14(14 + 1)]/4 = 52.5$ and $\text{Var}(W) = [14(14 + 1)(28 + 1)]/24 = 253.75$. Therefore,

$$w' = \frac{95.0 - 52.5}{\sqrt{253.75}} = 2.67$$

implying that we should *reject* H_0—it would appear that cyclazocine therapy *is* helpful in reducing heroin dependence.

Question 13.3.3 Two manufacturing processes are available for annealing a certain kind of copper tubing, the primary difference being in the temperature required. The critical response variable is the resulting tensile strength. To compare the methods, 15 pieces of tubing were broken into pairs. One piece from each pair was randomly selected to be annealed at a moderate temperature, the other piece at a high temperature. The resulting tensile strengths (in tons/sq in.) are listed below.

TENSILE STRENGTHS (TONS/SQ. IN.)

Pair	Moderate Temperature	High Temperature
1	16.5	16.9
2	17.6	17.2
3	16.9	17.0
4	15.8	16.1
5	18.4	18.2
6	17.5	17.7
7	17.6	17.9
8	16.1	16.0
9	16.8	17.3
10	15.8	16.1
11	16.8	16.5
12	17.3	17.6
13	18.1	18.4
14	17.9	17.2
15	16.4	16.5

Analyze these data with a Wilcoxon signed rank test. Use a two-sided alternative. Let $\alpha = 0.05$.

Question 13.3.4 To measure the effect on coordination associated with mild intoxication, 13 subjects were each given 15.7 ml of ethyl alcohol per square meter of body surface area and asked to write a certain phrase as many times as they could in the space of one minute (111). The number of correctly written letters was then counted and scaled, with a scale value of 0 representing the score a subject not under the influence of alcohol would be expected to achieve. Negative scores indicate *decreased* writing speeds; positive scores, *increased* writing speeds.

Subject	Score	Subject	Score
1	−6	8	0
2	10	9	−7
3	9	10	5
4	−8	11	−9
5	−6	12	−10
6	−2	13	−2
7	20		

Use the signed rank test to determine whether the level of alcohol provided in this study has any effect on writing speed. Let $\alpha = 0.05$. Omit Subject 8 from your calculations.

Question 13.3.5 Suppose that the population being sampled is symmetric and we wish to test $H_0: \tilde{\mu} = \tilde{\mu}_0$. Both the sign test and the signed rank test would be valid. Which procedure, if either, would you expect to have greater power? Why?

13.4 THE KRUSKAL–WALLIS TEST

In the final two sections of this chapter are the nonparametric counterparts for the two analysis-of-variance models introduced in Chapters 11 and 12. Neither of these procedures, the *Kruskal–Wallis test* and the *Friedman test*, will be derived. We will simply state the procedures and illustrate them with examples.

First, we consider the so-called *k-sample problem*. Suppose that k (≥ 2) independent random samples of sizes $n_1, n_2, \ldots, n_k$ are drawn, representing k populations having the same shape but possibly different locations. Our objective is to test whether the population medians are all the same—that is,

$$H_0: \quad \tilde{\mu}_1 = \tilde{\mu}_2 = \cdots = \tilde{\mu}_k$$

versus

$$H_1: \quad \text{not all the } \tilde{\mu}_j\text{'s are equal}$$

The Kruskal–Wallis procedure for testing H_0 is really quite simple, involving considerably fewer computations than the analysis of variance. The first step is to rank the entire set of $n = \sum_{i=1}^{k} n_i$ observations (from smallest to largest). Then the rank sum, $R_{\cdot j}$, is calculated for each sample. Table 13.4.1 shows the notation we will be using: it follows the same conventions as the dot notation of Chapter 11. The only real difference is the addition of r_{ij}, the symbol for the rank corresponding to y_{ij}.

TABLE 13.4.1 NOTATION FOR KRUSKAL–WALLIS PROCEDURE

	Sample		
1	2	$\cdots$	k
$y_{11}(r_{11})$	$y_{12}(r_{12})$		$y_{1k}(r_{1k})$
$y_{21}(r_{21})$			
$\vdots$	$\vdots$	$\cdots$	$\vdots$
$y_{n_1 1}(r_{n_1 1})$	$y_{n_2 2}(r_{n_2 2})$		$y_{n_k k}(r_{n_k k})$
Totals $\quad r_{\cdot 1}$	$r_{\cdot 2}$		$r_{\cdot k}$

The Kruskal–Wallis statistic, B, is defined as

$$B = \frac{12}{n(n+1)} \sum_{j=1}^{k} \frac{R_{\cdot j}^2}{n_j} - 3(n+1)$$

Notice how B resembles the computing formula for $SS_{\text{treatments}}$ in the analysis of variance. Here $\sum_{j=1}^{k} (R_{\cdot j}^2/n_j)$, and thus B, get larger and larger as the differences between the population medians increase. [Recall that a similar explanation was given for $SS_{\text{treatments}}$ and $\sum_{j=1}^{k} (T_{\cdot j}^2/n_j)$.] Theorem 13.4.1 gives the distribution of the Kruskal–Wallis statistic and the appropriate critical region for testing H_0 versus H_1.

CASE STUDY 13.4.1

On December 1, 1969, a lottery was held in Selective Service headquarters in Washington, D.C., to determine the draft status of all 19-year-old males. It was the first time such a procedure had been used since World War II. Priorities were established according to a person's birthday. Each of the 366 possible birthdates was written on a slip of paper and put into a small capsule. The capsules were then put into a large bowl, mixed, and drawn out one by one. By agreement, persons whose birthday corresponded to the first capsule drawn would have the highest draft priority; those whose birthday corresponded to the second capsule drawn, the second highest priority, and so on. Table 13.4.2 shows the order in which the 366 birthdates were drawn (151). The first date was September 14 (001); the last, June 8 (366).

We can think of the observed sequence of draft priorities as ranks from 1 to 366. If the lottery is random, the average of these ranks for each of the months should be approximately equal. If the lottery is *not* random, we would expect to see certain months having a preponderance of high ranks and other months a preponderance of low ranks. This suggests that we can test the lottery for randomness by using the Kruskal–Wallis statistic.

Substituting the $r_{\cdot j}$'s shown at the bottom of Table 13.4.2 into the formula for B gives

$$b = \frac{12}{366(367)} \left[\frac{(6236)^2}{31} + \cdots + \frac{(3768)^2}{31} \right] - 3(367)$$

$$= 25.95$$

By Theorem 13.4.1, B has approximately a χ^2 distribution with 11 degrees of freedom (when $H_0: \tilde{\mu}_{\text{Jan}} = \tilde{\mu}_{\text{Feb}} = \cdots = \tilde{\mu}_{\text{Dec}}$ is true). Thus, if we choose to test H_0 at the $\alpha = 0.01$ level of significance, we should reject the null hypothesis if $b \geq 24.7$ (see Table A.3 in the Appendix). But b did exceed that value, so the hypothesis of randomness would be rejected—looking at the data more closely, we can see that there was a tendency for birthdays late in the year to come early in the draft.

An even more resounding rejection of the randomness hypothesis can be gotten by dividing the 12 months into two half-years—the first, January

Date	Jan.	Feb.	Mar.	Apr.	May	June	July	Aug.	Sept.	Oct.	Nov.	Dec.
1	305	086	108	032	330	249	093	111	225	359	019	129
2	159	144	029	271	298	228	350	045	161	125	034	328
3	251	297	267	083	040	301	115	261	049	244	348	157
4	215	210	275	081	276	020	279	145	232	202	266	165
5	101	214	293	269	364	028	188	054	082	024	310	056
6	224	347	139	253	155	110	327	114	006	087	076	010
7	306	091	122	147	035	085	050	168	008	234	051	012
8	199	181	213	312	321	366	013	048	184	283	097	105
9	194	338	317	219	197	335	106	263	342	080	043
10	325	216	323	218	065	206	284	021	071	220	282	041
11	329	150	136	014	037	134	248	324	158	237	046	039
12	221	068	300	346	133	272	015	142	242	072	066	314
13	318	152	259	124	295	069	042	307	175	138	126	163
14	238	004	354	231	178	356	331	198	001	294	127	026
15	017	089	169	273	130	180	322	102	113	171	131	320
16	121	212	166	148	055	274	120	044	207	254	107	096
17	235	189	033	260	112	073	098	154	255	288	143	304
18	140	292	332	090	278	341	190	141	246	005	146	128
19	058	025	200	336	075	104	227	311	177	241	203	240
20	280	302	239	345	183	360	187	344	063	192	185	135
21	186	363	334	062	250	060	027	291	204	243	156	070
22	337	290	265	316	326	247	153	339	160	117	009	053
23	118	057	256	252	319	109	172	116	119	201	182	162
24	059	236	258	002	031	358	023	036	195	196	230	095
25	052	179	343	351	361	137	067	286	149	176	132	084
26	092	365	170	340	357	022	303	245	018	007	309	173
27	355	205	268	074	296	064	289	352	233	264	047	078
28	077	299	223	262	308	222	088	167	257	094	281	123
29	349	285	362	191	226	353	270	061	151	229	099	016
30	164		217	208	103	209	287	333	315	038	174	003
31	211		030		313		193	011		079		100
Totals:	6236	5886	7000	6110	6447	5872	5628	5377	4719	5656	4462	3768

through June; the second, July through December. Then the hypotheses to be tested are

$$H_0: \quad \tilde{\mu}_1 = \tilde{\mu}_2$$

versus

$$H_1: \quad \tilde{\mu}_1 \neq \tilde{\mu}_2$$

Table 13.4.3, derived from Table 13.4.2, summarizes the data appropriately. Substituting these values into the formula for the Kruskal–Wallis statistic gives a b value (with 1 degree of freedom) of 16.85:

$$b = \frac{12}{366(367)} \left[\frac{(37,551)^2}{182} + \frac{(29,610)^2}{184} \right] - 3(367)$$

$$= 16.85$$

TABLE 13.4.3 SUMMARY OF 1969 DRAFT LOTTERY BY SIX-MONTH PERIODS

	Jan–June (1)	July–Dec. (2)
$r_{.j}$	37,551	29,610
n_j	182	184

The significance of 16.85 can be gauged by recalling the moments of a chi square variable. If $B \sim \chi_1^2$, then $E(B) = 1$ and Var $(B) = 2$ (see Question 7.5.2). Therefore, the observed b value is more than 11 standard deviations away from its mean:

$$\frac{16.85 - 1}{\sqrt{2}} = 11.2$$

Analyzed this way, there can be little doubt that the lottery was not random!

Question 13.4.1 A civil rights group plans to file suit against a certain city because of its alleged non-compliance with a school busing order. The black-white ratio in this particular community is 20 : 80, and all the public schools are expected to reflect, as closely as possible, that ratio. The table below lists the actual percentages of black students in two randomly selected sets of schools—the first set is comprised of schools located in predominantly black neighborhoods; the second set, schools located in white neighborhoods.

PERCENTAGE OF BLACK STUDENTS

Schools in Black Neighborhoods	Schools in White Neighborhoods
36	21
28	14
41	11
32	30
46	29
39	6
24	18
32	25
45	23

Rank these percentages and use an appropriate procedure to test the civil rights group's claim. Let $\alpha = 0.05$.

Question 13.4.2 Recall Case Study 13.4.1. What might have accounted for the lack of randomness evident in the 1969 draft lottery?

Question 13.4.3 Analyze the serum binding data of Case Study 11.2.1 using the Kruskal–Wallis test. Let $\alpha = 0.05$.

Question 13.4.4 The production of a certain organic chemical requires the addition of ammonium chloride. The manufacturer can conveniently obtain the ammonium chloride in any one of three forms—powdered, moderately ground, and coarse. To see what effect, if any, the quality of the NH_4Cl has, they decide to run the reaction seven times with each form of ammonium chloride. The resulting yields (in pounds) are listed below.

ORGANIC CHEMICAL YIELDS (LB)

Powdered NH_4Cl	Moderately Ground NH_4Cl	Coarse NH_4Cl
146	150	141
152	144	138
149	148	142
161	155	146
158	154	139
149	150	145
154	148	137

Compare the yields with a Kruskal–Wallis test. Let $\alpha = 0.05$.

Question 13.4.5 Show that the Kruskal–Wallis statistic, B, as defined in Theorem 13.4.1 can also be written

$$B = \sum_{j=1}^{k} \left(\frac{n - n_j}{n} \right) Z_j^2$$

where

$$Z_j = \frac{\dfrac{R_{.j}}{n_j} - \dfrac{n + 1}{2}}{\sqrt{\dfrac{(n + 1)(n - n_j)}{12n_j}}}$$

[It can also be shown that (1) $E(R_{.j}/n_j) = (n + 1)/2$, (2) $\text{Var}(R_{.j}/n_j) = [(n + 1) \cdot (n - n_j)]/12n_j$, and (3) Z_j is approximately an $N(0, 1)$ variable if n_j is sufficiently large. These results, together with the expression showing B to be a function of the sum of squared Z_j's, and the fact that the Z_j's are subject to a constraint

$$\left(\sum_{j=1}^{k} R_{.j} = \frac{n(n + 1)}{2} \right)$$

—meaning the sum would have $k - 1$, rather than k, degrees of freedom—would be the basis for a proof of Theorem 13.4.1.]

13.5 THE FRIEDMAN TEST

The nonparametric analog of the analysis of variance for a randomized block design is *Friedman's test*, a procedure based on within-block *ranks*. Its form is similar to that of the Kruskal–Wallis statistic, and, like its predecessor, it has approximately a χ^2 distribution when H_0 is true. Theorem 13.5.1 gives a formal statement.

> **Theorem 13.5.1.** Suppose that k (≥ 2) treatments are ranked independently within b blocks. Let $r_{\cdot j}$, $j = 1, 2, \ldots, k$, be the rank sum of the jth treatment. The null hypothesis that the population medians of the k treatments are all equal is rejected at the α level of significance (approximately) if
>
> $$g = \frac{12}{bk(k+1)} \sum_{j=1}^{k} r_{\cdot j}^2 - 3b(k+1) \geq \chi^2_{1-\alpha, k-1}$$

CASE STUDY 13.5.1

Baseball rules allow a batter considerable leeway in how he is permitted to run from home plate to second base. Two of the possibilities are the narrow-angle and the wide-angle paths diagrammed in Figure 13.5.1. As a means of comparing the two, time trials were held involving 22 players (184). Each player ran both paths. Recorded for each runner was the time it took to go from a point 35 feet from home plate to a point 15 feet from second base.

Narrow-angle

Wide-angle

Figure 13.5.1 Batter's path from home plate to second base.

These are listed in the second and fourth columns of Table 13.5.1. In the third and fifth columns are the within-block ranks for each path. A rank of 1 was assigned to the better path, the one enabling the runner to round first base faster.

If $\tilde{\mu}_1$ and $\tilde{\mu}_2$ denote the true median rounding times associated with the narrow-angle and wide-angle paths, respectively, the hypotheses to be tested are

$$H_0: \quad \tilde{\mu}_1 = \tilde{\mu}_2$$

versus

$$H_1: \quad \tilde{\mu}_1 \neq \tilde{\mu}_2$$

TABLE 13.5.1 TIMES (SEC) REQUIRED TO ROUND FIRST BASE

Player	Narrow-Angle	Rank	Wide-Angle	Rank
1	5.50	1	5.55	2
2	5.70	1	5.75	2
3	5.60	2	5.50	1
4	5.50	2	5.40	1
5	5.85	2	5.70	1
6	5.55	1	5.60	2
7	5.40	2	5.35	1
8	5.50	2	5.35	1
9	5.15	2	5.00	1
10	5.80	2	5.70	1
11	5.20	2	5.10	1
12	5.55	2	5.45	1
13	5.35	1	5.45	2
14	5.00	2	4.95	1
15	5.50	2	5.40	1
16	5.55	2	5.50	1
17	5.55	2	5.35	1
18	5.50	1	5.55	2
19	5.45	2	5.25	1
20	5.60	2	5.40	1
21	5.65	2	5.55	1
22	6.30	2	6.25	1
		39		27

Let $\alpha = 0.05$. By Theorem 13.5.1, the Friedman statistic (under H_0) will have approximately a χ_1^2 distribution, and the decision rule will be

Reject H_0 if $g \geq 3.84$

But

$$g = \frac{12}{22(2)(3)} [(39)^2 + (27)^2] - 3(22)(3)$$

$$= 6.54$$

implying that the two paths are *not* equivalent, the wide-angle path being significantly better.

Question 13.5.1 The following data come from a field trial set up to assess the effects of different amounts of potash on the breaking strength of cotton fibers (25). The experiment was done in three blocks. The five treatment levels—36, 54, 72, 108, and 144 lbs. of potash per acre—were assigned randomly within each block. The variable recorded was the Pressley strength index.

		\multicolumn{5}{c}{Treatment (pounds of potash/acre)}				
		36	54	72	108	144
	1	7.62	8.14	7.76	7.17	7.46
Blocks	2	8.00	8.15	7.73	7.57	7.68
	3	7.93	7.87	7.74	7.80	7.21

Compare the effects of the different levels of potash applications using Friedman's test. Let $\alpha = 0.05$.

Question 13.5.2 Use Friedman's test to analyze the Transylvania-effect data given in Case Study 12.2.3.

Question 13.5.3 Until its recent indictment as a possible carcinogen, cyclamate was a widely used sweetener in soft drinks. The data below show a comparison of three laboratory methods for determining the percentage of sodium cyclamate in commercially produced orange drink. All three procedures were applied to each of 12 samples (145).

PERCENT SODIUM CYCLAMATE (W/W)

Sample	\multicolumn{3}{c}{Method}		
	Picryl Chloride	Davies	AOAC
1	0.598	0.628	0.632
2	0.614	0.628	0.630
3	0.600	0.600	0.622
4	0.580	0.612	0.584
5	0.596	0.600	0.650
6	0.592	0.628	0.606
7	0.616	0.628	0.644
8	0.614	0.644	0.644
9	0.604	0.644	0.624
10	0.608	0.612	0.619
11	0.602	0.628	0.632
12	0.614	0.644	0.616

Use Friedman's test to determine whether the three methods give significantly different results. Let $\alpha = 0.05$.

Question 13.5.4 Suppose that k treatments are to be applied within each of b blocks. Let $\bar{R}_{..}$ denote the average of the bk ranks and let $\bar{R}_{.j} = (1/b)R_{.j}$. Show that the Friedman statistic given in Theorem 13.5.1 can also be written

$$G = \frac{12b}{k(k+1)} \sum_{j=1}^{k} (\bar{R}_{.j} - \bar{R}_{..})^2$$

What analysis-of-variance expression does this resemble?

Sec. 13.5   The Friedman Test

Question 13.5.5 Use Friedman's test to analyze the triticale data given in Question 12.3.1. Let .05 be the level of significance.

Appendix

TABLE A.1 CUMULATIVE AREAS UNDER THE STANDARD NORMAL DISTRIBUTION

z	0	1	2	3	4	5	6	7	8	9
−3.	0.0013	0.0010	0.0007	0.0005	0.0003	0.0002	0.0002	0.0001	0.0001	0.0000
−2.9	0.0019	0.0018	0.0017	0.0017	0.0016	0.0016	0.0015	0.0015	0.0014	0.0014
−2.8	0.0026	0.0025	0.0024	0.0023	0.0023	0.0022	0.0021	0.0021	0.0020	0.0019
−2.7	0.0035	0.0034	0.0033	0.0032	0.0031	0.0030	0.0029	0.0028	0.0027	0.0026
−2.6	0.0047	0.0045	0.0044	0.0043	0.0041	0.0040	0.0039	0.0038	0.0037	0.0036
−2.5	0.0062	0.0060	0.0059	0.0057	0.0055	0.0054	0.0052	0.0051	0.0049	0.0048
−2.4	0.0082	0.0080	0.0078	0.0075	0.0073	0.0071	0.0069	0.0068	0.0066	0.0064
−2.3	0.0107	0.0104	0.0102	0.0099	0.0096	0.0094	0.0091	0.0089	0.0087	0.0084
−2.2	0.0139	0.0136	0.0132	0.0129	0.0126	0.0122	0.0119	0.0116	0.0113	0.0110
−2.1	0.0179	0.0174	0.0170	0.0166	0.0162	0.0158	0.0154	0.0150	0.0146	0.0143
−2.0	0.0228	0.0222	0.0217	0.0212	0.0207	0.0202	0.0197	0.0192	0.0188	0.0183
−1.9	0.0287	0.0281	0.0274	0.0268	0.0262	0.0256	0.0250	0.0244	0.0238	0.0233
−1.8	0.0359	0.0352	0.0344	0.0336	0.0329	0.0322	0.0314	0.0307	0.0300	0.0294
−1.7	0.0446	0.0436	0.0427	0.0418	0.0409	0.0401	0.0392	0.0384	0.0375	0.0367
−1.6	0.0548	0.0537	0.0526	0.0516	0.0505	0.0495	0.0485	0.0475	0.0465	0.0455
−1.5	0.0668	0.0655	0.0643	0.0630	0.0618	0.0606	0.0594	0.0582	0.0570	0.0559
−1.4	0.0808	0.0793	0.0778	0.0764	0.0749	0.0735	0.0722	0.0708	0.0694	0.0681
−1.3	0.0968	0.0951	0.0934	0.0918	0.0901	0.0885	0.0869	0.0853	0.0838	0.0823
−1.2	0.1151	0.1131	0.1112	0.1093	0.1075	0.1056	0.1038	0.1020	0.1003	0.0985
−1.1	0.1357	0.1335	0.1314	0.1292	0.1271	0.1251	0.1230	0.1210	0.1190	0.1170
−1.0	0.1587	0.1562	0.1539	0.1515	0.1492	0.1469	0.1446	0.1423	0.1401	0.1379
−0.9	0.1841	0.1814	0.1788	0.1762	0.1736	0.1711	0.1685	0.1660	0.1635	0.1611
−0.8	0.2119	0.2090	0.2061	0.2033	0.2005	0.1977	0.1949	0.1922	0.1894	0.1867
−0.7	0.2420	0.2389	0.2358	0.2327	0.2297	0.2266	0.2236	0.2206	0.2177	0.2148
−0.6	0.2743	0.2709	0.2676	0.2643	0.2611	0.2578	0.2546	0.2514	0.2483	0.2451
−0.5	0.3085	0.3050	0.3015	0.2981	0.2946	0.2912	0.2877	0.2843	0.2810	0.2776
−0.4	0.3446	0.3409	0.3372	0.3336	0.3300	0.3264	0.3228	0.3192	0.3156	0.3121
−0.3	0.3821	0.3783	0.3745	0.3707	0.3669	0.3632	0.3594	0.3557	0.3520	0.3483
−0.2	0.4207	0.4168	0.4129	0.4090	0.4052	0.4013	0.3974	0.3936	0.3897	0.3859
−0.1	0.4602	0.4562	0.4522	0.4483	0.4443	0.4404	0.4364	0.4325	0.4286	0.4247
−0.0	0.5000	0.4960	0.4920	0.4880	0.4840	0.4801	0.4761	0.4721	0.4681	0.4641

TABLE A.1 CUMULATIVE AREAS UNDER THE STANDARD NORMAL DISTRIBUTION (cont.)

z	0	1	2	3	4	5	6	7	8	9
0.0	0.5000	0.5040	0.5080	0.5120	0.5160	0.5199	0.5239	0.5279	0.5319	0.5359
0.1	0.5398	0.5438	0.5478	0.5517	0.5557	0.5596	0.5636	0.5675	0.5714	0.5753
0.2	0.5793	0.5832	0.5871	0.5910	0.5948	0.5987	0.6026	0.6064	0.6103	0.6141
0.3	0.6179	0.6217	0.6255	0.6293	0.6331	0.6368	0.6406	0.6443	0.6480	0.6517
0.4	0.6554	0.6591	0.6628	0.6664	0.6700	0.6736	0.6772	0.6808	0.6844	0.6879
0.5	0.6915	0.6950	0.6985	0.7019	0.7054	0.7088	0.7123	0.7157	0.7190	0.7224
0.6	0.7257	0.7291	0.7324	0.7357	0.7389	0.7422	0.7454	0.7486	0.7517	0.7549
0.7	0.7580	0.7611	0.7642	0.7673	0.7703	0.7734	0.7764	0.7794	0.7823	0.7852
0.8	0.7881	0.7910	0.7939	0.7967	0.7995	0.8023	0.8051	0.8078	0.8106	0.8133
0.9	0.8159	0.8186	0.8212	0.8238	0.8264	0.8289	0.8315	0.8340	0.8365	0.8389
1.0	0.8413	0.8438	0.8461	0.8485	0.8508	0.8531	0.8554	0.8577	0.8599	0.8621
1.1	0.8643	0.8665	0.8686	0.8708	0.8729	0.8749	0.8770	0.8790	0.8810	0.8830
1.2	0.8849	0.8869	0.8888	0.8907	0.8925	0.8944	0.8962	0.8980	0.8997	0.9015
1.3	0.9032	0.9049	0.9066	0.9082	0.9099	0.9115	0.9131	0.9147	0.9162	0.9177
1.4	0.9192	0.9207	0.9222	0.9236	0.9251	0.9265	0.9278	0.9292	0.9306	0.9319
1.5	0.9332	0.9345	0.9357	0.9370	0.9382	0.9394	0.9406	0.9418	0.9430	0.9441
1.6	0.9452	0.9463	0.9474	0.9484	0.9495	0.9505	0.9515	0.9525	0.9535	0.9545
1.7	0.9554	0.9564	0.9573	0.9582	0.9591	0.9599	0.9608	0.9616	0.9625	0.9633
1.8	0.9641	0.9648	0.9656	0.9664	0.9671	0.9678	0.9686	0.9693	0.9700	0.9706
1.9	0.9713	0.9719	0.9726	0.9732	0.9738	0.9744	0.9750	0.9756	0.9762	0.9767
2.0	0.9772	0.9778	0.9783	0.9788	0.9793	0.9798	0.9803	0.9808	0.9812	0.9817
2.1	0.9821	0.9826	0.9830	0.9834	0.9838	0.9842	0.9846	0.9850	0.9854	0.9857
2.2	0.9861	0.9864	0.9868	0.9871	0.9874	0.9878	0.9881	0.9884	0.9887	0.9890
2.3	0.9893	0.9896	0.9898	0.9901	0.9904	0.9906	0.9909	0.9911	0.9913	0.9916
2.4	0.9918	0.9920	0.9922	0.9925	0.9927	0.9929	0.9931	0.9932	0.9934	0.9936
2.5	0.9938	0.9940	0.9941	0.9943	0.9945	0.9946	0.9948	0.9949	0.9951	0.9952
2.6	0.9953	0.9955	0.9956	0.9957	0.9959	0.9960	0.9961	0.9962	0.9963	0.9964
2.7	0.9965	0.9966	0.9967	0.9968	0.9969	0.9970	0.9971	0.9972	0.9973	0.9974
2.8	0.9974	0.9975	0.9976	0.9977	0.9977	0.9978	0.9979	0.9979	0.9980	0.9981
2.9	0.9981	0.9982	0.9982	0.9983	0.9984	0.9984	0.9985	0.9985	0.9986	0.9986
3.	0.9987	0.9990	0.9993	0.9995	0.9997	0.9998	0.9998	0.9999	0.9999	1.0000

SOURCE: B. W. Lindgren, *Statistical Theory* (New York: Macmillan, 1962), pp. 392–393.

TABLE A.2 UPPER PERCENTILES OF STUDENT t DISTRIBUTIONS

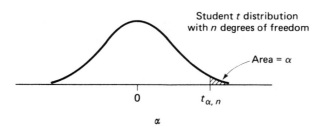

Student t distribution
with n degrees of freedom

Area = α

0       $t_{\alpha, n}$

α

df	0.20	0.15	0.10	0.05	0.025	0.01	0.005
1	1.376	1.963	3.078	6.3138	12.706	31.821	63.657
2	1.061	1.386	1.886	2.9200	4.3027	6.965	9.9248
3	0.978	1.250	1.638	2.3534	3.1825	4.541	5.8409
4	0.941	1.190	1.533	2.1318	2.7764	3.747	4.6041
5	0.920	1.156	1.476	2.0150	2.5706	3.365	4.0321
6	0.906	1.134	1.440	1.9432	2.4469	3.143	3.7074
7	0.896	1.119	1.415	1.8946	2.3646	2.998	3.4995
8	0.889	1.108	1.397	1.8595	2.3060	2.896	3.3554
9	0.883	1.100	1.383	1.8331	2.2622	2.821	3.2498
10	0.879	1.093	1.372	1.8125	2.2281	2.764	3.1693
11	0.876	1.088	1.363	1.7959	2.2010	2.718	3.1058
12	0.873	1.083	1.356	1.7823	2.1788	2.681	3.0545
13	0.870	1.079	1.350	1.7709	2.1604	2.650	3.0123
14	0.868	1.076	1.345	1.7613	2.1448	2.624	2.9768
15	0.866	1.074	1.341	1.7530	2.1315	2.602	2.9467
16	0.865	1.071	1.337	1.7459	2.1199	2.583	2.9208
17	0.863	1.069	1.333	1.7396	2.1098	2.567	2.8982
18	0.862	1.067	1.330	1.7341	2.1009	2.552	2.8784
19	0.861	1.066	1.328	1.7291	2.0930	2.539	2.8609
20	0.860	1.064	1.325	1.7247	2.0860	2.528	2.8453
21	0.859	1.063	1.323	1.7207	2.0796	2.518	2.8314
22	0.858	1.061	1.321	1.7171	2.0739	2.508	2.8188
23	0.858	1.060	1.319	1.7139	2.0687	2.500	2.8073
24	0.857	1.059	1.318	1.7109	2.0639	2.492	2.7969
25	0.856	1.058	1.316	1.7081	2.0595	2.485	2.7874
26	0.856	1.058	1.315	1.7056	2.0555	2.479	2.7787
27	0.855	1.057	1.314	1.7033	2.0518	2.473	2.7707
28	0.855	1.056	1.313	1.7011	2.0484	2.467	2.7633
29	0.854	1.055	1.311	1.6991	2.0452	2.462	2.7564
30	0.854	1.055	1.310	1.6973	2.0423	2.457	2.7500
31	0.8535	1.0541	1.3095	1.6955	2.0395	2.453	2.7441
32	0.8531	1.0536	1.3086	1.6939	2.0370	2.449	2.7385
33	0.8527	1.0531	1.3078	1.6924	2.0345	2.445	2.7333
34	0.8524	1.0526	1.3070	1.6909	2.0323	2.441	2.7284

df	0.20	0.15	0.10	0.05	0.025	0.01	0.005
35	0.8521	1.0521	1.3062	1.6896	2.0301	2.438	2.7239
36	0.8518	1.0516	1.3055	1.6883	2.0281	2.434	2.7195
37	0.8515	1.0512	1.3049	1.6871	2.0262	2.431	2.7155
38	0.8512	1.0508	1.3042	1.6860	2.0244	2.428	2.7116
39	0.8510	1.0504	1.3037	1.6849	2.0227	2.426	2.7079
40	0.8507	1.0501	1.3031	1.6839	2.0211	2.423	2.7045
41	0.8505	1.0498	1.3026	1.6829	2.0196	2.421	2.7012
42	0.8503	1.0494	1.3020	1.6820	2.0181	2.418	2.6981
43	0.8501	1.0491	1.3016	1.6811	2.0167	2.416	2.6952
44	0.8499	1.0488	1.3011	1.6802	2.0154	2.414	2.6923
45	0.8497	1.0485	1.3007	1.6794	2.0141	2.412	2.6896
46	0.8495	1.0483	1.3002	1.6787	2.0129	2.410	2.6870
47	0.8494	1.0480	1.2998	1.6779	2.0118	2.408	2.6846
48	0.8492	1.0478	1.2994	1.6772	2.0106	2.406	2.6822
49	0.8490	1.0476	1.2991	1.6766	2.0096	2.405	2.6800
50	0.8489	1.0473	1.2987	1.6759	2.0086	2.403	2.6778
51	0.8448	1.0471	1.2984	1.6753	2.0077	2.402	2.6758
52	0.8486	1.0469	1.2981	1.6747	2.0067	2.400	2.6738
53	0.8485	1.0467	1.2978	1.6742	2.0058	2.399	2.6719
54	0.8484	1.0465	1.2975	1.6736	2.0049	2.397	2.6700
55	0.8483	1.0463	1.2972	1.6731	2.0041	2.396	2.6683
56	0.8481	1.0461	1.2969	1.6725	2.0033	2.395	2.6666
57	0.8480	1.0460	1.2967	1.6721	2.0025	2.393	2.6650
58	0.8479	1.0458	1.2964	1.6716	2.0017	2.392	2.6633
59	0.8478	1.0457	1.2962	1.6712	2.0010	2.391	2.6618
60	0.8477	1.0455	1.2959	1.6707	2.0003	2.390	2.6603
61	0.8476	1.0454	1.2957	1.6703	1.9997	2.389	2.6590
62	0.8475	1.0452	1.2954	1.6698	1.9990	2.388	2.6576
63	0.8474	1.0451	1.2952	1.6694	1.9984	2.387	2.6563
64	0.8473	1.0449	1.2950	1.6690	1.9977	2.386	2.6549
65	0.8472	1.0448	1.2948	1.6687	1.9972	2.385	2.6537
66	0.8471	1.0447	1.2945	1.6683	1.9966	2.384	2.6525
67	0.8471	1.0446	1.2944	1.6680	1.9961	2.383	2.6513
68	0.8470	1.0444	1.2942	1.6676	1.9955	2.382	2.6501
69	0.8469	1.0443	1.2940	1.6673	1.9950	2.381	2.6491
70	0.8468	1.0442	1.2938	1.6669	1.9945	2.381	2.6480
71	0.8468	1.0441	1.2936	1.6666	1.9940	2.380	2.6470
72	0.8467	1.0440	1.2934	1.6663	1.9935	2.379	2.6459
73	0.8466	1.0439	1.2933	1.6660	1.9931	2.378	2.6450
74	0.8465	1.0438	1.2931	1.6657	1.9926	2.378	2.6640
75	0.8465	1.0437	1.2930	1.6655	1.9922	2.377	2.6431
76	0.8464	1.0436	1.2928	1.6652	1.9917	2.376	2.6421
77	0.8464	1.0435	1.2927	1.6649	1.9913	2.376	2.6413
78	0.8463	1.0434	1.2925	1.6646	1.9909	2.375	2.6406
79	0.8463	1.0433	1.2924	1.6644	1.9905	2.374	2.6396

TABLE A.2 UPPER PERCENTILES OF STUDENT t DISTRIBUTIONS (cont.)

				α			
df	0.20	0.15	0.10	0.05	0.025	0.01	0.005
80	0.8462	1.0432	1.2922	1.6641	1.9901	2.374	2.6388
81	0.8461	1.0431	1.2921	1.6639	1.9897	2.373	2.6380
82	0.8460	1.0430	1.2920	1.6637	1.9893	2.372	2.6372
83	0.8460	1.0430	1.2919	1.6635	1.9890	2.372	2.6365
84	0.8459	1.0429	1.2917	1.6632	1.9886	2.371	2.6357
85	0.8459	1.0428	1.2916	1.6630	1.9883	2.371	2.6350
86	0.8458	1.0427	1.2915	1.6628	1.9880	2.370	2.6343
87	0.8458	1.0427	1.2914	1.6626	1.9877	2.370	2.6336
88	0.8457	1.0426	1.2913	1.6624	1.9873	2.369	2.6329
89	0.8457	1.0426	1.2912	1.6622	1.9870	2.369	2.6323
90	0.8457	1.0425	1.2910	1.6620	1.9867	2.368	2.6316
91	0.8457	1.0424	1.2909	1.6618	1.9864	2.368	2.6310
92	0.8456	1.0423	1.2908	1.6616	1.9861	2.367	2.6303
93	0.8456	1.0423	1.2907	1.6614	1.9859	2.367	2.6298
94	0.8455	1.0422	1.2906	1.6612	1.9856	2.366	2.6292
95	0.8455	1.0422	1.2905	1.6611	1.9853	2.366	2.6286
96	0.8454	1.0421	1.2904	1.6609	1.9850	2.366	2.6280
97	0.8454	1.0421	1.2904	1.6608	1.9848	2.365	2.6275
98	0.8453	1.0420	1.2903	1.6606	1.9845	2.365	2.6270
99	0.8453	1.0419	1.2902	1.6604	1.9843	2.364	2.6265
100	0.8452	1.0418	1.2901	1.6602	1.9840	2.364	2.6260
∞	0.84	1.04	1.28	1.64	1.96	2.33	2.58

SOURCE: *Scientific Tables*, 6th ed. (Basel, Switzerland: J. R. Geigy, 1962), pp. 32–33.

TABLE A.3 UPPER AND LOWER PERCENTILES OF X^2 DISTRIBUTIONS

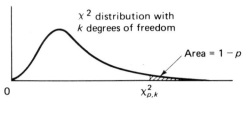

				p				
df	0.010	0.025	0.050	0.10	0.90	0.95	0.975	0.99
1	0.000157	0.000982	0.00393	0.0158	2.706	3.841	5.024	6.635
2	0.0201	0.0506	0.103	0.211	4.605	5.991	7.378	9.210
3	0.115	0.216	0.352	0.584	6.251	7.815	9.348	11.345
4	0.297	0.484	0.711	1.064	7.779	9.488	11.143	13.277
5	0.554	0.831	1.145	1.610	9.236	11.070	12.832	15.086
6	0.872	1.237	1.635	2.204	10.645	12.592	14.449	16.812
7	1.239	1.690	2.167	2.833	12.017	14.067	16.013	18.475
8	1.646	2.180	2.733	3.490	13.362	15.507	17.535	20.090
9	2.088	2.700	3.325	4.168	14.684	16.919	19.023	21.666
10	2.558	3.247	3.940	4.865	15.987	18.307	20.483	23.209
11	3.053	3.816	4.575	5.578	17.275	19.675	21.920	24.725
12	3.571	4.404	5.226	6.304	18.549	21.026	23.336	26.217
13	4.107	5.009	5.892	7.042	19.812	22.362	24.736	27.688
14	4.660	5.629	6.571	7.790	21.064	23.685	26.119	29.141
15	5.229	6.262	7.261	8.547	22.307	24.996	27.488	30.578
16	5.812	6.908	7.962	9.312	23.542	26.296	28.845	32.000
17	6.408	7.564	8.672	10.085	24.769	27.587	30.191	33.409
18	7.015	8.231	9.390	10.865	25.989	28.869	31.526	34.805
19	7.633	8.907	10.117	11.651	27.204	30.144	32.852	36.191
20	8.260	9.591	10.851	12.443	28.412	31.410	34.170	37.566
21	8.897	10.283	11.591	13.240	29.615	32.671	35.479	38.932
22	9.542	10.982	12.338	14.041	30.813	33.924	36.781	40.289
23	10.196	11.688	13.091	14.848	32.007	35.172	38.076	41.638
24	10.856	12.401	13.848	15.659	33.196	36.415	39.364	42.980
25	11.524	13.120	14.611	16.473	34.382	37.652	40.646	44.314
26	12.198	13.844	15.379	17.292	35.563	38.885	41.923	45.642
27	12.879	14.573	16.151	18.114	36.741	40.113	43.194	46.963
28	13.565	15.308	16.928	18.939	37.916	41.337	44.461	48.278
29	14.256	16.047	17.708	19.768	39.087	42.557	45.722	49.588
30	14.953	16.791	18.493	20.599	40.256	43.773	46.979	50.892
31	15.655	17.539	19.281	21.434	41.422	44.985	48.232	52.191
32	16.362	18.291	20.072	22.271	42.585	46.194	49.480	53.486
33	17.073	19.047	20.867	23.110	43.745	47.400	50.725	54.776
34	17.789	19.806	21.664	23.952	44.903	48.602	51.966	56.061

TABLE A.3 UPPER AND LOWER PERCENTILES OF X^2 DISTRIBUTIONS (cont.)

				p				
df	0.010	0.025	0.050	0.10	0.90	0.95	0.975	0.99
35	18.509	20.569	22.465	24.797	46.059	49.802	53.203	57.342
36	19.233	21.336	23.269	25.643	47.212	50.998	54.437	58.619
37	19.960	22.106	24.075	26.492	48.363	52.192	55.668	59.892
38	20.691	22.878	24.884	27.343	49.513	53.384	56.895	61.162
39	21.426	23.654	25.695	28.196	50.660	54.572	58.120	62.428
40	22.164	24.433	26.509	29.051	51.805	55.758	59.342	63.691
41	22.906	25.215	27.326	29.907	52.949	56.942	60.561	64.950
42	23.650	25.999	28.144	30.765	54.090	58.124	61.777	66.206
43	24.398	26.785	28.965	31.625	55.230	59.304	62.990	67.459
44	25.148	27.575	29.787	32.487	56.369	60.481	64.201	68.709
45	25.901	28.366	30.612	33.350	57.505	61.656	65.410	69.957
46	26.657	29.160	31.439	34.215	58.641	62.830	66.617	71.201
47	27.416	29.956	32.268	35.081	59.774	64.001	67.821	72.443
48	28.177	30.755	33.098	35.949	60.907	65.171	69.023	73.683
49	28.941	31.555	33.930	36.818	62.038	66.339	70.222	74.919
50	29.707	32.357	34.764	37.689	63.167	67.505	71.420	76.154

SOURCE: *Scientific Tables*, 6th ed. (Basel, Switzerland: J. R. Geigy, 1962), p. 36.

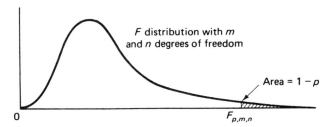

The figure above illustrates the percentiles of F distribution shown in Table A.4. Table A.4 is used with permission from Wilfrid J. Dixon and Frank J. Massey, Jr., *Introduction to Statistical Analysis*, 2nd. ed. (New York: McGraw-Hill, 1957), pp. 389–404.

PERCENTILES OF F DISTRIBUTIONS

n	p	1	2	3	4	5	6	7	8	9	10	11	12	p
1	.0005	$.0^662$	$.0^550$	$.0^338$	$.0^294$	.016	.022	.027	.032	.036	.039	.042	.045	.0005
	.001	$.0^525$	$.0^210$	$.0^260$	.013	.021	.028	.034	.039	.044	.048	.051	.054	.001
	.005	$.0^462$	$.0^251$	.018	.032	.044	.054	.062	.068	.073	.078	.082	.085	.005
	.010	$.0^325$	.010	.029	.047	.062	.073	.082	.089	.095	.100	.104	.107	.010
	.025	$.0^215$	.026	.057	.082	.100	.113	.124	.132	.139	.144	.149	.153	.025
	.05	$.0^262$	.054	.099	.130	.151	.167	.179	.188	.195	.201	.207	.211	.05
	.10	.025	.117	.181	.220	.246	.265	.279	.289	.298	.304	.310	.315	.10
	.25	.172	.389	.494	.553	.591	.617	.637	.650	.661	.670	.680	.684	.25
	.50	1.00	1.50	1.71	1.82	1.89	1.94	1.98	2.00	2.03	2.04	2.05	2.07	.50
	.75	5.83	7.50	8.20	8.58	8.82	8.98	9.10	9.19	9.26	9.32	9.36	9.41	.75
	.90	39.9	49.5	53.6	55.8	57.2	58.2	58.9	59.4	59.9	60.2	60.5	60.7	.90
	.95	161	200	216	225	230	234	237	239	241	242	243	244	.95
	.975	648	800	864	900	922	937	948	957	963	969	973	977	.975
	.99	405^1	500^1	540^1	562^1	576^1	586^1	593^1	598^1	602^1	606^1	608^1	611^1	.99
	.995	162^2	200^2	216^2	225^2	231^2	234^2	237^2	239^2	241^2	242^2	243^2	244^2	.995
	.999	406^3	500^3	540^3	562^3	576^3	586^3	593^3	598^3	602^3	606^3	609^3	611^3	.999
	.9995	162^4	200^4	216^4	225^4	231^4	234^4	237^4	239^4	241^4	242^4	243^4	244^4	.9995
2	.0005	$.0^650$	$.0^350$	$.0^242$	.011	.020	.029	.037	.044	.050	.056	.061	.065	.0005
	.001	$.0^520$	$.0^210$	$.0^268$	.016	.027	.037	.046	.054	.061	.067	.072	.077	.001
	.005	$.0^450$	$.0^250$	.020	.038	.055	.069	.081	.091	.099	.106	.112	.118	.005
	.01	$.0^320$	.010	.032	.056	.075	.092	.105	.116	.125	.132	.139	.144	.01
	.025	$.0^213$	.026	.062	.094	.119	.138	.153	.165	.175	.183	.190	.196	.025
	.05	$.0^250$	.053	.105	.144	.173	.194	.211	.224	.235	.244	.251	.257	.05
	.10	.020	.111	.183	.231	.265	.289	.307	.321	.333	.342	.350	.356	.10
	.25	.133	.333	.439	.500	.540	.568	.588	.604	.616	.626	.633	.641	.25
	.50	.667	1.00	1.13	1.21	1.25	1.28	1.30	1.32	1.33	1.34	1.35	1.36	.50
	.75	2.57	3.00	3.15	3.23	3.28	3.31	3.34	3.35	3.37	3.38	3.39	3.39	.75
	.90	8.53	9.00	9.16	9.24	9.29	9.33	9.35	9.37	9.38	9.39	9.40	9.41	.90
	.95	18.5	19.0	19.2	19.2	19.3	19.3	19.4	19.4	19.4	19.4	19.4	19.4	.95
	.975	38.5	39.0	39.2	39.2	39.3	39.3	39.4	39.4	39.4	39.4	39.4	39.4	.975
	.99	98.5	99.0	99.2	99.2	99.3	99.3	99.4	99.4	99.4	99.4	99.4	99.4	.99
	.995	198	199	199	199	199	199	199	199	199	199	199	199	.995
	.999	998	999	999	999	999	999	999	999	999	999	999	999	.999
	.9995	200^1	200^1	200^1	200^1	200^1	200^1	200^1	200^1	200^1	200^1	200^1	200^1	.9995
3	.0005	$.0^646$	$.0^350$	$.0^244$	.012	.023	.033	.043	.052	.060	.067	.074	.079	.0005
	.001	$.0^519$	$.0^210$	$.0^271$	.018	.030	.042	.053	.063	.072	.079	.086	.093	.001
	.005	$.0^446$	$.0^250$	.021	.041	.060	.077	.092	.104	.115	.124	.132	.138	.005
	.01	$.0^319$	.010	.034	.060	.083	.102	.118	.132	.143	.153	.161	.168	.01
	.025	$.0^212$	.026	.065	.100	.129	.152	.170	.185	.197	.207	.216	.224	.025
	.05	$.0^246$	.052	.108	.152	.185	.210	.230	.246	.259	.270	.279	.287	.05
	.10	.019	.109	.185	.239	.276	.304	.325	.342	.356	.367	.376	.384	.10
	.25	.122	.317	.424	.489	.531	.561	.582	.600	.613	.624	.633	.641	.25
	.50	.585	.881	1.00	1.06	1.10	1.13	1.15	1.16	1.17	1.18	1.19	1.20	.50
	.75	2.02	2.28	2.36	2.39	2.41	2.42	2.43	2.44	2.44	2.44	2.45	2.45	.75
	.90	5.54	5.46	5.39	5.34	5.31	5.28	5.27	5.25	5.24	5.23	5.22	5.22	.90
	.95	10.1	9.55	9.28	9.12	9.01	8.94	8.89	8.85	8.81	8.79	8.76	8.74	.95
	.075	17.4	16.0	15.4	15.1	14.9	14.7	14.6	14.5	14.5	14.4	14.4	14.3	.975
	.99	34.1	30.8	29.5	28.7	28.2	27.9	27.7	27.5	27.3	27.2	27.1	27.1	.99
	.995	55.6	49.8	47.5	46.2	45.4	44.8	44.4	44.1	43.9	43.7	43.5	43.4	.995
	.999	167	149	141	137	135	133	132	131	130	129	129	128	.999
	.9995	266	237	225	218	214	211	209	208	207	206	204	204	.9995

Read $.0^356$ as .00056, 200^1 as 2000, 162^4 as 1620000, etc.

p \ m	15	20	24	30	40	50	60	100	120	200	500	∞	p	n
.0005	.051	.058	062	.066	.069	.072	.074	.077	.078	.080	.081	.083	.0005	1
.001	.060	.067	.071	.075	.079	.082	.084	.087	.088	.089	.091	.092	.001	
.005	.093	.101	.105	.109	.113	.116	.118	.121	.122	.124	.126	.127	.005	
.01	.115	.124	.128	.132	.137	.139	.141	.145	.146	.148	.150	.151	.01	
.025	.161	.170	.175	.180	.184	.187	.189	.193	.194	.196	.198	.199	.025	
.05	.220	.230	.235	.240	.245	.248	.250	.254	.255	.257	.259	.261	.05	
.10	.325	.336	.342	.347	.353	.356	.358	.362	.364	.366	.368	.370	.10	
.25	.698	.712	.719	.727	.734	.738	.741	.747	.749	.752	.754	.756	.25	
.50	2.09	2.12	2.13	2.15	2.16	2.17	2.17	2.18	2.18	2.19	2.19	2.20	.50	
.75	9.49	9.58	9.63	9.67	9.71	9.74	9.76	9.78	9.80	9.82	9.84	9.85	.75	
.90	61.2	61.7	62.0	62.3	62.5	62.7	62.8	63.0	63.1	63.2	63.3	63.3	.90	
.95	246	248	249	250	251	252	252	253	253	254	254	254	.95	
.975	985	993	997	100^1	101^1	101^1	101^1	101^1	101^1	102^1	102^1	102^1	.975	
.99	616^1	621^1	623^1	626^1	629^1	630^1	631^1	633^1	634^1	635^1	636^1	637^1	.99	
.995	246^2	248^2	249^2	250^2	251^2	252^2	253^2	253^2	254^2	254^2	254^2	255^2	.995	
.999	616^3	621^3	623^3	626^3	629^3	630^3	631^3	633^3	634^3	635^3	636^3	637^3	.999	
.9995	246^4	248^4	249^4	250^4	251^4	252^4	252^4	253^4	253^4	253^4	254^4	254^4	.9995	
.0005	.076	.088	.094	.101	.108	.113	.116	.122	.124	.127	.130	.132	.0005	2
.001	.088	.100	.107	.114	.121	.126	.129	.135	.137	.140	.143	.145	.001	
.005	.130	.143	.150	.157	.165	.169	.173	.179	.181	.184	.187	.189	.005	
.01	.157	.171	.178	.186	.193	.198	.201	.207	.209	.212	.215	.217	.01	
.025	.210	.224	.232	.239	.247	.251	.255	.261	.263	.266	.269	.271	.025	
.05	.272	.286	.294	.302	.309	.314	.317	.324	.326	.329	.332	.334	.05	
.10	.371	.386	.394	.402	.410	.415	.418	.424	.426	.429	.433	.434	.10	
.25	.657	.672	.680	.689	.697	.702	.705	.711	.713	.716	.719	.721	.25	
.50	1.38	1.39	1.40	1.41	1.42	1.42	1.43	1.43	1.43	1.44	1.44	1.44	.50	
.75	3.41	3.43	3.43	3.44	3.45	3.45	3.46	3.47	3.47	3.48	3.48	3.48	.75	
.90	9.42	9.44	9.45	9.46	9.47	9.47	9.47	9.48	9.48	9.49	9.49	9.49	.90	
.95	19.4	19.4	19.5	19.5	19.5	19.5	19.5	19.5	19.5	19.5	19.5	19.5	.95	
.975	39.4	39.4	39.5	39.5	39.5	39.5	39.5	39.5	39.5	39.5	39.5	39.5	.975	
.99	99.4	99.4	99.5	99.5	99.5	99.5	99.5	99.5	99.5	99.5	99.5	99.5	.99	
.995	199	199	199	199	199	199	199	199	199	199	199	200	.995	
.999	999	999	999	999	999	999	999	999	999	999	999	999	.999	
.9995	200^1	200^1	200^1	200^1	200^1	200^1	200^1	200^1	200^1	200^1	200^1	200^1	.9995	
.0005	.093	.109	.117	.127	.136	.143	.147	.156	.158	.162	.166	.169	.0005	3
.001	.107	.123	.132	.142	.152	.158	.162	.171	.173	.177	.181	.184	.001	
.005	.154	.172	.181	.191	.201	.207	.211	.220	.222	.227	.231	.234	.005	
.01	.185	.203	.212	.222	.232	.238	.242	.251	.253	.258	.262	.264	.01	
.025	.241	.259	.269	.279	.289	.295	.299	.308	.310	.314	.318	.321	.025	
.05	.304	.323	.332	.342	.352	.358	.363	.370	.373	.377	.382	.384	.05	
.10	.402	.420	.430	.439	.449	.455	.459	.467	.469	.474	.476	.480	.10	
.25	.658	.675	.684	.693	.702	.708	.711	.719	.721	.724	.728	.730	.25	
.50	1.21	1.23	1.23	1.24	1.25	1.25	1.25	1.26	1.26	1.26	1.27	1.27	.50	
.75	2.46	2.46	2.46	2.47	2.47	2.47	2.47	2.47	2.47	2.47	2.47	2.47	.75	
.90	5.20	5.18	5.18	5.17	5.16	5.15	5.15	5.14	5.14	5.14	5.14	5.13	.90	
.95	8.70	8.66	8.63	8.62	8.59	8.58	8.57	8.55	8.55	8.54	8.53	8.53	.95	
.975	14.3	14.2	14.1	14.1	14.0	14.0	14.0	14.0	13.9	13.9	13.9	13.9	.975	
.99	26.9	26.7	26.6	26.5	26.4	26.3	26.3	26.2	26.2	26.2	26.1	26.1	.99	
.995	43.1	42.8	42.6	42.5	42.3	42.2	42.1	42.0	42.0	41.9	41.9	41.8	.995	
.999	127	126	126	125	125	125	124	124	124	124	124	123	.999	
.9995	203	201	200	199	199	198	198	197	197	197	196	196	.9995	

n	p	1	2	3	4	5	6	7	8	9	10	11	12	p
4	.0005	$.0^644$	$.0^350$	$.0^246$	.013	.024	.036	.047	.057	.066	.075	.082	.089	.0005
	.001	$.0^518$	$.0^210$	$.0^273$	.019	.032	.046	.058	.069	.079	.089	.097	.104	.001
	.005	$.0^444$	$.0^250$	.022	.043	.064	.083	.100	.114	.126	.137	.145	.153	.005
	.01	$.0^318$	.010	.035	.063	.088	.109	.127	.143	.156	.167	.176	.185	.01
	.025	$.0^211$	.026	.066	.104	.135	.161	.181	.198	.212	.224	.234	.243	.025
	05	$.0^244$	.052	.110	.157	.193	.221	.243	.261	.275	.288	.298	.307	.05
	.10	.018	.108	.187	.243	.284	.314	.338	.356	.371	.384	.394	.403	.10
	.25	.117	.309	.418	.484	.528	.560	.583	.601	.615	.627	.637	.645	.25
	.50	.549	.828	.941	1.00	1.04	1.06	1.08	1.09	1.10	1.10	1.11	1.13	.50
	.75	1.81	2.00	2.05	2.06	2.07	2.08	2.08	2.08	2.08	2.08	2.08	2.08	.75
	.90	4.54	4.32	4.19	4.11	4.05	4.01	3.98	3.95	3.94	3.92	3.91	3.90	.90
	.95	7.71	6.94	6.59	6.39	6.26	6.16	6.09	6.04	6.00	5.96	5.94	5.91	.95
	.975	12.2	10.6	9.98	9.60	9.36	9.20	9.07	8.98	8.90	8.84	8.79	8.75	.975
	.99	21.2	18.0	16.7	16.0	15.5	15.2	15.0	14.8	14.7	14.5	14.4	14.4	.99
	.995	31.3	26.3	24.3	23.2	22.5	22.0	21.6	21.4	21.1	21.0	20.8	20.7	.995
	.999	74.1	61.2	56.2	53.4	51.7	50.5	49.7	49.0	48.5	48.0	47.7	47.4	.999
	.9995	106	87.4	80.1	76.1	73.6	71.9	70.6	69.7	68.9	68.3	67.8	67.4	.9995
5	.0005	$.0^643$	$.0^350$	$.0^247$	.014	.025	.038	.050	.061	.070	.081	.089	.096	.0005
	.001	$.0^517$	$.0^210$	$.0^275$	.019	.034	.048	.062	.074	.085	.095	.104	.112	.001
	.005	$.0^443$	$.0^250$	.022	.045	.067	.087	.105	.120	.134	.146	.156	.165	.005
	.01	$.0^317$	.010	.035	.064	.091	.114	.134	.151	.165	.177	.188	.197	.01
	.025	$.0^211$	.025	.067	.107	.140	.167	.189	.208	.223	.236	.248	.257	.025
	.05	$.0^243$	.052	.111	.160	.198	.228	.252	.271	.287	.301	.313	.322	.05
	.10	.017	.108	.188	.247	.290	.322	.347	.367	.383	.397	.408	.418	.10
	.25	.113	.305	.415	.483	.528	.560	.584	.604	.618	.631	.641	.650	.25
	.50	.528	.799	.907	.965	1.00	1.02	1.04	1.05	1.06	1.07	1.08	1.09	.50
	.75	1.69	1.85	1.88	1.89	1.89	1.89	1.89	1.89	1.89	1.89	1.89	1.89	.75
	.90	4.06	3.78	3.62	3.52	3.45	3.40	3.37	3.34	3.32	3.30	3.28	3.27	.90
	.95	6.61	5.79	5.41	5.19	5.05	4.95	4.88	4.82	4.77	4.74	4.71	4.68	.95
	.975	10.0	8.43	7.76	7.39	7.15	6.98	6.85	6.76	6.68	6.62	6.57	6.52	.975
	.99	16.3	13.3	12.1	11.4	11.0	10.7	10.5	10.3	10.2	10.1	9.96	9.89	.99
	.995	22.8	18.3	16.5	15.6	14.9	14.5	14.2	14.0	13.8	13.6	13.5	13.4	.995
	.999	47.2	37.1	33.2	31.1	29.7	28.8	28.2	27.6	27.2	26.9	26.6	26.4	.999
	.9995	63.6	49.8	44.4	41.5	39.7	38.5	37.6	36.9	36.4	35.9	35.6	35.2	.9995
6	.0005	$.0^643$	$.0^350$	$.0^247$	.014	.026	.039	.052	.064	.075	.085	.094	.103	.0005
	.001	$.0^517$	$.0^210$	$.0^275$	.020	.035	.050	.064	.078	.090	.101	.111	.119	.001
	.005	$.0^443$	$.0^250$	.022	.045	.069	.090	.109	.126	.140	.153	.164	.174	.005
	.01	$.0^317$	.010	.036	.066	.094	.118	.139	.157	.172	.186	.197	.207	.01
	.025	$.0^211$	.025	.068	.109	.143	.172	.195	.215	.231	.246	.258	.268	.025
	.05	$.0^243$	.052	.112	.162	.202	.233	.259	.279	.296	.311	.324	.334	.05
	.10	.017	.107	.189	.249	.294	.327	.354	.375	.392	.406	.418	.429	.10
	.25	.111	.302	.413	.481	.524	.561	.586	.606	.622	.635	.645	.654	.25
	.50	.515	.780	.886	.942	.977	1.00	1.02	1.03	1.04	1.05	1.05	1.06	.50
	.75	1.62	1.76	1.78	1.79	1.79	1.78	1.78	1.78	1.77	1.77	1.77	1.77	.75
	.90	3.78	3.46	3.29	3.18	3.11	3.05	3.01	2.98	2.96	2.94	2.92	2.90	.90
	.95	5.99	5.14	4.76	4.53	4.39	4.28	4.21	4.15	4.10	4.06	4.03	4.00	.95
	.975	8.81	7.26	6.60	6.23	5.99	5.82	5.70	5.60	5.52	5.46	5.41	5.37	.975
	.99	13.7	10.9	9.78	9.15	8.75	8.47	8.26	8.10	7.98	7.87	7.79	7.72	.99
	.995	18.6	14.5	12.9	12.0	11.5	11.1	10.8	10.6	10.4	10.2	10.1	10.0	.995
	.999	35.5	27.0	23.7	21.9	20.8	20.0	19.5	19.0	18.7	18.4	18.2	18.0	.999
	.9995	46.1	34.8	30.4	28.1	26.6	25.6	24.9	24.3	23.9	23.5	23.2	23.0	.9995

TABLE A.4 PERCENTILES OF *F* DISTRIBUTIONS (cont.)

p \ m	15	20	24	30	40	50	60	100	120	200	500	∞	p	n
.0005	.105	.125	.135	.147	.159	.166	.172	.183	.186	.191	.196	.200	.0005	**4**
.001	.121	.141	.152	.163	.176	.183	.188	.200	.202	.208	.213	.217	.001	
.005	.172	.193	.204	.216	.229	.237	.242	.253	.255	.260	.266	.269	.005	
.01	.204	.226	.237	.249	.261	.269	.274	.285	.287	.293	.298	.301	.01	
.025	.263	.284	.296	.308	.320	.327	.332	.342	.346	.351	.356	.359	.025	
.05	.327	.349	.360	.372	.384	.391	.396	.407	.409	.413	.418	.422	.05	
.10	.424	.445	.456	.467	.478	.485	.490	.500	.502	.508	.510	.514	.10	
.25	.664	.683	.692	.702	.712	.718	.722	.731	.733	.737	.740	.743	.25	
.50	1.14	1.15	1.16	1.16	1.17	1.18	1.18	1.18	1.18	1.19	1.19	1.19	.50	
.75	2.08	2.08	2.08	2.08	2.08	2.08	2.08	2.08	2.08	2.08	2.08	2.08	.75	
.90	3.87	3.84	3.83	3.82	3.80	3.80	3.79	3.78	3.78	3.77	3.76	3.76	.90	
.95	5.86	5.80	5.77	5.75	5.72	5.70	5.69	5.66	5.66	5.65	5.64	5.63	.95	
.975	8.66	8.56	8.51	8.46	8.41	8.38	8.36	8.32	8.31	8.29	8.27	8.26	.975	
.99	14.2	14.0	13.9	13.8	13.7	13.7	13.6	13.6	13.6	13.5	13.5	13.5	.99	
.995	20.4	20.2	20.0	19.9	19.8	19.7	19.6	19.5	19.5	19.4	19.4	19.3	.995	
.999	46.8	46.1	45.8	45.4	45.1	44.9	44.7	44.5	44.4	44.3	44.1	44.0	.999	
.9995	66.5	65.5	65.1	64.6	64.1	63.8	63.6	63.2	63.1	62.9	62.7	62.6	.9995	
.0005	.115	.137	.150	.163	.177	.186	.192	.205	.209	.216	.222	.226	.0005	**5**
.001	.132	.155	.167	.181	.195	.204	.210	.223	.227	.233	.239	.244	.001	
.005	.186	.210	.223	.237	.251	.260	.266	.279	.282	.288	.294	.299	.005	
.01	.219	.244	.257	.270	.285	.293	.299	.312	.315	.322	.328	.331	.01	
.025	.280	.304	.317	.330	.344	.353	.359	.370	.374	.380	.386	.390	.025	
.05	.345	.369	.382	.395	.408	.417	.422	.432	.437	.442	.448	.452	.05	
.10	.440	.463	.476	.488	.501	.508	.514	.524	.527	.532	.538	.541	.10	
.25	.669	.690	.700	.711	.722	.728	.732	.741	.743	.748	.752	.755	.25	
.50	1.10	1.11	1.12	1.12	1.13	1.13	1.14	1.14	1.14	1.15	1.15	1.15	.50	
.75	1.89	1.88	1.88	1.88	1.88	1.88	1.87	1.87	1.87	1.87	1.87	1.87	.75	
.90	3.24	3.21	3.19	3.17	3.16	3.15	3.14	3.13	3.12	3.12	3.11	3.10	.90	
.95	4.62	4.56	4.53	4.50	4.46	4.44	4.43	4.41	4.40	4.39	4.37	4.36	.95	
.975	6.43	6.33	6.28	6.23	6.18	6.14	6.12	6.08	6.07	6.05	6.03	6.02	.975	
.99	9.72	9.55	9.47	9.38	9.29	9.24	9.20	9.13	9.11	9.08	9.04	9.02	.99	
.995	13.1	12.9	12.8	12.7	12.5	12.5	12.4	12.3	12.3	12.2	12.2	12.1	.995	
.999	25.9	25.4	25.1	24.9	24.6	24.4	24.3	24.1	24.1	23.9	23.8	23.8	.999	
.9995	34.6	33.9	33.5	33.1	32.7	32.5	32.3	32.1	32.0	31.8	31.7	31.6	.9995	
.0005	.123	.148	.162	.177	.193	.203	.210	.225	.229	.236	.244	.249	.0005	**6**
.001	.141	.166	.180	.195	.211	.222	.229	.243	.247	.255	.262	.267	.001	
.005	.197	.224	.238	.253	.269	.279	.286	.301	.304	.312	.318	.324	.005	
.01	.232	.258	.273	.288	.304	.313	.321	.334	.338	.346	.352	.357	.01	
.025	.293	.320	.334	.349	.364	.375	.381	.394	.398	.405	.412	.415	.025	
.05	.358	.385	.399	.413	.428	.437	.444	.457	.460	.467	.472	.476	.05	
.10	.453	.478	.491	.505	.519	.526	.533	.546	.548	.556	.559	.564	.10	
.25	.675	.696	.707	.718	.729	.736	.741	.751	.753	.758	.762	.765	.25	
.50	1.07	1.08	1.09	1.10	1.10	1.11	1.11	1.11	1.12	1.12	1.12	1.12	.50	
.75	1.76	1.76	1.75	1.75	1.75	1.75	1.74	1.74	1.74	1.74	1.74	1.74	.75	
.90	2.87	2.84	2.82	2.80	2.78	2.77	2.76	2.75	2.74	2.73	2.73	2.72	.90	
.95	3.94	3.87	3.84	3.81	3.77	3.75	3.74	3.71	3.70	3.69	3.68	3.67	.95	
.975	5.27	5.17	5.12	5.07	5.01	4.98	4.96	4.92	4.90	4.88	4.86	4.85	.975	
.99	7.56	7.40	7.31	7.23	7.14	7.09	7.06	6.99	6.97	6.93	6.90	6.88	.99	
.995	9.81	9.59	9.47	9.36	9.24	9.17	9.12	9.03	9.00	8.95	8.91	8.88	.995	
.999	17.6	17.1	16.9	16.7	16.4	16.3	16.2	16.0	16.0	15.9	15.8	15.7	.999	
.9995	22.4	21.9	21.7	21.4	21.1	20.9	20.7	20.5	20.4	20.3	20.2	20.1	.9995	

n	p	1	2	3	4	5	6	7	8	9	10	11	12	p
7	.0005	$.0^642$	$.0^350$	$.0^248$	.014	.027	.040	.053	.066	.078	.088	.099	.108	.0005
	.001	$.0^517$	$.0^210$	$.0^276$	.020	.035	.051	.067	.081	.093	.105	.115	.125	.001
	.005	$.0^442$	$.0^250$	.023	.046	.070	.093	.113	.130	.145	.159	.171	.181	.005
	.01	$.0^317$	.010	.036	.067	.096	.121	.143	.162	.178	.192	.205	.216	.01
	.025	$.0^210$	.025	.068	.110	.146	.176	.200	.221	.238	.253	.266	.277	.025
	.05	$.0^242$	.052	.113	.164	.205	.238	.264	.286	.304	.319	.332	.343	.05
	.10	.017	.107	.190	.251	.297	.332	.359	.381	.399	.414	.427	.438	.10
	.25	.110	.300	.412	.481	.528	.562	.588	.608	.624	.637	.649	.658	.25
	.50	.506	.767	.871	.926	.960	.983	1.00	1.01	1.02	1.03	1.04	1.04	.50
	.75	1.57	1.70	1.72	1.72	1.71	1.71	1.70	1.70	1.69	1.69	1.69	1.68	.75
	.90	3.59	3.26	3.07	2.96	2.88	2.83	2.78	2.75	2.72	2.70	2.68	2.67	.90
	.95	5.59	4.74	4.35	4.12	3.97	3.87	3.79	3.73	3.68	3.64	3.60	3.57	.95
	.975	8.07	6.54	5.89	5.52	5.29	5.12	4.99	4.90	4.82	4.76	4.71	4.67	.975
	.99	12.2	9.55	8.45	7.85	7.46	7.19	6.99	6.84	6.72	6.62	6.54	6.47	.99
	.995	16.2	12.4	10.9	10.0	9.52	9.16	8.89	8.68	8.51	8.38	8.27	8.18	.995
	.999	29.2	21.7	18.8	17.2	16.2	15.5	15.0	14.6	14.3	14.1	13.9	13.7	.999
	.9995	37.0	27.2	23.5	21.4	20.2	19.3	18.7	18.2	17.8	17.5	17.2	17.0	.9995
8	.0005	$.0^342$	$.0^350$	$.0^248$	.014	.027	.041	.055	.068	.081	.092	.102	.112	.0005
	.001	$.0^517$	$.0^210$	$.0^276$	.020	.036	.053	.068	.083	.096	.109	.120	.130	.001
	.005	$.0^442$	$.0^250$	.027	.047	.072	.095	.115	.133	.149	.164	.176	.187	.005
	.01	$.0^317$	.010	.036	.068	.097	.123	.146	.166	.183	.198	.211	.222	.01
	.025	$.0^210$	.025	.069	.111	.148	.179	.204	.226	.244	.259	.273	.285	.025
	.05	$.0^242$	.052	.113	.166	.208	.241	.268	.291	.310	.326	.339	.351	.05
	.10	.017	.107	.190	.253	.299	.335	.363	.386	.405	.421	.435	.445	.10
	.25	.109	.298	.411	.481	.529	.563	.589	.610	.627	.640	.654	.661	.25
	.50	.499	.757	.860	.915	.948	.971	.988	1.00	1.01	1.02	1.02	1.03	.50
	.75	1.54	1.66	1.67	1.66	1.66	1.65	1.64	1.64	1.64	1.63	1.63	1.62	.75
	.90	3.46	3.11	2.92	2.81	2.73	2.67	2.62	2.59	2.56	2.54	2.52	2.50	.90
	.95	5.32	4.46	4.07	3.84	3.69	3.58	3.50	3.44	3.39	3.35	3.31	3.28	.95
	.975	7.57	6.06	5.42	5.05	4.82	4.65	4.53	4.43	4.36	4.30	4.24	4.20	.975
	.99	11.3	8.65	7.59	7.01	6.63	6.37	6.18	6.03	5.91	5.81	5.73	5.67	.99
	.995	14.7	11.0	9.60	8.81	8.30	7.95	7.69	7.50	7.34	7.21	7.10	7.01	.995
	.999	25.4	18.5	15.8	14.4	13.5	12.9	12.4	12.0	11.8	11.5	11.4	11.2	.999
	.9995	31.6	22.8	19.4	17.6	16.4	15.7	15.1	14.6	14.3	14.0	13.8	13.6	.9995
9	.0005	$.0^641$	$.0^350$	$.0^248$	.015	.027	.042	.056	.070	.083	.094	.105	.115	.0005
	.001	$.0^517$	$.0^210$	$.0^277$	.021	.037	.054	.070	.085	.099	.112	.123	.134	.001
	.005	$.0^442$	$.0^250$	.023	.047	.073	.096	.117	.136	.153	.168	.181	.192	.005
	.01	$.0^317$	.010	.037	.068	.098	.125	.149	.169	.187	.202	.216	.228	.01
	.025	$.0^210$	.025	.069	.112	.150	.181	.207	.230	.248	.265	.279	.291	.025
	.05	$.0^240$	.052	.113	.167	.210	.244	.272	.296	.315	.331	.345	.358	.05
	.10	.017	.107	.191	.254	.302	.338	.367	.390	.410	.426	.441	.452	.10
	.25	.108	.297	.410	.480	.529	.564	.591	.612	.629	.643	.654	.664	.25
	.50	.494	.749	.852	.906	.939	.962	.978	.990	1.00	1.01	1.01	1.02	.50
	.75	1.51	1.62	1.63	1.63	1.62	1.61	1.60	1.60	1.59	1.59	1.58	1.58	.75
	.90	3.36	3.01	2.81	2.69	2.61	2.55	2.51	2.47	2.44	2.42	2.40	2.38	.90
	.95	5.12	4.26	3.86	3.63	3.48	3.37	3.29	3.23	3.18	3.14	3.10	3.07	.95
	.975	7.21	5.71	5.08	4.72	4.48	4.32	4.20	4.10	4.03	3.96	3.91	3.87	.975
	.99	10.6	8.02	6.99	6.42	6.06	5.80	5.61	5.47	5.35	5.26	5.18	5.11	.99
	.995	13.6	10.1	8.72	7.96	7.47	7.13	6.88	6.69	6.54	6.42	6.31	6.23	.995
	.999	22.9	16.4	13.9	12.6	11.7	11.1	10.7	10.4	10.1	9.89	9.71	9.57	.999
	.9995	28.0	19.9	16.8	15.1	14.1	13.3	12.8	12.4	12.1	11.8	11.6	11.4	.9995

p \ m	15	20	24	30	40	50	60	100	120	200	500	∞	p	n
.0005	.130	.157	.172	.188	.206	.217	.225	.242	.246	.255	.263	.268	.0005	7
.001	.148	.176	.191	.208	.225	.237	.245	.261	.266	.274	.282	.288	.001	
.005	.206	.235	.251	.267	.285	.296	.304	.319	.324	.332	.340	.345	.005	
.01	.241	.270	.286	.303	.320	.331	.339	.355	.358	.366	.373	.379	.01	
.025	.304	.333	.348	.364	.381	.392	.399	.413	.418	.426	.433	.437	.025	
.05	.369	.398	.413	.428	.445	.455	.461	.476	.479	.485	.493	.498	.05	
.10	.463	.491	.504	.519	.534	.543	.550	.562	.566	.571	.578	.582	.10	
.25	.679	.702	.713	.725	.737	.745	.749	.760	.762	.767	.772	.775	.25	
.50	1.05	1.07	1.07	1.08	1.08	1.09	1.09	1.10	1.10	1.10	1.10	1.10	.50	
.75	1.68	1.67	1.67	1.66	1.66	1.66	1.65	1.65	1.65	1.65	1.65	1.65	.75	
.90	2.63	2.59	2.58	2.56	2.54	2.52	2.51	2.50	2.49	2.48	2.48	2.47	.90	
.95	3.51	3.44	3.41	3.38	3.34	3.32	3.30	3.27	3.27	3.25	3.24	3.23	.95	
.975	4.57	4.47	4.42	4.36	4.31	4.28	4.25	4.21	4.20	4.18	4.16	4.14	.975	
.99	6.31	6.16	6.07	5.99	5.91	5.86	5.82	5.75	5.74	5.70	5.67	5.65	.99	
.995	7.97	7.75	7.65	7.53	7.42	7.35	7.31	7.22	7.19	7.15	7.10	7.08	.995	
.999	13.3	12.9	12.7	12.5	12.3	12.2	12.1	11.9	11.9	11.8	11.7	11.7	.999	
.9995	16.5	16.0	15.7	15.5	15.2	15.1	15.0	14.7	14.7	14.6	14.5	14.4	.9995	
.0005	.136	.164	.181	.198	.218	.230	.239	.257	.262	.271	.281	.287	.0005	8
.001	.155	.184	.200	.218	.238	.250	.259	.277	.282	.292	.300	.306	.001	
.005	.214	.244	.261	.279	.299	.311	.319	.337	.341	.351	.358	.364	.005	
.01	.250	.281	.297	.315	.334	.346	.354	.372	.376	.385	.392	.398	.01	
.025	.313	.343	.360	.377	.395	.407	.415	.431	.435	.442	.450	.456	.025	
.05	.379	.409	.425	.441	.459	.469	.477	.493	.496	.505	.510	.516	.05	
.10	.472	.500	.515	.531	.547	.556	.563	.578	.581	.588	.595	.599	.10	
.25	.684	.707	.718	.730	.743	.751	.756	.767	.769	.775	.780	.783	.25	
.50	1.04	1.05	1.06	1.07	1.07	1.07	1.08	1.08	1.09	1.09	1.09	1.09	.50	
.75	1.62	1.61	1.60	1.60	1.59	1.59	1.59	1.58	1.58	1.58	1.58	1.58	.75	
.90	2.46	2.42	2.40	2.38	2.36	2.35	2.34	2.32	2.32	2.31	2.30	2.29	.90	
.95	3.22	3.15	3.12	3.08	3.04	3.02	3.01	2.97	2.97	2.95	2.94	2.93	.95	
.975	4.10	4.00	3.95	3.89	3.84	3.81	3.78	3.74	3.73	3.70	3.68	3.67	.975	
.99	5.52	5.36	5.28	5.20	5.12	5.07	5.03	4.96	4.95	4.91	4.88	4.86	.99	
.995	6.81	6.61	6.50	6.40	6.29	6.22	6.18	6.09	6.06	6.02	5.98	5.95	.995	
.999	10.8	10.5	10.3	10.1	9.92	9.80	9.73	9.57	9.54	9.46	9.39	9.34	.999	
.9995	13.1	12.7	12.5	12.2	12.0	11.8	11.8	11.6	11.5	11.4	11.4	11.3	.9995	
.0005	.141	.171	.188	.207	.228	.242	.251	.270	.276	.287	.297	.303	.0005	9
.001	.160	.191	.208	.228	.249	.262	.271	.291	.296	.307	.316	.323	.001	
.005	.220	.253	.271	.290	.310	.324	.332	.351	.356	.366	.376	.382	.005	
.01	.257	.289	.307	.326	.346	.358	.368	.386	.391	.400	.410	.415	.01	
.025	.320	.352	.370	.388	.408	.420	.428	.446	.450	.459	.467	.473	.025	
.05	.386	.418	.435	.452	.471	.483	.490	.508	.510	.518	.526	.532	.05	
.10	.479	.509	.525	.541	.556	.568	.575	.588	.594	.602	.610	.613	.10	
.25	.687	.711	.723	.736	.749	.757	.762	.773	.776	.782	.787	.791	.25	
.50	1.03	1.04	1.05	1.05	1.06	1.06	1.07	1.07	1.07	1.08	1.08	1.08	.50	
.75	1.57	1.56	1.56	1.55	1.55	1.54	1.54	1.53	1.53	1.53	1.53	1.53	.75	
.90	2.34	2.30	2.28	2.25	2.23	2.22	2.21	2.19	2.18	2.17	2.17	2.16	.90	
.95	3.01	2.94	2.90	2.86	2.83	2.80	2.79	2.76	2.75	2.73	2.72	2.71	.95	
.975	3.77	3.67	3.61	3.56	3.51	3.47	3.45	3.40	3.39	3.37	3.35	3.33	.975	
.99	4.96	4.81	4.73	4.65	4.57	4.52	4.48	4.42	4.40	4.36	4.33	4.31	.99	
.995	6.03	5.83	5.73	5.62	5.52	5.45	5.41	5.32	5.30	5.26	5.21	5.19	.995	
.999	9.24	8.90	8.72	8.55	8.37	8.26	8.19	8.04	8.00	7.93	7.86	7.81	.999	
.9995	11.0	10.6	10.4	10.2	9.94	9.80	9.71	9.53	9.49	9.40	9.32	9.26	.9995	

n	p	1	2	3	4	5	6	7	8	9	10	11	12	p
10	.0005	$.0^641$	$.0^350$	$.0^249$	.015	.028	.043	.057	.071	.085	.097	.108	.119	.0005
	.001	$.0^517$	$.0^210$	$.0^277$	.021	.037	.054	.071	.087	.101	.114	.126	.137	.001
	.005	$.0^441$	$.0^250$	.023	.048	.073	.098	.119	.139	.156	.171	.185	.197	.005
	.01	$.0^317$	.010	.037	.069	.100	.127	.151	.172	.190	.206	.220	.233	.01
	.025	$.0^210$	.025	.069	.113	.151	.183	.210	.233	.252	.269	.283	.296	.025
	.05	$.0^241$	.052	.114	.168	.211	.246	.275	.299	.319	.336	.351	.363	.05
	.10	.017	.106	.191	.255	.303	.340	.370	.394	.414	.430	.444	.457	.10
	.25	.107	.296	.409	.480	.529	.565	.592	.613	.631	.645	.657	.667	.25
	.50	.490	.743	.845	.899	.932	.954	.971	.983	.992	1.00	1.01	1.01	.50
	.75	1.49	1.60	1.60	1.59	1.59	1.58	1.57	1.56	1.56	1.55	1.55	1.54	.75
	.90	3.28	2.92	2.73	2.61	2.52	2.46	2.41	2.38	2.35	2.32	2.30	2.28	.90
	.95	4.96	4.10	3.71	3.48	3.33	3.22	3.14	3.07	3.02	2.98	2.94	2.91	.95
	.975	6.94	5.46	4.83	4.47	4.24	4.07	3.95	3.85	3.78	3.72	3.66	3.62	.975
	.99	10.0	7.56	6.55	5.99	5.64	5.39	5.20	5.06	4.94	4.85	4.77	4.71	.99
	.995	12.8	9.43	8.08	7.34	6.87	6.54	6.30	6.12	5.97	5.85	5.75	5.66	.995
	.999	21.0	14.9	12.6	11.3	10.5	9.92	9.52	9.20	8.96	8.75	8.58	8.44	.999
	.9995	25.5	17.9	15.0	13.4	12.4	11.8	11.3	10.9	10.6	10.3	10.1	9.93	.9995
11	.0005	$.0^641$	$.0^350$	$.0^249$	.015	.028	.043	.058	.072	.086	.099	.111	.121	.0005
	.001	$.0^516$	$.0^210$	$.0^278$	.021	.038	.055	.072	.088	.103	.116	.129	.140	.001
	.005	$.0^440$	$.0^250$	.023	.048	.074	.099	.121	.141	.158	.174	.188	.200	.005
	.01	$.0^316$	.010	.037	.069	.100	.128	.153	.175	.193	.210	.224	.237	.01
	.025	$.0^210$	.025	.069	.114	.152	.185	.212	.236	.256	.273	.288	.301	.025
	.05	$.0^241$	.052	.114	.168	.212	.248	.278	.302	.323	.340	.355	.368	.05
	.10	.017	.106	.192	.256	.305	.342	.373	.397	.417	.435	.448	.461	.10
	.25	.107	.295	.408	.481	.529	.565	.592	.614	.633	.645	.658	.667	.25
	.50	.486	.739	.840	.893	.926	.948	.964	.977	.986	.994	1.00	1.01	.50
	.75	1.47	1.58	1.58	1.57	1.56	1.55	1.54	1.53	1.53	1.52	1.52	1.51	.75
	.90	3.23	2.86	2.66	2.54	2.45	2.39	2.34	2.30	2.27	2.25	2.23	2.21	.90
	.95	4.84	3.98	3.59	3.36	3.20	3.09	3.01	2.95	2.90	2.85	2.82	2.79	.95
	.975	6.72	5.26	4.63	4.28	4.04	3.88	3.76	3.66	3.59	3.53	3.47	3.43	.975
	.99	9.65	7.21	6.22	5.67	5.32	5.07	4.89	4.74	4.63	4.54	4.46	4.40	.99
	.995	12.2	8.91	7.60	6.88	6.42	6.10	5.86	5.68	5.54	5.42	5.32	5.24	.995
	.999	19.7	13.8	11.6	10.3	9.58	9.05	8.66	8.35	8.12	7.92	7.76	7.62	.999
	.9995	23.6	16.4	13.6	12.2	11.2	10.6	10.1	9.76	9.48	9.24	9.04	8.88	.9995
12	.0005	$.0^641$	$.0^350$	$.0^249$	.015	.028	.044	.058	.073	.087	.101	.113	.124	.0005
	.001	$.0^516$	$.0^210$	$.0^278$	.021	.038	.056	.073	.089	.104	.118	.131	.143	.001
	.005	$.0^439$	$.0^250$	.023	.048	.075	.100	.122	.143	.161	.177	.191	.204	.005
	.01	$.0^316$	.010	.037	.070	.101	.130	.155	.176	.196	.212	.227	.241	.01
	.025	$.0^210$	.025	.070	.114	.153	.186	.214	.238	.259	.276	.292	.305	.025
	.05	$.0^241$	.052	.114	.169	.214	.250	.280	.305	.325	.343	.358	.372	.05
	.10	.016	.106	.192	.257	.306	.344	.375	.400	.420	.438	.452	.466	.10
	.25	.106	.295	.408	.480	.530	.566	.594	.616	.633	.649	.662	.671	.25
	.50	.484	.735	.835	.888	.921	.943	.959	.972	.981	.989	.995	1.00	.50
	.75	1.46	1.56	1.56	1.55	1.54	1.53	1.52	1.51	1.51	1.50	1.50	1.49	.75
	.90	3.18	2.81	2.61	2.48	2.39	2.33	2.28	2.24	2.21	2.19	2.17	2.15	.90
	.95	4.75	3.89	3.49	3.26	3.11	3.00	2.91	2.85	2.80	2.75	2.72	2.69	.95
	.975	6.55	5.10	4.47	4.12	3.89	3.73	3.61	3.51	3.44	3.37	3.32	3.28	.975
	.99	9.33	6.93	5.95	5.41	5.06	4.82	4.64	4.50	4.39	4.30	4.22	4.16	.99
	.995	11.8	8.51	7.23	6.52	6.07	5.76	5.52	5.35	5.20	5.09	4.99	4.91	.995
	.999	18.6	13.0	10.8	9.63	8.89	8.38	8.00	7.71	7.48	7.29	7.14	7.01	.999
	.9995	22.2	15.3	12.7	11.2	10.4	9.74	9.28	8.94	8.66	8.43	8.24	8.08	.9995

p	15	20	24	30	40	50	60	100	120	200	500	∞	p	n
.0005	.145	.177	.195	.215	.238	.251	.262	.282	.288	.299	.311	.319	.0005	**10**
.001	.164	.197	.216	.236	.258	.272	.282	.303	.309	.321	.331	.338	.001	
.005	.226	.260	.279	.299	.321	.334	.344	.365	.370	.380	.391	.397	.005	
.01	.263	.297	.316	.336	.357	.370	.380	.400	.405	.415	.424	.431	.01	
.025	.327	.360	.379	.398	.419	.431	.441	.459	.464	.474	.483	.488	.025	
.05	.393	.426	.444	.462	.481	.493	.502	.518	.523	.532	.541	.546	.05	
.10	.486	.516	.532	.549	.567	.578	.586	.602	.605	.614	.621	.625	.10	
.25	.691	.714	.727	.740	.754	.762	.767	.779	.782	.788	.793	.797	.25	
.50	1.02	1.03	1.04	1.05	1.05	1.06	1.06	1.06	1.06	1.07	1.07	1.07	.50	
.75	1.53	1.52	1.52	1.51	1.51	1.50	1.50	1.49	1.49	1.49	1.48	1.48	.75	
.90	2.24	2.20	2.18	2.16	2.13	2.12	2.11	2.09	2.08	2.07	2.06	2.06	.90	
.95	2.85	2.77	2.74	2.70	2.66	2.64	2.62	2.59	2.58	2.56	2.55	2.54	.95	
.975	3.52	3.42	3.37	3.31	3.26	3.22	3.20	3.15	3.14	3.12	3.09	3.08	.975	
.99	4.56	4.41	4.33	4.25	4.17	4.12	4.08	4.01	4.00	3.96	3.93	3.91	.99	
.995	5.47	5.27	5.17	5.07	4.97	4.90	4.86	4.77	4.75	4.71	4.67	4.64	.995	
.999	8.13	7.80	7.64	7.47	7.30	7.19	7.12	6.98	6.94	6.87	6.81	6.76	.999	
.9995	9.56	9.16	8.96	8.75	8.54	8.42	8.33	8.16	8.12	8.04	7.96	7.90	.9995	
.0005	.148	.182	.201	.222	.246	.261	.271	.293	.299	.312	.324	.331	.0005	**11**
.001	.168	.202	.222	.243	.266	.282	.292	.313	.320	.332	.343	.353	.001	
.005	.231	.266	.286	.308	.330	.345	.355	.376	.382	.394	.403	.412	.005	
.01	.268	.304	.324	.344	.366	.380	.391	.412	.417	.427	.439	.444	.01	
.025	.332	.368	.386	.407	.429	.442	.450	.472	.476	.485	.495	.503	.025	
.05	.398	.433	.452	.469	.490	.503	.513	.529	.535	.543	.552	.559	.05	
.10	.490	.524	.541	.559	.578	.588	.595	.614	.617	.625	.633	.637	.10	
.25	.694	.719	.730	.744	.758	.767	.773	.780	.788	.794	.799	.803	.25	
.50	1.02	1.03	1.03	1.04	1.05	1.05	1.05	1.06	1.06	1.06	1.06	1.06	.50	
.75	1.50	1.49	1.49	1.48	1.47	1.47	1.47	1.46	1.46	1.46	1.45	1.45	.75	
.90	2.17	2.12	2.10	2.08	2.05	2.04	2.03	2.00	2.00	1.99	1.98	1.97	.90	
.95	2.72	2.65	2.61	2.57	2.53	2.51	2.49	2.46	2.45	2.43	2.42	2.40	.95	
.975	3.33	3.23	3.17	3.12	3.06	3.03	3.00	2.96	2.94	2.92	2.90	2.88	.975	
.99	4.25	4.10	4.02	3.94	3.86	3.81	3.78	3.71	3.69	3.66	3.62	3.60	.99	
.995	5.05	4.86	4.76	4.65	4.55	4.49	4.45	4.36	4.34	4.29	4.25	4.23	.995	
.999	7.32	7.01	6.85	6.68	6.52	6.41	6.35	6.21	6.17	6.10	6.04	6.00	.999	
.9995	8.52	8.14	7.94	7.75	7.55	7.43	7.35	7.18	7.14	7.06	6.98	6.93	.9995	
.0005	.152	.186	.206	.228	.253	.269	.280	.305	.311	.323	.337	.345	.0005	**12**
.001	.172	.207	.228	.250	.275	.291	.302	.326	.332	.344	.357	.365	.001	
.005	.235	.272	.292	.315	.339	.355	.365	.388	.393	.405	.417	.424	.005	
.01	.273	.310	.330	.352	.376	.391	.401	.422	.428	.441	.450	.458	.01	
.025	.337	.374	.394	.416	.437	.450	.461	.481	.487	.498	.508	.514	.025	
.05	.404	.439	.458	.478	.499	.513	.522	.541	.545	.556	.565	.571	.05	
.10	.496	.528	.546	.564	.583	.595	.604	.621	.625	.633	.641	.647	.10	
.25	.695	.721	.734	.748	.762	.771	.777	.789	.792	.799	.804	.808	.25	
.50	1.01	1.02	1.03	1.03	1.04	1.04	1.05	1.05	1.05	1.05	1.06	1.06	.50	
.75	1.48	1.47	1.46	1.45	1.45	1.44	1.44	1.43	1.43	1.43	1.42	1.42	.75	
.90	2.11	2.06	2.04	2.01	1.99	1.97	1.96	1.94	1.93	1.92	1.91	1.90	.90	
.95	2.62	2.54	2.51	2.47	2.43	2.40	2.38	2.35	2.34	2.32	2.31	2.30	.95	
.975	3.18	3.07	3.02	2.96	2.91	2.87	2.85	2.80	2.79	2.76	2.74	2.72	.975	
.99	4.01	3.86	3.78	3.70	3.62	3.57	3.54	3.47	3.45	3.41	3.38	3.36	.99	
.995	4.72	4.53	4.43	4.33	4.23	4.17	4.12	4.04	4.01	3.97	3.93	3.90	.995	
.999	6.71	6.40	6.25	6.09	5.93	5.83	5.76	5.63	5.59	5.52	5.46	5.42	.999	
.9995	7.74	7.37	7.18	7.00	6.80	6.68	6.61	6.45	6.41	6.33	6.25	6.20	.9995	

n	p \ m	1	2	3	4	5	6	7	8	9	10	11	12	p
15	.0005	$.0^641$	$.0^350$	$.0^249$	.015	.029	.045	.061	.076	.091	.105	.117	.129	.0005
	.001	$.0^516$	$.0^210$	$.0^279$	.021	.039	.057	.075	.092	.108	.123	.137	.149	.001
	.005	$.0^439$	$.0^250$	.023	.049	.076	.102	.125	.147	.166	.183	.198	.212	.005
	.01	$.0^316$	.010	.037	.070	.103	.132	.158	.181	.202	.219	.235	.249	.01
	.025	$.0^210$	.025	.070	.116	.156	.190	.219	.244	.265	.284	.300	.315	.025
	.05	$.0^241$	.051	.115	.170	.216	.254	.285	.311	.333	.351	.368	.382	.05
	.10	.016	.106	.192	.258	.309	.348	.380	.406	.427	.446	.461	.475	.10
	.25	.105	.293	.407	.480	.531	.568	.596	.618	.637	.652	.667	.676	.25
	.50	.478	.726	.826	.878	.911	.933	.948	.960	.970	.977	.984	.989	.50
	.75	1.43	1.52	1.52	1.51	1.49	1.48	1.47	1.46	1.46	1.45	1.44	1.44	.75
	.90	3.07	2.70	2.49	2.36	2.27	2.21	2.16	2.12	2.09	2.06	2.04	2.02	.90
	.95	4.54	3.68	3.29	3.06	2.90	2.79	2.71	2.64	2.59	2.54	2.51	2.48	.95
	.975	6.20	4.76	4.15	3.80	3.58	3.41	3.29	3.20	3.12	3.06	3.01	2.96	.975
	.99	8.68	6.36	5.42	4.89	4.56	4.32	4.14	4.00	3.89	3.80	3.73	3.67	.99
	.995	10.8	7.70	6.48	5.80	5.37	5.07	4.85	4.67	4.54	4.42	4.33	4.25	.995
	.999	16.6	11.3	9.34	8.25	7.57	7.09	6.74	6.47	6.26	6.08	5.93	5.81	.999
	.9995	19.5	13.2	10.8	9.48	8.66	8.10	7.68	7.36	7.11	6.91	6.75	6.60	.9995
20	.0005	$.0^640$	$.0^350$	$.0^250$	.015	.029	.046	.063	.079	.094	.109	.123	.136	.0005
	.001	$.0^516$	$.0^210$	$.0^279$	.022	.039	.058	.077	.095	.112	.128	.143	.156	.001
	.005	$.0^439$	$.0^250$	.023	.050	.077	.104	.129	.151	.171	.190	.206	.221	.005
	.01	$.0^316$	.010	.037	.071	.105	.135	.162	.187	.208	.227	.244	.259	.01
	.025	$.0^210$	.025	.071	.117	.158	.193	.224	.250	.273	.292	.310	.325	.025
	.05	$.0^240$	.051	.115	.172	.219	.258	.290	.318	.340	.360	.377	.393	.05
	.10	.016	.106	.193	.260	.312	.353	.385	.412	.435	.454	.472	.485	.10
	.25	.104	.292	.407	.480	.531	.569	.598	.622	.641	.656	.671	.681	.25
	.50	.472	.718	.816	.868	.900	.922	.938	.950	.959	.966	.972	.977	.50
	.75	1.40	1.49	1.48	1.47	1.45	1.44	1.43	1.42	1.41	1.40	1.39	1.39	.75
	.90	2.97	2.59	2.38	2.25	2.16	2.09	2.04	2.00	1.96	1.94	1.91	1.89	.90
	.95	4.35	3.49	3.10	2.87	2.71	2.60	2.51	2.45	2.39	2.35	2.31	2.28	.95
	.975	5.87	4.46	3.86	3.51	3.29	3.13	3.01	2.91	2.84	2.77	2.72	2.68	.975
	.99	8.10	5.85	4.94	4.43	4.10	3.87	3.70	3.56	3.46	3.37	3.29	3.23	.99
	.995	9.94	6.99	5.82	5.17	4.76	4.47	4.26	4.09	3.96	3.85	3.76	3.68	.995
	.999	14.8	9.95	8.10	7.10	6.46	6.02	5.69	5.44	5.24	5.08	4.94	4.82	.999
	.9995	17.2	11.4	9.20	8.02	7.28	6.76	6.38	6.08	5.85	5.66	5.51	5.38	.9995
24	.0005	$.0^640$	$.0^350$	$.0^250$	.015	.030	.046	.064	.080	.096	.112	.126	.139	.0005
	.001	$.0^516$	$.0^210$	$.0^279$	.022	.040	.059	.079	.097	.115	.131	.146	.160	.001
	.005	$.0^440$	$.0^250$	.023	.050	.078	.106	.131	.154	.175	.193	.210	.226	.005
	.01	$.0^316$	.010	.038	.072	.106	.137	.165	.189	.211	.231	.249	.264	.01
	.025	$.0^210$	.025	.071	.117	.159	.195	.227	.253	.277	.297	.315	.331	.025
	.05	$.0^240$	.051	.116	.173	.221	.260	.293	.321	.345	.365	.383	.399	.05
	.10	.016	.106	.193	.261	.313	.355	.388	.416	.439	.459	.476	.491	.10
	.25	.104	.291	.406	.480	.532	.570	.600	.623	.643	.659	.671	.684	.25
	.50	.469	.714	.812	.863	.895	.917	.932	.944	.953	.961	.967	.972	.50
	.75	1.39	1.47	1.46	1.44	1.43	1.41	1.40	1.39	1.38	1.38	1.37	1.36	.75
	.90	2.93	2.54	2.33	2.19	2.10	2.04	1.98	1.94	1.91	1.88	1.85	1.83	.90
	.95	4.26	3.40	3.01	2.78	2.62	2.51	2.42	2.36	2.30	2.25	2.21	2.18	.95
	.975	5.72	4.32	3.72	3.38	3.15	2.99	2.87	2.78	2.70	2.64	2.59	2.54	.975
	.99	7.82	5.61	4.72	4.22	3.90	3.67	3.50	3.36	3.26	3.17	3.09	3.03	.99
	.995	9.55	6.66	5.52	4.89	4.49	4.20	3.99	3.83	3.69	3.59	3.50	3.42	.995
	.999	14.0	9.34	7.55	6.59	5.98	5.55	5.23	4.99	4.80	4.64	4.50	4.39	.999
	.9995	16.2	10.6	8.52	7.39	6.68	6.18	5.82	5.54	5.31	5.13	4.98	4.85	.9995

p	15	20	24	30	40	50	60	100	120	200	500	∞	p	n
.0005	.159	.197	.220	.244	.272	.290	.303	.330	.339	.353	.368	.377	.0005	**15**
.001	.181	.219	.242	.266	.294	.313	.325	.352	.360	.375	.388	.398	.001	
.005	.246	.286	.308	.333	.360	.377	.389	.415	.422	.435	.448	.457	.005	
.01	.284	.324	.346	.370	.397	.413	.425	.450	.456	.469	.483	.490	.01	
.025	.349	.389	.410	.433	.458	.474	.485	.508	.514	.526	.538	.546	.025	
.05	.416	.454	.474	.496	.519	.535	.545	.565	.571	.581	.592	.600	.05	
.10	.507	.542	.561	.581	.602	.614	.624	.641	.647	.658	.667	.672	.10	
.25	.701	.728	.742	.757	.772	.782	.788	.802	.805	.812	.818	.822	.25	
.50	1.00	1.01	1.02	1.02	1.03	1.03	1.03	1.04	1.04	1.04	1.04	1.05	.50	
.75	1.43	1.41	1.41	1.40	1.39	1.39	1.38	1.38	1.37	1.37	1.36	1.36	.75	
.90	1.97	1.92	1.90	1.87	1.85	1.83	1.82	1.79	1.79	1.77	1.76	1.76	.90	
.95	2.40	2.33	2.29	2.25	2.20	2.18	2.16	2.12	2.11	2.10	2.08	2.07	.95	
.975	2.86	2.76	2.70	2.64	2.59	2.55	2.52	2.47	2.46	2.44	2.41	2.40	.975	
.99	3.52	3.37	3.29	3.21	3.13	3.08	3.05	2.98	2.96	2.92	2.89	2.87	.99	
.995	4.07	3.88	3.79	3.69	3.59	3.52	3.48	3.39	3.37	3.33	3.29	3.26	.995	
.999	5.54	5.25	5.10	4.95	4.80	4.70	4.64	4.51	4.47	4.41	4.35	4.31	.999	
.9995	6.27	5.93	5.75	5.58	5.40	5.29	5.21	5.06	5.02	4.94	4.87	4.83	.9995	
.0005	.169	.211	.235	.263	.295	.316	.331	.364	.375	.391	.408	.422	.0005	**20**
.001	.191	.233	.258	.286	.318	.339	.354	.386	.395	.413	.429	.441	.001	
.005	.258	.301	.327	.354	.385	.405	.419	.448	.457	.474	.490	.500	.005	
.01	.297	.340	.365	.392	.422	.441	.455	.483	.491	.508	.521	.532	.01	
.025	.363	.406	.430	.456	.484	.503	.514	.541	.548	.562	.575	.585	.025	
.05	.430	.471	.493	.518	.544	.562	.572	.595	.603	.617	.629	.637	.05	
.10	.520	.557	.578	.600	.623	.637	.648	.671	.675	.685	.694	.704	.10	
.25	.708	.736	.751	.767	.784	.794	.801	.816	.820	.827	.835	.840	.25	
.50	.989	1.00	1.01	1.01	1.02	1.02	1.02	1.03	1.03	1.03	1.03	1.03	.50	
.75	1.37	1.36	1.35	1.34	1.33	1.33	1.32	1.31	1.31	1.30	1.30	1.29	.75	
.90	1.84	1.79	1.77	1.74	1.71	1.69	1.68	1.65	1.64	1.63	1.62	1.61	.90	
.95	2.20	2.12	2.08	2.04	1.99	1.97	1.95	1.91	1.90	1.88	1.86	1.84	.95	
.975	2.57	2.46	2.41	2.35	2.29	2.25	2.22	2.17	2.16	2.13	2.10	2.09	.975	
.99	3.09	2.94	2.86	2.78	2.69	2.64	2.61	2.54	2.52	2.48	2.44	2.42	.99	
.995	3.50	3.32	3.22	3.12	3.02	2.96	2.92	2.83	2.81	2.76	2.72	2.69	.995	
.999	4.56	4.29	4.15	4.01	3.86	3.77	3.70	3.58	3.54	3.48	3.42	3.38	.999	
.9995	5.07	4.75	4.58	4.42	4.24	4.15	4.07	3.93	3.90	3.82	3.75	3.70	.9995	
.0005	.174	.218	.244	.274	.309	.331	.349	.384	.395	.416	.434	.449	.0005	**24**
.001	.196	.241	.268	.298	.332	.354	.371	.405	.417	.437	.455	.469	.001	
.005	.264	.310	.337	.367	.400	.422	.437	.469	.479	.498	.515	.527	.005	
.01	.304	.350	.376	.405	.437	.459	.473	.505	.513	.529	.546	.558	.01	
.025	.370	.415	.441	.468	.498	.518	.531	.562	.568	.585	.599	.610	.025	
.05	.437	.480	.504	.530	.558	.575	.588	.613	.622	.637	.649	.659	.05	
.10	.527	.566	.588	.611	.635	.651	.662	.685	.691	.704	.715	.723	.10	
.25	.712	.741	.757	.773	.791	.802	.809	.825	.829	.837	.844	.850	.25	
.50	.983	.994	1.00	1.01	1.01	1.02	1.02	1.02	1.02	1.02	1.03	1.03	.50	
.75	1.35	1.33	1.32	1.31	1.30	1.29	1.29	1.28	1.28	1.27	1.27	1.26	.75	
.90	1.78	1.73	1.70	1.67	1.64	1.62	1.61	1.58	1.57	1.56	1.54	1.53	.90	
.95	2.11	2.03	1.98	1.94	1.89	1.86	1.84	1.80	1.79	1.77	1.75	1.73	.95	
.975	2.44	2.33	2.27	2.21	2.15	2.11	2.08	2.02	2.01	1.98	1.95	1.94	.975	
.99	2.89	2.74	2.66	2.58	2.49	2.44	2.40	2.33	2.31	2.27	2.24	2.21	.99	
.995	3.25	3.06	2.97	2.87	2.77	2.70	2.66	2.57	2.55	2.50	2.46	2.43	.995	
.999	4.14	3.87	3.74	3.59	3.45	3.35	3.29	3.16	3.14	3.07	3.01	2.97	.999	
.9995	4.55	4.25	4.09	3.93	3.76	3.66	3.59	3.44	3.41	3.33	3.27	3.22	.9995	

n	p	m 1	2	3	4	5	6	7	8	9	10	11	12	p
30	.0005	$.0^6 40$	$.0^3 50$	$.0^2 50$	.015	.030	.047	.065	.082	.098	.114	.129	.143	.0005
	.001	$.0^5 16$	$.0^2 10$	$.0^2 80$	.022	.040	.060	.080	.099	.117	.134	.150	.164	.001
	.005	$.0^4 40$	$.0^2 50$	.024	.050	.079	.107	.133	.156	.178	.197	.215	.231	.005
	.01	$.0^3 16$	.010	.038	.072	.107	.138	.167	.192	.215	.235	.254	.270	.01
	.025	$.0^2 10$	.025	.071	.118	.161	.197	.229	.257	.281	.302	.321	.337	.025
	.05	$.0^2 40$	.051	.116	.174	.222	.263	.296	.325	.349	.370	.389	.406	.05
	.10	.016	.106	.193	.262	.315	.357	.391	.420	.443	.464	.481	.497	.10
	.25	.103	.290	.406	.480	.532	.571	.601	.625	.645	.661	.676	.688	.25
	.50	.466	.709	.807	.858	.890	.912	.927	.939	.948	.955	.961	.966	.50
	.75	1.38	1.45	1.44	1.42	1.41	1.39	1.38	1.37	1.36	1.35	1.35	1.34	.75
	.90	2.88	2.49	2.28	2.14	2.05	1.98	1.93	1.88	1.85	1.82	1.79	1.77	.90
	.95	4.17	3.32	2.92	2.69	2.53	2.42	2.33	2.27	2.21	2.16	2.13	2.09	.95
	.975	5.57	4.18	3.59	3.25	3.03	2.87	2.75	2.65	2.57	2.51	2.46	2.41	.975
	.99	7.56	5.39	4.51	4.02	3.70	3.47	3.30	3.17	3.07	2.98	2.91	2.84	.99
	.995	9.18	6.35	5.24	4.62	4.23	3.95	3.74	3.58	3.45	3.34	3.25	3.18	.995
	.999	13.3	8.77	7.05	6.12	5.53	5.12	4.82	4.58	4.39	4.24	4.11	4.00	.999
	.9995	15.2	9.90	7.90	6.82	6.14	5.66	5.31	5.04	4.82	4.65	4.51	4.38	.9995
40	.0005	$.0^6 40$	$.0^3 50$	$.0^2 50$	.016	.030	.048	.066	.084	.100	.117	.132	.147	.0005
	.001	$.0^5 16$	$.0^2 10$	$.0^2 80$	.022	.042	.061	.081	.101	.119	.137	.153	.169	.001
	.005	$.0^4 40$	$.0^2 50$	.024	.051	.080	.108	.135	.159	.181	.201	.220	.237	.005
	.01	$.0^3 16$	.010	.038	.073	.108	.140	.169	.195	.219	.240	.259	.276	.01
	.025	$.0^3 99$	.025	.071	.119	.162	.199	.232	.260	.285	.307	.327	.344	.025
	.05	$.0^2 40$	.051	.116	.175	.224	.265	.299	.329	.354	.376	.395	.412	.05
	.10	.016	.106	.194	.263	.317	.360	.394	.424	.448	.469	.488	.504	.10
	.25	.103	.290	.405	.480	.533	.572	.603	.627	.647	.664	.680	.691	.25
	.50	.463	.705	.802	.854	.885	.907	.922	.934	.943	.950	.956	.961	.50
	.75	1.36	1.44	1.42	1.40	1.39	1.37	1.36	1.35	1.34	1.33	1.32	1.31	.75
	.90	2.84	2.44	2.23	2.09	2.00	1.93	1.87	1.83	1.79	1.76	1.73	1.71	.90
	.95	4.08	3.23	2.84	2.61	2.45	2.34	2.25	2.18	2.12	2.08	2.04	2.00	.95
	.975	5.42	4.05	3.46	3.13	2.90	2.74	2.62	2.53	2.45	2.39	2.33	2.29	.975
	.99	7.31	5.18	4.31	3.83	3.51	3.29	3.12	2.99	2.89	2.80	2.73	2.66	.99
	.995	8.83	6.07	4.98	4.37	3.99	3.71	3.51	3.35	3.22	3.12	3.03	2.95	.995
	.999	12.6	8.25	6.60	5.70	5.13	4.73	4.44	4.21	4.02	3.87	3.75	3.64	.999
	.9995	14.4	9.25	7.33	6.30	5.64	5.19	4.85	4.59	4.38	4.21	4.07	3.95	.9995
60	.0005	$.0^6 40$	$.0^3 50$	$.0^2 51$	.016	.031	.048	.067	.085	.103	.120	.136	.152	.0005
	.001	$.0^5 16$	$.0^2 10$	$.0^2 80$	.022	.041	.062	.083	.103	.122	.140	.157	.174	.001
	.005	$.0^4 40$	$.0^2 50$	.024	.051	.081	.110	.137	.162	.185	.206	.225	.243	.005
	.01	$.0^3 16$	.010	.038	.073	.109	.142	.172	.199	.223	.245	.265	.283	.01
	.025	$.0^3 99$	.025	.071	.120	.163	.202	.235	.264	.290	.313	.333	.351	.025
	.05	$.0^2 40$	.051	.116	.176	.226	.267	.303	.333	.359	.382	.402	.419	.05
	.10	.016	.106	.194	.264	.318	.362	.398	.428	.453	.475	.493	.510	.10
	.25	.102	.289	.405	.480	.534	.573	.604	.629	.650	.667	.680	.695	.25
	.50	.461	.701	.798	.849	.880	.901	.917	.928	.937	.945	.951	.956	.50
	.75	1.35	1.42	1.41	1.38	1.37	1.35	1.33	1.32	1.31	1.30	1.29	1.29	.75
	.90	2.79	2.39	2.18	2.04	1.95	1.87	1.82	1.77	1.74	1.71	1.68	1.66	.90
	.95	4.00	3.15	2.76	2.53	2.37	2.25	2.17	2.10	2.04	1.99	1.95	1.92	.95
	.975	5.29	3.93	3.34	3.01	2.79	2.63	2.51	2.41	2.33	2.27	2.22	2.17	.975
	.99	7.08	4.98	4.13	3.65	3.34	3.12	2.95	2.82	2.72	2.63	2.56	2.50	.99
	.995	8.49	5.80	4.73	4.14	3.76	3.49	3.29	3.13	3.01	2.90	2.82	2.74	.995
	.999	12.0	7.76	6.17	5.31	4.76	4.37	4.09	3.87	3.69	3.54	3.43	3.31	.999
	.9995	13.6	8.65	6.81	5.82	5.20	4.76	4.44	4.18	3.98	3.82	3.69	3.57	.9995

p \ m	15	20	24	30	40	50	60	100	120	200	500	∞	p	n
.0005	.179	.226	.254	.287	.325	.350	.369	.410	.420	.444	.467	.483	.0005	**30**
.001	.202	.250	.278	.311	.348	.373	.391	.431	.442	.465	.488	.503	.001	
.005	.271	.320	.349	.381	.416	.441	.457	.495	.504	.524	.543	.559	.005	
.01	.311	.360	.388	.419	.454	.476	.493	.529	.538	.559	.575	.590	.01	
.025	.378	.426	.453	.482	.515	.535	.551	.585	.592	.610	.625	.639	.025	
.05	.445	.490	.516	.543	.573	.592	.606	.637	.644	.658	.676	.685	.05	
.10	.534	.575	.598	.623	.649	.667	.678	.704	.710	.725	.735	.746	.10	
.25	.716	.746	.763	.780	.798	.810	.818	.835	.839	.848	.856	.862	.25	
.50	.978	.989	.994	1.00	1.01	1.01	1.01	1.02	1.02	1.02	1.02	1.02	.50	
.75	1.32	1.30	1.29	1.28	1.27	1.26	1.26	1.25	1.24	1.24	1.23	1.23	.75	
.90	1.72	1.67	1.64	1.61	1.57	1.55	1.54	1.51	1.50	1.48	1.47	1.46	.90	
.95	2.01	1.93	1.89	1.84	1.79	1.76	1.74	1.70	1.68	1.66	1.64	1.62	.95	
.975	2.31	2.20	2.14	2.07	2.01	1.97	1.94	1.88	1.87	1.84	1.81	1.79	.975	
.99	2.70	2.55	2.47	2.39	2.30	2.25	2.21	2.13	2.11	2.07	2.03	2.01	.99	
.995	3.01	2.82	2.73	2.63	2.52	2.46	2.42	2.32	2.30	2.25	2.21	2.18	.995	
.999	3.75	3.49	3.36	3.22	3.07	2.98	2.92	2.79	2.76	2.69	2.63	2.59	.999	
.9995	4.10	3.80	3.65	3.48	3.32	3.22	3.15	3.00	2.97	2.89	2.82	2.78	.9995	
.0005	.185	.236	.266	.301	.343	.373	.393	.441	.453	.480	.504	.525	.0005	**40**
.001	.209	.259	.290	.326	.367	.396	.415	.461	.473	.500	.524	.545	.001	
.005	.279	.331	.362	.396	.436	.463	.481	.524	.534	.559	.581	.599	.005	
.01	.319	.371	.401	.435	.473	.498	.516	.556	.567	.592	.613	.628	.01	
.025	.387	.437	.466	.498	.533	.556	.573	.610	.620	.641	.662	.674	.025	
.05	.454	.502	.529	.558	.591	.613	.627	.658	.669	.685	.704	.717	.05	
.10	.542	.585	.609	.636	.664	.683	.696	.724	.731	.747	.762	.772	.10	
.25	.720	.752	.769	.787	.806	.819	.828	.846	.851	.861	.870	.877	.25	
.50	.972	.983	.989	.994	1.00	1.00	1.01	1.01	1.01	1.01	1.02	1.02	.50	
.75	1.30	1.28	1.26	1.25	1.24	1.23	1.22	1.21	1.21	1.20	1.19	1.19	.75	
.90	1.66	1.61	1.57	1.54	1.51	1.48	1.47	1.43	1.42	1.41	1.39	1.38	.90	
.95	1.92	1.84	1.79	1.74	1.69	1.66	1.64	1.59	1.58	1.55	1.53	1.51	.95	
.975	2.18	2.07	2.01	1.94	1.88	1.83	1.80	1.74	1.72	1.69	1.66	1.64	.975	
.99	2.52	2.37	2.29	2.20	2.11	2.06	2.02	1.94	1.92	1.87	1.83	1.80	.99	
.995	2.78	2.60	2.50	2.40	2.30	2.23	2.18	2.09	2.06	2.01	1.96	1.93	.995	
.999	3.40	3.15	3.01	2.87	2.73	2.64	2.57	2.44	2.41	2.34	2.28	2.23	.999	
.9995	3.68	3.39	3.24	3.08	2.92	2.82	2.74	2.60	2.57	2.49	2.41	2.37	.9995	
.0005	.192	.246	.278	.318	.365	.398	.421	.478	.493	.527	.561	.585	.0005	**60**
.001	.216	.270	.304	.343	.389	.421	.444	.497	.512	.545	.579	.602	.001	
.005	.287	.343	.376	.414	.458	.488	.510	.559	.572	.602	.633	.652	.005	
.01	.328	.383	.416	.453	.495	.524	.545	.592	.604	.633	.658	.679	.01	
.025	.396	.450	.481	.515	.555	.581	.600	.641	.654	.680	.704	.720	.025	
.05	.463	.514	.543	.575	.611	.633	.652	.690	.700	.719	.746	.759	.05	
.10	.550	.596	.622	.650	.682	.703	.717	.750	.758	.776	.793	.806	.10	
.25	.725	.758	.776	.796	.816	.830	.840	.860	.865	.877	.888	.896	.25	
.50	.967	.978	.983	.989	.994	.998	1.00	1.00	1.01	1.01	1.01	1.01	.50	
.75	1.27	1.25	1.24	1.22	1.21	1.20	1.19	1.17	1.17	1.16	1.15	1.15	.75	
.90	1.60	1.54	1.51	1.48	1.44	1.41	1.40	1.36	1.35	1.33	1.31	1.29	.90	
.95	1.84	1.75	1.70	1.65	1.59	1.56	1.53	1.48	1.47	1.44	1.41	1.39	.95	
.975	2.06	1.94	1.88	1.82	1.74	1.70	1.67	1.60	1.58	1.54	1.51	1.48	.975	
.99	2.35	2.20	2.12	2.03	1.94	1.88	1.84	1.75	1.73	1.68	1.63	1.60	.99	
.995	2.57	2.39	2.29	2.19	2.08	2.01	1.96	1.86	1.83	1.78	1.73	1.69	.995	
.999	3.08	2.83	2.69	2.56	2.41	2.31	2.25	2.11	2.09	2.01	1.93	1.89	.999	
.9995	3.30	3.02	2.87	2.71	2.55	2.45	2.38	2.23	2.19	2.11	2.03	1.98	.9995	

n	p	1	2	3	4	5	6	7	8	9	10	11	12	p
120	.0005	$.0^640$	$.0^350$	$.0^251$	.016	.031	.049	.067	.087	.105	.123	.140	.156	.0005
	.001	$.0^516$	$.0^210$	$.0^281$	.023	.042	.063	.084	.105	.125	.144	.162	.179	.001
	.005	$.0^439$	$.0^250$	.024	.051	.081	.111	.139	.165	.189	.211	.230	.249	.005
	.01	$.0^316$	.010	.038	.074	.110	.143	.174	.202	.227	.250	.271	.290	.01
	.025	$.0^399$	.025	.072	.120	.165	.204	.238	.268	.295	.318	.340	.359	.025
	.05	$.0^239$	.051	.117	.177	.227	.270	.306	.337	.364	.388	.408	.427	.05
	.10	.016	.105	.194	.265	.320	.365	.401	.432	.458	.480	.500	.518	.10
	.25	.102	.288	.405	.481	.534	.574	.606	.631	.652	.670	.685	.699	.25
	.50	.458	.697	.793	.844	.875	.896	.912	.923	.932	.939	.945	.950	.50
	.75	1.34	1.40	1.39	1.37	1.35	1.33	1.31	1.30	1.29	1.28	1.27	1.26	.75
	.90	2.75	2.35	2.13	1.99	1.90	1.82	1.77	1.72	1.68	1.65	1.62	1.60	.90
	.95	3.92	3.07	2.68	2.45	2.29	2.18	2.09	2.02	1.96	1.91	1.87	1.83	.95
	.975	5.15	3.80	3.23	2.89	2.67	2.52	2.39	2.30	2.22	2.16	2.10	2.05	.975
	.99	6.85	4.79	3.95	3.48	3.17	2.96	2.79	2.66	2.56	2.47	2.40	2.34	.99
	.995	8.18	5.54	4.50	3.92	3.55	3.28	3.09	2.93	2.81	2.71	2.62	2.54	.995
	.999	11.4	7.32	5.79	4.95	4.42	4.03	3.77	3.55	3.38	3.24	3.12	3.02	.999
	.9995	12.8	8.10	6.34	5.39	4.79	4.37	4.07	3.82	3.63	3.47	3.34	3.22	.9995
∞	.0005	$.0^639$	$.0^350$	$.0^251$	.016	.032	.050	.069	.088	.108	.127	.144	.161	.0005
	.001	$.0^516$	$.0^210$	$.0^281$	.023	.042	.063	.085	.107	.128	.148	.167	.185	.001
	.005	$.0^439$	$.0^250$	.024	.052	.082	.113	.141	.168	.193	.216	.236	.256	.005
	.01	$.0^316$	.010	.038	.074	.111	.145	.177	.206	.232	.256	.278	.298	.01
	.025	$.0^398$	.025	.072	.121	.166	.206	.241	.272	.300	.325	.347	.367	.025
	.05	$.0^239$	.051	.117	.178	.229	.273	.310	.342	.369	.394	.417	.436	.05
	.10	.016	.105	.195	.266	.322	.367	.405	.436	.463	.487	.508	.525	.10
	.25	.102	.288	.404	.481	.535	.576	.608	.634	.655	.674	.690	.703	.25
	.50	.455	.693	.789	.839	.870	.891	.907	.918	.927	.934	.939	.945	.50
	.75	1.32	1.39	1.37	1.35	1.33	1.31	1.29	1.28	1.27	1.25	1.24	1.24	.75
	.90	2.71	2.30	2.08	1.94	1.85	1.77	1.72	1.67	1.63	1.60	1.57	1.55	.90
	.95	3.84	3.00	2.60	2.37	2.21	2.10	2.01	1.94	1.88	1.83	1.79	1.75	.95
	.975	5.02	3.69	3.12	2.79	2.57	2.41	2.29	2.19	2.11	2.05	1.99	1.94	.975
	.99	6.63	4.61	3.78	3.32	3.02	2.80	2.64	2.51	2.41	2.32	2.25	2.18	.99
	.995	7.88	5.30	4.28	3.72	3.35	3.09	2.90	2.74	2.62	2.52	2.43	2.36	.995
	.999	10.8	6.91	5.42	4.62	4.10	3.74	3.47	3.27	3.10	2.96	2.84	2.74	.999
	.9995	12.1	7.60	5.91	5.00	4.42	4.02	3.72	3.48	3.30	3.14	3.02	2.90	.9995

p \ *m*	15	20	24	30	40	50	60	100	120	200	500	∞	*p*	*n*
.0005	.199	.256	.293	.338	.390	.429	.458	.524	.543	.578	.614	.676	.0005	**120**
.001	.223	.282	.319	.363	.415	.453	.480	.542	.568	.595	.631	.691	.001	
.005	.297	.356	.393	.434	.484	.520	.545	.605	.623	.661	.702	.733	.005	
.01	.338	.397	.433	.474	.522	.556	.579	.636	.652	.688	.725	.755	.01	
.025	.406	.464	.498	.536	.580	.611	.633	.684	.698	.729	.762	.789	.025	
.05	.473	.527	.559	.594	.634	.661	.682	.727	.740	.767	.785	.819	.05	
.10	.560	.609	.636	.667	.702	.726	.742	.781	.791	.815	.838	.855	.10	
.25	.730	.765	.784	.805	.828	.843	.853	.877	.884	.897	.911	.923	.25	
.50	.961	.972	.978	.983	.989	.992	.994	1.00	1.00	1.00	1.01	1.01	.50	
.75	1.24	1.22	1.21	1.19	1.18	1.17	1.16	1.14	1.13	1.12	1.11	1.10	.75	
.90	1.55	1.48	1.45	1.41	1.37	1.34	1.32	1.27	1.26	1.24	1.21	1.19	.90	
.95	1.75	1.66	1.61	1.55	1.50	1.46	1.43	1.37	1.35	1.32	1.28	1.25	.95	
.975	1.95	1.82	1.76	1.69	1.61	1.56	1.53	1.45	1.43	1.39	1.34	1.31	.975	
.99	2.19	2.03	1.95	1.86	1.76	1.70	1.66	1.56	1.53	1.48	1.42	1.38	.99	
.995	2.37	2.19	2.09	1.98	1.87	1.80	1.75	1.64	1.61	1.54	1.48	1.43	.995	
.999	2.78	2.53	2.40	2.26	2.11	2.02	1.95	1.82	1.76	1.70	1.62	1.54	.999	
.9995	2.96	2.67	2.53	2.38	2.21	2.11	2.01	1.88	1.84	1.75	1.67	1.60	.9995	
.0005	.207	.270	.311	.360	.422	.469	.505	.599	.624	.704	.804	1.00	.0005	**∞**
.001	.232	.296	.338	.386	.448	.493	.527	.617	.649	.719	.819	1.00	.001	
.005	.307	.372	.412	.460	.518	.559	.592	.671	.699	.762	.843	1.00	.005	
.01	.349	.413	.452	.499	.554	.595	.625	.699	.724	.782	.858	1.00	.01	
.025	.418	.480	.517	.560	.611	.645	.675	.741	.763	.813	.878	1.00	.025	
.05	.484	.543	.577	.617	.663	.694	.720	.781	.797	.840	.896	1.00	.05	
.10	.570	.622	.652	.687	.726	.752	.774	.826	.838	.877	.919	1.00	.10	
.25	.736	.773	.793	.816	.842	.860	.872	.901	.910	.932	.957	1.00	.25	
.50	.956	.967	.972	.978	.983	.987	.989	.993	.994	.997	.999	1.00	.50	
.75	1.22	1.19	1.18	1.16	1.14	1.13	1.12	1.09	1.08	1.07	1.04	1.00	.75	
.90	1.49	1.42	1.38	1.34	1.30	1.26	1.24	1.18	1.17	1.13	1.08	1.00	.90	
.95	1.67	1.57	1.52	1.46	1.39	1.35	1.32	1.24	1.22	1.17	1.11	1.00	.95	
.975	1.83	1.71	1.64	1.57	1.48	1.43	1.39	1.30	1.27	1.21	1.13	1.00	.975	
.99	2.04	1.88	1.79	1.70	1.59	1.52	1.47	1.36	1.32	1.25	1.15	1.00	.99	
.995	2.19	2.00	1.90	1.79	1.67	1.59	1.53	1.40	1.36	1.28	1.17	1.00	.995	
.999	2.51	2.27	2.13	1.99	1.84	1.73	1.66	1.49	1.45	1.34	1.21	1.00	.999	
.9995	2.65	2.37	2.22	2.07	1.91	1.79	1.71	1.53	1.48	1.36	1.22	1.00	.9995	

TABLE A.5 UPPER PERCENTILES OF STUDENTIZED RANGE DISTRIBUTIONS

Studentized range distribution with k and ν degrees of freedom

Area = α

$Q_{\alpha,k,\nu}$

ν	k / $1-\alpha$	2	3	4	5	6	7	8	9	10	11	12	13	14	15	16
1	0.95	18.0	27.0	32.8	37.1	40.4	43.1	45.4	47.4	49.1	50.6	52.0	53.2	54.3	55.4	56.3
	0.99	90.0	135	164	186	202	216	227	237	246	253	260	266	272	277	282
2	0.95	6.09	8.3	9.8	10.9	11.7	12.4	13.0	13.5	14.0	14.4	14.7	15.1	15.4	15.7	15.9
	0.99	14.0	19.0	22.3	24.7	26.6	28.2	29.5	30.7	31.7	32.6	33.4	34.1	34.8	35.4	36.0
3	0.95	4.50	5.91	6.82	7.50	8.04	8.48	8.85	9.18	9.46	9.72	9.95	10.2	10.4	10.5	10.7
	0.99	8.26	10.6	12.2	13.3	14.2	15.0	15.6	16.2	16.7	17.1	17.5	17.9	18.2	18.5	18.8
4	0.95	3.93	5.04	5.76	6.29	6.71	7.05	7.35	7.60	7.83	8.03	8.21	8.37	8.52	8.66	8.79
	0.99	6.51	8.12	9.17	9.96	10.6	11.1	11.5	11.9	12.3	12.6	12.8	13.1	13.3	13.5	13.7
5	0.95	3.64	4.60	5.22	5.67	6.03	6.33	6.58	6.80	6.99	7.17	7.32	7.47	7.60	7.72	7.83
	0.99	5.70	6.97	7.80	8.42	8.91	9.32	9.67	9.97	10.2	10.5	10.7	10.9	11.1	11.2	11.4
6	0.95	3.46	4.34	4.90	5.31	5.63	5.89	6.12	6.32	6.49	6.65	6.79	6.92	7.03	7.14	7.24
	0.99	5.24	6.33	7.03	7.56	7.97	8.32	8.61	8.87	9.10	9.30	9.49	9.65	9.81	9.95	10.1
7	0.95	3.34	4.16	4.68	5.06	5.36	5.61	5.82	6.00	6.16	6.30	6.43	6.55	6.66	6.76	6.85
	0.99	4.95	5.92	6.54	7.01	7.37	7.68	7.94	8.17	8.37	8.55	8.71	8.86	9.00	9.12	9.24
8	0.95	3.26	4.04	4.53	4.89	5.17	5.40	5.60	5.77	5.92	6.05	6.18	6.29	6.39	6.48	6.57
	0.99	4.74	5.63	6.20	6.63	6.96	7.24	7.47	7.68	7.87	8.03	8.18	8.31	8.44	8.55	8.66
9	0.95	3.20	3.95	4.42	4.76	5.02	5.24	5.43	5.60	5.74	5.87	5.98	6.09	6.19	6.28	6.36
	0.99	4.60	5.43	5.96	6.35	6.66	6.91	7.13	7.32	7.49	7.65	7.78	7.91	8.03	8.13	8.23
10	0.95	3.15	3.88	4.33	4.65	4.91	5.12	5.30	5.46	5.60	5.72	5.83	5.93	6.03	6.11	6.20
	0.99	4.48	5.27	5.77	6.14	6.43	6.67	6.87	7.05	7.21	7.36	7.48	7.60	7.71	7.81	7.91
11	0.95	3.11	3.82	4.26	4.57	4.82	5.03	5.20	5.35	5.49	5.61	5.71	5.81	5.90	5.99	6.06
	0.99	4.39	5.14	5.62	5.97	6.25	6.48	6.67	6.84	6.99	7.13	7.25	7.36	7.46	7.56	7.65
12	0.95	3.08	3.77	4.20	4.51	4.75	4.95	5.12	5.27	5.40	5.51	5.62	5.71	5.80	5.88	5.95
	0.99	4.32	5.04	5.50	5.84	6.10	6.32	6.51	6.67	6.81	6.94	7.06	7.17	7.26	7.36	7.44
13	0.95	3.06	3.73	4.15	4.45	4.69	4.88	5.05	5.19	5.32	5.43	5.53	5.63	5.71	5.79	5.86
	0.99	4.26	4.96	5.40	5.73	5.98	6.19	6.37	6.53	6.67	6.79	6.90	7.01	7.10	7.19	7.27
14	0.95	3.03	3.70	4.11	4.41	4.64	4.83	4.99	5.13	5.25	5.36	5.46	5.55	5.64	5.72	5.79
	0.99	4.21	4.89	5.32	5.63	5.88	6.08	6.26	6.41	6.54	6.66	6.77	6.87	6.96	7.05	7.12

TABLE A.5 UPPER PERCENTILES OF STUDENTIZED RANGE DISTRIBUTIONS (cont.)

v	k / $1-\alpha$	2	3	4	5	6	7	8	9	10	11	12	13	14	15	16
15	0.95	3.01	3.67	4.08	4.37	4.60	4.78	4.94	5.08	5.20	5.31	5.40	5.49	5.58	5.65	5.72
	0.99	4.17	4.83	5.25	5.56	5.80	5.99	6.16	6.31	6.44	6.55	6.66	6.76	6.84	6.93	7.00
16	0.95	3.00	3.65	4.05	4.33	4.56	4.74	4.90	5.03	5.15	5.26	5.35	5.44	5.52	5.59	5.66
	0.99	4.13	4.78	5.19	5.49	5.72	5.92	6.08	6.22	6.35	6.46	6.56	6.66	6.74	6.82	6.90
17	0.95	2.98	3.63	4.02	4.30	4.52	4.71	4.86	4.99	5.11	5.21	5.31	5.39	5.47	5.55	5.61
	0.99	4.10	4.74	5.14	5.43	5.66	5.85	6.01	6.15	6.27	6.38	6.48	6.57	6.66	6.73	6.80
18	0.95	2.97	3.61	4.00	4.28	4.49	4.67	4.82	4.96	5.07	5.17	5.27	5.35	5.43	5.50	5.57
	0.99	4.07	4.70	5.09	5.38	5.60	5.79	5.94	6.08	6.20	6.31	6.41	6.50	6.58	6.65	6.72
19	0.95	2.96	3.59	3.98	4.25	4.47	4.65	4.79	4.92	5.04	5.14	5.23	5.32	5.39	5.46	5.53
	0.99	4.05	4.67	5.05	5.33	5.55	5.73	5.89	6.02	6.14	6.25	6.34	6.43	6.51	6.58	6.65
20	0.95	2.95	3.58	3.96	4.23	4.45	4.62	4.77	4.90	5.01	5.11	5.20	5.28	5.36	5.43	5.49
	0.99	4.02	4.64	5.02	5.29	5.51	5.69	5.84	5.97	6.09	6.19	6.29	6.37	6.45	6.52	6.59
24	0.95	2.92	3.53	3.90	4.17	4.37	4.54	4.68	4.81	4.92	5.01	5.10	5.18	5.25	5.32	5.38
	0.99	3.96	4.54	4.91	5.17	5.37	5.54	5.69	5.81	5.92	6.02	6.11	6.19	6.26	6.33	6.39
30	0.95	2.89	3.49	3.84	4.10	4.30	4.46	4.60	4.72	4.83	4.92	5.00	5.08	5.15	5.21	5.27
	0.99	3.89	4.45	4.80	5.05	5.24	5.40	5.54	5.65	5.76	5.85	5.93	6.01	6.08	6.14	6.20
40	0.95	2.86	3.44	3.79	4.04	4.23	4.39	4.52	4.63	4.74	4.82	4.91	4.98	5.05	5.11	5.16
	0.99	3.82	4.37	4.70	4.93	5.11	5.27	5.39	5.50	5.60	5.69	5.77	5.84	5.90	5.96	6.02
60	0.95	2.83	3.40	3.74	3.98	4.16	4.31	4.44	4.55	4.65	4.73	4.81	4.88	4.94	5.00	5.06
	0.99	3.76	4.28	4.60	4.82	4.99	5.13	5.25	5.36	5.45	5.53	5.60	5.67	5.73	5.79	5.84
120	0.95	2.80	3.36	3.69	3.92	4.10	4.24	4.36	4.48	4.56	4.64	4.72	4.78	4.84	4.90	4.95
	0.99	3.70	4.20	4.50	4.71	4.87	5.01	5.12	5.21	5.30	5.38	5.44	5.51	5.56	5.61	5.66
∞	0.95	2.77	3.31	3.63	3.86	4.03	4.17	4.29	4.39	4.47	4.55	4.62	4.68	4.74	4.80	4.85
	0.99	3.64	4.12	4.40	4.60	4.76	4.88	4.99	5.08	5.16	5.23	5.29	5.35	5.40	5.45	5.49

SOURCE: Olive Jean Dunn and Virginia A. Clark, *Applied Statistics: Analysis of Variance and Regression* (New York: Wiley, 1974), pp. 371–372.

TABLE A.6 UPPER AND LOWER PERCENTILES OF THE WILCOXON SIGNED RANK STATISTIC, W

	w_1^*	w_2^*	$P(W \leq w_1^*) = P(W \geq w_2^*)$
$n = 4$	0	10	0.062
	1	9	0.125
$n = 5$	0	15	0.031
	1	14	0.062
	2	13	0.094
	3	12	0.156
$n = 6$	0	21	0.016
	1	20	0.031
	2	19	0.047
	3	18	0.078
	4	17	0.109
	5	16	0.156
$n = 7$	0	28	0.008
	1	27	0.016
	2	26	0.023
	3	25	0.039
	4	24	0.055
	5	23	0.078
	6	22	0.109
	7	21	0.148
$n = 8$	0	36	0.004
	1	35	0.008
	2	34	0.012
	3	33	0.020
	4	32	0.027
	5	31	0.039
	6	30	0.055
	7	29	0.074
	8	28	0.098
	9	27	0.125
$n = 9$	1	44	0.004
	2	43	0.006
	3	42	0.010
	4	41	0.014
	5	40	0.020
	6	39	0.027
	7	38	0.037
	8	37	0.049
	9	36	0.064
	10	35	0.082
	11	34	0.102
	12	33	0.125

SOURCE: Wilfrid J. Dixon and Frank J. Massey, Jr., *Introduction to Statistical Analysis*, 2nd. ed. (New York: McGraw-Hill, 1957), pp. 443–444.

TABLE A.6 UPPER AND LOWER PERCENTILES OF THE WILCOXON SIGNED RANK STATISTIC, W (cont.)

	w_1^*	w_2^*	$P(W \leq w_1^*) = P(W \geq w_2^*)$
$n = 10$	3	52	0.005
	4	51	0.007
	5	50	0.010
	6	49	0.014
	7	48	0.019
	8	47	0.024
	9	46	0.032
	10	45	0.042
	11	44	0.053
	12	43	0.065
	13	42	0.080
	14	41	0.097
	15	40	0.116
	16	39	0.138
$n = 11$	5	61	0.005
	6	60	0.007
	7	59	0.009
	8	58	0.012
	9	57	0.016
	10	56	0.021
	11	55	0.027
	12	54	0.034
	13	53	0.042
	14	52	0.051
	15	51	0.062
	16	50	0.074
	17	49	0.087
	18	48	0.103
	19	47	0.120
	20	46	0.139
$n = 12$	7	71	0.005
	8	70	0.006
	9	69	0.008
	10	68	0.010
	11	67	0.013
	12	66	0.017
	13	65	0.021
	14	64	0.026
	15	63	0.032
	16	62	0.039
	17	61	0.046
	18	60	0.055
	19	59	0.065
	20	58	0.076
	21	57	0.088
	22	56	0.102
	23	55	0.117
	24	54	0.133

Answers to Odd-Numbered Exercises

CHAPTER 2

2.2.1 $S = \{(s,s,s),\ (s,s,f),\ (s,f,s),\ (f,s,s),\ (s,f,f),\ (f,s,f),\ (f,f,s),\ (f,f,f)\};\ A = \{(s,f,s),\ (f,s,s)\};\ B = \{(f,f,f)\}$

2.2.3 $(1,3,4),(2,3,4),(1,3,5),(2,3,5),(1,3,6),(2,3,6)$

2.2.5 Let $I_i = i$th innocent suspect, $i = 1,2,3$, and $G_i = i$th guilty suspect, $i = 1,2$. Then $S = \{(I_1,I_2),\ (I_1,I_3),\ (I_1,G_1),\ (I_1,G_2),\ (I_2,I_3),\ (I_2,G_1),\ (I_2,G_2),\ (I_3,G_1),\ (I_3,G_2),\ (G_1,G_2)\};\ A = \{(I_1,I_2),(I_1,I_3),(I_1,G_1),(I_1,G_2),(I_2,I_3),(I_2,G_1),(I_2,G_2),(I_3,G_1),(I_3,G_2)\}.$

2.2.7 $(9,9),(9,\text{no }9\text{ or }7,9),(9,\text{no }9\text{ or }7,\text{no }9\text{ or }7,9),\ldots,(9,\text{no }9\text{ or }7,\ldots,\text{no }9\text{ or }7,9)$

2.2.11 $A = (A_{11} \cap A_{21}) \cup (A_{12} \cap A_{22})$

2.2.15 270

2.2.19 $N(A \cup B \cup C) = N(A) + N(B) + N(C) - N(A \cap B) - N(A \cap C) - N(B \cap C) + N(A \cap B \cap C)$

2.3.1 (a) $\frac{1}{12}$ (b) $\frac{11}{12}$ (c) $\frac{5}{12}$

2.3.3 $(-1 + \sqrt{5})/2$

2.3.5 *Hint*: Use the fact that $P(A \cup B) \le 1$.

2.3.7 $P(A \cup B) = P(A) + P(B) - 1 + P(A^C \cup B^C)$

2.4.1 With ace–six flats, $P(\text{sum} = 7) = \frac{3}{16}$; with fair dice, $P(\text{sum} = 7) = \frac{1}{6}$.

2.4.3 $\frac{3}{10}$ **2.4.5** $\frac{3}{10}$ **2.4.7** $\frac{12}{216}$

2.5.1 $P(A) = \frac{3}{4}$ **2.5.3** 0.82   **2.5.5** $(1 - e^{-4})/(1 - e^{-9})$

2.5.7 $\sqrt{\pi/4}$ **2.6.1** (a) 0.40   (b) 0.30   (c) 0.23

2.6.3 Intuitively, $P(80 \le t \le 85 \mid t \ge 70)$ should be greater than $P(80 \le t \le 85)$. In Example 2.6.3, it is: $0.19 > 0.0313$.

2.6.5. *Hint*: Use Theorem 2.3.4.

2.6.7 *Hint*: Compare the definitions of $P(A \mid B)$ and $P(B \mid A)$.

2.6.9 $\frac{5}{8}$ **2.6.11** $\frac{7}{15}$ **2.6.13** $0.005;\ \frac{1}{360},\ 360$

2.6.15 0.13   **2.6.17** $\frac{2}{3}$ **2.6.19** $\frac{1}{13}$

2.6.21 $\frac{11}{16}$ **2.6.23** $\frac{4}{7}$ **2.6.25** 0.015

2.6.27 $\frac{7}{16}$ **2.6.29** 0.078   **2.6.31** 14

2.7.1 No, because $P(A \mid B) = 0 \ne P(A)$.

2.7.3 No, because $(0.35)(0.40) = 0.14 \ne 0.12;\ 0.37$.

2.7.5 $P(A \cap B) = \frac{6}{36} = (\frac{1}{2})(\frac{1}{3}) = P(A) \cdot P(B);\qquad P(A \cap C) = \frac{3}{36} = (\frac{1}{2})(\frac{1}{6}) = P(A) \cdot P(C);$ $P(B \cap C) = \frac{2}{36} = (\frac{1}{3})(\frac{1}{6}) = P(B) \cdot P(C); P(A \cap B \cap C) = \frac{1}{36} = (\frac{1}{2})(\frac{1}{3})(\frac{1}{6}) = P(A) \cdot P(B) \cdot P(C)$

2.7.7 *Hint*: $P(A_1 \cup A_2 \cup \cdots \cup A_n) = 1 - P(A_1^C \cap A_2^C \cap \cdots \cap A_n^C)$.

2.7.9 $\frac{29}{90}$

2.7.11 There is no reason to assume that the six characteristics listed in Table 2.7.1 are necessarily independent. If they are not, the prosecutor's probability is incorrect. Also, given the large population of Los Angeles, the likelihood of two people fitting the eyewitness's descriptions is not as small as the "1 in 12 million" argument would suggest (see Case Study 2.11.1).

2.7.13 0.952   **2.7.15** $1 - (\frac{3}{7})^{rm}$ **2.7.17** 0.20

2.8.1 25   **2.8.3** 7   **2.8.5** $P(\text{fair}) = 1/(1 + 2^n)$

2.8.7 $(\frac{5}{6})^{k-1} \cdot (\frac{1}{6});\ \frac{5}{11}$ **2.9.1** 1680   **2.9.3** 45; aeu, cdx

2.9.5 (a) 6,760,000   (b) 3,407,040   (c) 6,759,324

2.9.7 256   **2.9.9** 4   **2.9.11** 24; 4

2.9.13 4.6

2.9.17 $2 \cdot 6 \cdot 8!$

2.9.25 95,550

2.10.1 $n/2^{n-1}$

2.10.5 $13 / \binom{15}{3}$

2.9.15 (a) 8!   (b) $2 \cdot (4!)^2$

2.9.23 10

2.9.27 $\binom{11}{2} \cdot \binom{8}{3}$

2.10.3 $2^{50} / \binom{100}{50}$

2.10.7 $4! \cdot 6!/10!$

2.10.9 $7!/7^7; 7/7^7$. We are assuming that each person is equally likely to want to get off at any of the floors 2 through 8. In some cases this would be an unreasonable assumption; persons living on the lower floors might frequently walk up to their apartments, implying that a random person on the elevator would have a greater than one-seventh chance of getting off at, say, the eighth floor.

2.10.11 See Table 2.10.1.

2.10.13 $\frac{10}{19}$

2.10.15 $\left[\binom{2}{1} \cdot \binom{2}{1} \right]^4 \cdot \binom{32}{4} / \binom{48}{12}$

2.10.17 $\binom{n+r}{2}^n / 2^n$

2.11.1 $\binom{54}{7} / \binom{98}{7}; \binom{54}{6}\binom{44}{1} / \binom{98}{7}; \binom{54}{5}\binom{44}{2} / \binom{98}{7}$

2.11.3 $\left[\binom{6}{2}\binom{26}{4} + \binom{6}{1}\binom{26}{5} + \binom{6}{0}\binom{26}{6} \right] / \binom{32}{6}$

2.11.5 4

2.11.7 A "three/one" distribution is more likely.

2.11.9 $\binom{25}{3} (0.06)^3 (0.94)^{22}$

2.11.11 $\frac{1}{4}$

2.11.13 0.004

2.11.15 0.61

CHAPTER 3

3.2.1 $P(Y = 3) = \frac{1}{3}, P(Y = 6) = \frac{2}{3}$

3.2.3 (a) $P((0,0)) = \frac{24}{315}, \quad P((0,1)) = \frac{96}{315}, \quad P((0,2)) = \frac{60}{315}, \quad P((1,0)) = \frac{18}{315}, \quad P((1,1)) = \frac{72}{315},$
$P((1,2)) = \frac{45}{315}$
(b) $f_Y(0) = \frac{198}{315}, f_Y(1) = \frac{72}{315}, f_Y(2) = \frac{45}{315}$

3.2.5 $f_Y(y) = (0.30)^{y-1}(0.70), y = 1, 2, \ldots$

3.2.7 2

3.2.9 $F_Y(y) = 0, \ y < 0; \ F_Y(y) = 0.008, \ 0 \le y < 1; \ F_Y(y) = 0.104, \ 1 \le y < 2; \ F_Y(y) = 0.488,$
$2 \le y < 3; \ F_Y(y) = 1, \ y \ge 3$. When graphed, $F_Y(y)$ is a step function, similar to Figure 3.2.9.

3.2.11 $F_Y(y) = 0, \ y < 0; \ F_Y(y) = 0.659, \ 0 \le y < 1; \ F_Y(y) = 0.958, \ 1 \le y < 2; \ F_Y(y) = 0.998,$
$2 \le y < 3; \ F_Y(y) = 1.000, \ y \ge 3$. The graph is a step function.

3.2.13 $F_Y(y) = 0, y < a; F_Y(y) = (y - a)/(b - a), a \le y < b; F_Y(y) = 1, y \ge b$

3.2.15 $F_Y(y) = 0, \quad y < -1; \quad F_Y(y) = y + y^2/2 + \frac{1}{2}, \quad -1 \le y < 0; \quad F_Y(y) = y - y^2/2 + \frac{1}{2},$
$0 \le y < 1; F_Y(y) = 1, y \ge 1$

3.2.17 $\frac{5}{16}$

3.2.19 (a) ln 2   (b) 0.223   (c) 0.223   (d) $f_Y(y) = 1/y, 1 < y < e$

3.2.21 $1 - 25e^{-6}$

3.2.23 $h(y) = 1/\lambda$, $y > 0$. The exponential model assumes no "wear-out," as evidenced by the fact that $h(y)$ is a constant for all $y > 0$.

3.3.1 $\frac{1}{10}$

3.3.3 $f_{X,Y}(x, y) = \binom{4}{x}\binom{4}{y}\binom{44}{4 - x - y} \bigg/ \binom{52}{4}$;

$x = 0, 1, \ldots, 4$; $y = 0, 1, \ldots, 4$; $x + y = 0, 1, \ldots, 4$

3.3.5

		y		
		0	1	2
	0	$\frac{16}{36}$	$\frac{8}{36}$	$\frac{1}{36}$
x	1	$\frac{8}{36}$	$\frac{2}{36}$	0
	2	$\frac{1}{36}$	0	0

$f_Z(0) = \frac{16}{36}$, $f_Z(1) = \frac{16}{36}$, $f_Z(2) = \frac{4}{36}$

3.3.7 $\frac{1}{2}$

3.3.9 (a) $S = \{(0, 1), (0, 2), (0, 3), (0, 4), (0, 5), (0, 6), (1, 1), (1, 2), (1, 3), (1, 4), (1, 5), (1, 6)\}$
(b) $\frac{1}{3}$ (c) $F_{X,Y}(x, y)$ is a set of nonoverlapping planes parallel to the xy-axis. Conceptually, $F_{X,Y}(x, y)$ is the two-dimensional extension of the step function associated with $F_Y(y)$ when Y is discrete.

3.3.11 $F_{X,Y}(x, y) = 0$, $x < 5$, $y < 2$; $F_{X,Y}(x, y) = \frac{1}{2}$, $5 \le x < 6$, $2 \le y < 3$; $F_{X,Y}(x, y) = \frac{2}{3}$, $5 \le x < 6$, $y \ge 3$; $F_{X,Y}(x, y) = \frac{2}{3}$, $x \ge 6$, $2 \le y < 3$; $F_{X,Y}(x, y) = 1$, $x \ge 6$, $y \ge 3$

3.3.13 $f_{X,Y}(x, y) = 1$, $0 < x < 1$, $0 < y < 1$. Graphed, $f_{X,Y}(x, y)$ is a plane of height one over the unit square.

3.3.15 $f_X(x) = 1$, $0 < x < 1$; $f_Y(y) = -\ln y$, $0 < y < 1$

3.3.17 $45/512$; $15/64$

3.3.19 $e^{-4.2}$

3.3.21 (a) $f_{X,Y}(x, y) = x + y$, $0 < x < 1$, $0 < y < 1$
(b) $f_{Y,Z}(y, z) = (y + \frac{1}{2})e^{-z}$, $0 < y < 1$, $z > 0$
(c) $f_Z(z) = e^{-z}$, $z > 0$

3.4.1 $f_{X_1, \ldots, X_n}(x_1, \ldots, x_n) = (1/\lambda)^n e^{-(1/\lambda) \cdot \Sigma_{i=1}^n x_i} > 0$, $i = 1, \ldots, n$

3.4.3 Yes; $f_{X_1, X_2}(x_1, x_2) = 12x_1 x_2(1 - x_2) = 2x_1 \cdot 6x_2(1 - x_2) = f_{X_1}(x_1) \cdot f_{X_2}(x_2)$, $0 < x_1 < 1$, $0 < x_2 < 1$

3.4.5 The region in the xy-plane over which $f_{X,Y}(x, y) > 0$ is not rectangular: knowing the value of x tells us something about the value of y.

3.5.1 $f_Y(-1) = \frac{3}{21}$, $f_Y(1) = \frac{12}{21}$, $f_Y(3) = \frac{6}{21}$

3.5.3 $f_Y(y) = \frac{1}{3}$, $-7 \le y \le -4$

3.5.5 $f_Y(y) = -3y^2/4 - 3y - \frac{9}{4}$, $-3 < y < -1$

3.5.7 (a) $f_Y(y) = (1/y^2) \cdot e^{-1/y}$, $y > 0$
(b) $f_Y(y) = e^{y - e^y}$, $-\infty < y < \infty$

3.5.9 $f_Y(y) = (81y/25 + 288/5)e^{-(9y/5 + 32)^2/2}$, $y > -160/9$

3.5.11 $F_Z(z) = z^2/2$, $0 < z \le 1$; $F_Z(z) = 2z - z^2/2 - 1$, $1 < z \le 2$

3.5.13 (a) $F_Z(z) = 0$, $z \le 0$; $F_Z(z) = z/2$, $0 < z < 1$; $F_Z(z) = 1 - \frac{1}{2z}$, $z \ge 1$

(b) $f_Z(z) = \frac{1}{2}$, $0 < z < 1$; $f_Z(z) = \dfrac{1}{2z^2}$, $z \ge 1$

3.5.15 $f_Z(z) = e^{-(r+s)}(r + s)^z/z!$, $z = 0, 1, 2, \ldots$; yes

3.5.17 $f_Z(z) = ze^{-z}$, $z > 0$

3.5.21 $f_Y(y) = (\frac{2}{9}) \cdot (y + 1)$, $-1 < y < 2$

3.5.23 $f_Z(z) = \frac{1}{2}z^2 e^{-z}$, $z > 0$

Answers to Odd-Numbered Exercises

3.6.1 0.82; 0.45   **3.6.3** $1 - p^2$ **3.7.1**   0.015

3.7.3 $f_{Y|x}(y) = \binom{6}{y}\binom{4}{3-x-y} \Big/ \binom{10}{3-x}, \; y = 0, 1, \ldots, 3 - x$

3.7.5 $f_{X,Y|1}(1,1) = \frac{1}{5}, f_{X,Y|1}(2,2) = \frac{4}{5}; f_{X,Y|2}(2,1) = \frac{1}{4}, f_{X,Y|2}(1,2) = \frac{1}{4}, f_{X,Y|2}(2,2) = \frac{1}{2}$

3.7.9 $f_{Y|x}(y) = (x + y)/(x + \frac{1}{2}), \; 0 \le y \le 1$

3.7.11 $f_Y(y) = (\frac{1}{3}) \cdot (2y + 2), \; 0 < y < 1$

3.7.13 $\frac{2}{3}$ **3.8.1** $\frac{5}{3}$ **3.8.5**   1.6

3.8.7 $\displaystyle\int_1^\infty (1/x^2)\,dx = 1$, but $E(X) = \displaystyle\int_1^\infty (1/x)\,dx = \ln x \Big|_1^\infty$ is not finite.

3.8.9   (a) $c/(2 - c)$ (b) $2 \log 2$

3.8.11   (a) $\frac{6}{15}$ (b) $\frac{4}{3}$ **3.8.13** $\frac{91}{35}$ **3.9.1**   35

3.9.3 $E(X)$ **3.9.5**   112   **3.9.7**   9.6

3.9.9 $r(n + 1)/2$; no   **3.9.11**   \$50,000   **3.9.13**   2

3.9.15 $\frac{1}{4}$ **3.9.17**   12.25   **3.10.1** $\frac{12}{25}$

3.10.3   0.75   **3.10.5** $\frac{3}{80}$

3.10.7   Johnny should pick $(a + b)/2$.

3.10.9 $10°C$

3.10.11   0.49 in.; We are assuming that the 50 brick lengths and 49 mortar thicknesses are independent.

3.10.13   0.16 ohm

3.10.15 $E(T) = np(n + 1)/2$, where $p = P(\text{gambler wins } i\text{th hand})$, $i = 1, 2, \ldots, n$; Var $(T) = np(1 - p)(n + 1)(2n + 1)/6$

3.11.1 $E(X^r) = 2^r/(r + 1)$; $\frac{8}{7}$

3.11.5 $E(X^{2k}) = n^k \Gamma(k + \frac{1}{2})\Gamma(n/2 - k)/\Gamma(\frac{1}{2})\Gamma(n/2)$, $2k < n$

3.12.3 $[(e^t - 1)/t]^2$

3.12.5 $E(X) = 1/p$; Var $(X) = q/p^2$

3.12.9   Let $Z = X + Y$. Then $f_Z(z) = e^{-2\lambda}(2\lambda)^z/z!$, $z = 0, 1, \ldots$

3.13.1   (a) 0.050   (b) 0.25

3.13.3   *Hint:* $P(X > 148 \text{ or } X < 52) = 1 - P(|X - \mu| \le 48)$

CHAPTER 4

4.2.1   0.865; binomial   **4.2.3**   0.10

4.2.5 6.9×10^{-12} **4.2.9** $f_Z(z) = e^{-10}10^z/z!$, $z = 0, 1, \ldots$

4.2.11   0.36; yes   **4.2.13**   0.195; 0.430

4.2.15

Number of hits, x	Expected number
0	227.3
1	211.4
2	98.3
3	30.5
4	7.1
5+	1.3

4.2.19   *Hint:* Consider the interval $(a, a + y)$. Let Y denote the interval between consecutive occurrences. Let X denote the number of occurrences in an interval of length y. What does $P(X = 0)$ equal? What must be the relationship between y and Y if $X = 0$?

4.2.21 0.24; 0.47

4.2.23 *Hint:* Consider $M_X(t) \cdot M_Y(t)$.

4.2.25 $f_Y(0) = 163.6, f_Y(1) = 266.7, f_Y(2) = 259.4, f_Y(3) = 192.8, P(Y \geq 4) = 213.5$

4.3.1 (a) 0.0994   (b) 0.1210

4.3.3

n	Exact binomial	DeMoivre-Laplace limit
2	0.500	0.6826
5	0.625	0.6826
8	0.711	0.6826
10	0.656	0.6826
12	0.612	0.6826
15	0.698	0.6826

4.3.5 0.001   **4.3.7** 0.92

4.3.9 (a) $\sum_{k=75}^{80} \binom{100}{k}(0.70)^k(0.30)^{100-k}$ (b) 0.1525

4.3.11 (a) 0.0668   (b) 0.1587

4.3.13 16%   **4.3.15** 0.7745

4.3.17 $P(X \geq 9.0) \doteq 0.04$. The data suggest that the baboon does not belong to the genus *Papio*, but the evidence is far from overwhelming.

4.3.19 0.0062; 0.3312; 0.5934   **4.3.21** (a) 583.8   (b) 9.5

4.3.23

Rate	Observed frequency	Expected frequency
< 2.9	1	1.8
3.0–3.9	4	5.7
4.0–4.9	16	12.2
5.0–5.9	15	14.9
6.0–6.9	7	10.3
7.0–7.9	4	4.1
> 8.0	3	1.0
	$\overline{50}$	$\overline{50.0}$

4.4.1 0.06; *Hint:* Let X = game during which the third-string quarterback is injured. Then $P(\text{team starts fourth QB}) = \sum_{x=3}^{11} P(X = x)$.

4.4.5 $\left(\frac{5}{6}\right)^3 \cdot \left(\frac{1}{6}\right)$

4.4.7 *Hint:* $M_X(t) = \sum_{x=1}^{\infty} e^{xt} q^{x-1} p = (p/q) \cdot \sum_{x=1}^{\infty} (qe^t)^x$

4.5.1 0.65   **4.5.5** 0.05; 0.029; 0.021   **4.5.7** 0.988; 20

4.7.1 *Hint:* $\Gamma\left(\frac{7}{2}\right) = \Gamma(3.5) = \Gamma(2.5 + 1) = 2.5\Gamma(2.5)$

4.7.9 $f_Y(y) = \frac{1}{2}(1/1000)^3 y^2 e^{-y/1000}, \, y > 0$

CHAPTER 5

5.3.1 0.41

5.3.3 Var $(W_1) = \theta^2/n$; Var $(W_2) = \theta^2$

5.4.1 1.33   **5.4.3** $(n + 1)Y_{\min}$

5.4.5 $W = \left(\frac{5}{3}\right) \cdot Y'_3$; $w = 30$; w would be inappropriate if $w < y'_4$.

5.4.7 $a + b = 1$ **5.4.11** $W = 3X^2$

5.4.15 Yes; $\lim_{n \to \infty} E(W_n) = \lim_{n \to \infty} [n/(n + 1)]\theta = \theta$.

5.4.17 Median $= [(n + 1)/(n2^{1/n})]\theta \neq \theta$, except when $n = 1$.

5.5.1 $n; 1$ **5.5.3** $3/(n(n + 2))$

5.5.5 $\frac{8}{9}$ **5.5.7** $(2n + 1)(2n + 3)/2n$

5.6.1 Cramer–Rao lower bound $= \lambda^2/n$; Var $(\bar{Y}) = \lambda^2/n$

5.6.3 Both sides of the first equation in Definition 5.6.2 equal σ^2/n.

5.6.5 (a) θ^2/n (b) $\theta^2/[n(n + 2)]$

5.7.1 74

5.7.3 *Hint:* Does $1 - \text{Var}(W)/\varepsilon^2$ converge to 1 as n increases?

5.7.5 *Hint:* Recall that Var $(\bar{Y}) = \sigma^2/n$.

5.7.13 *Hint:* $L(p) = (1 - p)^{\sum_{i=1}^n x_i - n} p^n = (1 - p)^{w-n} p^n$.

5.8.1 $w = y/n$ **5.8.3** $w = y_{min}$

5.8.5 1.69   **5.8.7** $\sqrt{\bar{y}}$

5.8.9 $\hat{\theta}_1 = 5:14; \hat{\theta}_2 = 5:29; \hat{\mu} = 5:21\frac{1}{2}$

5.8.11 $\hat{\theta} = -1 - n/\ln \prod_{i=1}^n x_i$; MLE $= 0.75$; moments estimate $= 0.50$

5.8.13 2.17 (per 100,000)

5.8.15 $\bar{y}; \sqrt{(3/n) \cdot \sum_{i=1}^n y_i^2 - 3\bar{y}^2}$

5.9.1 $(w_1/[(n + 1)(1 - \sqrt[n]{0.05})], w_1/[(n + 1)(1 - \sqrt[n]{0.95})])$; intervals based on $(n + 1) \cdot Y_{min}$ will be wider.

5.9.3 Yes, if the confidence interval is not symmetric around the point estimate.

5.10.1 (0.76, 0.86)   **5.10.3** (0.251, 0.379)   **5.10.5** (0.008, 0.91)

5.10.7 543   **5.10.9** 1024

CHAPTER 6

6.2.1 13

6.2.3 Let $p = P$(any particular psychotic–normal pair will be correctly identified). Test H_0: $p = \frac{1}{2}$ vs. $H_1: p > \frac{1}{2}$. For $\alpha = 0.05$, $y^* = 9$. No.

6.2.5 Let $p = P$(person dies in three-month period preceding birthday). Test $H_0: p = \frac{1}{4}$ vs. $H_1: p < \frac{1}{4}$. Reject H_0 if $(y - np_0)/\sqrt{np_0(1 - p_0)} \leq -2.33$. Conclusion: Reject H_0: obs. $Z = -10.8$.

6.2.7 (a) $H_0: p = 0.40; H_1: p > 0.40$
(b) Accept H_0: obs. $Z = 0.91$; critical value is 2.33.

6.2.9 Reject H_0. The observed test statistic (6.41) exceeds the 0.05 critical value (1.64).

6.2.11 0.0087   **6.3.1** No.

6.3.3 $\alpha = 0.064; \beta = 0.107$. Type I errors would be considered more serious than Type II errors.

6.3.5 $\alpha = 0.0188; \beta = 0.806$

6.3.7 $\alpha = 0.5; \beta = 0.33$ when urn is 60% white; $\beta = 0.18$ when urn is 70% white.

6.3.9 (a) 0.0808   (b) 0.8413   (c) Reject H_0 if $y \leq 0.84$.   (d) 0.9726

6.4.1 $\Lambda = \max_\omega L(p)/\max_\Omega L(p)$, where $\max_\omega L(p) = p_0^n(1 - p_0)^{\sum_{i=1}^n X_i - n}$ and $\max_\Omega L(p) = (n/\sum_{i=1}^n X_i)^n [1 - (n/\sum_{i=1}^n X_i)]^{\sum_{i=1}^n X_i - n}$

6.4.3 $\lambda = \{(2\pi)^{-n/2} e^{-(1/2)\sum_{i=1}^n (y_i - \mu_0)^2}\}/\{(2\pi)^{-n/2} e^{-(1/2)\sum_{i=1}^n (y_i - \bar{y})^2}\} = e^{-(1/2)\{(\bar{y} - \mu_0)/(1/\sqrt{n})\}^2}$. Base the test on $Z = (\bar{y} - \mu_0)/(1/\sqrt{n})$.

CHAPTER 7

7.2.1   14.7

7.2.3   The 1-standard deviation interval is (7.3, 20.3) and contains 65% of the observations; the 2-standard deviation interval is (0.8, 26.8) and contains 96% of the observations.

7.2.5   15%; the shape of the data's histogram suggests that the normality assumption may not be appropriate.

7.2.7 $\hat{\mu} = \bar{y}$; $\hat{\sigma}^2 = [(n-1)/n]s^2$

7.2.9   Let $K(x) = (x - \mu)^2$, $p(\sigma) = -1/(2\sigma^2)$, $S(x) = 0$, and $q(\sigma) = -\ln(\sqrt{2\pi}\,\sigma)$; sufficient statistic $= \sum_{i=1}^{n}(x_i - \mu)^2$.

7.3.3   0.29; 0.43; 0.23   **7.3.5**   0.027; $\sigma = 0.224$

7.3.7   2663   **7.3.9** $P(Z > 3.16) \doteq 0.0008$

7.3.11   0.885   **7.4.1**   0.0418

7.4.3   0.121   **7.4.5**   0.0062

7.4.7   Using the continuity correction, $P(Y \geq 9) = P(Y \geq 8.5) \doteq P(Z \geq 0.78) = 0.22$; based on the Poisson pdf, $P(Y \geq 9) = 0.21$.

7.4.9

Salary	Observed frequency	Expected frequency
≤ 9	160.39	263.91
9–12	1079.20	999.23
12–15	1337.35	1310.69
15–18	959.24	953.47
18–21	519.83	522.50
21–24	234.15	236.81
24–27	94.19	97.75
27–30	33.77	36.88
30–35	17.33	17.33
≥ 35	7.55	4.43

7.5.3   16

7.5.5   Test H_0: $\sigma^2 = 1.1$ vs. H_1: $\sigma^2 < 1.1$. Observed $\chi^2 = 2.22$. Critical value $= 1.145$. Accept H_0.

7.5.7 $(-\infty, 82.5), (78.3, +\infty)$

7.5.9   9   **7.5.11**   0.592

7.5.13 $-2 \ln \lambda = 4.0$; reject H_0 (critical value $= 3.841$)

7.5.17   (a) $M_Y(t) = M_{2n\bar{X}/\theta}(t) = M_{\bar{X}}[(2n/\theta) \cdot t] = [M_{X_i}(2t/\theta)]^n = \{1/[1 - \theta(2t/\theta)]\}^n$
        $= [1/(1 - 2t)]^n$
        (b) $(2n\bar{x}/\chi^2_{1-\alpha/2,\,2n},\ 2n\bar{x}/\chi^2_{\alpha/2,\,2n})$

7.6.3 $m/(n-2)$, for $n > 2$

7.6.5 $E(Y) = r/(r+s)$; $\text{Var}(Y) = rs/(r+s+1)(r+s)^2$

7.6.7   1.059   **7.6.9**   0.70

7.6.11 $E(X^{2k}) = n^k \Gamma(k + \frac{1}{2})\Gamma(n/2 - k)/\Gamma(\frac{1}{2})\Gamma(n/2)$, for $2k < n$; for $2k \geq n$, $E(X^{2k})$ does not exist.

7.6.13   (5.14, 6.78)   **7.6.15**   (0.237, 0.731)

7.6.17 $(\bar{y} - 1.96 \cdot \sigma/\sqrt{n},\ \bar{y} + 1.96 \cdot \sigma/\sqrt{n})$; $n = 15.37\sigma^2/L^2$

7.6.19   No. Theorem 7.6.4 is a more "practical" result. An experimenter who does not know μ (and is trying to estimate it with a confidence interval) is not likely to know σ^2.

7.6.21 *Hint:* μ and σ^2 must jointly satisfy the two inequalities $(80.4 - \mu)^2 < (1.96)^2 \sigma^2 / 18$ and $7.564 < (450.3/\sigma^2) < 30.191$.

7.7.1 Let μ = workers' true average FEV_1/VC ratio. Test $H_0: \mu = 0.80$ vs. $H_1: \mu < 0.80$. Accept H_0: obs. $t = -1.723$, which is larger than $-t_{0.025, 18} = -2.110$.

7.7.3 Test $H_0: \mu = 94$ vs. $H_1: \mu < 94$. Observed $t = -1.05$; critical value $= -1.8331$. Accept H_0.

7.7.7 Reject H_0. The value of the test statistic (-3.27) is less than the 0.05 critical value (-1.64). The coating appears to be beneficial.

7.7.9 9

CHAPTER 8

8.2.1 The numerator is a linear combination of independent normals (see Theorem 7.3.1).

8.2.3 Observed t ratio $= 0.93$ (with 14 degrees of freedom). Accept the null hypothesis that carpeting has no effect on levels of airborne bacteria, since $\pm t_{0.025, 14} = \pm 2.1448$.

8.2.5 Observed t ratio $= -1.28$ (with 10 degrees of freedom). Since $\pm t_{0.025, 10} = \pm 2.2281$, accept $H_0: \mu_A = \mu_B$.

8.2.7 Observed t ratio $= 1.72$ (with 17 degrees of freedom). Accept $H_0: \mu_X = \mu_Y$, since $\pm t_{0.025, 17} = \pm 2.1098$.

8.2.9 *Hint:* See Question 8.2.1 to establish normality. Compute $E(Z)$ and Var (Z).

8.3.1 Observed F ratio $= 0.29$ (with 8 and 5 degrees of freedom). Accept $H_0: \sigma_X^2 = \sigma_Y^2$. Critical values are 0.208 and 6.76. Yes.

8.3.3 Observed F ratio $= 0.31$ (with 9 and 9 degrees of freedom). Accept $H_0: \sigma_X^2 = \sigma_Y^2$, since $F_{0.025, 9, 9} = 0.248$ and $F_{0.975, 9, 9} = 4.03$.

8.4.1 Observed Z ratio $= 1.76$. Accept $H_0 (\pm z_{0.025} = \pm 1.96)$.

8.4.3 Observed Z ratio $= -7.92$. Reject $H_0 (\pm z_{0.025} = \pm 1.96)$.

8.4.5 $-2 \ln \lambda = 10.37$. Reject $H_0: p_H = p_W$ at the $\alpha = 0.01$ level of significance since $-2 \ln \lambda$ is greater than or equal to $\chi^2_{0.99, 1} (= 6.635)$.

8.4.7 0.52

8.5.1 $(0.06, 1.19)$; $H_0: \mu_X = \mu_Y$ would be rejected.

8.5.5 $(0.01, 0.14)$ **8.5.7** $(0.67, 8.10)$ **8.5.9** $(-0.020, 0.084)$

CHAPTER 9

9.2.1 $(7!/2! \, 4! \, 1!) \cdot (0.266)^3 (0.468)^4 = 0.095$

9.2.3 0.006

9.2.5 $[12!/(2!)^6] \cdot (1/21)^2 (2/21)^2 (3/21)^2 (4/21)^2 (5/21)^2 (6/21)^2 = 0.0005$

9.2.9 20/9

9.3.1 Reject H_0; obs. $\chi^2 = 7.43$, which exceeds $\chi^2_{0.95, 1} = 3.841$.

9.3.3 Observed $\chi^2 = 2.46$ (with 4 degrees of freedom). Accept H_0, since $\chi^2_{0.95, 4} = 9.488$.

9.3.5 Reject H_0. Observed $\chi^2 = 17.6$; $\chi^2_{0.95, 7} = 14.067$.

9.3.7 Reject H_0 since the test statistic equals 8.5.4, which exceeds the critical value $\chi^2_{0.95, 1} = 3.841$.

9.3.9 Accept H_0. Observed test statistic $= 2.10$ (with 2 degrees of freedom). *Note:* If the original five classes are taken to be "0–0.19," "0.20–0.39," and so on, they need to be redefined as "0–0.39," "0.40–0.59," and "0.60–1.00," because of the small expected frequencies associated with "0–0.19" and "0.80–1.00."

9.3.11 Observed $\chi^2 = (9 - 6.25)^2/6.25 + (11 - 12.5)^2/12.5 + (8 - 15.625)^2/15.625 + (22 - 15.625)^2/15.625 = 7.7$; $\chi^2_{0.90, 3} = 6.251$. Reject H_0.

9.4.1 Observed $\chi^2 = 1.14$ (with 2 degrees of freedom). Accept H_0.

9.4.3 Observed $\chi^2 = 1.98$ (with 4 degrees of freedom). Accept H_0, since $\chi^2_{0.99, 4} = 13.277$.

9.4.5 Test $H_0: f_X(x) = 1/\theta$; MLE for $\theta = X_{max} = 115$. Divide the numbers 1–115 into five classes, "1–23," "24–46," and so on (each class has expected frequency 10). Observed $\chi^2 = 2.6$ (with 3 degrees of freedom). Accept H_0.

9.4.7 $\hat{\theta} = 0.65$; expected frequencies for the four phenotypes listed are 1441.6, 190.4, 190.4, and 353.6, respectively.

9.5.1 Accept the null hypothesis that the criteria are independent; the observed $\chi^2 = 2.13$, which is less than the critical value $\chi^2_{0.95, 1} = 3.841$.

9.5.3 Time of infection and risk of pregnancy complication are independent: the test statistic equals 50.2, which exceeds the critical value $\chi^2_{0.99, 1} = 6.635$.

9.5.5 Accept H_0. Observed $\chi^2 = 3.90$ (with 4 degrees of freedom); $\chi^2_{0.95, 4} = 9.488$.

CHAPTER 10

10.2.1 105/72

10.2.7 8/450; 0.492

10.2.17 0.825

10.3.9 2/9; 9/4

10.4.3 $y = 96.6 - 4.92x$

10.4.11 log-log

10.4.15 $y = 3498.19e^{-0.298x}$

10.5.7 (1.99, 4.59)

10.2.5 $-8/484$; $-2/15\sqrt{14}$

10.2.15 -0.453

10.3.5 $E(Y \mid x) = 2(x + 1)$

10.4.1 $y = 25.2 + 3.3x$; $84.5°F$

10.4.9 5.0 days

10.4.13 $a = 2243.7$; $b = -1.1428$

10.5.5 $\frac{16}{25}$

10.5.9 Accept H_0 since the observed t equals 1.631, which is less than the critical value $t_{0.05, 4} = 2.132$.

10.5.11 Obs. $t = \sqrt{n - 2}\,(\hat{A} - a)\,\sqrt{n \sum_{i=1}^{n} (x_i - \bar{x})^2}/\sqrt{n\hat{\sigma}^2 \sum_{i=1}^{n} (x_i^2)} \sim t_{n-2}$

10.5.13 Obs. $t = (\hat{b} - \hat{a})/s_{\hat{b} - \hat{a}} \sim t_{n+m-4}$, where $s^2_{\hat{b} - \hat{a}} = \hat{\sigma}^2_b + \hat{\sigma}^2_a = \hat{\sigma}^2/\sum_{i=1}^{n} (x_i - \bar{x})^2 + \hat{\sigma}^2/\sum_{i=1}^{m} (u_i - \bar{u})^2$ and $\hat{\sigma}^2 = [(n - 2)\hat{\sigma}^2_1 + (m - 2)\hat{\sigma}^2_2]/(n + m - 4)$

10.5.15 (110.52, 135.06); committee should be more interested in confidence interval, which concerns industry averages.

10.5.17 (a) $y = 9.406 + 0.081x$ (b) Reject H_0 since the test statistic equals 3.407, which is greater than $t_{0.05, 10} = 1.812$.   (c) (0.028, 0.135)   (d) (23.51, 27.84)   (e) (25.57, 42.03)   (f) $(-2.358, 21.17)$

10.5.21 Coefficient of determination $= r^2$

10.6.1 0.942; 0.209

10.6.3 $c\mu_X + d\mu_Y$; $c^2\sigma^2_X + d^2\sigma^2_Y + 2cd\rho(X, Y)\sigma_X\sigma_Y$

10.6.5 $E(X) = E(Y) = 0$; Var $(X) = 4$; Var $(Y) = 1$; $\rho(X, Y) = \frac{1}{2}$; $k = 1/2\pi\sqrt{3}$

10.7.1 Reject H_0 since the test statistic equals -4.247, which is less than $-t_{0.025, 7} = -2.365$.

10.7.3 Observed $t = 4.66$; reject H_0 ($t_{0.05, 19} = 1.7291$).

CHAPTER 11

11.2.1 Observed F ratio $= 22.36$ (with 1 and 14 degrees of freedom). Reject H_0.

11.2.5 Observed F ratio $= 0.002185/0.0001453 = 15.0$. Reject H_0. Within the limits of rounding error, obs. $F = ($obs. $t)^2$.

11.2.7 $E(Y) = 2\gamma + r$

11.3.1

Source	df	SS	MS	F
Varieties	4	270.27	67.57	6.39
Error	10	105.68	10.57	
Total	14	375.95		

Reject H_0 ($F_{0.95, 4, 10} = 3.48$).

11.4.1

Source	df	SS	MS	F
Tubes	2	510.71	255.36	11.56
$\begin{pmatrix} C_1 \\ C_2 \end{pmatrix}$	$\begin{pmatrix} 1 \\ 1 \end{pmatrix}$	$\begin{pmatrix} 264.033 \\ 246.678 \end{pmatrix}$	$\begin{pmatrix} 264.033 \\ 246.678 \end{pmatrix}$	11.95 11.17
Error	42	927.73	22.09	
Total	44	1438.44		

The second contrast is defined by $C_2 = \frac{1}{2}\mu_A - \mu_B + \frac{1}{2}\mu_C$. It tests the subhypothesis $H_0'': (\mu_A + \mu_C)/2 = \mu_B$. At the 0.01 level, the original null hypothesis as well as the two subhypotheses are all rejected.

11.4.3 Let $C = \mu_1 - \frac{1}{2}\mu_{II} - \frac{1}{2}\mu_{III}$. Observed F ratio $= (113.64/1) \div (276.50/15) = 6.17$. Reject $H_0: \mu_1 = (\mu_{II} + \mu_{III})/2$, since $F_{0.95, 1, 15} = 4.54$.

11.4.5 $\hat{\mu} = 22.9$, $\hat{\tau}_1 = 1.3$, $\hat{\tau}_2 = 3.3$, $\hat{\tau}_3 = -4.6$; yes; a χ^2 goodness-of-fit test for normality

11.5.1 Consider 95% Tukey intervals. Then $D = 3.67/\sqrt{6}$ and $MS_{error} = 18.43$.

Comparison	95% Tukey interval	Conclusion
I vs. II	$(-8.10, \quad 4.76)$	NS
I vs. III	$(-15.43, \; -2.57)$	Reject
II vs. III	$(-13.76, \; -0.90)$	Reject

11.5.3

Source	df	SS	MS	F
Companies	2	113,646.33	56,823.16	5.81*
Error	15	146,753.67	9,783.58	
Total	17	260,400.00		

$D = 3.67/\sqrt{6}$

Pairwise difference	95% Tukey interval	Conclusion
Cutter vs. Abbott	$(-78.9, \quad 217.5)$	NS
Cutter vs. McGaw	$(-271.0, \quad 25.4)$	NS
Abbott vs. McGaw	$(-340.4, \; -44.0)$	Reject

11.6.1 Take the square root of each observation before doing the analysis of variance.

Source	df	SS	MS	F
Developers	1	1.8330	1.8330	6.20*
Error	10	2.9578	0.2958	
Total	11	4.7908		

11.6.3 *Hint:* Use the arc sine transformation.

CHAPTER 12

12.2.1 -0.7

12.2.3

Source	df	SS	MS	F
States	1	61.63	61.63	7.20*
Students	14	400.80	28.63	
Error	14	119.87	8.56	
Total	29	582.30		

12.2.9 $C_1 = \mu_P - \mu_R$; $SS_{C_1} = 18.225$; observed F ratio $= 7.35$; $C_2 = \mu_{BV} - \mu_B$; $SS_{C_2} = 38.025$; observed F ratio $= 15.33$; $C_3 = \mu_P - \mu_{BV} + \mu_R - \mu_B$; $SS_{C_3} = 0.128$; observed F ratio $= 0.05$

Reject at the $\alpha = 0.10$ level of significance the null subhypotheses associated with C_1 and C_2. C_3 tests whether the average effect of the Plain and Roast Beef flavors is significantly different from the average effect of the Butter Vanilla and Bread flavors.

12.3.1

Source	df	SS	MS	F
Varieties	1	0.61	0.61	4.58 NS
Locations	3	0.55		
Error	3	0.40	0.1333	
Total	7	1.56		

12.3.3 $\bar{d} \pm t_{\alpha/2, \, b-1}(s_D/\sqrt{b})$. For the data of Case Study 12.3.1, the 95% confidence interval for μ_D is $(-0.21, 0.43)$.

CHAPTER 13

13.2.1 Let $p = P(Y_i > X_i)$. For $\alpha = 0.05$, we should reject $H_0: p = \frac{1}{2}$ in favor of $H_1: p < \frac{1}{2}$ if $y_+ \leq 2$, since $P(Y_+ \leq 2) = 0.033$ but $P(Y_+ \leq 3) > 0.05$. Here, $y_+ = 2$, so reject H_0.

13.2.3 Let Y_+ denote the number of Y_i's for which $Y_i - 7.39$ is positive. Omit the observation that *equals* 7.39. Then $y_+ = 4$, based on $n = 43$ observations. Observed Z ratio $= -5.34$. Reject H_0.

13.2.5

y_+	$P(Y_+ = y_+)$
0	$\frac{1}{128}$
1	$\frac{7}{128}$
2	$\frac{21}{128}$
3	$\frac{35}{128}$
4	$\frac{35}{128}$
5	$\frac{21}{128}$
6	$\frac{7}{128}$
7	$\frac{1}{128}$
	1

Possible α levels for one-sided test: $\frac{1}{128}, \frac{8}{128}, \frac{29}{128}$, and so on.

13.2.7 The approximate, large sample observed Z ratio is 1.89. With ± 1.96 being the 0.05 critical values, we should accept H_0: the blades have not been shown to be significantly different.

13.3.1 $W = 9$. Accept H_0.

13.3.3 Observed Z ratio $= 0.99$. Accept H_0.

13.3.5 Signed rank test, because it makes use of more of the information in the data.

13.4.1 $B = 9.83$. Reject H_0, since $\chi^2_{0.95,\,1} = 3.841$.

13.4.3 $B = 15.3$ (with 4 degrees of freedom). Reject H_0.

13.5.1 $G = 8.8$ (with 4 degrees of freedom). Accept H_0.

13.5.3 $G = 17.0$ (with 2 degrees of freedom). Reject H_0.

13.5.5 Accept H_0.

Bibliography

1. Alexander, S. A. "Price Movements in Speculative Markets: Trends or Random Walks," in *The Random Character of Stock Market Prices*. Edited by Paul H. Cootner. Cambridge, Mass.: The M.I.T. Press, 1964.

2. Allen, Raymond L., and Doubleday, Lawrence W. "The Deterrence of Unauthorized Intrusion in Residences, a Systems Analysis," in *First International Electronic Crime Countermeasures Conference*. Lexington, Ky.: ORES Publications, 1973.

3. Ball, J. A. C., and Taylor, A. R. "The Effect of Cyclandelate on Mental Function and Cerebral Blood Flow in Elderly Patients," in *Research on the Cerebral Circulation*. Edited by John Stirling Meyer, Helmut Lechner, and Otto Eichhorn. Springfield, Ill.: Thomas, 1969.

4. Barnicot, N. A., and Brothwell, D. R. "The Evaluation of Metrical Data in the Comparison of Ancient and Modern Bones," in *Medical Biology and Etruscan Origins*. Edited by G. E. W. Wolstenholme and Cecilia M. O'Connor. Boston: Little, Brown, and Company, 1959.

5. Barnothy, Jeno M. "Development of Young Mice," in *Biological Effects of Magnetic Fields*. Edited by Madeline F. Barnothy. New York: Plenum Press, 1964.

6. Bartle, Robert G. *The Elements of Real Analysis*, 2nd ed. New York: John Wiley & Sons, 1976.

7. Bellany, Ian. "Strategic Arms Competition and the Logistic Curve." *Survival*, 16 (1974), 228–30.

8. Bennett, W. R., Jr. "How Artificial is Intelligence?" *American Scientist*, 65, no. 6 (1977), 694–702.

9. Berger, R. J., and Walker, J. M. "A Polygraphic Study of Sleep in the Tree Shrew." *Brain, Behavior and Evolution*, 5 (1972), 62.

10. *Biometrika Tables for Statisticians*, vol. 1. Edited by E. S. Pearson and H. O. Hartley. London: Cambridge University Press, 1954.

11. Blackman, Sheldon, and Catalina, Don. "The Moon and the Emergency Room." *Perceptual and Motor Skills*, 37 (1973), 624–26.

12. Boneau, Alan C. "The Effects of Violations of Assumptions Underlying the *t* Test." *Psychological Bulletin*, 57 (1960), 49–64.

13. Bortkiewicz, L. *Das Gesetz der Kleinen Zahlen*. Leipzig: Teubner, 1898.

14. Boyd, Edith. "The Specific Gravity of the Human Body." *Human Biology*, 5 (1933), 651–52.

15. Brinegar, Claude S. "Mark Twain and the Quintus Curtius Snodgrass Letters: A Statistical Test of Authorship." *Journal of the American Statistical Association*, 58 (1963), 85–96.

16. Bullard, Roger W., and Shumake, Stephen A. "Food Temperature Preference Response of *Desmodus rotundus*." *Journal of Mammalogy*, 54 (1973), 299–302.

17. Camner, Per, and Philipson, Klas. "Urban Factor and Tracheobronchial Clearance." *Archives of Environmental Health*, 27 (1973), 82.

18. Carver, W. A. "A Genetic Study of Certain Chlorophyll Deficiencies in Maize." *Genetics*, 12 (1927), 415–40.

19. Casler, Lawrence. "The Effects of Hypnosis on GESP." *Journal of Parapsychology*, 28 (1964), 126–34.

20. Chow, G. C. "Statistical Demand Functions and their Use in Forecasting," in *The Demand for Durable Goods*. Edited by A. C. Harberger. Chicago: University of Chicago Press, 1960.

21. Clarke, R. D. "An Application of the Poisson Distribution." *Journal of the Institute of Actuaries*, 22 (1946), 48.

22. Clason, Clyde B. *Exploring the Distant Stars*. New York: G. P. Putnam's Sons, 1958, p. 337.

23. Clemens, S. L. *Life on the Mississippi*. New York: Harper & Row, 1917, p. 156.

24. Cochran, W. G. "Approximate Significance Levels of the Behrens-Fisher Test." *Biometrics*, 20 (1964), 191–95.

25. Cochran, W. G., and Cox, Gertrude M. *Experimental Designs*, 2nd ed., New York: John Wiley & Sons, 1957, p. 108.

26. Collins, Robert L. "On the Inheritance of Handedness." *Journal of Heredity*, 59, no. 1 (1968).

27. Coulson, J. C. "The Significance of the Pair-bond in the Kittiwake," in *Parental Behavior in Birds*. Edited by Rae Silver. Stroudsburg, Pa.: Dowden, Hutchinson, & Ross, 1977.

28. Cramer, Harald. *Mathematical Methods of Statistics*. Princeton, N. J.: Princeton University Press, 1946.

29. Daily Stock Price Record, NYSE, January 1978–March 1978. New York: Standard and Poor, 1978.

30. Das, S. C. "Fitting Truncated Type III Curves to Rainfall Data." *Australian Journal of Physics*, 8 (1955), 298–304.

31. David, F. N. *Games, Gods, and Gambling*. New York: Hafner, 1962, p. 168.

32. Diaz, Jose Luis, and Huttunen, Matti O. "Persistent Increase in Brain Serotonin Turnover after Chronic Administration of LSD in the Rat." *Science*, 174 (1971), 62–63.

33. Dixon, Wilfrid, and Massey, Frank J., Jr. *Introduction to Statistical Analysis*, 2nd ed. New York: McGraw-Hill, 1957.

34. Dubois, Cora, ed. *Lowie's Selected Papers in Anthropology*. Berkeley, Calif.: University of California Press, 1960, pp. 137–42.

35. Fadeley, Robert Cunningham. "Oregon Malignancy Pattern Physiographically Related to Hanford, Washington, Radioisotope Storage." *Journal of Environmental Health*, 27 (1965), 883–97.

36. Fagen, Robert M. "Exercise, Play, and Physical Training in Animals," in *Perspectives in Ethology*. Edited by P. P. G. Bateson and Peter H. Klopfer. New York: Plenum Press, 1976.

37. Fairley, William B. "Evaluating the 'Small' Probability of a Catastrophic Accident from the Marine Transportation of Liquefied Natural Gas," in *Statistics and Public Policy*. Edited by William B. Fairley and Frederick Mosteller. Reading, Mass.: Addison-Wesley, 1977.

38. Fairley, William B., and Mosteller, Frederick. "A Conversation about Collins," in *Statistics and Public Policy*. Edited by William B. Fairley and Frederick Mosteller. Reading, Mass.: Addison-Wesley, 1977.

39. Fears, Thomas; Scotts, Joseph; and Scheiderman, Marvin A. "Skin Cancer, Melanoma, and Sunlight." *American Journal of Public Health*, 66 (1976), 461–64.

40. Feller, W. "Statistical Aspects of ESP." *Journal of Parapsychology*, 4 (1940), 271–98.

41. ———. *An Introduction to Probability Theory and Its Applications*, vol. 1. 2nd ed. New York: John Wiley & Sons, 1957.

42. Finkbeiner, Daniel T. *Introduction to Matrices and Linear Transformations*. San Francisco: W. H. Freeman, 1960.

43. Fishbein, Morris. *Birth Defects*. Philadelphia: Lippincott, 1962, p. 177.

44. Fisher, R. A. "On the 'Probable Error' of a Coefficient of Correlation Deduced from a Small Sample." *Metron*, 1 (1921), 3–32.

45. ———. "The Negative Binomial Distribution." *Annals of Eugenics*, 11 (1941), 182–87.

46. Fisz, Marek. *Probability Theory and Mathematical Statistics*, 3rd ed. New York: John Wiley & Sons, 1963, pp. 43–44.

47. Fraser, D. A. S. *Probability and Statistics: Theory and Applications*. North Scituate, Mass.: Duxbury, 1976, pp. 62–63.

48. Freund, John E. *Mathematical Statistics*, 2nd ed. Englewood Cliffs, N.J.: Prentice-Hall, Inc., 1971, p. 226.

49. Furuhata, Tanemoto, and Yamamoto, Katsuichi. *Forensic Odontology*. Springfield, Ill.: Thomas, 1967, p. 84.

50. Gabriel, K. R., and Neumann, J. "On a Distribution of Weather Cycles by Length." *Quarterly Journal of The Royal Meteorological Society*, 83 (1957), 375–80.

51. Galton, Francis. *Memories of My Life*. London: Methuen, 1908, p. 302.

52. ———. *Natural Inheritance*. London: Macmillan, 1908.

53. Gendreau, Paul, et al. "Changes in EEG Alpha Frequency and Evoked Response Latency During Solitary Confinement." *Journal of Abnormal Psychology*, 79 (1972), 54–59.

54. Gerber, Robert C., et al. "Kinetics of Aurothiomalate in Serum and Synovial Fluid." *Arthritis and Rheumatism*, 15 (1972), 626.

55. Gibbons, Jean Dickinson. *Nonparametric Statistical Inference*. New York: McGraw-Hill, 1971.

56. Gluckson, R., and Leone, T. "The *Sports Illustrated* Cover Jinx." Preprint, University of Southern California, 1984.

57. Goldman, Malcolm. *Introduction to Probability and Statistics*. New York: Harcourt, Brace & World, 1970, pp. 399–403.

58. Goodman, Leo A. "Serial Number Analysis." *Journal of the American Statistical Association*, 47 (1952), 622–34.

59. Graybill, Franklin A. *An Introduction to Linear Statistical Models*, vol. 1. New York: McGraw-Hill, 1961.

60. Griffin, Donald R.; Webster, Frederic A.; and Michael, Charles R. "The Echolocation of Flying Insects by Bats." *Animal Behavior*, 8 (1960), 148.

61. Grover, Charles A. "Population Differences in the Swell Shark *Cephaloscyllium ventriosum*." *California Fish and Game*, 58 (1972), 191–97.

62. Haggard, William H.; Bilton, Thaddeus H.; and Crutcher, Harold L. "Maximum Rainfall from Tropical Cyclone Systems which Cross the Appalachians." *Journal of Applied Meteorology*, 12 (1973), 50–61.

63. Haight, F. A. "Group Size Distributions, with Applications to Vehicle Occupancy," in *Random Counts in Physical Science, Geological Science, and Business*, vol. 3. Edited by G. P. Patil. University Park, Pa.: Pennsylvania State University Press, 1970.

64. Hansel, C. E. M. *ESP: A Scientific Evaluation*. New York: Scribner's, 1966, pp. 86–89.

65. Hare, Edward; Price, John; and Slater, Eliot. "Mental Disorder and Season of Birth: A National Sample Compared with the General Population." *British Journal of Psychiatry*, 124 (1974), 81–86.

66. Hasselblad, V. "Estimation of Finite Mixtures of Distributions from the Exponential Family." *Journal of the American Statistical Association*, 64 (1969), 1459–71.

67. Hastings, N. A. J., and Peacock, J. B. *Statistical Distributions*. London: Butterworth, 1975.

68. Hazel, W. M., and Eglof, W. K. "Determination of Calcium in Magnesite and Fused Magnesia." *Industrial and Engineering Chemistry, Analytical Edition*, 18 (1946), 759–60.

69. Heath, Clark W., and Hasterlik, Robert J. "Leukemia among Children in a Suburban Community." *The American Journal of Medicine*, 34 (1963), 796–812.

70. Hendy, M. F., and Charles, J. A. "The Production Techniques, Silver Content and Circulation History of the Twelfth-Century Byzantine Trachy." *Archaeometry*, 12 (1970), 13–21.

71. Hennekens, C., et al. "Coffee Drinking and Death Due to Coronary Heart Disease." *New England Journal of Medicine*, 294 (1976), 633–36.

72. Hersen, Michel. "Personality Characteristics of Nightmare Sufferers." *Journal of Nervous and Mental Diseases*, 153 (1971), 29–31.

73. Hogg, Robert V., and Craig, Allen T. *Introduction to Mathematical Statistics*, 3rd ed. New York: Macmillan, 1970.

74. Hollander, Myles, and Wolfe, Douglas A. *Nonparametric Statistical Methods*. New York: John Wiley & Sons, 1973.

75. Horvath, Frank S., and Reid, John E. "The Reliability of Polygraph Examiner Diagnosis of Truth and Deception." *Journal of Criminal Law, Criminology, and Police Science*, 62 (1971), 276–81.

76. Hudgens, Gerald A.; Denenberg, Victor H.; and Zarrow, M. X. "Mice Reared with Rats: Effects of Preweaning and Postweaning Social Interactions upon Adult Behaviour." *Behaviour*, 30 (1968), 259–74.

77. Hulbert, Roger H., and Krumbiegel, Edward R. "Synthetic Flavors Improve Acceptance of Anticoagulant-Type Rodenticides." *Journal of Environmental Health*, 34 (1972), 407–11.

78. Huxtable, J.; Aitken, M. J.; and Weber, J. C. "Thermoluminescent Dating of Baked Clay Balls of the Poverty Point Culture." *Archaeometry*, 14 (1972), 269–75.

79. Hynek, Joseph Allen. *The UFO Experience: A Scientific Inquiry*. Chicago: Rognery, 1972.

80. Ibrahim, Michel A., et al. "Coronary Heart Disease: Screening by Familial Aggregation." *Archives of Environmental Health*, 16 (1968), 235–40.

81. Imai, Yoshituka. "Linkage Groups of the Japanese Morning Glory." *Genetics*, 14 (1929), 223–55.

82. Jacobson, Eugene, and Kossoff, Jerome. "Self-percept and Consumer Attitudes Toward Small Cars," in *Consumer Behavior in Theory and in Action*. Edited by Steuart Henderson Britt. New York: John Wiley & Sons, 1970.

83. James, Andrew, and Moncada, Robert. "Many Set Color TV Lounges Show Highest Radiation." *Journal of Environmental Health*, 31 (1969), 359–60.

84. Johnson, Norman L., and Kotz, Samuel. *Continuous Univariate Distributions*, vol. 1. Boston: Houghton-Mifflin, 1970.

85. ———. *Continuous Univariate Distributions*, vol. 2. Boston: Houghton-Mifflin, 1970.

86. ———. *Discrete Distributions*. Boston: Houghton-Mifflin, 1969.

87. Kneafsey, James T. *Transportation Economic Analysis*. Lexington, Mass.: Heath, 1975.

88. Kronoveter, Kenneth J., and Somerville, Gordon W. "Airplane Cockpit Noise Levels and Pilot Hearing Sensitivity." *Archives of Environmental Health*, 20 (1970), 498.

89. Kulldorff, Gunnar. "Estimation of One or Two Parameters of the Exponential Distribution on the Basis of Suitably Chosen Order Statistics." *The Annals of Mathematical Statistics*, 34 (1963), 1419–31.

90. Larsen, Richard J., and Marx, Morris L. *An Introduction to Probability and Its Applications.* Englewood Cliffs, N.J.: Prentice-Hall, Inc., 1985.

91. Lathem, Edward Connery, ed. *The Poetry of Robert Frost.* New York: Holt, Rinehart and Winston, 1969, p. 362.

92. Lavalle, Irving H. *An Introduction to Probability, Decision, and Inference.* New York: Holt, Rinehart and Winston, 1970.

93. Li, Frederick P. "Suicide Among Chemists." *Archives of Environmental Health*, 19 (1969), 519.

94. Liu, C. L. *Introduction to Combinatorial Mathematics.* New York: McGraw-Hill, 1968.

95. Lottenbach, K. "Vasomotor Tone and Vascular Response to Local Cold in Primary Raynaud's Disease." *Angiology*, 32 (1971), 4–8.

96. Maguire, B. A.; Pearson, E. S.; and Wynn, A. H. A. "The Time Intervals between Industrial Accidents." *Biometrika*, 39 (1952), 168–80.

97. Maistrov, L. E. *Probability Theory—A Historical Sketch.* New York: Academic Press, 1974.

98. Malina, Robert M. "Comparison of the Increase in Body Size between 1899 and 1970 in a Specially Selected Group with that in the General Population." *Physical Anthropology*, 37 (1972), 135–41.

99. Mann, H. B. *Analysis and Design of Experiments.* New York: Dover, 1949.

100. Matui, Isamu. "Statistical Study of the Distribution of Scattered Villages in Two Regions of the Tonami Plain, Toyama Prefecture," in *Spatial Patterns*. Edited by Berry and Marble. Englewood Cliffs, N.J.: Prentice-Hall, Inc., 1968.

101. McIntyre, Donald B. "Precision and Resolution in Geochronometry," in *The Fabric of Geology*. Edited by Claude C. Albritton, Jr. Stanford, Calif.: Freeman, Cooper, and Co., 1963.

102. "Medical News." *Journal of the American Medical Association*, 219 (1972), 981.

103. Miettinen, Jorma K. "The Accumulation and Excretion of Heavy Metals in Organisms," in *Heavy Metals in the Aquatic Environment*. Edited by P. A. Krenkel. Oxford: Pergamon Press, 1975.

104. Miller, Rupert G., Jr. *Simultaneous Statistical Inference.* New York: McGraw-Hill, 1966.

105. Miller, Russell R. "Drug Surveillance Utilizing Epidemiologic Methods." *American Journal of Hospital Pharmacy*, 30 (1973), 584–92.

106. Minkoff, Eli C. "A Fossil Baboon from Angola, with a Note on *Australopithecus*." *Journal of Paleontology*, 46 (1972), 836–44.

107. Morgan, Peter J. "A Photogrammetric Survey of Hoseason Glacier, Kemp Coast, Antarctica." *Journal of Glaciology*, 12 (1973), 113–20.

108. Moriarity, Shane. Personal communication.

109. Mulcahy, Risteard; McGilvray, J. W.; and Hickey, Noel. "Cigarette Smoking Related to Geographic Variations in Coronary Heart Disease Mortality and to Expectation of Life in the Two Sexes." *American Journal of Public Health*, 60 (1970), 1516.

110. Munford, A. G. "A Note on the Uniformity Assumption in the Birthday Problem." *American Statistician*, 31 (1977), 119.

111. Nash, Harvey. *Alcohol and Caffeine*. Springfield, Ill.: Charles C. Thomas, 1962, p. 96.

112. Neter, John; Wasserman, William; and Kutner, Michael H. *Applied Linear Regression Models*. Homewood, Ill.: Richard D. Irwin, Inc., 1983, pp. 132–40.

113. Newsweek. March 6, 1978, p. 78.

114. ———. March 21, 1983, p. 65.

115. Nye, Francis Iven. *Family Relationships and Delinquent Behavior*. New York: John Wiley & Sons, 1958, p. 37.

116. Olvin, J. F. "Moonlight and Nervous Disorders." *American Journal of Psychiatry*, 99 (1943), 578–84.

117. Orringer, Eugene, et al. "Splenectomy in Chronic Thrombocytopenic Purpura." *Journal of Chronic Diseases*, 23 (1970), 117–22.

118. Papoulis, Athanasios. *Probability, Random Variables, and Stochastic Processes*. New York: McGraw-Hill, 1965, pp. 206–07.

119. Pascal, Gerald, and Suttell, Barbara. "Testing the Claims of a Graphologist." *Journal of Personality*, 16 (1947), 192–97.

120. Passingham, R. E. "Anatomical Differences between the Neocortex of Man and Other Primates." *Brain, Behavior, and Evolution*, 7 (1973), 337–59.

121. Pearson, E. S., and Kendall, M. G. *Studies in the History of Statistics and Probability*. London: Griffin, 1970.

122. Pearson, Karl. "Notes on the History of Correlation." *Biometrika*, 13, 1920–1921.

123. Phillips, David P. "Deathday and Birthday: An Unexpected Connection," in *Statistics: A Guide to the Unknown*. Edited by Judith M. Tanur, et al. San Francisco: Holden-Day, 1972.

124. Phillips, Lawrence D. *Bayesian Statistics for Social Scientists*. London: Thomas Nelson & Sons, Ltd., 1973.

125. Pierce, George W. *The Songs of Insects*. Cambridge, Mass.: Harvard University Press, 1949, pp. 12–21.

126. Porter, John W., et al. "Effect of Hypnotic Age Regression on the Magnitude of the Ponzo Illusion." *Journal of Abnormal Psychology*, 79 (1972), 189–94.

127. Quetelet, L. A. J. *Lettres sur la Theorie des Probabilites, appliquee aux Sciences Morales et Politiques*. Bruxelles: M. Hayez, Imprimeur de L'Academie Royal des Sciences, des Lettres et des Beaux-Arts de Belgique, 1846, p. 400.

128. Ragsdale, A. C., and Brody, S. *Journal of Dairy Science*, 5 (1922), 214.

129. Rahman, N. A. *Practical Exercises in Probability and Statistics*. New York: Hafner, 1972.

130. Resnick, Richard B.; Fink, Max; and Freedman, Alfred M. "A Cyclazocine Typology in Opiate Dependence." *American Journal of Psychiatry*, 126 (1970), 1256–60.

131. Rich, Clyde L. "Is Random Digit Dialing Really Necessary?" *Journal of Marketing Research*, 14 (1977), 300–05.

132. Richardson, Lewis F. "The Distribution of Wars in Time." *Journal of The Royal Statistical Society*, 107 (1944), 242–50.

133. Ritter, Brunhilde. "The Use of Contact Desensitization, Demonstration-plus-participation and Demonstration-alone in the Treatment of Acrophobia." *Behaviour Research and Therapy*, 7 (1969), 157–64.

134. Roberts, Charlotte A. "Retraining of Inactive Medical Technologists—Whose Responsibility?" *American Journal of Medical Technology*, 42 (1976), 115–23.

135. Rochat, Roger W. "Cervical Cancer Screening: The Effect of Infrequently Occurring Disease on the Accuracy of Diagnosis." Presented to the Society for Epidemiological Research. Toronto: 1976.

136. Rohatgi, V. K. *An Introduction to Probability Theory and Mathematical Statistics*. New York: John Wiley & Sons, 1976, p. 81.

137. Rosenthal, Robert, and Jacobson, Lenore F. "Teacher Expectations for the Disadvantaged." *Scientific American*, 218 (1968), 19–23.

138. Roth, Lewis F. "Juvenile Susceptibility of Ponderosa Pine to Dwarf Mistletoe." *Phytopathology*, 64 (1974), 689–92.

139. Roulette, Amos. "An Assessment of Unit Dose Injectable Systems." *American Journal of Hospital Pharmacy*, 29 (1972), 61.

140. Rowley, Wayne A. "Laboratory Flight Ability of the Mosquito, *Culex Tarsalis Coq.*" *Journal of Medical Entomology*, 7 (1970), 713–16.

141. Ruckman, Joseph E.; Zscheile, Frederick P., Jr.; and Qualset, Calvin O. "Protein, Lysine, and Grain Yields of Triticale and Wheat as Influenced by Genotype and Location." *Journal of Agricultural and Food Chemistry*, 21 (1973), 697–700.

142. Rutherford, Sir Ernest; Chadwick, James; and Ellis, C. D. *Radiations from Radioactive Substances*. London: Cambridge University Press, 1951, p. 172.

143. Sagan, Carl. *Cosmos*. New York: Random House, 1980, pp. 298–302.

144. Salvosa, Carmencita B.; Payne, Philip R.; and Wheeler, Erica F. "Energy Expenditure of Elderly People Living Alone or in Local Authority Homes." *American Journal of Clinical Nutrition*, 24 (1971), 1468.

145. Saturley, B. A. "Colorimetric Determination of Cyclamate in Soft Drinks, Using Picryl Chloride." *Journal of the Association of Official Analytical Chemists*, 55 (1972), 892–94.

146. Scheffe, Henry. *The Analysis of Variance*. New York: John Wiley & Sons, 1959.

147. Schell, Emil D. "Samuel Pepys, Isaac Newton, and Probability." *The American Statistician*, 14 (1960), 27–30.

148. Schoeneman, Robert L.; Dyer, Randolph H.; and Earl, Elaine M. "Analytical Profile of Straight Bourbon Whiskies." *Journal of the Association of Official Analytical Chemists*, 54 (1971), 1247–61.

149. *Scientific Tables*, 6th ed. Basel: Geigy, 1962.

150. Searle, Shayle R. *Linear Models*. New York: John Wiley & Sons, 1971.

151. Selective Service System. Office of the Director. Washington, D.C., 1969.

152. Sen, Nrisinha, et al. "Effect of Sodium Nitrite Concentration on the Formation of Nitrosopyrrolidine and Dimethylnitrosamine in Fried Bacon." *Journal of Agricultural and Food Chemistry*, 22 (1974), 540–41.

153. Shahidi, Syed A., et al. "Celery Implicated in High Bacteria Count Salads." *Journal of Environmental Health*, 32 (1970), 669.

154. Sharpe, Roger S., and Johnsgard, Paul A. "Inheritance of Behavioral Characters in F_2 Mallard × Pintail (*Anas Platyrhynchos L.* × *Anas Acuta L.*) Hybrids." *Behaviour*, 27 (1966), 259–72.

155. Shaw, G. B. *The Doctor's Dilemma, with a Preface on Doctors*. New York: Brentano's 1911, p. lxiv.

156. Shore, Neil S.; Greene, Reginald; and Kazemi, Homayoun. "Lung Dysfunction in

Workers Exposed to *Bacillus subtilis* Enzyme. " *Environmental Research*, 4 (1971), 512–19.

157. Smith, W. Lynn. "Facilitating Verbal-Symbolic Functions in Children with Learning Problems and 14–6 Positive Spike EEG Patterns with Ethosuximide (Zarontin)," in *Drugs and Cerebral Function*. Edited by Wallace Smith. Springfield, Ill.: Charles C. Thomas, 1970.

158. Srb, Adrian M.; Owen, Ray D.; and Edgar, Robert S. *General Genetics*, 2nd ed. San Francisco: W. H. Freeman, 1965.

159. Sukhatme, P. V. "On Fisher and Behren's Test of Significance for the Difference in Means of Two Normal Samples." *Sankhya*, 4 (1938), 39–48.

160. Sutton, D. H. "Gestation Period." *Medical Journal of Australia*, 1 (1945), 611–13.

161. Szalontai, S., and Timaffy, M. "Involutional Thrombopathy," in *Age with a Future*. Edited by P. From Hansen. Philadelphia: F. A. Davis, 1964.

162. *Tables of the Incomplete Γ-Function.* Edited by Karl Pearson. London: Cambridge University Press, 1922.

163. *Tennessean* (Nashville). Jan. 20, 1973.

164. ———. Aug. 30, 1973.

165. Thatcher, A. R. "The Distribution of Earnings of Employees in Great Britain." *Journal of the Royal Statistical Society*, Series A, 131 (1968), 133–80.

166. Thorndike, Frances. "Applications of Poisson's Probability Summation." *Bell System Technical Journal*, 5 (1926), 604–24.

167. Treuhaft, Paul S., and McCarty, Daniel J. "Synovial Fluid pH, Lactate, Oxygen and Carbon Dioxide Partial Pressure in Various Joint Diseases." *Arthritis and Rheumatism*, 14 (1971), 476–77.

168. Trugo, Luiz C.; Macrae, Robert; and Dick, James. "Determination of Purine Alkaloids and Trigonelline in Instant Coffee and Other Beverages Using High Performance Liquid Chromatography." *Journal of the Science of Food and Agriculture*, 34 (1983), 300–06.

169. Turco, Salvatore, and Davis, Neil. "Particulate Matter in Intravenous Infusion Fluids— Phase 3." *American Journal of Hospital Pharmacy*, 30 (1973), 612.

170. Vidins, Eva I.; Fox, Jo Ann E.; and Beck, Ivan T. "Transmural Potential Difference (PD) in the Body of the Esophagus in Patients with Esophagitis, Barrett's Epithelium and Carcinoma of the Esophagus." *American Journal of Digestive Diseases*, 16 (1971), 991–99.

171. Vincent, Pauline. "Factors Influencing Patient Noncompliance: A Theoretical Approach." *Nursing Research*, 20 (1971), 514.

172. Vogel, John H. K.; Horgan, John A.; and Strahl, Cheryl L. "Left Ventricular Dysfunction in Chronic Constrictive Pericarditis." *Chest*, 59 (1971), 489.

173. Walter, William G., and Stober, Angie. "Microbial Air Sampling in a Carpeted Hospital." *Journal of Environmental Health*, 30 (1968), 405.

174. *Weather Atlas of the United States.* U.S. Department of Commerce, 1968, p. 61.

175. Weiss, William. "Cigarette Smoke Gas Phase and *Paramecium* Survival." *Archives of Environmental Health*, 17 (1968), 63.

176. Werner, Martha; Stabenau, James R.; and Pollin, William. "Thematic Apperception Test Method for the Differentiation of Families of Schizophrenics, Delinquents, and 'Normals'." *Journal of Abnormal Psychology*, 75 (1970), 139–45.

177. Whitworth, William Allen. *Choice and Chance*. New York: Hafner, 1965.

178. Wilks, Samuel S. *Mathematical Statistics.* New York: John Wiley & Sons, 1962.

179. Wilton, D. P., and Fay, R. W. "Response of Adult *Anopheles Stephensi* to Light of Various Wavelengths." *Journal of Medical Entomology,* 9 (1972), 301–04.

180. Winslow, Charles. *The Conquest of Epidemic Disease.* Princeton, N.J.: Princeton University Press, 1943, p. 303.

181. Wiorkowski, John J. "A Curious Aspect of Knockout Tournaments of Size 2^m." *The American Statistician,* 26 (1972), 28–30.

182. Wolf, Stewart, ed. *The Artery and the Process of Arteriosclerosis: Measurement and Modification.* (Proceedings of an Interdisciplinary Conference on Fundamental Data on Reactions of Vascular Tissue in Man, April 19–25, 1970, Lindau, West Germany.) New York: Plenum Press, 1972, p. 116.

183. Wood, Robert M. "Giant Discoveries of Future Science." *Virginia Journal of Science,* 21 (1970), 169–77.

184. Woodward, W. F. "A Comparison of Base Running Methods in Baseball." M.Sc. Thesis, Florida State University (1970).

185. Wrightman, Lawrence S. "Wallace Supporters and Adherence to 'Law and Order'," in *Human Social Behavior.* Edited by Robert A. Baron and Robert M. Liebert. Homewood, Ill.: Dorsey Press, 1971.

186. Wyler, Allen R.; Minoru, Masuda; and Holmes, Thomas H. "Magnitude of Life Events and Seriousness of Illness." *Psychosomatic Medicine,* 33 (1971), 115–22.

187. Yochem, Donald, and Roach, Darrell. "Aspirin: Effect on Thrombus Formation Time and Prothrombin Time of Human Subjects." *Angiology,* 22 (1971), 72.

188. Yule, G. "Why Do We Sometimes Get Nonsense-Correlations between Time Series?— A Study in Sampling and the Nature of Time Series." *Journal of the Royal Statistical Society,* 89 (1926), 1–69.

189. Yule, G. U., and Kendall, M. G. *An Introduction to the Theory of Statistics,* 14th ed. London: Charles Griffin, 1965.

190. Zaret, Thomas M. "Predators, Invisible Prey, and the Nature of Polymorphism in the *Cladocera* (Class *Crustacea*)." *Limnology and Oceanography,* 17 (1972), 171–84.

191. Zelazo, Philip R.; Zelazo, Nancy Ann; and Kolb, Sarah. "'Walking' in the Newborn." *Science,* 176 (1972), 314–15.

192. Zelinsky, Daniel. *A First Course in Linear Algebra,* 2nd ed. New York: Academic Press, 1973.

193. Ziv, G., and Sulman, F. G. "Binding of Antibiotics to Bovine and Ovine Serum." *Antimicrobial Agents and Chemotherapy,* 2 (1972), 206–13.

Index